The
GENE
WARS

The
GENE
WARS

*Science, Politics,
and the Human Genome*

ROBERT COOK-DEEGAN

W · W · Norton & Company · New York · London

FIRST EDITION

The text of this book is composed in Galliard with the display set in Garamond Old style and Stradivarius. Composition and manufacturing by the Maple-Vail Book Manufacturing Group. Book design by Marjorie J. Flock.

Library of Congress Cataloging-in-Publication Data

Cook-Deegan, Robert M.
    The gene wars: science, politics, and the human genome / Robert Cook-Deegan.
        p.   cm.
    Includes bibliographical references.
    1. Human Genome Project.   2. Human gene mapping.
QH445.2.C66   1994
573.2'12—dc20   93-10762

ISBN 0-393-03572-7

W. W. Norton & Company, Inc., 500 Fifth Avenue, New York, N.Y. 10110
W. W. Norton & Company Ltd., 10 Coptic Street, London WC1A PU

1 2 3 4 5 6 7 8 9 0

To Kathryn, Patrick, and Maeve

Contents

PART FOUR

Genome Gone Global

PART FIVE

Ethical, Legal, and Social Issues

Preface

ACENTRAL TENET OF MODERN BIOLOGY holds that the long-chain organic molecule deoxyribonucleic acid (DNA) encodes all the instructions needed to create life. The recipe for making a human being is written out sequentially in a four-letter digital code, embodied in the six feet of DNA coiled inside virtually every human cell. Amassing the scientific tools to decode that instruction set has been a major preoccupation of molecular biology ever since 1953, when James D. Watson and Francis Crick first described DNA's double helical structure. The Human Genome Project is a natural outgrowth of this effort.

The genome project emerged from several independent sources in the mid-1980s. Once it welled up in the scientific community, the project quickly became a contentious political issue as well as a scientific one. Among its friends, enthusiasm for the project inspired rhetorical exuberance. The project became the "Holy Grail of biology" and a quest for the all-revealing "Book of Man." These sincere, but nonetheless hyperbolic, conceptions overshadowed what is, in essence, an extensive road-building project for genetic exploration.

One factor that incited passion about the genome project among scientists was its price tag. Very early in the project's evolution, the figure of $3 billion began to be bandied about. This estimate of the project's eventual cost was made long before any reliable projections could be made about its constituent parts; indeed, even the cost of getting started was highly uncertain. The $3 billion was to be spread out over ten or twenty years. If the National Institutes of Health continued spending at its current rate, however, it would expend a total of $150 to $200 billion during that time on all of its activities. In that case the genome project might account for between 1 and 2 percent of the total NIH expenditures for many years to come. But the larger debate was never cast in these terms. Rather the debate became one of "big" science versus "small" science. The reliance on systematic technology development and goal-directed gene-mapping efforts presaged a new style for biology, one that elicited excitement from those attracted to whiz-bang technologies but drew gasps of revulsion from those who aspired to cultivate biology on a more modest scale and with a decentralized organization. The battle was, among other things, over whose vision would control the budget and which scientific aesthetic would prevail.

The purpose of the genome project is to explore the territory discovered in 1865, when Gregor Mendel first described hereditary "elements." The early genetic pioneers could only glimpse this new world indirectly, by studying how traits were inherited. They could not guess what the substance of genes might be. That awaited elucidation in 1944, when Oswald T. Avery, Colin McLeod, and Maclyn McCarty, working at the Rockefeller Institute in New York, demonstrated that DNA in bacteria contained the inherited genetic information. Subsequent biology showed that DNA is indeed the universal language of life. All organisms, with a few minor variations, use the same DNA code.

Maps of this biological frontier could be drawn only after the structural basis of inheritance had been worked out. The study of genetics was the study of how characters are inherited—how knowing something about parents tells us something about their children. In the 1950s and 1960s, molecular biology spawned molecular genetics. The central target of this new approach to genetics became explaining how DNA was copied so that it could be passed on to subsequent generations, on one hand, and how its instructions were translated into cellular function, on the other. The structural analysis of DNA function became a dominant theme of modern biology, including research aimed at conquering human diseases.

A somewhat separate science of human genetics developed in parallel. Human genetics long emphasized the study of pedigrees—the transmission of traits from one generation to another—and the genetic differences among human populations. Medical genetics, the study of how some diseases were passed on as genetic traits, constituted a once abstruse spinoff of human genetics. During the 1970s and 1980s, genetics was drawn out of the backwater and entered the mainstream of biomedical research, emerging as a dominant strategy to understand mysterious diseases. This development resulted from the confluence of this tradition of human genetics with the *molecular* genetics practiced in bacteria, yeast, fruit flies, nematodes, mice, and other organisms.

A revisionist history has already crept into the reconstruction of the genome project's genesis. Specialists in human and medical genetics have, in some quarters, attempted to claim the origin of the genome project as their own. This could well have occurred, indeed perhaps should have. It would have been logical for the Human Genome Project to emerge from within human genetics or medical genetics. The progression from gene mapping to the Human Genome Project is so natural that many within the field believe it must have arisen so. But that is not the way it happened.

None of the three main instigators of the Human Genome Project, as it actually developed, was a human geneticist. Each used genetic techniques and engaged in biomedical research on human diseases, but none was engaged in studying human genetic disease or human pedigrees as a primary focus. The genome project began, instead, as a technological vision. Its name came from thinking about systematically applying the tools of molecular genetics to the entire genome—the full complement of DNA in human cells. In this view, gene mapping was a subsidiary step, with determination of the complete DNA code as the primary goal. Human genetics later recaptured some pieces of the genome project by redefining its goals, but the genome project clearly began outside of the mainstream in human genetics.

Advances in methods for analyzing the structure of DNA rapidly, in small amounts, and with great precision allowed some of the power of genetics, so well demonstrated in other organisms, to be applied to the study of human disease and normal human physiology. The real roots of the genome project were in yeast genetics, nematode genetics, and bacterial genetics. A few pioneers had long used molecular methods to dissect human diseases, and with some success. But the emergence of new technologies, demonstrably useful for studying other organisms, made the direct analysis of DNA much more tractable in humans. The genome project grew from thinking boldly about how to apply emerging technologies of DNA analysis to the study of human biology and, ultimately, to the task of tackling the entire human genome.

This book is an account of the origins of the Human Genome Project. The scientific ideas took hold only after they were publicly aired, provoked a vigorous debate, and were then repackaged to make them politically palatable. The main story line is the creation of a bureaucratic structure to carry those ideas to fruition. The ideas derived from science and technology; the genome project as a sociological phenomenon, however, came from the actions of many people, often working without knowledge of others treading on convergent paths.

The book describes two kinds of history: the technological advances that predated the project and the events that followed, once the idea of a genome project came to the surface. After an introductory chapter, the technical developments that led to discussions of a concerted genome project are described in Chapters 2 through 4. Those who want to go straight to the politics, or who are put off by the technical arcana, can skip these and go directly from Chapter 1 to Chapter 5. Chapters 5 through 20 and the Epilogue describe the political and historical events built on the technical foundation.

Throughout its early history, I was positioned as a close observer of the genome project, and was at times a minor participant. From 1986 through 1988, I was part of a team that prepared a report on the genome project for Congress. I then directed a small congressional bioethics commission, which died in the crossfire of abortion politics. Upon its death, I briefly joined the National Center for Human Genome Research at the National Institutes of Health as an outside consultant, beginning my association two months after it was created. Along the way, it became clear that the policy story needed to be told. The Alfred P. Sloan Foundation, and subsequently the National Science Foundation, provided generous support to gather information and to begin writing the book. I conducted interviews with the main characters in this story beginning in 1986, and continued for six years. This is their story, the genesis of the Human Genome Project—a case study in the politics of modern science.

PART ONE

The Scientific Foundation

1

Why Genetics?

THE OKLAHOMA SUNSHINE hit me right in the eyes, ending a long and tearful night. It was on odd way to celebrate the fourth of July, 1976, the bicentennial of the signing of the Declaration of Independence. I awoke in the farmhouse built by Robert Ross.* It was completed some thirty years before, when he was in his early thirties. The house and nearby barn were monuments to Robert's once considerable skill in carpentry—premortem tombstones for the man he had been before Alzheimer's disease destroyed his mind.

Robert's wife, Emma, and daughter, Ellie May, took me to see the corral he had tried to build at age forty. It was a rail fence, vastly simpler that the beautiful barn and farmhouse from which it extended. The fence was mis-shapen, boards wopperjawed and nails askew—an external embodiment of the decay in his cerebral cortex. Emma dated the onset of his illness to the day he came to her, tears of bitter frustration in his eyes, when he realized he did not know how to build the fence. He became aware he had lost the ability to think. It was the beginning of a devastating travail, destined to last twenty-five years. The awareness of dwindling capacities tortured him less as the years passed, because his capacity for any kind of awareness dissipated. The pain slowly shifted from Robert Ross to his family, and particularly to Emma, his partner for life.

*Names have been changed to protect the privacy of family members.

My first night in Oklahoma was absorbed in recapitulating the course of Robert's disease. It was a story of relentless loss. Over the first ten years, Robert slowly deteriorated, progressively losing his ability to do farm work, then to help with even the most menial chores. He became a ward of his wife and children, and by age fifty he could not even recognize them. Robert's blank stare was the most painful torture for his wife and children. These were the central characters of his life, its most cherished rewards. Now they were meaningless, literally beyond recognition, and they knew it even if Robert did not. Such a fire sears even those emotionally thick of skin, leaving permanent scars.

The Ross home country lay in the middle of the Oklahoma panhandle. This was land that at one time no state had wanted, and so it was here that North America's native populations were displaced in the face of the European migration westward. It later became a capital in the Dust Bowl. The Ross family weathered the Depression only to succumb to another catastrophe. Ten of Robert's thirteen siblings also developed Alzheimer's disease. The disease struck a half-dozen cousins. This toll of disease left the family reeling, attempting to deal with a remorseless foe they could not see, whose advance they could not stop.

I went to Oklahoma with Jeanie, a technician from a University of Colorado genetics laboratory, to meet the family and to collect blood and saliva samples. Our mission was to refine the already considerable mass of pedigree data we had assembled on this gifted but star-crossed family, and to bring samples back to Denver for analysis. I was twenty-three, fresh from my first year in medical school; Jeanie was in her early forties. I had taken one elective course in neurology, and had read seventy or eighty articles on Alzheimer's disease, the bulk of world literature at that time. (Contrast this with the more than four thousand items found in a search of the medical literature over a thirty month period in the early 1990s.) In those days, Alzheimer's disease was not yet a household word; it was a scientific backwater. A decade later, Alzheimer's disease got "hot" as a research topic, attaining that status in part because of a 1976 editorial by Robert Katzman in a leading neurology journal, which described its ravages on the population.[1]

I first met Robert Ross in the spring of 1976, at a Veterans Administration long-term care facility in Fort Lyon, in southeastern Colorado. The VA hospital was initially constructed to house those afflicted with tuberculosis, hence its isolated location on the arid plains. By sheer coincidence, my grandfather had been clinical director there in the 1930s, and my father, now a Denver physician, was born there. As hygiene and antibiotics conquered tuberculosis, the hospital was transformed into a mental health facility, accepting patients from a multistate area. It included a few special wards for long-staying patients. Robert had been there for several months when I first encountered him, flagged for further workup by James Austin, chairman of the department of neurology at the University of Colorado. Austin and other neurologists from the university periodically visited the Fort Lyon facility. I came on this particular visit specifically to try to find cases of Alzheimer's disease that might have a genetic origin.

Robert was sixty when I met him, completely mute, with his arms contracted permanently into the fetal position. Robert's only responses to the outside world were myoclonic (involuntary) jerks provoked by loud noises or bright lights. Robert was the shell of a man who had been dearly loved. His story was typical of the clinical history of Alzheimer's disease, unusual only in having started at such a young age (forty) and having followed such a protracted course (ultimately twenty-five years). There were several notes in his files indicating that others in his family had Alzheimer's disease. Hence our interest. At the time, there was an active debate about whether Alzheimer's disease could be inherited. The most popular British textbook stated flatly, "It is not inherited,"[2] yet there were twenty or so papers in the literature, dating back to German research between the wars,[3-5] suggesting that some families carried an Alzheimer's gene.[6;7]

Austin was convinced that there was indeed an inherited form of the disease, and thought that a genetic research strategy was likely to be productive. The basic idea was to isolate a gene associated with the heritable form of Alzheimer's disease and then to determine the gene's function. It was a conceptually sound strategy that had been widely discussed for many diseases, although never successfully carried out at the time. The problem was that the tools to implement the strategy were primitive, and might well be inadequate for the task. The first step was to trace the disease through a large family, to see if there was a pattern suggesting inheritance of a single gene.

I met Emma Ross in Denver a few weeks after first seeing her husband, Robert. She had driven from Oklahoma to Denver, bringing a homemade pedigree with hundreds of individuals. The pedigree was written on pieces of blank school paper taped together. When unfolded, the collage covered a conference table. Some symbols had the wrong shapes, the data were incomplete, and the connections between some parts of the family were unclear, but the essentials were sound. The extended family tree represented years of diligent effort. The hundreds of hours that went into constructing the pedigree were far beyond what we had any right to expect. It was immediately obvious that the simplest explanation for who got Alzheimer's disease in the Ross family was a gene that caused Alzheimer's disease in a single dose; in other words, one copy of the bad gene from either parent was enough to trigger the disease. The Ross family might contain the information necessary to find the gene. Emma and I agreed that the next step was for me to meet the family.

The Ross pedigree spanned six generations, from Robert's great-grandfather to his grandchildren. The great-grandfather, of German extraction, immigrated from near the Volga River in Russia to the Midwestern plains of the United States in the decades following the American Civil War. The great-grandfather reportedly suffered from severe confusion late in life. Robert's grandfather and father had similarly been said to have some mental illness, although it was not clear what it was. Their families had to take care of them, beginning in their late forties. Given the small population base and the tendency to marry within ethnic groups, it was possible that Robert's father and mother both carried the Alzheimer's gene, explaining why eleven of fourteen children were affected. (At the time, only nine of the eleven cases were clear-cut; the other two cases were confirmed years later.) It was also possible this family was just extraordinarily unlucky, having rolled the genetic dice and gotten bad rolls in all but three cases. The basis for inheriting Alzheimers's disease in the one group of brothers and sisters was not entirely clear, but the family was so large that it seemed likely to be a fruitful source of information. We needed to document the clinical histories and to obtain samples for biochemical and genetic analysis.

Jeanie and I met the Ross family at a schoolhouse reunion. It was a hot Oklahoma Saturday. Emma and Ellie May had contacted the branches and twigs of the family tree, and had arranged a family picnic to celebrate the Fourth of July. There were fifty to sixty people present. Jeanie and I already had a foretaste of the havoc Alzheimer's disease had wrought on this family. En route to Oklahoma, we stopped to see a cousin afflicted with the disease in nearby Kansas; before the meeting, we visited the family of a brother who had the disease but who had not yet admitted he was affected. The rest of the family already knew nonetheless. (Two years later, the symptoms were much worse, and the disease openly accepted, but this patient's immediate family remained estranged from the rest of the family.)

The Ross family grew up as Thomas Jefferson would have urged, as stalwart citizen farmers—sturdy stoics whose lives were centered on church, family, and farm. Moods swung with the weather, dependent on prospects for that year's crop. Members of the Ross family bore a heavy additional burden. Children of those who developed Alzheimer's disease stood an even chance of developing, or escaping from, the disease before they turned fifty.

Those born to an affected parent were constantly on watch for signs of early Alzheimer's disease in themselves. They had seen the wreckage of the

---

Pedigree of the Rosses shows the devastating impact Alzheimer's disease can have on a family. (Family members affected by the disease are indicated by gray symbols.) The disease was traced back four generations to Robert Ross's great-grandfather, who immigrated from Russia to the Oklahoma panhandle more than a century ago. The pedigree was initially constructed by Emma Ross (as she is known in this book). It was then corrected and amended as other family members were systematically contacted and clinical records and autopsy reports were checked on every person recorded as having Alzheimer's disease, on the brothers and sisters of those affected, and on those who married into the family. In this version, the pedigree has been altered somewhat to protect the confidentiality of family members. Nevertheless, the essentials have been retained, showing the four generations of affected family members, and many more at risk in the succeeding generations.

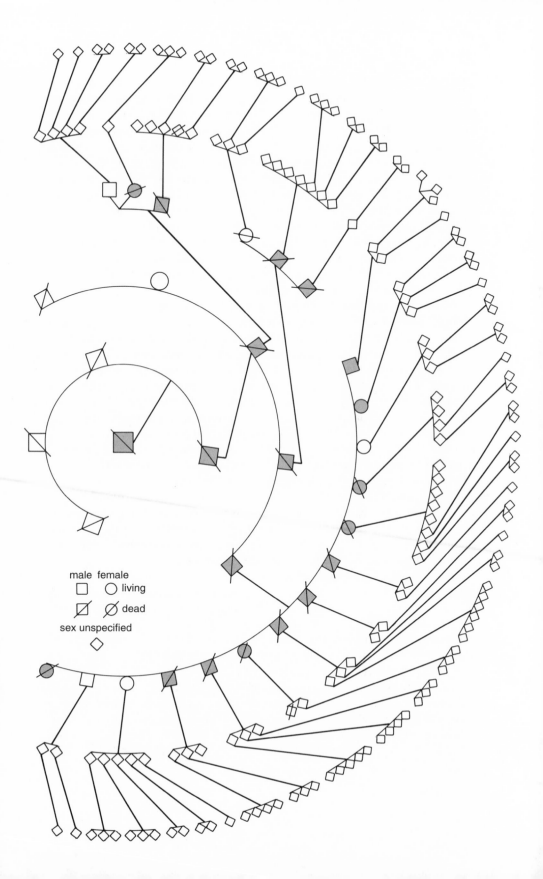

male female
☐ ○ living
⬔ ⊘ dead
sex unspecified
◇

disease, and loathed foisting it on their children and spouses. Those not directly at risk dreaded the day they would notice the first symptoms in loved brothers or sisters, uncles or aunts. Every car key forgotten, every light left on, every meeting missed became the focus of great distress. Was the disease beginning? This intensified the family dynamics, with denial, anger, and valiant acceptance constantly ripening on some branch of the family tree.

Emma was committed to shedding light on the disease. She assembled her pedigree, and kept records on deaths and births. Some members angrily resented Emma's meddling in their affairs. They protected their privacy against the incursions of scientists like me who sought to study the family in hopes of uncovering some clue to the causes of the dread disease. For some people at some times, the pain was just too overwhelming to let others near.

The stories of those who opened themselves to our inquiries were diverse, but they had a common tragic theme: the slow death of a mind in an otherwise healthy body. Each family described a period of grief that long preceded death and stretched on until death came, often years or even decades later. In many ways, death was a release. There was immense strife as once robust men and women became abject dependents. Every time a marriage was in prospect, there was a debate about when and how much to tell the prospective family member. The new spouse might someday bear responsibility for taking care of an Alzheimer's victim. Those who married into the family knowing the risks faced a difficult initiation. Others were not warned, and their resentment at having been kept in the dark haunted family gatherings.

As a twenty-three-year-old neophyte bearing the tools of science, I was eventually welcomed into the family as an intimate observer and archivist, privy to the most private family stories. I learned many details not known by others within the family—who had been adopted, who had artificial insemination, who was illegitimate. In the long tradition of medical research, with this intimacy I was handed the responsibility to guard the information. I was admitted to the inner sanctum because my art, molecular biology, was a source of hope—if not for the afflicted, then for their children.

In the schoolhouse, we ate a bounteous meal, replete with the Midwestern staples of beef, fresh corn, pie, and chocolate cake. I gave a short talk about Alzheimer's disease. The talk ended with an explanation of why Jeanie and I were there and how we hoped to locate the gene causing Alzheimer's disease by looking for other genes that might be near it, inherited along with it. Their large family and well-documented medical histories, I told them, would give us the best chance yet to get close to the Alzheimer's gene.

Genes are stretches of deoxyribonucleic acid (DNA) that contain the instructions to make a biological molecule. There are roughly six feet of DNA tightly coiled in each of the trillions of cells in the human body (with the exception of a few cells, like red blood cells, that lose their DNA as they mature). DNA is packaged with proteins into chromosomes, microscopic "colored bodies" in the cell's nucleus.

We were searching for a molecular handle on a mysterious disease. If we could find the gene's approximate location, its position among the chromosomes, we would take a step toward finding the gene itself, the actual DNA encoding some faulty molecule responsible for Alzheimer's disease. Through an extremely tedious but logical series of investigations, we might be able to find the molecule produced by the errant gene, and thus discover at least one molecular defect underlying the disease. We might even be lucky and find that the gene causing Alzheimer's disease was already known, but not yet associated with disease. There might well be other ways to develop Alzheimer's disease besides having a bad gene, but studying a clearly genetic form, in families such as the Rosses, was a logical scientific strategy. It was a relatively "clean" way to study a disease otherwise so immensely difficult to approach.

DNA, I explained, consists of long strings of chemical building blocks. There are four constituent chemicals, or nucleotide bases, abbreviated as A, C, G, and T (for adenine, cytosine, guanine, and thymine). These bases are attached to a backbone that is wound into the famous double helix. The bases form the steps in the spiral staircase of life. The DNA code is expressed in the order of the A's, C's, G's, and T's going up (or down) the spiral staircase.

Genes are stretches of DNA that produce something. Usually they contain the instructions for making a protein. The process of going from gene to protein involves several steps. First, the order of A's, C's, G's, and T's in chromosomal DNA is transferred to a molecule of ribonucleic acid (RNA) by cellular machinery. RNA is quite similar to DNA, but it is chemically less stable. For most genes, RNA is, in turn, translated into the order of amino acids that make up proteins. Proteins make up many of the complex structures in and between cells, and they mediate most chemical reactions within the body.

Proteins are the workhorses of the biological world. They can become cellular structures themselves, or the precursors of other structures (such as the components of membranes that surround cells, or sugars that bond to proteins). A large family of proteins called enzymes catalyzes biochemical reactions within cells. Proteins are made up of strings of amino acids, of which there are twenty common varieties. All twenty have a common structural element that allows them to be knitted into protein strings. The structural backbone element is linked to a diverse range of chemical structures. These chemical differences allow amino acids to perform widely differing functions. Some amino acids are best at linking to others. Some are especially useful in catalyzing reactions of a particular type, such as removal of a water molecule or rupture of a chemical bond. Others fit smoothly into membranes. The twenty common amino acids present an enormously variable repertoire of chemical functions when strung together by the hundreds into proteins. The chemical properties of proteins are determined by the order and type of these twenty amino acids. The linear sequence of A's, C's, G's, and T's in DNA is thus translated, as a rule, into chemical function by determining the linear

Chromosomes, the repositories of genetic information, consist of extremely long DNA molecules bound to proteins and other cellular components and wound into the distinctive "supercoiled" shapes seen in the microscope. The micrograph below shows the full complement of 46 chromosomes in a cell of a normal human male during the metaphase stage of cell division. The chromosomes have been stained to reveal their characteristic banding patterns. The diagram at right shows the detailed banded structure of the 22 autosomes (or nonsex chromosomes, present in pairs in the micrograph) and the one pair of sex chromosomes (X and Y). The chromosomes are aligned along their centromeres, or constricted regions (gray areas). The total amount of DNA incorporated into a complete chromosome set of this type is the human genome. *Photo courtesy Department of Clinical Cytogentics, Addenbrookes Hospital, Cambridge, England / Science Photo Library / Photo Researchers, Inc.*

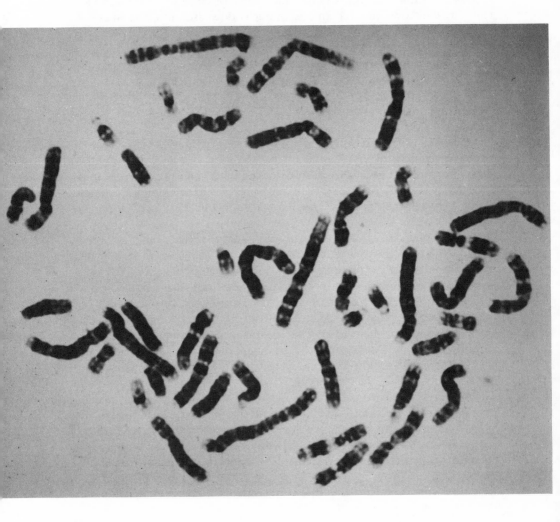

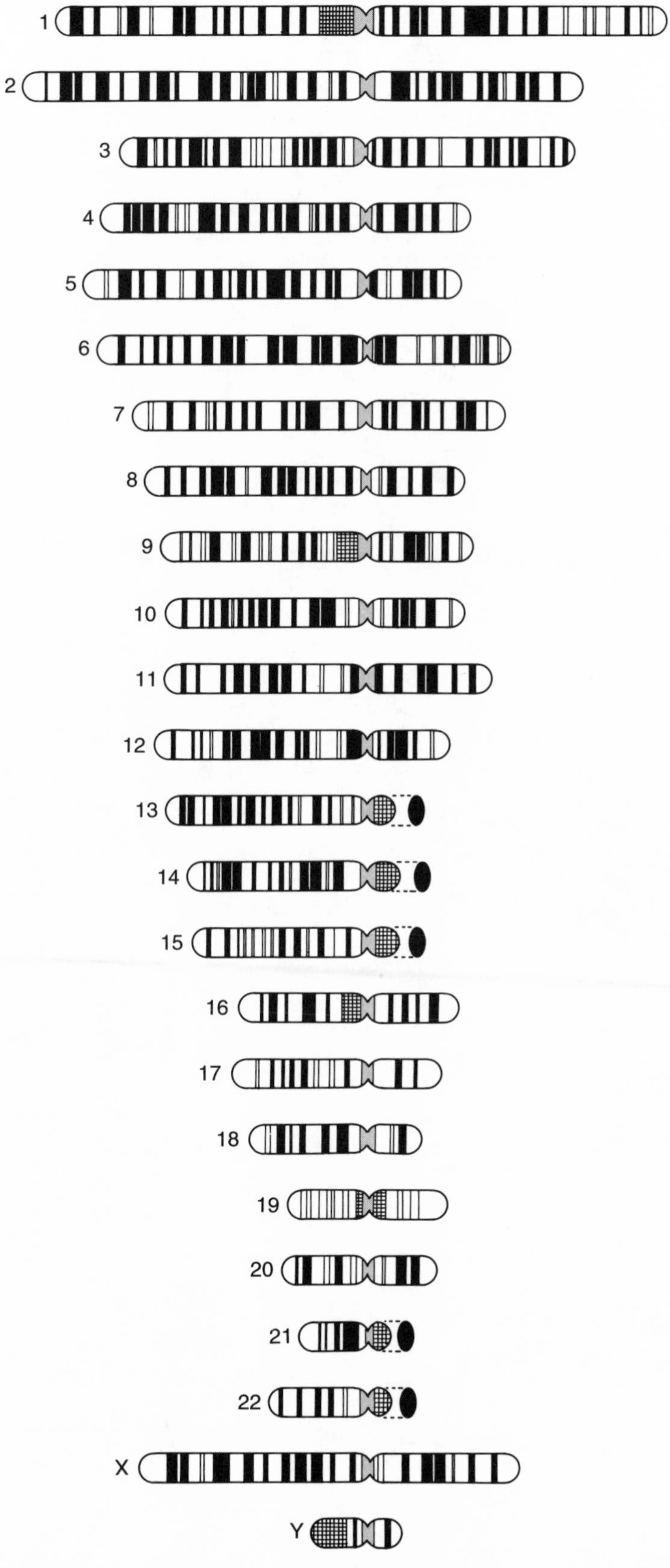
1
2
3
4
5
6
7
8
9
10
11
12
13
14
15
16
17
18
19
20
21
22
X
Y

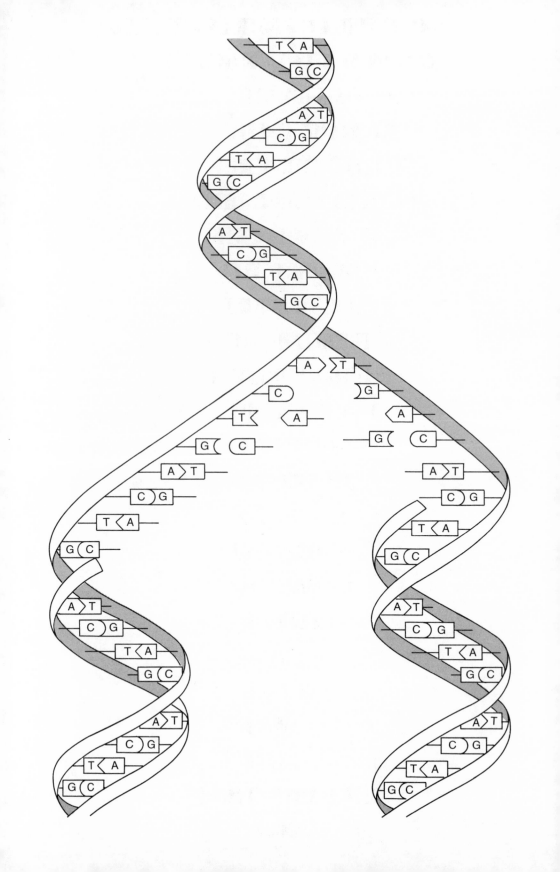

sequence of amino acids in proteins. A one-dimensional digital code in DNA is translated into a string of amino acids, which in turn folds into a three-dimensional functioning molecule.

Somewhere in the six feet of DNA was a gene that caused nerve cells to die prematurely, and Alzheimer's disease to develop in the Ross family. The smallest human chromosomes were estimated to contain about fifty million DNA bases strung together. The largest chromosome was roughly five times longer. One set of human chromosomes (one of each pair of chromosomes) contained an estimated three billion base pairs. Fishing the Alzheimer's gene out of this vast ocean of DNA was an awesome task. We needed a navigational chart—a map.

In 1976, there were only seventy or so "markers," genes whose location was known on the twenty-two pairs of nonsex human chromosomes.[8] These markers were the reference points by which to navigate on a genetic voyage in search of an unknown gene—an undiscovered island. Many chromosomes had only one or two markers each, so the signposts were few indeed. The tools of human molecular genetics were exceedingly imprecise; we as investigators were frustratingly impotent. But it was worth a shot. Any action was better than hopeless waiting. The seventy markers were useful, but far fewer than we needed to have a good chance of locating the gene.

I returned to Oklahoma and Nebraska and Kansas and Texas several times over the next decade, always to a family reunion followed by an assembly line to gather more samples. We found other families in Colorado, in California, and in the Midwest and began to study them as well. Over the years, we could apply methods developed in the rapidly expanding area of human molecular genetics. Each year, we obtained more clinical records on family members. Each year, there was another seminar. At every family meeting, I could report progress, but nothing even vaguely resembling a major breakthrough. Twice there was a newly discovered victim whose misfortune cast a pall over the meeting, making our scientific progress seem paltry by comparison. Certainly our own work was inconclusive. It was the usual story of medical science— pushing inadequate analytical tools to the limit in search of some clue about how the body works. It was awful, but it seemed important to persist in the face of long odds.

A geneticist can work for years in a laboratory, never seeing an affected patient or commiserating with an afflicted family. The daily laboratory routine is relatively stable, if intense and demanding. Once the impact of a disease is directly experienced—the pain and devastation it causes for specific people—

DNA replicates itself by "unzipping" its two helical strands and incorporating new nucleotide building blocks from the surrounding medium in precisely the right order to form two identical copies of the original double helix. Each strand contains all the information needed to make the opposite strand, because the nucleotide bases represented by A's bind specifically to T's (and vice versa), while G's bind specifically to C's.

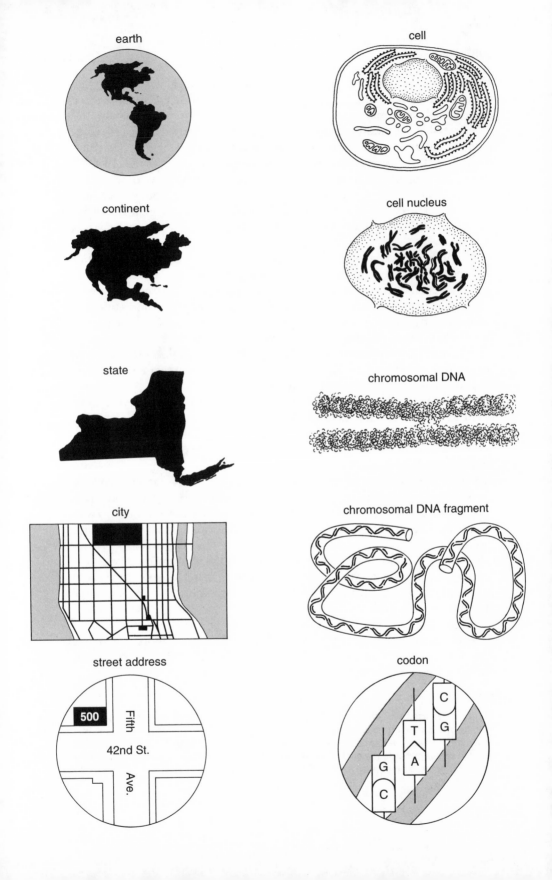

earth

cell

continent

cell nucleus

state

chromosomal DNA

city

chromosomal DNA fragment

street address

codon

laboratory work acquires new meaning. It demands greater urgency. The stakes go up; the room for excuses and tolerance of delay go dramatically down. Laboratory manipulations become less an exercise in abstract problem-solving and more a holy crusade against a common enemy. Disease becomes evil; eradicating it a primary need. Medical research differs from other scientific fields in this respect. It is driven by this passion for life—the hunger to understand life in order to preserve it.

Robert Ross died in 1981, at age sixty-five, of pneumonia. Two hours later, I cut the spinal cord and lifted his brain out of the cranium. Robert's brain was now a pound of gelid mush in my hands. I weighed the brain on a scale. I will always remember this bizarre act, the culmination of years of work. I knew Emma and Robert's children. I had seen the farm he built. This was the brain that felt emotions and thought thoughts whose objects I had myself encountered. I knew that beloved wife, that farm, that family. The frustration of a mangled fence began here.

The ravages of Alzheimer's disease had reduced Robert's brain mass by a third. It was grossly abnormal, and the severity of the deterioration was apparent as soon as the brain tipped the scale. We sent the tissue off to D. Carleton Gajdusek's laboratory at the National Institutes of Health for analysis, along with tissue frozen from various organs. Under the microscope, Robert's brain was riddled with microscopic plaques, craters left by the bombs in his genes. His was roughly the sixtieth autopsy I performed during my internship year, but the most memorable by far. Studying Robert's brain was another small step toward ridding the world of Alzheimer's disease. If the tools of genetics had been more powerful at the time, it could have been a longer stride.

In 1987, two groups of investigators linked the inheritance of Alzheimer's disease to a region of chromosome 21.[9, 10] It appeared likely that somewhere in the DNA on that chromosome was a gene that caused the disease in some families, although apparently not the Rosses.[11-13] In 1991, several groups identified a mutation correlated with Alzheimer's disease in two families,[14; 15] but this mutation proved quite rare. It occurred in only a few of the families with Alzheimer's disease mapped to chromosome 21.[16] In other families, there is evidence of a gene on chromosome 19,[17] and in yet other families, on chromosome 14.[18] The chromosome 14 gene seemed likely to account for most cases of early-onset familial Alzheimer's disease. Alzheimer's disease has become one of the clearest examples of "genetic heterogeneity" in medicine—clinically similar disorders caused by different gene defects.[7]

The genetics of the Ross family are still obscure. The Rosses are among a

---

Geographic analogy gives a rough sense of the relative sizes of the subcellular entities involved in genome research. In the world of the cell, the information encoded in a triplet of nucleotide bases, or codon, corresponds loosely to the address of a single building.

Genetic maps differ in scale. The enlarged drawing of human chromosome 21 at left shows the region where the first linkage to Alzheimer's disease was identified in 1987. The scale shows the approximate length of this region in centimorgans, a measure of how often chromosomal segments are inherited together. (A distance of one centimorgan between genes indicates that they are likely to be separated only once in a hundred times in the process of meiosis, the kind of cell division that produces new sperm and egg cells.) The diagram at center is a physical map of a region of DNA, with an array of overlapping DNA fragments spanning the region. The scale is given in kilobases, or thousands of base pairs. Each fragment represents a length of DNA that has been cloned in yeast cells, so that large amounts can be copied and studied directly. By identifying clones containing adjacent chromosomal fragments, DNA from the region can be systematically scanned in search of genes. The diagram at right shows DNA mapping at its ultimate resolution, with the sequence of individual base pairs constituting the genetic information.

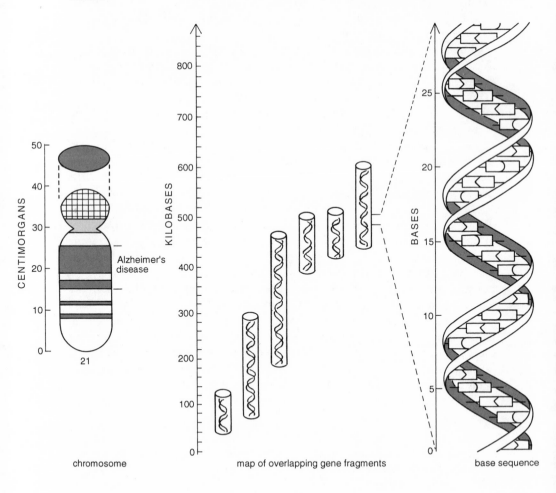

chromosome   map of overlapping gene fragments   base sequence

group now known as "Volga German" families. These families left Germany to live in the Volga River valley in prerevolutionary Russia. They suffered religious persecution, and many emigrated to the United States. Alzheimer's disease runs through a few of these families, which have been studied in the American Midwest. As noted above, the pattern of inheritance suggests a single gene, but none has yet been mapped.

This complex and confusing genetic story may well be a prototype of how researchers can use genetics to study neurological and psychiatric disorders. Diseases once thought to be coherent clinical syndromes may turn out to have several different causes. The clinical and anatomic similarities obscure underlying biological differences that the precision of molecular genetics can distinguish. In the case of Alzheimer's, the very tentative conclusion seems to be that there is at least one identified gene, the amyloid precursor protein gene, on chromosome 21, and perhaps another site on chromosome 21. There is strong evidence of another gene on chromosome 14, accounting for most familial cases with early onset, and a gene on chromosome 19 associated with later onset cases. There is yet another unmapped gene, or perhaps more than one, in the Volga German families. Four or more genes may cause what has until very recently been considered a single genetic form of Alzheimer's disease.

To add yet another complication, it remains unclear how many cases of Alzheimer's disease are genetic in origin, as opposed to other unknown causes (head trauma, viruses, environmental toxins, and other postulated agents). This is a matter of considerable controversy in neurology and genetics. The conventional wisdom is that only 10 to 15 percent of cases are associated with a single gene of major effect, but some researchers argue that the vast majority of cases may actually be due to genes whose effect is obscured because the disease begins so late in life that many affected people die before they develop symptoms.[19; 20]

In many cases, some would say a majority, genes may be far less important, perhaps even irrelevant. In the absence of firm data, opinion runs rampant, and no one opinion is any better or worse than the others until it is proved right or wrong. The hope is that genetic studies will provide a tool to trace the causal path, leading to further progress and suggestions of other possible causes. The ultimate goals are prevention, treatment, and possibly even cure.

Molecular genetics is a short cut to understanding mechanism through structure. The great appeal of the genetic approach is its immense explanatory power. If a gene is part of a causal chain, then there is something concrete to study—how the gene turns on and off, what it produces, what the gene product might do. That is why it is so attractive to study Alzheimer's disease, cancer, arthritis, diabetes, and other major killing and disabling diseases through genetics. Genes do not do everything, and the genetic approach must be wedded to biochemistry and physiology to complete understanding of a causal chain, but molecular genetics has been advancing more rapidly than these other fields.

Technology emanating from molecular genetics will continue to shift the conceptual foundations of biology and medicine toward the study of DNA.

A familial form of Alzheimer's disease was known in the 1930s, but success in finding even an approximate chromosomal address for a gene causing it came only fifty years later. The discovery was made possible by new tools and techniques. Success awaited the construction of genetic maps, sets of markers on all the human chromosomes that could be used to trace the inheritance of regions of chromosomes through families. As the methods of molecular biology became more powerful, they were applied to problems of increasing scale and complexity. In the 1980s, a group of scientist-administrators independently spawned the idea of systematically mapping the human chromosomes and spelling out the molecular detail of the DNA they contain. The idea for a genetic map of the human chromosomes combined with technological innovations in other fields and eventually jelled into what became known as the Human Genome Project. The assault on genetic disease was but one of many historical roots of the genome project, but it was this root that lent the project special urgency. Alzheimer's disease was but one of hundreds of scourges at which molecular genetics took aim.

The Human Genome Project emerged as a unifying force, focusing the full intensity of molecular biology on the development of tools to crack open the diseases that eluded understanding. The tools were maps and methods; the genome project was a political package in which to present them to policymakers and the public.

2
Mapping Our Genes

BETWEEN THE TIME I first met Robert Ross and his death, a revolution began in human genetics. The manifesto of this revolution was a 1980 joint paper published in the *American Journal of Human Genetics*.[1] The paper, by David Botstein of the Massachusetts Institute of Technology (MIT), Raymond White and Mark Skolnick of the University of Utah, and Ronald Davis of Stanford University, proposed a systematic approach to finding and organizing markers on the human chromosomes. A map consisting of such markers spaced throughout the chromosomes could then be used to locate genes by correlating the inheritance of the markers with the inheritance of traits (including genetic diseases) in families. Human geneticists could trace the inheritance of small chromosomal fragments through families for the first time. The 1980 paper drew on techniques first developed for yeast genetics and extended ideas just then emerging in human genetics. The work leading up to this landmark 1980 paper began in 1978, when yeast geneticists Botstein and Davis were presented with a novel problem in human genetics.

David Botstein had a background eminently suitable for making this conceptual breakthrough. He had trained initially at the University of Michigan, one of the world's centers of human genetics after World War II, and had gone into yeast genetics at MIT. He was therefore imbued with human genetics, but worked on one of the most genetically tractable organisms, the source of many new molecular genetic techniques. Ronald Davis was long known to have a flair for devising new technologies in molecular genetics. Any time there was a new way to cut, insert, separate, purify, or otherwise manipulate DNA fragments, Davis's laboratory at Stanford was likely to be involved. Botstein and Davis crystallized a molecular marking strategy out of a complex brew of methods for DNA analysis.

The strategy of narrowing the region in which to search for genes was long used in experimental organisms, where controlled breeding and powerful genetic techniques simplified the task. The conceptual breakthrough in 1978 was to show how existing methods could be applied to the *human* genome. Botstein and Davis's technique was the conceptual engine that drove human genetics from the era of the horse-drawn carriage into the age of the automobile. With these new maps, studying families like the Rosses changed from an improbable quest fueled by hope to a simple matter of persistence. To the extent that particular genes caused disease, their technique was a reliable way to find them.

Genetics became a major scientific field early in the twentieth century. As early as 1865, the Austrian monk Gregor Mendel had noted that the simplest way to explain the inheritance of certain characteristics of peas and other plants was to postulate "factors" donated by each parent. Exactly what physical structure conferred inheritance was not immediately clear, however. Chromosomes were first observed inside cells in 1877. Walter S. Sutton, a medical student working with Edmund B. Wilson at Columbia University, proposed in 1902 that chromosomes carried Mendel's hereditary factors.[2-4] Three years later, Nettie M. Stevens of Bryn Mawr College, also working with Wilson, explained how factors on the X and Y chromosomes could explain the inheritance of gender, independently corroborating her earlier work on insects.[5; 6]

In 1906, the English scientist William Bateson, a champion of Mendelism, christened the study of inheritance "genetics."[5; 7-9] In an independent coinage, Mendel's hereditary factors became "genes."[10] Thus, by 1910, the field had a name, and specific elementary objects to study.

Mendel's work, published in 1866, was largely ignored for thirty-five years, not because it was obscure or unavailable, but because its relevance to the dominant biological controversy of its day—evolution—was not immediately apparent.[5; 9] His work was rediscovered independently in 1900 by three scientists in Holland, Germany, and Austria,[11; 12] spawning the birth of Mendelian genetics. Genetic mechanisms to explain variation among generations immediately became the focal point in a protracted dispute about mechanisms of evolution. The controversy was ultimately resolved with the emergence of theoretical population genetics in the 1920s and 1930s. This new field combined statistical analysis of variations with the study of inheritance to explain how small genetic changes—mutations—could work with natural selection to explain evolution.[9]

Genetics grew rapidly with the study of the fruit fly *Drosophila melanogaster* and other species. Thomas Hunt Morgan, first at Columbia and then at the California Institute of Technology, blazed a path through the chromosomes of *Drosophila,* creating the paradigm for genetics in other organisms. The idea of looking for clusters of genetic traits, or characters, that were often inherited together—a phenomenon called genetic linkage—emerged from this group in a series of brilliant investigations.[12]

Applying the concepts of heredity emerging from the study of plants and other organisms, the British physician Archibald Garrod laid the foundation for medical genetics during the first decade of the twentieth century. At the Hospital for Sick Children in London, he studied the disease alcaptonuria. This condition caused a child's urine to turn dark, and later resulted in discolored cartilage and arthritis. Garrod studied the chemical abnormalities of the affected children's urine, finding an excess of homogenistic acid, a metabolic by-product. This by-product accumulated like water behind a dam because the enzyme did not function. After conferring with Bateson, Garrod deduced that the inheritance of alcaptonuria could be explained by Mendel's hereditary factors; the enzyme defect must be determined by a gene.[13; 14] Garrod further generalized his theory to the notion of "inborn errors of metabolism" for many other disorders.[15] Garrod thus established a firm link between genes and some human diseases.

Human gene mapping began in 1911, when Morgan's Columbia colleague Edmund B. Wilson deduced that the gene for color blindness must lie on the X chromosome because of its distinctive pattern of inheritance—fathers did not pass it on to sons, and it was rare among women.[16] The X chromosome could be distinguished by its size; females had two copies and males only a single copy. For five decades, study of the characteristic inheritance patterns of X-linked disease remained the most reliable gene-mapping method.

The first mapping of a human disease trait on another chromosome was published in 1968.[17] Genetic linkage to a human nonsex chromosome was first established in 1951,[18] but the nonsex chromosomes could not then be readily distinguished, so just which chromosome contained it was not known. It was possible to distinguish nonsex chromosomes in an occasional family when a particular chromosome had an unusual shape, or when chromosome fragments were rearranged and caused detectable clinical features, but vast regions of chromosomes other than X and Y resisted mapping. (At the time, geneticists erroneously believed there were forty-eight human chromosomes, further evidence of the technical limitations of the day.) In the late 1960s, two technical developments freed mapping from dependency on rare anomalies.

Somatic-cell hybridization mixed chromosomes from different organisms, fusing together cells from humans and other organisms. The mixed chromosomes fragmented and reorganized into metastable cell lines that retained various amounts of human DNA. It turned out that most rodent-human cell lines, after a few generations, kept mainly rodent DNA and only a small amount of human DNA, and were relatively stable over time.[19] If two genes were located near each other on the same chromosome, they would be expressed together in hybrid cell lines. By assembling large numbers of such cell lines, and devising ways to select only those cells containing genes of interest, it became possible to map genes by finding which genes were expressed together from different bits of chromosomes.[20] This was a laborious way to study the linkage of different known genes, suggesting their location near one another.

Linking genes to one another did not necessarily mean knowing which chromosomes contained them.[21] The largest and smallest chromosomes could be distinguished, but a large group of intermediate size generally could not. Geneticists needed a method to distinguish *all* of the chromosomes. Torbjörn

same gene to find a change in a single DNA base pair. A gene was thus located by linkage to DNA marker variants, isolated by mapping the region in detail, and its mutant identified by looking at DNA sequence.

As the genetic studies of yeast and other organisms progressed, new techniques also fed a growth spurt in the emerging specialty of medical genetics. According to a leading authority in the field, the ability to distinguish chromosomes and to map at least some genes "gave the clinical geneticist his (or her) organ; just as the cardiologist had the heart, the neurologist the nervous system, the gastroenterologist the GI tract, the clinical geneticist had the genome."[41] Clinical genetics focused on lesions of the genome just as surgeons dealt with tumors or cardiologists dealt with damaged heart muscle. The first item on the agenda of clinical genetics was to define the lesions and characterize the diseases they caused.

In 1978, Yuet Wai Kan and A. M. Dozy at the University of California, San Francisco, found that a particular variant was commonly associated with the sickle-cell gene in families of North African origin. In 87 percent of cases, cutting DNA with a restriction enzyme generated a DNA fragment of a distinctive length, associated with the sickle-cell gene.[42] The sequence difference detected by the restriction enzyme was not in the gene itself, but could nonetheless mark the chromosome whence it came. In those families of Northern African descent, it was possible to tell whether a child inherited the chromosome associated with sickle-cell variant or the chromosome usually carrying a normal gene. If children had only the sickle-cell variant, then they were likely to have gotten the gene from both parents, and prone to develop sickle-cell disease. By tracking a DNA marker, one could indirectly track the gene in those families where the DNA sequence variants held. The marker variant, itself of no functional significance, was used to establish linkage with the mutant sickle-cell gene. Kan and Dozy noted that such markers could be quite useful for determining linkage with genes.

Two British groups also noted that normal variations among individuals could be used as markers to trace inheritance, and for linkage to genetic diseases and other traits.[43; 44] Alec Jeffreys at the University of Leicester was most interested in studying variations among human populations; Ellen Solomon

---

Physical mapping of chromosomal DNA is done in stages. First the DNA is cleaved into various lengths by special enzymes called restriction enzymes, which cut the DNA at specific sites (1). The resulting fragments are then combined with other fragments of DNA (vectors), typically forming circular loops of DNA that can be cloned, or copied in large numbers, in yeast or bacteria (2,3,4). The collection of cloned fragments, known as a clone library, provides a large enough supply of the original chromosomal DNA to analyze directly. The clones are next cleaved by enzymes and the fragments are separated according to size by running them through an agarose gel (5). By looking at the fragments common to different clones, researchers can piece together the original order of the fragments in the chromosomal DNA (6,7). A complete physical map can be assembled by correctly ordering the overlapping DNA fragments from one end of the chromosome to the other.

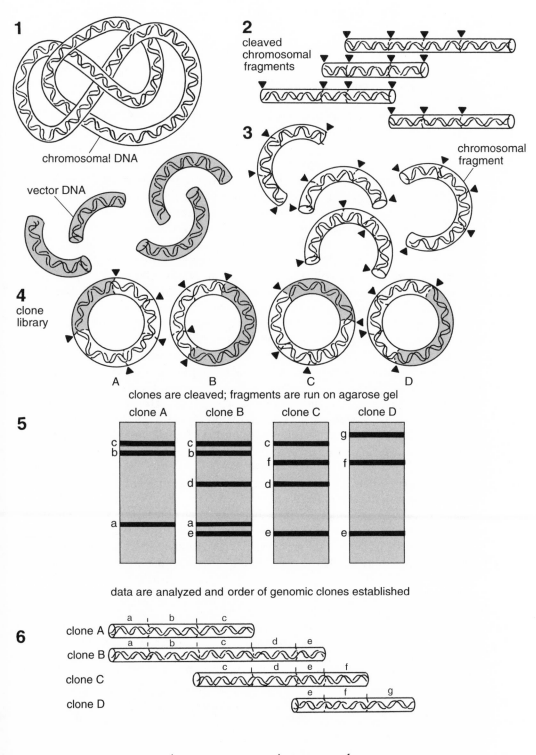

1 chromosomal DNA

2 cleaved chromosomal fragments

3 chromosomal fragment

4 clone library

A   B   C   D

clones are cleaved; fragments are run on agarose gel

5

clone A   clone B   clone C   clone D

clone A: c, b, a
clone B: c, b, d, a, e
clone C: c, f, d, e
clone D: g, f, e

data are analyzed and order of genomic clones established

6

clone A   a | b | c
clone B   a | b | c | d | e
clone C   c | d | e | f
clone D   e | f | g

7

a | b | c | d | e | f | g

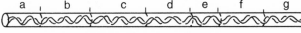

map of chromosomal DNA

and Walter Bodmer at the Imperial Cancer Research Fund in London, two highly regarded population geneticists, were also in the thick of many searches for human disease genes. The idea of a genetic linkage map based on DNA markers was in the air, and the 1980 paper by Botstein, White, Skolnick and Davis was the key that explained to those outside the inner sanctum of human genetics how such a linkage map might work.

In April 1978, Mark Skolnick and his colleagues at the University of Utah were trying to solve the mystifying inheritance patterns of the disease hemo-chromatosis. Hemochromatosis, resulting from toxic accumulations of iron, most often caused cirrhosis of the liver, orange-green coloration of the skin, diabetes, and insidious deterioration of heart muscle. The Utah pedigrees were meticulously documented, but difficult to interpret. Variations in the expression of symptoms made it difficult to ascertain how the disorder was inherited (depending among other things on the amount of iron ingested). Women often escaped symptoms longer than men did because they lost iron through menstruation (the blood lost each month carried iron with it).

In the mid-1970s, a French team noted a linkage between hemochromatosis and highly diverse proteins on cell surfaces, used by the immune system to distinguish "self" from foreign cells.[45–47] In studying organ transplantation, surgeons and immunologists had long known about the need to match donor and recipient according to cell surface markers, known as Human Leukocyte Antigen (HLA) types, to avoid organ rejection. HLA surface proteins were encoded by a gene complex known to reside on chromosome 6 in humans. The association between hemochromatosis and HLA markers suggested that by studying the nearby markers, one could dissect the inheritance of hemo-chromatosis.

Kerry Kravitz, then a graduate student at the University of Utah, worked with Skolnick to study the large pedigrees. Kravitz and Skolnick used the classic "boot-strapping" process of human genetics—finding families with many cases of a disease, establishing a rigorous clinical definition of the condition, and then systematically finding more cases (and excluding those that did not fit the clinical definition). They identified individuals with hemochromatosis, then tested iron levels in the patients' relatives, thus identifying new cases that had not been diagnosed. Better clinical information allowed another round of case finding, and so on. This had the added benefit of enabling treatment for the newly identified patients, including several women who had not yet shown symptoms. Kravitz and Skolnick kept testing for linkage between clinical hem-ochromatosis and HLA type. HLA markers were proteins, so this was not a case of direct DNA analysis, but the proteins were indirect indicators of genetic diversity. (The genes coding for the proteins were different.) The Utah group eventually accumulated enough cases to conclude that the disease was reces-sive—it required two copies of the gene, from both father and mother, to develop the disease.[48] Further analysis suggested ominously that the disease was forty times more common than previously believed.[49]

Kravitz presented his initial results at the annual review of graduate students, at the Wasatch Mountain ski resort Alta in April 1978. Botstein from MIT and Davis from Stanford were invited as outside reviewers. After Kravitz presented his results of statistical associations between HLA variants and hemochromatosis, the group discussed how to use statistical associations to map genes. Botstein and Davis, both of whom were familiar with how restriction fragment patterns had been used in yeast, supported the notion of using co-inheritance of a disease and some nearby marker as a mapping tool.

Botstein tends to think and talk excessively fast, and often at the same time. In one of his characteristic verbal explosions, he realized that correlating genetic differences with disease—the general approach used by Kravitz to track hemochromatosis—could be generalized and made much more powerful by direct analysis of DNA variations, using techniques Botstein and Davis were both familiar with in yeast.[50-55] If there were enough differences among individuals in families, the technique could be used to locate genes by dint of inheritance alone, with no knowledge of gene function and no particular candidate genes in hand.[56]

Botstein later recounted a vivid memory of looking up at Davis, both knowing that this was a conceptual breakthrough.[53-55] If one could only find enough genetic linkage markers spanning the chromosomes, a full-blown map should be possible. Genetic linkage in humans had been sporadically successful. The only markers available were generally genes, and these were usually not variable enough among individuals within a family to be able to trace their inheritance unambiguously. This limited their usefulness for genetic linkage analysis. The dearth of good markers hampered linkage analysis. Investigators searching for a gene were unlikely to detect a linked marker because the odds of finding a variable marker near the gene of interest was low. It was a crapshoot, with the odds stacked heavily in favor of the house. Moreover, the markers were distant from one another, so that their relative order and the distance between them were hard to determine. This complicated the process of determining the size of the chromosomal region containing a gene, and the gene's orientation with respect to different markers, even if a nearby marker was linked to the gene. The 1980 paper noted that "no method of systematically mapping human genes has been devised, largely because of the paucity" of markers that varied frequently among individuals.[1] Finding a large collection of such markers dispersed throughout the genome would be an enormous task, but it should theoretically work. Once these markers were ordered relative to one another, they could anchor a map, and be used to search for genes expeditiously, *even if one knew nothing more than the pattern of inheritance in a family*.

The importance of large families was the reason Skolnick was in Utah in the first place. Mormon families trace their pedigrees in great detail for religious reasons, searching for distant relatives who can be guided to salvation. The Church of Latter-Day Saints supports an elaborate research center for genealogy. Members of the church tend to have large families, again to increase the number of those redeemed by the faith. The scrupulous attention to family history has created a gold mine for human genetics. Indeed, something like mining is involved, as some of the genealogical records are carefully preserved in a mountain vault, to prevent their destruction in the event of war.

Skolnick set out to computerize large pedigrees for genetic analysis. He and others devised computer algorithms to do the tedious computations of probabilities for genetic linkage. He recognized the importance of Botstein and Davis's discussion, but he was not a molecular biologist. Botstein and Davis were intrigued, but they had many other projects more directly related to their past interests. The idea of a concerted effort to construct a genetic linkage map of DNA markers awaited another investigator's initiative.

Enter Raymond L. White. White, who was then at the University of Massachusetts at Worcester, came to the project that would establish his scientific reputation through a triangular MIT connection. White had been a graduate student with Maurice Fox at MIT. Soon after the fateful 1978 meeting at Alta, Fox bumped into Skolnick at a National Institutes of Health (NIH) meeting in Bethesda, Maryland, on breast cancer. Skolnick conveyed Botstein's idea to Fox, who then spoke with Botstein when he got back to MIT. Fox called White to tell him about Botstein's idea. White, becoming claustrophobic about prospects for a career studying the genetics of blowflies,

---

DNA Marker near the gene for Alzheimer's disease on chromosome 14 can be used to trace the course of the disease through a family. As the drawing at the top of the opposite page shows, the marker is not coincident with the Alzheimer's gene, but is close to it and thus is highly likely to be inherited with it. In this hypothetical family, there are five variations of the DNA marker. Repeated DNA sequences have been inserted one, two, three, four, and five times into DNA fragments cut out of each family member's DNA, differentially increasing the length of the respective fragments. Since each person inherits two copies of chromosome 14, he or she can have makers of two different lengths (known technically as restriction fragment length polymorphisms, or RFLPs). In this case, the grandmother exhibits RFLP pattern 1,5 and has Alzheimer's disease. Her son shows pattern 1,4 and is also affected. This indicates two things: (1) that he inherited a copy of chromosome 14 with pattern 1 from his mother and pattern 4 from his father; (2) that Alzheimer's disease is associated with pattern 1, since this son must have gotten the disease along with pattern 1. The unaffected sister in this generation inherited the other copy of chromosome 14, bearing pattern 5 from her mother and pattern 2 from her father. With this information, it is possible to predict that two siblings in the next generation are likely to carry the Alzheimer's gene, and thus to develop the disease (if they live long enough), because they also inherited pattern 1, while the two other siblings will be spared. The use of genetic markers such as RFLP's to track which copy of a chromosome was inherited by different family members depends on finding a marker that differs among family members. It was by repeating this process in many different families, and in different cases in the same family, that investigators were able to build up the evidence that a specific region of chromosome 14 is associated with Alzheimer's disease. The precise location of the gene and the nature of the gene defect can be determined only when the gene is found in the region.

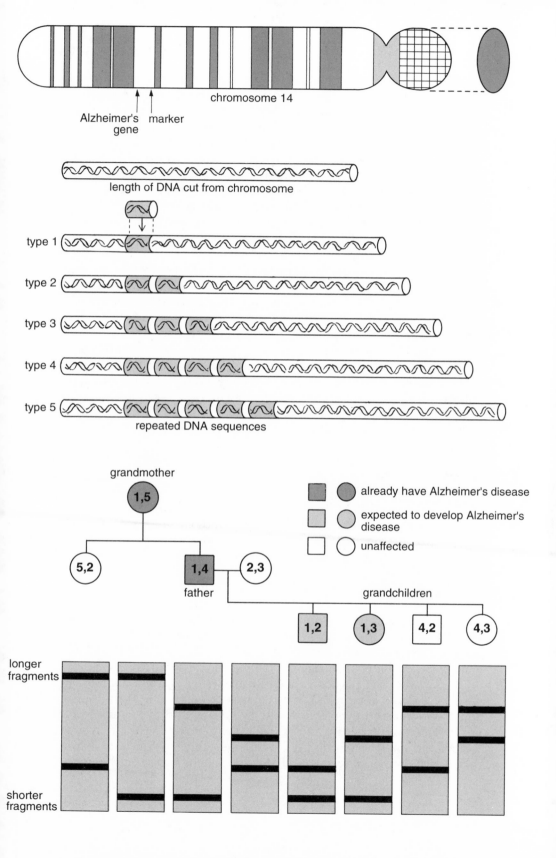

chromosome 14

Alzheimer's gene   marker

length of DNA cut from chromosome

type 1

type 2

type 3

type 4

type 5

repeated DNA sequences

grandmother

1,5

already have Alzheimer's disease

expected to develop Alzheimer's disease

unaffected

5,2

1,4
father

2,3

grandchildren

1,2   1,3   4,2   4,3

longer fragments

shorter fragments

was looking for an exciting new project to pursue. Botstein had severe space constraints at MIT, and could not immediately embark on a new venture.[57] White called Botstein, with whom he had worked for a year while Fox was on sabbatical; White had found his project. They began collaborating on a research proposal to seek grant funds from NIH.[52–54; 58; 59]

White took the lead drafting a proposal to seek NIH grant funds. The application was sent to NIH on February 27, 1979. Its first goal was "to develop a new set of genetic markers for the human genome based on DNA restriction fragment length polymorphisms (RFLPs)."[60] These RFLPs were markers of human genetic variation that could be used to trace the inheritance of chromosome regions through family pedigrees. The grant proposal mentioned Skolnick's HLA linkage work, and recounted the prior use of RFLPs in analysis of viruses and yeast. The proposal then laid out the great advantages for studying human inheritance if "one does not have to 'isolate the gene' in order to do mapping." By tracing the inheritance of naturally occurring variants, one could look for associations with disease genes, without knowing which protein the disease gene produced. An RFLP map would be just the tool needed to solve the biggest problem confronting most genetic diseases— to find unknown genes. If enough RFLPs could be mapped to the chromosomes, they would "provide a new horizon in genetics. . . . If successful, this endeavor will transform research with a low probability of success—familial linkage studies—into a legitimate endeavor. . . . the new tool will also [permit analysis of] syndromes which seem to 'run in families' but are too difficult to characterize genetically."[60]

White got his funding to start work at Worcester, and by November 1979, White and his colleague Arlene Wyman had isolated their first human RFLP. The probe from the first RFLP was named pAW101 (for "plasmid, Arlene Wyman 101").[61] The RFLP showed almost as much genetic variation as the HLA locus, making it likely that the two chromosomes in any given person would often be distinguishable, and that parents would have different markers as well. This heterogeneity made it a wonderful tool to find a gene near it on the tip of chromosome 14, where Wyman eventually mapped it.

If RFLP differences similar to the one associated with the sickle-cell gene could be found throughout the chromosomes, then the inheritance of chromosomal regions could be similarly traced through families and correlated with the presence or absence of disease. If an unknown disease gene, say the one causing cystic fibrosis, was located near the site detected by an RFLP, then it might be possible to figure out which chromosome from each parent carried the disease gene. The inheritance of RFLP markers on different chromosomes could be analyzed in many individuals in many families. If markers from one region were consistently inherited with the disease, then there was statistical evidence that the gene was in that region. To locate a gene of unknown location and unknown function, one would keep studying markers from many different chromosomal regions and look for regions consistently associated with the disease. Given enough markers, one would eventually find one near the gene. Having found the approximate location on the chromosomes, DNA from that region could be isolated and studied in greater detail in hopes of finding the mutation itself—the DNA change that caused a genetic disease. The concept was elegant and powerful, and the first step was to develop enough RFLP markers.

White soon moved to Utah, to take advantage of the Mormon pedigrees and because the Howard Hughes Medical Institute was willing to give him substantial funding to construct a map. Over the next few years, the Utah group systematically searched for RFLP markers, refining their techniques. They made DNA from members of more than forty families available for genetic typing by other groups, and contributed more markers to the genetic map of humans than any other group.

Another team dedicated its efforts to finding genetic linkage markers and to constructing systematically a complete genetic linkage map. Helen Donis-Keller led a group at Collaborative Research, Inc., a private firm located near Boston. During 1979, 1980, and 1981, Botstein had several conversations with NIH staff about assembling a complete RFLP map. He finally despaired of enticing NIH into the ring, but he believed an RFLP map was centrally important. He and Davis were both on Collaborative Research's scientific advisory board. White and, for an even shorter period, Skolnick were also initially consultants to the company. The possibility of RFLP mapping was much discussed among the company's scientific advisers. Collaborative Research was casting about for new markets and technical avenues, wanting to create a limited research and development partnership (a tax-favored investment tool in vogue until it was dismembered in the 1986 tax-reform law).

Donis-Keller had come to gene mapping through molecular biology. She worked as a graduate student with Nobelist Walter Gilbert at Harvard. She then moved to Harvard Medical School to work on viral molecular biology with Bernard Fields. Gilbert called her in 1981 to see if she would like to work for the newly forming biotechnology company Biogen, and she was hired as its third U.S. employee. Biogen grew rapidly, and working there became chaotic. Donis-Keller left Biogen for Collaborative Research in the spring of 1983. Nobelist David Baltimore headed the scientific advisory group to Collaborative Research. He, Botstein, and others prevailed on Donis-Keller to join Collaborative Research, to do strategic planning. Genetic linkage mapping with RFLP's became one of several projects under discussion as priorities for the company.

Donis-Keller applied for an NIH Small Business Innovation Research award, but was told the work was "not sufficiently innovative." She and James Wimbush then went to Wall Street, looking for $50 million. They came close to securing venture capital under a limited-partnership arrangement, but ultimately failed. They did presentations for Johnson & Johnson, Union Carbide, and other large companies. Donis-Keller laid out the strategy of completing a genetic linkage map and using it to locate genes, thereby making available a new method to detect genetic conditions. Having the tools for genetic linkage would not only spin off diagnostic tests, but would also give the company an edge in finding genes more quickly. Other methods could be used to find the gene itself, providing a target for drug development and gene therapy. Donis-Keller and the upper management at Collaborative Research were repeatedly rebuffed on Wall Street. She was appalled at the lack of vision among American corporations. They simply did not believe that genetics would be critical to cancer, heart disease, or other major health problems that bred diagnostic and therapeutic markets.

In 1984, Tom Oesterling became president of Collaborative Research, and the company decided to go ahead with RFLPs, using internal funds. The genetic linkage mapping team at Collaborative Research grew from four in April 1983 to twenty-four by the end of 1984. Work progressed steadily for three years. By the summer of 1987, things began "to come together."[62] The Collaborative Research team decided to push for a complete genetic linkage map in time for the human gene mapping conference in Paris that September. Rumors that the Utah team was planning to publish such a map were rampant, and they spurred Collaborative Research's efforts. Jean-Marc Lalouel, from the Utah Group, told one of the company's researchers at the Paris meeting that Utah had "won the war," and the Boston group feared it might be true.[63] The Utah group did distribute a pamphlet at the Paris conference, but the Donis-Keller team had submitted their publication to the journal *Cell* just before leaving for Paris.

Collaborative Research's announcement of a genetic linkage map caused quite a stir in the fall of 1987.[64-66] The corporate office decided to hold a press conference at the American Society of Human Genetics meeting in October, and announced the impending publication of a genetic linkage map containing markers from the public domain and from the company's own collection.[62; 67] Following the press conference, other groups opened fire. The map was said to be incomplete and the spacing between some markers was indeterminate, but the group at Collaborative Research calculated that their map was suffi-cient to locate 95 percent of any new genes and markers. White reported that the Utah group would publish maps one chromosome at a time when there were no gaps. Some of the controversy derived from the fact that about a fourth of the markers in the map had been discovered elsewhere, and the family resources necessary to do the mapping were contributed by a variety of other groups. Someone had to publish the first map, however, and no one group could ever claim full credit for the pooled resources. Ambivalence about a for-profit company's sponsorship of work in the *Cell* paper further complicated professional rivalries that were already intense. Despite the professional ten-sions, or perhaps abetted by them, the genetic linkage map beginning to coalesce around the efforts of the groups at Utah, Collaborative Research, and elsewhere became an enormously powerful tool.

The number, pace, and scale of hunts for human disease genes increased dramatically in the late 1980s. RFLP mapping reached a fever pitch, often flashing a sharp, competitive edge. In musing on the history of their field, geneticists James Crow and William Dove compared the 1987 "map flap" with publication of the first genetic linkage map, Alfred Sturtevant's 1913 paper on *Drosophila:* "These quiet beginnings stand in abrupt contrast to the current hubbub over the human linkage map and the proper definition of a map. With its rival factions and the glare of publicity, the mapping race is almost a genetic Olympics."[68] Crow and Dove betrayed a tinge of nostalgia, even tacit disapproval, of the style among the new upstarts. But the world had changed.

With maps in hand, gene hunts became highly competitive races. Teams led by Francis Collins and Ray White crossed the line to the neurofibromatosis (type I) gene in a dead heat, and leveraged favors out of editors eager to publish hot new findings. Their articles came out the same day in *Science* and *Cell*[69] The glare of publicity made genetic linkage mapping and gene hunting a high-stakes game, a national sport with guaranteed coverage in the *New York Times* and other major newspapers. Within a decade, genetic linkage mapping had gone from stepchild to celebrity, a scientific Cinderella story. But Cinderella needed an escort to protect her from her stepsisters.

The glue holding the various genetic linkage efforts together, despite the rivalries and tensions, was the Centre d'Etude du Polymorphisme Humain (CEPH), a Paris organization founded by Nobelist Jean Dausset with funds from a scientific award and gifts from a private French donor.[70; 71] CEPH was formed with the express purpose of enabling groups to pool their efforts in constructing a complete genetic linkage map. The first meeting to piece together the coalition took place in November 1984, when major mapping groups from Europe and North America descended on Paris, or rather ascended to a hallowed terrain where common good transcended rivalry. Cell cultures from members of large reference families around the world were collected by CEPH. Twenty-seven families came from the Utah Mormon pedigrees. One family came from the thoroughly studied Amish pedigrees of Pennsylvania Dutch country. Two families were small branches of an enormous Huntington's disease cluster in Venezuela, and ten French families were contributed by Dausset. Together they constituted forty families well suited for genetic linkage analysis. The idea was to test members of the CEPH family panel and to use the resulting RFLP marker data to make linkage maps of all the chromosomes. Each group that participated in CEPH agreed to genotype all informative families in the panel (to use their markers on the DNA taken from individuals in those families) and to send the data back to a shared database.

While many of the CEPH families were initially found by studying pedigrees for specific diseases (Huntington's, bipolar disorder, and other conditions), gene-hunting was *not* the thrust of the CEPH collaboration. It was instead to orient DNA markers through systematic analysis of *reference* families. The families were not used because their pedigrees showed the inheritance of genetic diseases, but because there were sufficient numbers of living members in well-defined pedigrees from whom DNA was readily available. Once the markers were mapped in the reference families, then the same markers could be used to hunt for genes in pedigrees containing any variety of genetic diseases or other traits. Government funding agencies appeared oblivious to this distinction between gene-hunting and systematic mapping. Map construction would take a massive effort to find informative markers, map them, and order them relative to one another. Gene hunts might eventually produce a map, but the bedrock of gene hunts was family pedigrees constructed with scrupulous attention to the accuracy of diagnosis. Those hunting for specific genes continued to be supported, and many groups hunted by finding clusters of genetic linkage markers, but those attempting to produce global genetic linkage maps got little help from the government.

The world of science was the beneficiary of the competition among Donis-Keller, White, and the other genetic linkage mappers. Between them, the Utah and Collaborative Research groups performed a great service. Wyman and White's first marker was found under a grant from the National Institute of General Medical Sciences, but for the complex and vast effort needed to find sufficient markers and to orient them on a map of the human genome, funding came almost entirely from the Howard Hughes Medical Institute and corporate sources. This government policy failure played a major role in later debates about the need for a systematic genome project. NIH's rejection of overtures to support the construction of genetic linkage maps was mentioned repeatedly in debates about whether it would not similarly spurn other mapping ventures.

Mapping by genetic linkage harked back to a style of mathematical genetics developed late in the last century and early in this century, some of which preceded the term "genetics." The approach was fundamentally classical genetics—the study of the inheritance of observable differences among individuals—supplemented by clinical observation to define the genetic characters under study, as augmented by the modern tools of molecular marking. The process relied on the mathematics of probabilities to make correlations. Those who studied evolutionary biology and population genetics immediately understood the significance of genetic linkage mapping. They were joined by a few medical geneticists comfortable with the statistical techniques of linkage.

When RFLP markers helped locate the genes responsible for Huntington's disease[72] and Duchenne muscular dystrophy in 1983,[73] human geneticists took notice. The Duchenne gene was already known to reside on the X chromosome, by dint of its inheritance pattern, but the location of the Huntington's gene was a complete mystery, and there was little prospect of finding it by traditional methods. (The Huntington's story is told in greater detail in Chapter 16.) The first spectacular successes were followed in short order by polycystic kidney disease,[74] retinoblastoma,[75] and cystic fibrosis[76–79] in 1985. These were each major prizes, and genetic linkage mapping with RFLPs took center stage.

According to a *Newsweek* feature in August 1987, a disease a week was being mapped by genetic linkage.[80] Technical advances further extended the ability to work backward from approximate gene location, determined by linkage to a marker, to find the gene itself and identify its product (in most cases, a protein). The first successful search for a gene of unknown function, starting from chromosomal location, ended in 1987 with the cloning of a gene causing chronic granulomatous disease.[81] This was soon followed by the genes for Duchenne muscular dystrophy[82] and retinoblastoma.[75; 83; 84] In each of these cases, however, the gene's approximate location (on the X chromosome for chronic granulomatous disease and Duchenne; on chromosome 13 for retinoblastoma) was already known from patterns of inheritance, human-hamster cell hybrids, and the study of patients with small chromosomal deletions. RFLPs played a role in narrowing the search, but the thinness of the RFLP map and limits inherent in using only pedigree studies to locate genes required additional strategies.

The sea change came with discovery of the gene for cystic fibrosis (CF).[85–87] The CF story propelled molecular genetics to the fore. Huntington's disease was the first mapped by linkage to an RFLP marker,[72] the first great triumph of RFLP linkage. But knowing the gene's location did not lead to the Huntington's gene itself, which remained elusive until 1993. In contrast, CF was mapped by RFLP in 1985, and the gene found in 1989.

CF was one of the most common seriously disabling single-gene diseases in Europe and North America. It became the first disease for which a gene of completely unknown location was mapped by genetic linkage, and then the regional DNA studied until a gene was found and the protein product identified. The CF gene was first located on chromosome 7 by Lap-Chee Tsui of the University of Toronto, collaborating with Collaborative Research.[76; 88; 89] Rumors of the chromosome 7 location early in 1985 induced other groups to look intensively for other markers nearby. The Utah group and Robert Williamson's group at St. Mary's Hospital in London both found linked markers even more tightly linked to CF (meaning their markers were closer to the gene).[77; 78; 88; 89] The CF linkage studies were highly competitive and were covered closely by the scientific press.[88; 90] Competition produced quick results. By locating the CF gene in a region of chromosome 7, RFLP mapping solved one of the highest priority problems, and one of the knottiest in human genetics. The race did not stop with locating the gene, but continued with sustained intensity for four more years until the gene was isolated. Indeed, it did not end even then, as there were still prospects of gene therapy and targeted drug development to pursue.

In a special issue of *Science* published on October 12, 1990, or eleven days after the genome project officially began, scientists took stock of their accomplishments to date. The hugeness of the task before human geneticists was starkly apparent. There were inconsistencies in nomenclature, the genetic distances measured for the same regions differed markedly, depending on analytical assumptions, and little of the genome had been mapped in detail.[91; 92] The human genome was indeed still "Terra Incognita."[93] In another special issue of *Science* just two years later, the tone was far more upbeat. A news piece noted that the genome project "hit its stride even sooner than its most ardent enthusiasts had predicted. Data are pouring out of the genome centers, new technologies are coming on line, and perhaps most notably, the first two high-resolution maps of human chromosomes are now complete."[94] *Science* also published a linkage map of the human genome incorporating more than sixteen hundred markers, including many markers of far greater usefulness than those in published in the 1987 Collaborative Research map.[95] The map's authorship also changed in interesting ways from its 1987 counterpart. The authors were referred to as the NIH / CEPH Collaborative Mapping Group, with many collaborators listed for each chromosome, and Helen Donis-Keller served as overall coordinating editor. Four weeks later, *Nature* published another second-generation linkage map, arising from the prodigious efforts of a French collaboration. This map contained 814 markers spanning an estimated 90 percent of the genome, and most of the markers were far more useful for tracing inheritance than their 1987 counterparts.[96; 97]

In retrospect, the systematic search for chromosomal markers and the construction of linkage maps were among the most significant accomplishments of human genetics in the 1980s and into the 1990s. Their full impact was not to be felt for several years. Maps not only made gene-hunting easier but also opened entirely new possibilities for tracing the inheritance of multiple genes and the study of how genes in one region influenced those in another. The groups in Utah, at Collaborative Research, at CEPH, in the high-technology French collaboration, and at many other laboratories throughout the world forged a genetic linkage map of a human being, a practical tool never seen before.

Genetic linkage mappers blazed a trail for those who would use human genetics to crack the tough nuts of human disease—diseases that had resisted assault by traditional research methods. The scientific strategy to answer the question "What gene is at fault?" was beautifully laid out by science writer Maya Pines, in a report prepared for the Howard Hughes Medical Institute.[98] The construction of genetic linkage maps, ironically, got little aid at first from the government agencies charged with supporting the overall biomedical research effort. This was partly because private institutions, most notably the Howard Hughes Medical Institute and CEPH, were quicker to respond and partly because the visionaries took matters into their own hands when they encountered obstacles to government support.

The formation of CEPH proved a watershed in human genetics. Dausset's idea, the commitment of the Donis-Keller, White, and other laboratories to share data, and the agreement to make DNA from common families generally available represented a considerable commitment from each participating group. Each family pedigree took immense effort to construct and to check. Agreeing to share access to these pedigrees entailed a degree of cooperation in map construction often overlooked in the race to find individual disease-associated genes by *using* the map. CEPH promised a coherent approach to map the human genome, and it was called a "human genome mapping project" as early as 1985.[71] The collaborative arrangement had its weaknesses and tensions, but it outlived the public clashes to unify the efforts. When NIH joined the effort in 1990, and with the emergence of a high-technology French collaboration at more or less the same time, progress was even more rapid. Despite being recognized as a genome mapping project, however, genetic linkage mapping efforts did not grow into "the" Human Genome Project. Those constructing genetic linkage maps could stake a legitimate claim on the Human Genome Project, but the bureaucratic edifice bearing that title grew from different sources, from three independent proposals to determine a reference DNA sequence of the entire genome.

Genetic linkage mapping was eventually folded into the genome project as it evolved, but it was at best an afterthought in the earliest genome project proposals. The history of the genome project would have been a more logical progression had human genetics spawned it. The focus on DNA sequencing that gave rise to the genome project was more than just a matter of emphasis— the sequencing proposals formulated in 1985 and 1986 came from a different group of individuals not directly engaged in RFLP mapping. There was some overlap of interests, but the impetus for DNA sequencing arose from those who contemplated the structural study of DNA, not from classical genetics and the study of inherited characters. The confluence of structural and classical genetics was delayed, but it was inevitable.

It was a long way from determining chromosomal location by RFLP mapping to isolating a gene. The work was tedious, methods were unreliable, and they often proved inadequate to the task. Finding genes required better methods to study DNA from a given chromosomal region. To move expeditiously from approximate chromosomal location to actual gene required a different kind of map—a physical map. Physical mapping bridged the gap between genetic linkage and DNA sequence, an important intermediate step. The techniques for making physical maps were first developed in other organisms, and only later applied to humans.

changed while producing sperm and egg cells. Genes only a few thousand base pairs apart in one region could be separated as often as those hundreds of thousands of base pairs apart in another region, and it differed between males and females for most regions as well. On average, a 1 percent change of recombination translated to a million base pairs (in humans). Both kinds of maps were important, as they served different functions.

Genetic linkage maps provide a bridge from studying how a feature was inherited in an organism to locating the genes for that feature; physical maps are ways to directly catalog DNA by region. The problem of how to make a genetic linkage map in humans had, in principle, been solved by RFLP markers. A parallel problem was how to make a physical map of the human genome and other genomes of interest. As discussion of the Human Genome Project began in 1985, physical mapping of large genomes was just beginning through work on yeasts and nematode worms.

Two groups began independently to apply the cloning and ordering strategy to make maps of this pair of model organisms. Maynard Olson's laboratory at Washington University began to map the chromosomes of *Saccharomyces cerevisiae,* or baker's yeast, which is used not only in baking but also in wine and beer fermentation. John Sulston and Alan Coulson at the Medical Research Council laboratory in Cambridge, England—later joined by Robert Waterston at Washington University in St. Louis—worked toward a physical map of *Caenorhabditis elegans,* a soil-dwelling nematode about a millimeter long. Genome-scale physical mapping of both organisms began in the early 1980s and showed promising results by 1986.[3; 4] Both projects focused on organisms central to biological understanding.

Yeast was emerging as the core model for eukaryotic genetics, that is, the genetics of organisms whose cells have a separate nucleus containing the chromosomes. (Bacteria and many other lower organisms do not sequester their chromosomes in a separate compartment, or nucleus; they are called prokaryotes, meaning "before nucleus.") The 12.5 million base pairs in the genome of *S. cerevisiae* were a logical early target for physical mapping. Having a set of ordered genomic DNA clones would be an extremely powerful addition to the already formidable armamentarium assembled to conquer yeast genetics. Botstein, together with Gerald Fink of the Whitehead Institute for Biomedical Research, noted several features marking yeast as a model, most important "the facility with which the relation between gene structure and protein function can be established."[5] The wealth of data on yeast mutants from classical genetics, usually bred by selecting for those yeast cells that could survive under stressful conditions, combined with the immense power of DNA exchange within yeast cells, made it possible to introduce mutations into known genes. The effects of these mutations could be quickly assessed because of the short generation time. By introducing mutations, it was generally possible to snare genes in the genome. By studying what happened to the protein or RNA gene product when a mutation occurred and correlating it to how the organism's biology changed, it was possible to draw inferences from gene structure to protein function, and thence to physiology. The speed and precision of yeast biology expedited efforts to link the triad of gene, protein, and function, so that "proteins first discovered elsewhere but present in yeast may best be studied first in yeast."[5]

The cooperative sociology of the yeast research community was another important factor noted by Botstein and Fink: "newcomers find themselves in an atmosphere that encourages cooperation. . . . not only are the published strains and mutants generally made available, but many (if not quite all) laboratories in the field routinely exchange strains, protocols, and ideas long before publication." Yeast genetics was ripe for a structural approach, and stocks of ordered DNA clones representing the entire genome would be an immensely useful tool. Olson's proposal to make a physical map from bacterial clones like those pioneered by Maniatis was thus enthusiastically greeted by his peers, and his grant was approved.

Olson planned to start by making a library, but then to try to put the books in order. The ordering strategy was to find DNA clones that overlapped one another. By finding next neighbors, then the next, and so on, eventually the order of the cloned DNA fragments would be established and they could be assigned to their chromosomal region of origin.

Yeasts were wonderful experimental models for many aspects of eukaryotic genetics, but these single-celled organisms were unsuitable for studying the complex interactions found in more complex organisms. Yeasts do not have brains or adrenal glands, for example, and so they do not fabricate intricate connections between brain cells or develop specialized hormone-secreting cells to communicate to other organs widely separated in the body. Yeasts are far from simple, but large animals are immensely more complex, with trillions of cells somehow coordinated into a whole.

The ideal organism to address such questions, it turned out, was the nematode *Caenorhabditis elegans*. This worm was an unlikely candidate to win a beauty contest, but its three-day generation time and penchant for self-fertilization, thus automatically establishing pairs of identical chromosomes, made it an excellent choice for genetic inquiry.[6] Like many other basic lines of inquiry in molecular biology, the foundation for *C. elegans* biology was laid predominantly at the Medical Research Council (MRC) laboratory in Cambridge, England.

Sydney Brenner of the MRC selected *C. elegans* in the early 1960s as a model to study multicellular phenomena, especially the nervous system.[7–9] Brenner wanted the smallest animal possible that was nonetheless complicated enough "to study the effects of mutations in single genes . . . to isolate mutants affecting the behavior of an animal and see what changes have been produced in the nervous system."[7] Brenner outlined his first ideas about how a research program might go forward in an October 1963 proposal to the MRC:

The *new major problem* in molecular biology is the genetics and biochemistry of control mechanisms in cellular development. . . . Part of the success of molecular genetics is due to the use of extremely simple organisms which could be handled in large numbers: bacteria and bacterial viruses. . . . We should like to attack the problem of cellular development in a similar fashion, choosing the simplest possible differentiated organism and subjecting it to the analytical methods of microbial genetics. Thus we want a multicellular organism which has a short life cycle, is easily cultivated, and is small enough to be handled in large numbers, like a microorganism. It should have relatively few cells, so that exhaustive studies of lineage and patterns can be made, and should be amenable to genetic analysis.[9]

Brenner went on to tout the virtues of *Caenorhabditis briggsiae,* which was the initial focus until supplanted by another species, *C. elegans.* Brenner's logic

Sydney Brenner, a researcher at the Medical Research Council's molecular biology laboratory in Cambridge, England, was an influential early champion of genome mapping. His experimental animal of choice, beginning in the 1960s, was the nematode worm *Caenorhabditis elegans. Courtesy Sydney Brenner*

followed that being pursued by Seymour Benzer at Caltech to understand neural function and other complex phenomena in *Drosophila,* the fruit fly. To carry out Brenner's agenda, it was necessary to find mutant nematodes, with observable differences, and also to assemble an awesome mass of structural information—the lines of descent of all cells in the worm's body, and the connections between them. Several laboratories at the MRC laboratory in Cambridge dedicated themselves to doing just that.

John Sulston, through a monumental effort, traced the development of the more than nine hundred somatic cells in the nematode's body, by watching the worms develop under a microscope with special optics that enabled him to observe every cell in the transparent body of the nematode.[6; 7; 10–13] Sulston made meticulous records of which cells produced which, and thus created a "pedigree" of all 959 nongerm cells. This was an incredible feat, producing the basic information necessary to trace what happened when the development of specific cells was disrupted. With this set of lineages, it was possible to observe directly the effects of killing a particular cell and to observe how its death affected the nematode's behavior and ability to survive.

In another *tour de force,* John White, Eileen Southgate, and Nichol Thompson of the MRC group reconstructed the "wiring diagram" of the worm's nervous system, analyzing twenty thousand photographs taken by an electron microscope.[6; 10; 14] They traced the main connections linking all 302 nerve cells. These two efforts are mind-boggling in their detail. The nematode was a reductionist's delight. It was conceivable that with these basic tools it would become possible to understand the entire organism's biology in all its mechanistic detail. The notion was not that all the biology would be explained by structural details, but that the structure was the best scientific strategy to try to get at both genetic and environmental factors influencing development.

The cell lineages and connectivity maps were the starting points to study what happened when something disrupted normal structure. Structural genetics was the crucial missing element. Even as the work began on *C. elegans,* the purpose was to correlate behavior with structure; DNA was the conceptual starting point. Sulston and Brenner estimated the size of the genome early on, to see what they were up against. They estimated the genome size by grinding up nematodes and seeing how much DNA their cells contained, and also by observing how long it took for separated DNA strands to reassemble into double helices. (The more complex the genome, the longer this took.) These methods suggested that the genome consisted of eighty million base pairs, giving it "the smallest value of any animal."[15] (This estimate was increased to 100 million base pairs in the late 1980s. Corrections came because the physical map then nearing completion gave more accurate data, and it became clear that the genome of the bacterium *Escherichia coli,* whose size had been a scaling factor for the *C. elegans* calculation, was larger than originally thought.)[16] The genome was segmented into six chromosomes containing seven hundred genes that were mapped by 1988.[6] A physical map of the worm's genome was the critical next step. Sulston and Coulson took it.

Extending the length of DNA in each clone simplified the physical mapping process. Dozens of laboratories helped improve cloning vectors that could consistently contain DNA inserts thirty thousand to forty thousand base pairs long, and these quickly became standard fare. This reduced the number of DNA fragments that had to be sorted through and ordered to make a library. With these advances, it became possible to take DNA from chromosomes, clone it, and reconstruct the order of cloned DNA fragments. Eventually a complete map of the *C. elegans* genome could be assembled. This kind of physical map had an enormous advantage—the chromosomal DNA would be not only mapped but also cloned and stored in the freezer for further analysis. If one wanted to study DNA from a region known to contain a gene, for example, one could go the freezer and pull it out, or call the scientist who had it in his or her freezer.

Coulson and Sulston labored for several years to make collections of *C. elegans* DNA fragment clones and then "fingerprint" the fragments.[4] DNA fragment patterns were stored in a computer, which looked for other clones that might exhibit a similar pattern, thus indicating overlap. Coulson and Sulston had over 80 percent of the *C. elegans* genome covered by sixteen thousand clones in 1986. When ordered into overlapping clusters, they fell into seven hundred groups. This was a great boon to the close-knit nematode research community, but the problem remained of how to close the gaps. Those seven hundred clusters should eventually resolve into the six chromosomes. David Burke, Georges Carle, and Maynard Olson of Washington University in St. Louis helped solve the problem with a gift from yeast.

Burke, a student in Olson's laboratory, became interested in the goings-on in chromosomal structure. Burke and Carle constructed cloning vectors that yeast cells would recognize as chromosomes. Because they were treated as chromosomes, the fragments of DNA being copied could be enormously large. The idea was to co-opt the normal cellular machinery that copied and distributed DNA to new cells. Burke succeeded in making artificial chromosomes containing DNA fragments more than ten times larger than could be cloned in bacteria. The products of Burke and Carle's artifice became known as yeast artificial chromosomes, or YACs for short.[17]

The length of DNA fragments cloned in YACs helped solve several serious problems at once. First, far fewer clones were needed to span a chromosomal region. Second, the problems in cloning some genes in bacteria might be overcome in yeast, whose biology was more similar to that of higher organisms.[18] Third, the longer fragments improved prospects for detecting overlap, making it easier to find next-neighbor clones. The fraction of the clone needed to detect overlap was smaller, dramatically improving the speed of making a complete physical map.[19] Using YACs, the MRC and Washington University groups were able to span many gaps, reducing the number of contiguously mapped regions (nicknamed contigs) from 700 to 346 in seven months.[20] By late 1989, the number of gaps in the *C. elegans* map was down to 190;[21] by 1992, more than ninety million base pairs of the genome were physically mapped, with only forty remaining gaps.[22] A new refrain was heard in nematode laboratories: "The gaps in maps are filled mainly with the YACs."

Physical mapping was also greatly assisted by the ability to separate much longer DNA fragments in the laboratory. In 1984, David Schwartz and Charles Cantor, then at Columbia University, developed the DNA separation technique known as pulsed-field electrophoresis. The method was an elegant, if somewhat slow, way to distinguish DNA fragments millions of base pairs in length. Dealing with such enormous fragments entailed special handling to minimize inadvertent fragmentation and applying electric fields that changed direction periodically.[23] Many embellishments of this technique soon followed, and it became possible to make maps of large chromosomal regions by this technique relatively quickly. Although initially the technique did not result in sets of *clones* of DNA from the region for further study, it did make mapping much faster, and it could help establish landmarks for subsequent analytical steps. With YAC-sized clones, pulsed-field electrophoresis was necessary for separation. The method became integral to physical mapping involving very large DNA fragment clones, until supplanted by faster techniques.

The physical maps of yeasts and nematodes became extremely powerful tools for genetics and for general understanding of biology. They eliminated the need for each researcher to develop a clone library laboriously and to screen it independently. The nematode physical map was put in a computer database available to all laboratories. Those who discovered genes or mutant organisms fed into the system, thus linking physical and genetic maps in a living network of cooperating laboratories. This well-stocked toolbox—containing genetic linkage maps, cell lineage maps, physical maps, and mutant strains with known characteristics—enabled biologists to approach an understanding of *C. elegans* unequaled by understanding of any other organism of comparable complexity.[24] The wealth of structural detail, the quality of the researchers, and the persistent pursuit of the newest technologies proved the prescience of Brenner's insight.

The problem of physically mapping the human chromosomes was greatly underestimated. It had been blithely assumed that yeast and nematode maps would be completed quickly and that the same techniques would translate quickly and readily to the far larger and more complex human genome. The estimates of complete human chromosome maps within two or three years, proffered in 1987 and 1988, proved too optimistic, but only by a half decade or so. Despite misgivings as the genome project was launched in 1990, the first physical maps of human chromosomes were published in 1992. David Page and his group at the Whitehead Institute produced a map of the Y chromosome,[25; 26] and a team led by Daniel Cohen of France reported a map of chromosome 21.[27; 28]

The road to physical maps of yeast and nematode genomes was not yet at an end. With physical maps nearing completion, the next step was to determine the entire DNA sequence of the yeast and *C. elegans* genomes. A consortium of European laboratories began to sequence yeast chromosomes in 1988, joined by a group at Stanford University in 1990. The Washington University and MRC groups began a transatlantic joint project to sequence the *C. elegans* genome in 1990.[22] The road of a billion nucleotides began with a single base.

4

Sequence upon Sequence

THE FAR-REACHING SIGNIFICANCE of the discovery of DNA's double-helical structure was immediately apparent to many scientists. The physicist George Gamow, for example, was one of the first to realize that the information stored in DNA must be in the form of a four-letter digital code. Writing ten months after Watson and Crick first described the structure of DNA, he noted:

The hereditary properties of any given organism could be characterized by a long number written in a four-digital system. . . . the enzymes (proteins), the composition of which must be completely determined by the deoxyribonucleic acid molecule, are long peptide chains formed by about twenty different kinds of amino acids, and can be considered as long "words" based on a twenty-letter alphabet."[1]

Like the order of 0's and 1's, the two-letter digital code in computer software, the order of the chemical subunits of DNA contained the instructions not only for the assembly of proteins but also for their biological function. The software was useless without the hardware to translate it, but a great deal could be learned by looking at the software code. If one was trying to understand what a computer was doing or the nature of an error that disrupted it, then the software was a good place to start.

Determining the order of bases in the entire genome is the core idea that started the genome project. Each nucleated cell of the human body contains forty-six chromosomes: twenty-two pairs of nonsex chromosomes and a pair of sex chromosomes (two X's for females, and X and a Y for males). The idea, as initially posed, was to determine a reference sequence for each chromosome. Each chromosome is an extraordinarily long DNA molecule, from fifty million to hundreds of millions of base pairs in length, bundled with proteins and RNA. The total number of base pairs in a complete reference sequence of the human genome was estimated at more than three billion, necessarily a rough estimate until physical maps are completed.

Given that no more than a few hundred thousand base pairs had ever been sequenced in a contiguous region in the mid-1980s, sequencing the genome was a largely impractical idea when first posed. The idea nonetheless initiated a debate within science that broadened the definition of the Human Genome

Project; DNA sequence information remained a central, if no longer exclusive, objective. Having the map and sequence information would not impart all knowledge about biology, because most interesting questions are about function rather than structure. DNA sequence information was, however, enormously useful for several reasons.

The DNA code is shared among organisms. Genes with similar functions are often historically related, having sprung from a common ancestral gene. Their DNA sequences are similar, and hunting for such similarities is a powerful way to get hints about the function of a new gene. If a newly discovered disease gene is similar in sequence to a yeast gene that codes for a cell surface receptor, for example, experiments to look for receptor proteins are in order. A cancer-associated gene whose sequence is similar to a growth-regulating molecular switch in yeast is a clue to the origins of cancer.

DNA sequence is, in this sense, the *lingua franca* of biology, because all organisms speak it. Most genes are conserved through evolution or built from bits and pieces of existing genes, as tweaked and tuned by evolutionary history. Examining sequence similarities discloses the historical relationships between genes, and hence the relatedness of the proteins they produce. Structural similarity suggests (but does not establish) functional similarity. The sequence of amino acids in a protein can provide the same kind of information about relatedness, but it requires having sufficient amounts of pure protein to analyze. It is much easier to examine sequence at the level of the gene, because DNA can be spliced and copied, providing enough material to sequence. By using DNA sequence to suggest the function of the protein it produces, scientists can shortcut the long and tedious process of purifying the protein.

DNA sequence is also a natural way to catalog genetic information. What better way to keep track of genetic information than by storing its digital code? The cataloging process can lead to surprises. The first regions selected in the project to sequence the *C. elegans* genome, for example, revealed far more genes than expected. The region was selected, in part, because it was known to be gene-rich, but the number of genes was twice as large as projected. The project also turned up several genes of known function that had not been previously found.[2]

Sequence data also promised to serve as a starting point for biology in a way that had not until then been systematically pursued. In early 1992, a massive European collaboration succeeded in sequencing chromosome 3 of baker's yeast, the first chromosome of a nucleated cell to be sequenced. The sequence was achieved by thirty-five laboratories coordinated through a European Community project, culminating a complex three-year collaboration. An even larger and more complex collaboration than the *C. elegans* sequencing effort, it was a major stride forward. The publication in *Nature* had 147 authors from thirty-seven institutions.[3]

The complete sequence contained 182 apparent genes, only thirty-four of which had ever been mapped; this was somewhat higher than a theoretical estimate of 160 genes made by a Japanese group.[4] It showed the promise of sequencing for discovering genes, as "even in a genome as small and as intensively studied as that of yeast, only a minor fraction of the genes has been identified by classical means."[3] The authors of the *Nature* article concluded that systematic sequencing projects could "reveal new functions that have been missed by more traditional approaches and also illuminate the mechanisms of genome evolution."[3] With the sequence in hand, an obvious goal was to determine the function of these 182 genes. This was much more feasible in yeast than most other organisms because, in the authors' words, "the functional analysis of novel genes discovered from the sequencing is facilitated by the easy methods for gene disruption and replacement . . . available in yeast."[3] It turned out that only 20 percent of the newly sequenced genes were similar to those found in various databases, and that "yeast molecular geneticists are working on only a small subset of the problems presented by their organism."[3] Once this functional catalog expanded, it would serve as the reference book for studying function in other organisms. The yeast sequencing project was a major boost to European genetics. The battle was joined to sequence the remainder of the yeast genome, with groups from Canada, the UK, Japan, and the United States mounting projects on one or more chromosomes, in hopes of having a complete reference sequence by year 2002.[5] The *C. elegans,* yeast, and other large sequencing projects began the slow process of turning biology on its head—starting from DNA sequence information and working toward function rather than the other way around.

As noted in Chapter 1, DNA is transcribed into RNA, usually on the way to protein. RNA serves several functions, some of it involved in editing and splicing the genetic code. RNA can be a messenger—the vehicle to translate from DNA code to strings of amino acids that become proteins. Some RNA becomes part of the cellular machinery and is never translated into protein. Proteins, however, make up most cellular structures and mediate the vast bulk of chemical reactions within cells.

The flow of information is usually from DNA through RNA to protein— what Crick called the "central dogma," as elaborated in the late 1950s and early 1960s.[6] We now know that information can also go from RNA to DNA when cells are infected with some viruses, reversing the flow, as occurs with AIDS infection. Some cellular RNA is occasionally copied into DNA and then inserted into chromosomes as well, but these counterexamples are small eddy currents in the torrential outflow of information that begins with DNA.

Chromosomal DNA is the terrain to be mapped by the genome project, and DNA sequencing provides the map with the highest possible resolution. The order of base pairs in DNA is the raw information. Getting at that order is consequently of central importance. In finding the specific defect for cystic fibrosis, for example, several laboratories engaged in massive sequencing efforts. Indeed, the Du Pont company, which developed an automated DNA sequencing instrument and sold it for a few years, crowed that its technology had uncovered the gene.[7] A few humans were also involved, but the importance of technology was nonetheless worth noting. Two methods for determining DNA sequence were developed independently at Cambridges separated by the Atlantic Ocean.

Methods to sequence DNA were conceptual extensions of work on sequencing proteins. In 1945, Frederick Sanger in Cambridge, England, set out to determine the order of amino acids in insulin, the protein hormone used to treat diabetes.[8] The sequence of insulin was a landmark in protein chemistry and earned Sanger a Nobel Prize in 1958. Protein sequencing was given a major boost in 1950, when Pehr Edman of the University of Lund in Sweden discovered a way to chop one amino acid at a time from the end of a protein.[9] By removing one amino acid at a time, the sequence was directly determined. Previous methods had required breaking proteins into small fragments, analyzing the order of amino acids in the fragments by a variety of methods, and then reconstructing the overall order of the original molecule. Edman's method was not only easier to understand but also proved well suited to automation. By 1967, instruments to determine amino acid sequence for proteins were on the market. These evolved into rapid and reliable protein sequencing instruments.[10; 11]

The next step up from protein sequencing was RNA. It took Robert W. Holley and his colleagues at Cornell University seven years to determine the order of seventy-seven bases in one form of RNA. Like the first protein-sequencing methods, they broke the RNA molecule into small fragments and reconstructed the order.[12; 13] For several years, DNA sequencing was done by transcribing it to RNA and then deducing the RNA sequence of short segments. This was the strategy used to determine the first DNA sequence, published in 1971: the short "sticky ends" of bacteriophage lambda (λ).[8; 14] The methods were too slow and tedious to be scaled up for large DNA molecules. Sanger recognized this, and the importance of more efficient DNA sequencing.

Sanger's DNA sequencing method took advantage of how cells make linear strings of DNA. Cell enzymes start from DNA in a chromosome and use it as the template to make copies, preserving the order of base pairs. The chemical backbone of the base-pair unit (nucleotide) in the DNA molecule is the same, a sugar and a phosphate group. The only variable is which base is inserted. The outer backbone is thus a monotonous repeat of sugar-phosphate-sugar-phosphate . . . The information is contained in the order of bases in the "rungs" of the DNA ladder. Each base has only one complementary base. A binds to T and C binds to G. DNA in chromosomes is stored as two strands of DNA bound to each other, with one strand exactly complementary to the other. In making copies of the chromosomes, the strands are unzipped and new copies made of each strand. A new pair of identical double-stranded DNAs results.

Sanger's idea was to adapt the cell's natural machinery, but to introduce chemical tricks to produce the DNA sequence. Sanger's initial method was to supply all of the components, but to "starve" the reaction of one of the four bases needed to make DNA. The cell would copy strands of DNA, but would run out of one of the bases. All the chains would run out just before the same base, starved of that one component. In the sequence ACGTCGGTGC, for example, starving for T would produce ACGTCGG(blank) and ACG(blank). The T in the longer fragment was produced before it ran out of T precursors;

Frederick Sanger pioneered work on the sequencing of DNA, RNA, and proteins at the MRC's Cambridge laboratory. His emphasis on understanding biological function through molecular structure has guided British participation in genome research from the beginning. *Courtesy Frederick Sanger*

the shorter fragment, perhaps because the reaction started just a bit later, ran out before it got there. By separating the resulting molecules by length, it would be obvious that the eighth and fourth positions should be T, because the chains of seven and three bases (one less than the "T" positions) were present. Starving for G would produce AC(end), ACGTC(end), ACGTCG(end), and ACGTCGGT(end), meaning G was in positions 3, 6, 7, and 9. The whole sequence would follow directly from starving for each of the four nucleotide precursors.

Sanger's next trick was to find chemicals that were inserted in place of A's, C's, T's, and G's, but caused a growing DNA chain to end. In the above example, the fragments for the "false" T (T*) would be ACGT* and ACGTCGGT*. With one terminator for each base type, the sequence could again be read directly by just measuring how long they were. Sanger presented his first partial DNA sequence to an awestruck audience in May 1975[15–17] and published the simpler chain-terminator method in 1977.[18]

Sanger noted in a marvelous autobiographical article reviewing his career, "Sequences, Sequences, and Sequences," that "of the three main activities involved in scientific research, thinking, talking, and doing, I much prefer the last and am probably best at it."[8] He also gave some insight into why the MRC laboratory in Cambridge played such a central role in the development of modern molecular biology.

I was in the fortunate position of having a permanent research appointment with the (British) Medical Research Council, and was not under the usual obligation of having to produce a regular output of publishable material, with the result that I could afford to attack problems that were more "way out" and longer-term: in fact, as few others could adopt this approach, I felt under some obligation to do so. . . . I like the idea of doing something that nobody else is doing rather than racing to be the first to complete a project.[8]

Several thousand miles away, in the other Cambridge, Allan Maxam and Walter Gilbert of Harvard developed an entirely different sequencing method, based on chemical disruption of DNA. They were studying the regulation of a bacterial gene, a region of DNA that served as a "switch" to turn the gene on and off. In the off state, no RNA was transcribed; when it was on, RNA was produced copiously. A protein that stuck directly to DNA appeared to throw the switch. When the protein, dubbed the repressor, bound to its DNA target, it blocked RNA transcription of genes nearby. Maxam and Gilbert wanted to study the specific DNA site recognized by the repressor protein. They isolated a DNA fragment from the region and showed that a short stretch of DNA was "protected" from degradation when the repressor was present. When bound to DNA, the repressor protein protected the DNA from digestive enzymes. They made RNA from this region and spent two years laboriously determining the sequence of a twenty-four-base-pair region of DNA, using the fracture-and-reconstruct methods.[19]

A Soviet scientist, Andrei Mirzabekov, visited the laboratory twice during this period. His first visit, late in 1974 or early in 1975, was brief. He was finding ways to break DNA at specific base pairs by selectively adding methyl groups to specific DNA bases. Mirzabekov found that dimethyl sulfate destabilized the DNA, leading to breakage specifically at adenine (A) and guanine (G) bases. Mirzabekov, Gilbert, Maxam, and graduate student Jay Gralla discussed the possible use of such DNA-fragmenting reactions to study the repressor-binding region. They already knew the DNA sequence from the region, but wanted to know precisely where the protein bound. The idea was that protein binding would block not only enzymes but also chemical methylation—the addition of methyl (CH_3) groups to the bases in DNA. If they compared DNA fragments fractured in the presence and absence of repressor, there would theoretically be a stretch of DNA that would break without repressor, but would be protected with it.

The first experiment was a total mess, but the second showed the expected pattern. The results were reported at a Danish symposium in the summer of

1975.[20] This was independent verification of the digestion experiments, but direct chemical fragmentation of DNA gave a much more specific binding profile.

The new method had another extremely appealing feature. If reaction conditions could be found to fracture DNA selectively at each of the four bases constituting any DNA, it would be possible to read the DNA sequence directly by separating fragments according to length. Maxam adjusted reaction conditions until he could fragment DNA at G only, or at both A and G.[21] If a fragment appeared in both reactions, it was a G; if it appeared in the A + G reaction but not the G, it was an A. Maxam then found a similar method to break DNA at cytosine (C) and thymine (T). The base occupying each position in DNA could thus be inferred from the fragmentation in four separate chemical reactions. A DNA sequencing method was born.

Late in that summer of 1975, Maxam gave a talk at the annual New Hampshire Gordon Conference on nucleic acids, while Gilbert was in the Soviet Union. Maxam distributed a protocol for Maxam-Gilbert sequencing at that meeting.[22] Their method of DNA sequencing was published in 1977.[23]

Sanger later confided, "I cannot pretend that I was altogether overjoyed by the appearance of a competitive method [for DNA sequencing],"[8] although the two methods proved to have complementary strengths. The preferred approach varied according to what was sequenced and how the DNA was prepared.

Thus, between 1974 and 1976, two independent techniques for sequencing DNA were developed, each an elegant solution to a central methodological problem. The capacity to sequence DNA opened up an enormous range of experiments and complemented the other major technical triumph of molecular biology during this period—artificially recombinant DNA. Embellishments of these techniques were used to determine the sequence of progressively larger fragments, and eventually whole genomes. The first sequence of twelve base pairs in 1968 grew to the 5,386-base-pair genome of the bacterial virus phi-X, achieved in 1977 with the new Sanger method. The DNA sequence of the small chromosome within a human mitochondrion, the cell's energy pack, was determined in 1981. It consists of more than sixteen thousand base pairs.

Genome of a virus, one of the first—and smallest—genomes ever to be decoded, was worked out initially in 1977 by Frederick Sanger and his colleagues at the Medical Research Council Laboratory of Molecular Biology in Cambridge, England. The genome is represented here by a sequence of more than 5,000 letters, corresponding to the four nucleotides (adenine, thymine, cytosine, and guanine) of the viral DNA molecule, which has only a single circular strand during part of its life cycle. Starting from the top, the sequence runs from left to right on odd-numbered lines, and from right to left on even-numbered lines. (In its circular form, the two ends of the molecule are connected.) The particular sequence shown, representing the entire genetic inheritance of the extremely small bacterial virus designated phi-X174, is divided into nine genes, which code in turn for the amino acid sequences of nine different proteins. More than 500,000 such pages would be needed to similarly display the human genome.

GAGTTTTATCGCTTCCATGACGCAGAAGTTAACACTTTCGGATATTTCTGATGAGTCGAAAAATTATCTTGATAAACGAGGAATTACTACTGCTTGTTTA
TCAACTACCGCTTTCCAGCGTTTCATTCTCGAAGAGCTCGACGCGTTCCTATCCAGCTTAAAAGAGTAAAAGGCGGTCGTCAGGTGAAGCTAAATTAAGC
AACGATTCTGTCAAAAACTGACGCGTTGGATGAGGAGAAGTGGCTTAATATGCTTGGCACGTTCGTCAAGGACTGGTTTAGATATGACTCACATTTTGTT
TGAACTGAGTACTAAAGAATGGATAATCACCAACTTGTCGTAGCCTGAGTCTATCATTAGGTGCGAGAAAATTTTACAGTTGTTCTCTTAGAGATGGTAC
TACTGAACAATCCGTACGTTTCCAGACCGCTTTGGCCTCTATTAAGCTCATTCAGGCTTCTGCCGTTTTGGATTTAACCGAAGATGATTTCGATTTTCTG
TCGCTCCCATAGGATGTTTCAGGTCGCATGGTATTTGCGTTCGGAGTTGCGTCGCTGCTCGTGCTCTCGCCAGTCATCGTTAGGTTTGAAACAATGAGCA
TTCCTGCTCCTGTTGAGTTTATTGCTGCCGTCATTGCTTATTATGTTCATCCCGTCAACATTCAAACGGCCTGTCTCATCATGGAAGGCGCTGAATTTAC
CAGTCATTCTTGCAGTCACAAAGGACGCGCATGTGCGTTCCATTTGCGCTTGTTAAGTCGCCGAAATTGGCCTGCGAGCTGCGGTAATTATTACAAAAGG
GCAGAAGAAAACGTGCGTCAAAAATTACGTGCGGAAGGAGTGATGTAATGTCTAAAGGTAAAAAACGTTCTGGCGCTCGCCCTGGTCGTCCGCAGCCGTT
TTAAATAGGAGTTCATTCCCCGGCTTCGGGGACGTTATTTTTTAACAACTGGTGGATGTATGGTTTCTGCTGCCGGAAATGCGAACGGAAATCATGGAGCA
ATGTCTAATATTCAAACTGGCGCCGAGCGTATGCCGCATGACCTTTCCCATCTTGGCTTCCTTGCTGGTCAGATTGGTCGTCTTATTACCATTTCAACTA
TACAGATGTCATCTCAGTTATCGTTCCGGTGCTGCGTTACCTCTTTCTGCCTCTCGCGGTTGCCGCAGGTAGAGCTTCCTCAGCGGTCGCTATTGGCCTC
TTTTTACTTTTTATGTCCCTCATCGTCACGTTTATGGTGAACAGTGGATTAAGTTCATGAAGGATGGTGTTAATGCCACTCCTCTCCCGACTGTTAACCAA
GATATCTATAGTTTATTGGGACTTTGTTTACGAATCCCTAAAATAACCATAGTCCCAATTAGCACGGTTCTTTTCGCCGTACCAGTTATATTGGTCATCA
CGTATTTTAAAGCGCCGTGG...ATGCCTGACCGTACCGAGGCTAACCCTAATGAGCTTAATCAAGATGATGCTCGTTATGGTTTCCGTTGCTGCCATCT
TCGTCGAACGTCTGGGTATTACAGTTATCTACACCATCTTCAGCAGTAAACCGCTCTTTCGAGTCAGAGTCCTCCTTCGCCTCGTCAGGTTTACAAAAAC
TATGCTAATTTGCATACTGACCAAGAACGTGATTACTTCATGCAGCGTTACCATGA.GTTATTTCTTCATTTGGAGGTAAAACCTCATATGACGCTGACA
ACTTGTGCTGGTCTTTTGACCGGATTGCTGCAAACCAGTCAAGGTAGTTGTAGTATCGGTCTACGGGTCTCTAATCTCGCGTACTGTTCATTTCCTGCCA
ACAGACCTATAAACATTCTGTGCCGCGTTTCTTTGTTCCTGAGCATGGCACTATGTTTACTCTTGCGCTGGTTCGTTTTCCGCCTACTGCGACTAAAGAG
GGAAGTATCTTTAAAGTGCGCCGCCGTTCAACGGTATGTTTTGTCCCAGCGGTCGTTATAGCCATATTCAGTTTCGTGGAAATCGCAATTCCATGACTTA
ATGTTTTCCGTTCTGGTGATTCGTCTAAGAAGTTTAAGATTGCTGAGGGTCAGTGGTATCGTTATGCGCCTTCGTATGTTTCTCCTGCTTATCACCTTCT
TTGACTTGCTGACTTTGTGACCAGTATTAGTACCAACGCTTATTCATGCGCAAGAACCTTTAGTGGTCTTCCGCCAAGGACTTACTTACCCTTCGGAAGT
GTTGCAGTGGATAGTCTTACCTCATGTGACGTTTATCGCAATCTGCCGACCACTCGCGATTCAATCATGACTTCGTGATAAAAGATTGAGTGTGAGGTTA
TTTCAGACTTTGTACTAATTTGAGGATTCGTCTTTTGGATGGCGCGAAGCGAACCAGTTGGGGAGTCGCCGTTTTTAATTTTAAAAATGGCGAACGCAAT
TATTTCTCGCCACAATTCAAACTTTTTTTTCTGATAAGCTGGTTCTCACTTCTGTTACTCCAGCTTCTTCGGCACCTGTTTTACAGACACCTAAAGCTACA
GGACTAATCGCCGCAACTGTCTACATAGGTAGACTTACGTTACTTCTTTTGGTGGTAATGGTCGTATTTGGCAGTTTGATAGTTTTATATTGCAACTGCT
TTGTTTCAGTTGGTGCTGATATTGCTTTTGATGCCGACCCTAAATTTTTTGCCTGTTTGGTTCGCTTTGAGTCTTCTTCGGTTCCGACTACCCTCCCGAC
TAACGGCCCGCAAGCCCCTTCCTGCAGTTATCAGTGTGTCAGGAACTGCCATATTATTGGTGGTAGTACCGCTGGTAGGTTTCCTATTTGTAGTATCCGT
AACGTCTACGTTGGTTTCATGGTTTGGTCTAACTTTACCGCTACTAAATGCCGCGGATTGGTTTCGCTGAATCAGGTTATTAAAGAGATTATTTGTCTCC
AACTGGCGGAGGTTTGTTAAATCTGTACCGCGGTGGTCGTTCTCGTCTTCGTTATGGCGGTCGTTATCGTGGTTTGTATTTAGTGGAGTGAATTCACCGA
AAAGCCGCCTCCGGTGGCATTCAAGGTGATGTGCTTGCTACCGATAACAATACTGTAGGCATGGGTGATGCTGGTATTAAATCTGCCATTCAAGGCTCTA
CTTCACGGTCGGACGTTGCATGGAAGTTCTTCAGGAAATGGTCGAAATCGTTATCGTGTGCTTTGTTTTGATCCCCGCCGGAGTAGTCCCAATCCTTGTA
TGCCGTTTCTGATAAGTTGCTTGATTTGGTTGGACTTGGTGGCAAGTCTGCCGCTGATAAAGGAAAGGATACTCGTGATTATCTTGCTGCTGCATTTCCT
GGTCAACGTAAAATCATTCGAGAAAAACTAAGAGTTTAGGCCGCAGTTGGTATGGTCGTCTCCTTCGTAGTCGTGGTCGTGCGAGGGTTCGTAATTCGAG
ACAATCAGAAAGAGATTGCCGAGATGCAAAATGAGACTCAAAAAGAGATTGCTGGCATTCAGTCGGCGACTTCACGCCAGAATACGAAAGACCAGGTATA
CCTTTGGACGACAACGAACCTTTCTAACCACAAAAGGTATTATCTGCGTTGCGTCGTCATCTGAGGAAGACA.CTTATTCGTTCGTAGAGTAAAACACGT
GAGATTATGCGCCAAATGCTTACTCAAGCTCAAACGGCTGGTCAGTATTTTACCAATGACCAAATCAAAGAAATGACTCGCAAGGTTAGTGCTGAGGTTG
GGTCTTCGTCGTAGTCACTGCTGTAATCTTTATAGGAAACGTCATCGCGGTTATACTCTTCTCGGTATGGCGACTAAGACGCAAACGACTACTTGATTCA
TGTGGTTGATATTTTTCATGGTATTGATAAAGCTGTTGCCGATACTTGGAACAATTTCTGGAAAGACGGTAAAGCTGATGGTATTGGCTCTAATTTGTCT
TAAGTCTTCCCATTATTCTTGCTTGGTATTTTTTCGGAGGTTCTAAACCTCCGTACTTTTGTATGTTAACCCTCCCACAGTTAGGACTGCCAATAAAGGA
GTCACGCTGATTATTTTGACTTTGAGCGTATCGAGGCTCTTAAACCTGCTATTGAGGCTTGTGGCATTTCTACTCTTTCTCAATCCCCAATGCTTGGCTT
CGGCAGTTGTATGTATAGTGGTAATAGCTTGAGTTGCGGGACGTATGCTTTTCTGTCTTAGAGAAGGTTCTCGAACTACGACAATAGGTAGACGAATACC
CATAAGGCTGCTTCTGACGTTCGTGATGAGTTTGTATCTGTTACTGAGAAGTTAATGGATGAATTGGCACAATGCTACAATGTGCTCCCCCAACTTGATA
CCCGCAAGTCGTCGGTCGAACGTTTTGACGCATTGGCAGAAGAGCAAGAGATTTTTGGTAAAAAGCAGGGGAAGCCCCGCCACCAGATATCACAATAATT
TCTTAAGGATATTCGCGATGAGTATAATTACCCCAAAAAGAAAGGTATTAAGGATGAGTGTTCAAGATTGCTGGAGGCCTCCACTAAGTATCGCGTAGA
GATTAGCCAGCAGTCGGTTGCACTCTCACAGTTTTTGCTATTTGGTTGGTAGTCGTACTCGGACAGCGTAACGTAAGTAGTTTGCGACTTATCGTTTCGG
AGGCGTTTTATGATAATCCCAATGGTTTGCGTGACTATTTTCGTGATATTGGTCGTATGGTTCTTGCTGCCGAGGGTCGCAAGGCTAATGATTCACACGC
CGATGGACATCCTTCACAGGCGTATTTCACGTGGCGTACCTTTACTTCTGCCGGTAATCGACATGGTATGAGTCCGTGTGTTTTTATGACTATCGTCAGC
GTTGACCCTAATTTTGGTTGTCGGGTACGCAATCGCCGCCAGTTAAATAGCTTGCAAAATACGTGGCCTTATGGTTACAGTATGCCCATCGCAGTTCGCT
TGTATCTTTGGTTGTCGGTATATTGACCATCGAAATTCGCCGAGTGGAAATCGTAGTTGTCCGGTGTTGGTTGGTCTTGCACTTTTTCGCAGGACGCACA
GGCGAAATACGTTAACAAAAAGTCAGATATGGACCTTGCTGCTAAAGGTCTAGGAGCTAAAGAATGGAACAACTCACTAAAAACCAAGCTGTCGCTACTT
CGAACCATTCAACCTAATTCGTGAGGCACCTGTCTAAACAGTAACACTCGTAAAAGTAGGGCTTCAACGCCGAGTAAGACTAAGACTTGTCGAAGAACCC
TGGGTTACGACGCGACGCCGTTCAACCAGATATTGAAGCAGAACGCAAAAAGAGAGATGAGATTGAGGCTGGGAAAAGTTACTGTAGCCGACGTTTTGGC
ACGTCCAACCTATGCGGTTAGTAAAAATAGCTTCGCGCGTATTTAAACTCGTCTAAACAGCAGTCTCCAACGCGG

In 1984, the MRC Cambridge group sequenced the 172,000-base-pair genome of the Epstein-Barr virus, the cause of infectious mononucleosis and other conditions. The final landmark of the 1980s was the human cytomegalovirus, whose genome has more than 229,000 base pairs.[24] These achievements were accomplished without much automation. Computers were used to assemble the data into a coherent whole, but the vast bulk of the effort was done by human hands, eyes, and minds.

The next major step for DNA sequencing was to make it faster, cheaper, and more accurate. With enormous stretches of DNA whose sequence was yet to be determined, and with sequencing broken down to a series of standard procedures in repetitive steps, DNA sequencing was a natural target for automation.

Automation of biochemical reactions entailed mixing diverse reagents in small volumes, generally running a reaction inside a single small droplet. Molecular biologists and biochemists frequently ran dozens of reactions at once, each in its own test tube. Getting machines to do some of these mindless tasks would free postdoctoral fellows and graduate students to do more creative things, to move their minds from their hands to their science. By augmenting the duration, reliability, and speed of laboratory work, automation made possible experiments too large and complex to do manually. In the words of one molecular biologist, a robot was the "ultimate postdoc," an endorsement of the new instrumentation but also an acknowledgment of the tedious tasks routinely relegated to young and eager minds in molecular biology.

Devising instruments to automate processes used in molecular biology was pursued at only a few universities. Most of the work was concentrated in companies that sold analytical instruments to laboratories. A few companies were formed to develop instruments, usually growing out of academic centers to fill market niches left vacant by larger companies. Procedures to determine the order of amino acids in proteins were first automated in the late 1960s, followed by instruments to synthesize short proteins from amino acids. Analysis of DNA was next in line. Serious efforts to *synthesize* short segments of DNA, essential to developing highly sensitive probes for analyzing genetic experiments, began in the late 1970s and proved successful by the early 1980s.

Automation of DNA sequence determination began around this time in both Japan and the United States. In the United States, the first successful efforts leading to the current generation of fluorescence-based DNA sequencers began in the late 1970s at Caltech, one of the few academic centers interested in both molecular biology and instrument development.

Leroy Hood was at the center of efforts to marry high-technology instrumentation to molecular biology. Hood grew up near Shelby, Montana, a high school quarterback who directed his team to successive state championships. He went to Caltech as an undergraduate, and upon graduation joined an accelerated M.D. program at Johns Hopkins. He returned to Caltech to pursue a Ph.D., working in the laboratory of William J. Dreyer, whose interests centered on protein structure. In 1967, Dreyer's group was involved in automating the Edman reaction for protein sequence determination. The Caltech group worked with the Beckman Corporation to make an instrument. It became the first in a long line of triumphs, establishing Caltech as the capital of biotechnology instrumentation. Over the next two decades, Hood and his group at Caltech emerged as preeminent innovators.

Hood's main contribution to instrumentation was to create a laboratory environment with a breadth of expertise ranging from engineering through organic chemistry to molecular biology. His contributions to biology were broader than instrumentation. He was one of the leaders in molecular immunology, aimed at understanding the fundamental mechanisms at play in the body's principal defense system. He commanded one of the largest molecular biology groups in the world,[25; 26] with a legion of young faculty, postdoctoral fellows, graduate students, technicians, and others that generally hovered around sixty-five, and edged over one hundred in the summers.[27]

Developing instruments was a respectable enterprise at Caltech.[28] An accomplished fabrication shop helped build prototype machines, and the Hood group promulgated a philosophy that tied instrument development to solution of pressing problems in molecular biology, especially in immunology but also in several other areas. The group stayed near the forefront of molecular biology and defined the cutting edge of instrumentation development.

The Caltech group discussed ways to automate DNA sequencing procedures as soon as the Sanger and Maxam-Gilbert procedures became known, but there were several difficulties to be overcome. Henry Huang labored for several years to automate the standard DNA sequencing methods, using funds donated to Hood's group by Monsanto and money obtained from the Caltech president's fund.[29] Huang worked to build an apparatus that could detect DNA fragments, but the problem of detecting a very small signal from the DNA amid a very noisy signal, compounded by the primitive computers of the day, doomed the enterprise. Huang's efforts sparked a continuing interest in DNA sequencing, however, and it was picked up by Lloyd Smith when he joined the laboratory in April 1982.

Smith came to the Hood group from Stanford, where he had applied lasers and fluorescent methods to the study of biological questions.[30; 31] While he did not initially go to Caltech to automate DNA sequencing, Smith became intrigued by the problem. Hood had several times urged Huang to try fluorescent labeling instead of ultraviolet absorption, but Huang pointed out the difficulties.[29] Huang also considered measuring molecular mass directly to detect DNA, but again the technical problems were daunting. The Caltech group lacked the expertise in organic chemistry needed to attach fluorescent dyes to DNA. Smith filled that critical gap.

Huang left Caltech in September 1982, and work on DNA sequencing moved to Smith. A complex enterprise grew up, involving the talents of a team of people at Caltech and the biotechnology instrumentation firm Applied Biosystems. Smith's expertise combined organic chemistry and instrumentation, including the use of fluorescence and lasers. Smith worked to find appropriate dyes, then worked on the tedious and artful task of finding ways to attach them to DNA without perturbing the detection of DNA fragment size. He and the rest of the Caltech team also had to devise a practical way to activate the dyes by laser and to detect fluorescent emissions. Over two years, the Caltech team pushed inexorably forward.[32; 33]

Hood secured the necessary funding from the Weingart Institute and also used funds from corporate donations to the Caltech group from Baxter-Travenol, Monsanto, and Upjohn. Hood's ability to cultivate enthusiasm went well beyond academic circles, undergirding his prodigious fund-raising abilities. Priming the money pump was necessary to lay the foundation supporting a legion of young investigators who fiddled with expensive hardware.

While a DNA-sequencing prototype was being assembled at Caltech, a parallel effort moved forward at the nascent Applied Biosystems. Kip Connell from Hewlett-Packard directed the team there. Steve Fung worked on different ways to link fluorescent dyes to DNA. From 1983 through 1985, there was much give and take between Applied Biosystems, which focused on detection and slab gel techniques, and Caltech, where the emphasis was on fluorescent dye chemistry. The Caltech group announced its prototype in a June 1986 presentation.[32] By May 1986, Applied Biosystems had a commercial instrument ready for testing. Its DNA sequencer hit the market in 1987, and was soon joined by a rival fluorescence-based machine manufactured by Du Pont[34] and another machine based on detecting radioactive phosphorus.[35; 36]

Both the Caltech and Applied Biosystems projects got a big boost from two brothers and a fateful car ride. One day late in 1982, Tim Hunkapiller was driving his older brother Mike to the airport. Both brothers worked in Hood's group at Caltech. The Hunkapiller brothers grew up in Oklahoma and attended university there. Mike made his way to Caltech, where he studied physical chemical methods to understand enzyme-mediated reactions. In 1976, two years after getting his Ph.D., Mike planned to return to Oklahoma. Hood talked him into staying at Caltech, and Mike got involved in designing an instrument to sequence proteins in much smaller amounts than possible with the instruments developed previously at Caltech and elsewhere. Mike devised a new method that was far more sensitive and could be done on much smaller samples.[10] The result was a prototype instrument to do protein sequencing that could be used for a wide array of problems unapproachable with the previous generation of instruments.

The Caltech group probed several companies for interest in manufacturing the protein sequencer. They approached Becton-Dickinson, Du Pont, and Beckman Instruments. Beckman middle managers saw little prospect for expansion of the protein sequencing market they already dominated. They seemed to believe a new instrument might compete with their existing one. Du Pont nibbled but did not bite.

Applied Biosystems was founded to build instruments for the new biotechnology. Marvin Carruthers and Hood were on the board of AmGen, a biotechnology company. Carruthers, from the University of Colorado, worked with Hood's group to devise chemistry for a different machine—one that made short stretches of DNA with specified base sequences. The Carruthers collaboration with Hood's group at Caltech culminated in a machine brought to market by Applied Biosystems in 1982.

The group of venture capitalists who helped found AmGen spoke with Hood and Carruthers, who talked enthusiastically about an emerging instrumentation market for molecular biology. The venture capital sponsors were willing to try to tap this market, so they provided capital to start the company that became Applied Biosystems.

A group from Applied Biosystems visited Caltech in the spring of 1981, just after the first gas-phase protein sequencer had been developed. Applied Biosystems later picked up the license for the protein sequencer and evinced interest in the protein synthesizer, DNA synthesis machine, DNA sequencer, and other instruments under development at Caltech. With this suite of four instruments, a laboratory could break down proteins and DNA into their component sequences or build up a specified protein or DNA sequence from scratch. These were essential steps in a wide range of molecular biology experiments. Instruments to sequence and synthesize proteins and DNA formed the technological quartet for a new approach to biological research.

Applied Biosystems became a successful startup company, beating its larger rivals in the instrumentation business. It first marketed the Caltech-inspired protein sequencer in February 1982, and followed in December with the DNA synthesizer based on improved chemistry developed in Caruthers's laboratory. (Beckman had a sublicense to this chemistry as well.) Arnold Beckman, whose foundation was a generous funder of research at Caltech, hit the roof. His managers might not have been interested, but, he felt, Hood should have notified him directly that Beckman Instrument's middle managers had tunnel vision. While Beckman himself no longer had line authority, his name was still on the company, and he believed he could have taken remedial action. Beckman's fury intensified while Applied Biosystems displaced Beckman Instruments as the dominant force in protein sequencing within a year of introducing its instrument. Hood learned that selling a new idea required approaches to top management and also convincing middle management and corporate technical experts. Individuals at many levels could block a new idea; progress required all the gates to be open. Tensions between the Hood group and Beckman were ultimately healed over, after a few years' delay in securing the donation for the new Beckman Institute.

In 1982, Mike Hunkapiller was consulting with Applied Biosystems. Mike followed Applied Biosystems' early history with great interest, but was more interested in several biology projects at Caltech. Within a year, he changed his mind. He left Caltech to join the firm in July 1983, later becoming vice president for science and technology.[37] Applied Biosystems ultimately succeeded in developing the full set of four instruments for DNA and protein sequencing and synthesis and then began to work on yet other projects, such as the development of robots able to perform diverse biochemical reactions. Mike Hunkapiller was head of the development team.

Tim Hunkapiller had an entirely different scientific background and personal style. He was studying evolutionary biology and doing fieldwork in Oklahoma when he applied to Caltech for graduate school. He applied to Caltech in part because the application was free. He was rejected, but with the suggestion that he might be accepted if he took more physical chemistry. He agreed to take a special tutorial and was accepted at Caltech.

The Hunkapiller brothers' different styles led to a fruitful scientific collaboration. The DNA-sequencing project was one beneficiary. Tim's eye was caught by an article about separating molecules in gels within very thin capillary tubes, lodged in silicon wafers like those used for computer chips. Tim and Mike were speculating about how capillary separation techniques might be used to determine DNA sequences when Tim brought up the possibility of using four different color dyes, one for each nucleotide of DNA. This resuscitated Huang's idea in a new form. They discussed the idea with Smith, who was receptive. Smith pointed out that color dyes would not work because they would not be sufficiently sensitive to reveal the minuscule amounts of DNA in a gel. He suggested fluorescent dyes. With the Hunkapillers and Smith all enthusiastic, the balance tipped. Mike's support was especially critical, since he was Hood's trusted lieutenant.

It was a very long road, however, from idea to instrument. Smith donned the yoke at Caltech, and Applied Biosystems mounted its separate but interwoven effort. Tim Hunkapiller did not work directly in the group that Smith forged to make the DNA sequencer. Tim instead went on to focus on the use of computers to analyze information derived from molecular biology and became a prominent national figure in discussions on the subject.

While Applied Biosystems was hard at work in California, several other companies were also developing DNA sequencing machines. In 1987, the

---

Automated DNA sequencing technique is used to determine the order of the A's, C's, G's, and T's in a cloned sample of DNA. The DNA is first heated to separate the strands. The single strands are then cut at various points, and the resulting fragments are added to a mixture of fluorescent dyes in which strands that end in a specific letter are all tagged with the same fluorescent dye. The mixture is next run through an agarose gel, which in turn is passed through a laser beam. The shorter fragments, which travel farther in the gel, are exposed to the beam first, followed by successively longer fragments. Each base (A,C,G, or T) emits a different fluorescent color corresponding to its position in the DNA sequence. The DNA sequence can be inferred from the sequence of fluorescing colors emitted as the fragments in the gel go past the laser beam.

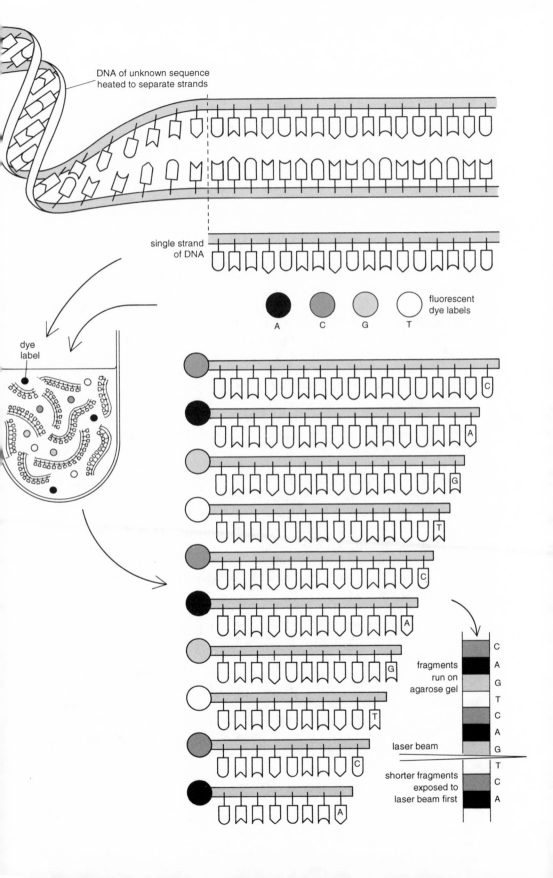

DNA of unknown sequence
heated to separate strands

single strand
of DNA

fluorescent
dye labels

A C G T

dye
label

fragments
run on
agarose gel

C
A
G
T
C
A
G

laser beam

shorter fragments
exposed to
laser beam first

T
C
A

chemical and pharmaceutical giant Du Pont announced a different method to use fluorescent dyes for DNA sequencing[34] and began to market its Genesis 2000 instrument within a year. EG&G Biomolecular, one of scores of EG&G companies that originally spun off from MIT in the postwar period, marketed another machine that was based on detecting radioactive phosphorus. The EG&G machine cost considerably less than the fluorescent instruments and was based on methods quite familiar to molecular geneticists.[35] It was aimed at sequencing projects on the scale undertaken by the average laboratory, however, not on the mega-sequencing scale envisioned by genome enthusiasts.[36]

Piecing together the history of the DNA sequencer revealed the tensions between science and industry in a highly competitive environment. Such tensions over priority and who had which idea first permeate science, and money intensified the conflict. The DNA sequencer became a source of great pride in collaboration, but also of some friction between Applied Biosystems and Caltech and among the major collaborators who made the first successful machines. Control over the underlying patents and royalties hung in the balance, in addition to scientific credit for originating an important new technology. The final truth was that no individual could take full credit.

Just as Nobel selection committees were perpetually unfair in conferring a prize on "winners" in science—ignoring the way science had changed so that most major advances required the efforts of hundreds, not one or two— choosing among the contributors to a technological development was not just perilous, it was nonsense. Lee Hood clearly created the working environment that harnessed the talents of those interested in technology and biology. Henry Huang kept the idea of automated DNA sequencing alive long enough to pass the baton. Lloyd Smith directed the team that fabricated the Caltech prototype DNA sequencer, as Mike Hunkapiller had done before him on protein sequencing. Tim Hunkapiller linked the analytical instruments to their computer interpretation. An Applied Biosystems team built its own prototype instrument that was quickly and successfully commercialized and spread throughout the world. No villains misappropriated the property of others, but competition for priority—and perhaps money—loosened the bonds of trust. Success shattered the collegium of science.

The commercial overlay of technology development also provoked international tensions. In this regard, the Caltech–Applied Biosystems group was regarded collectively as a national treasure. Norman Anderson, who had worked at several national laboratories over the years and had himself developed several instruments for biological research, chatted with me about U.S.–Japan trade tensions at a human genome meeting in July 1986. Referring to Japanese economic competition, he saluted Hood by saying, "Thank God he's on our side."[38] The United States was hardly alone in developing DNA sequencing machines and other instrumentation for molecular biology. There was movement aplenty on the other sides. The United States had a lead, but Europe and Japan were in the contest.

In Japan, the Science and Technology Agency (STA) began in 1981 to support a project to automate DNA sequencing. This program was the brainchild of Akiyoshi Wada, who was handed a mantle from STA to improve the analysis of DNA. He chose to focus on instrumentation and to automate well-established techniques, rather than simultaneously to develop new methods and automation technologies. Wada enticed several corporate sponsors (Fuji Photo, Seiko, and Matsui Knowledge Industries) into the project, which was eventually housed at the RIKEN Institute in Tsukuba Science City.[39-46] An independent automation effort at Hitachi culminated in a DNA sequencer that in 1989 was marketed only in Japan. (The Japanese genome program is discussed in detail in Chapter 15.)

The automation effort at the European Molecular Biology Laboratory in Heidelberg began in the early 1980s, supported by several European governments. Wilhelm Ansorge directed a team that produced an instrument that used fluorescent-dye detection. The EMBL instrument employed a scheme of fluor-labeled bases somewhat different from both the Caltech and the Du Pont designs.[47] The EMBL prototype served as the basis for the ALF DNA sequencing system marketed by LKB-Pharmacia, a Swedish company, beginning in 1989. The ALF system had complementary strengths to the Applied Biosystems approach, and large-scale sequencing projects used both.[2]

The ability to synthesize and sequence proteins and DNA revolutionized molecular biology; automating these tasks promised to consolidate the revolution. Hood and others saw automation as enabling assaults on larger and harder problems. He, Mike Hunkapiller, and Lloyd Smith preached that sequencing would continue to accelerate and that dedication to massive sequencing initiatives should await the development of better technology and better strategies for employing it. Automation would make currently impractical goals more practical.[48-50] Hood projected impacts in the practice of diagnosis and treatment, well beyond basic research applications.[51]

As the Human Genome Project shifted from a topic of debate to an ongoing research program in the early 1990s, it was becoming clear that large-scale DNA sequencing efforts would prove to be more than mere quantitative extensions of existing technology. After several years of research a second generation of automated "sequenators" appeared on the verge of development into instruments. New techniques seemed capable of increasing the speed of sequencing, reducing the amount of DNA needed to derive a sequence, and extending the number of bases that could be sequenced at a time.[52] Improvements seemed likely to be ten times or so more efficient at deriving sequence information from DNA prepared for sequencing. The steps for preparing DNA and the computer algorithms to piece sequences together began to loom as the main obstacles. Dedicated sequencing on a grand scale would entail much greater attention to the accuracy of data, quality control of laboratory practices, quantitative assessments of the validity of base pair assignments, sophisticated algorithms to weave long stretches of sequence data together, mathematical and computer methods to make comparisons of sequence data, and generally a more systematic production mode of operation.[53]

DNA sequencing caught the fancy of those who saw a new way to do biology. A simple but revolutionary new technique for producing enormous amounts of short stretches of DNA, called the polymerase chain reaction (PCR), promised yet another revolution.

Kary Mullis invented the PCR technique while working for Cetus Corporation. Mullis was prone to flights of fancy and emotive explosions. Of the three qualities needed to excel in corporate research and development—creativity, willingness to work hard, and an ability to work with others—he was off the charts on the first, passable on the second, and encountered difficulties with number three. He conceived of PCR in 1983, and it eventually worked. Mullis's Ph.D. was in chemistry, and he was hired to work on DNA synthesis at the University of California, San Francisco Medical Center, during the late 1970s. In 1979, he was hired by Cetus, a company named for the whale in its logo, in Emeryville across the Bay. At Cetus, Mullis eventually headed the group that made short stretches of DNA for experiments. Synthesizing DNA and using it to detect specific sequences led him to think about ways to copy DNA without having to clone it in bacteria, yeast, or other organisms. PCR was a powerful and simple technique to do just that.

PCR was an enormously powerful technique. By 1989, its revolutionary implications had begun to emerge. In genome mapping, the rapidity and simplicity of the technique enabled genome researchers to contemplate using the common language—short stretches of defined DNA sequence taken from PCR reactions—to merge genetic linkage maps and a variety of physical maps.[54] C. Thomas Caskey, a genome research director from Baylor University, introduced Mullis at a 1990 DNA sequencing conference, asserting that the genome project itself had become practical only in the wake of Mullis's discovery.[55] PCR was a godsend.

While heaven-sent, PCR did not arrive by direct descent. It came by a more circuitous route, namely Route 128 overlooking the Anderson Valley in Mendocino County, a "moonlit mountain road into northern California's redwood country."[56] Mullis struck upon the idea while driving along it with his coworker one Friday evening, for a weekend respite at his cabin.[56; 57] Later, back at Cetus, "I ran my favorite kind of experiment: one involving a single test tube and producing a yes or no answer. Would the PCR amplify the DNA sequence I had selected? The answer was yes."[56]

Mullis presented his idea as a scientific poster at Cetus's scientific retreat in June 1984. It was a troubled period in his personal life, and Mullis blew his fuse. He got into a late-night shouting match with another Cetus scientist, and was so combative that he kept many awake until past three in the morning with phone calls and bouts of yelling. Someone finally called a private security officer to walk Mullis along the beach until he calmed down. His tenure at Cetus was seriously in question: Mullis was creating havoc—shouting, threatening a coworker who was going out with his erstwhile girlfriend, arguing with Cetus's evening guards when he didn't have his badge to enter the building after hours. Rather than fire Mullis, Cetus ended his duties as a head of the DNA synthesis laboratory and let him pursue the PCR idea full-time.[58]

By June 1985, Mullis and many other Cetus scientists and technicians had made it work. PCR had gone from an idea to a demonstrated technique of great promise. It was beginning to be used throughout the company. The question of when and whether to publish it was discussed at several meetings. Mullis was to write up the basic idea promptly, and another team was to work on a paper using PCR to detect the sickle-cell mutation, as the first application to a practical problem. Mullis was slow to write up the basic method, becoming obsessed instead with producing fractal images on the company computer, and Cetus decided to let the sickle-cell paper go ahead, with Randall Saiki as the first author, because he had generated the data. Saiki presented the results in October. The paper was published in the December 20 issue of *Science*.[59] Mullis's paper, however, was rejected—first by *Nature* and then by *Science*. The importance of the technique and the numerous variations not covered in the Saiki paper were apparently lost on the reviewers. Like art critics in the time of Cézanne, they missed the point.

Mullis was not listed as senior author of the work cited as the standard first paper on PCR, an unfortunate consequence of his wanting to include additional experiments, his procrastination, and the short-sighted reviews at *Science* and *Nature*. Mullis had not been eager to publish the technique in the first place, and now he was cut out of the traditional standard of due credit. Most molecular biologists first heard of his technique only when he presented it at the June 1986 symposium, "The Molecular Biology of *Homo sapiens*" at Cold Spring Harbor Laboratory on Long Island, where his talk duly impressed the cognoscenti.[60] This presentation was arranged when a Cetus scientist called James Watson, who was organizing the meeting, to alert him to this important new technique. In his Cold Spring Harbor paper, Mullis noted the process of copying DNA repeatedly "seems not a little boring until the realization occurs that this procedure is catalyzing a doubling with each cycle in the amount of the fragment defined by the positions" of the short synthetic DNA stretches inserted into the reaction.[60] With the laws of exponential arithmetic, a series of doublings quickly amounted to a lot of DNA copies. The audience included many of the most esteemed molecular biologists in the world, and Mullis got center stage.

The original description of the method, along the lines of the paper rejected by *Science* and *Nature,* was finally published in a specialized journal only in late 1987.[61] In 1989, *Science* hailed PCR and its molecular accouterments as the first exemplars in an odd new annual ritual of declaring a "molecule of the year."[62; 63] *Science* thus belatedly recognized the fundamental importance of the technique, while neglecting to mention its earlier editorial mistake.

The publication experience caused the simmering tempers at Cetus to boil over. Mullis left Cetus in the summer of 1986, within months of his triumphant talk at Cold Spring Harbor.[58]

Copying DNA became possible in the 1970s with the advent of DNA cloning, and was a major advance, but PCR avoided the limitations and extra steps involved in molecular cloning. If what you wanted was information from a short stretch of DNA, rather than a long piece of DNA to study, PCR was the answer. PCR made it possible to amplify DNA fragments from much smaller samples of DNA, even down to the theoretical limit, a single molecule. PCR also made the copying process faster and cheaper and avoided the rearrangements and cell-culture manipulations in a bacterium, yeast, or other cell.

The PCR reaction was carried out in a test tube, using well-defined ingredients. These included the DNA to be analyzed, the nucleotide precursors to make new DNA, and a DNA polymerase enzyme. The first DNA polymerase was discovered by Arthur Kornberg at Stanford in 1955, in work that earned him a Nobel Prize,[64] and the intervening years had turned up a host of details about the process.[65] The PCR reaction built on what was known about how enzymes synthesized new DNA strands from existing ones.

The critical element for PCR was a set of DNA primers, short stretches of DNA of defined sequence at the ends of the DNA region to be copied. When bound to sample DNA, the primers created short regions of double-stranded DNA. The polymerase enzymes could not start making DNA strands from anywhere, but had to start from one end of a double-stranded region. Given such a starting point, as provided by primer that bound to native DNA, the enzyme could proceed to make a second strand complementary to the first. Twin primers pointing in opposite directions bracketed the DNA to be copied, thus defining the DNA region to be copied by PCR.

Starting with a very small amount of sample DNA and appropriate primers, a fragment could be copied (amplified) thousands, millions, or even hundreds of billions of times. The PCR reaction entailed heating the reaction mixture in cycles, to separate the DNA strands from each other. The original reaction

---

Polymerase chain reaction is used to rapidly amplify a specific DNA region. The region to be amplified is first bracketed by a pair of DNA primers, which set the starting points for making new strands of DNA, in opposite directions. The new strands of DNA are then separated, and copies are made again with a new pair of primers having the same sequence, so that only the same region of DNA is consistently copied. Each cycle leads to a near-doubling of DNA from that region. After dozens of repetitions, the process can yield billions of copies of the original DNA. The inventor of the PCR technique, Kary B. Mullis, was awarded the 1993 Nobel Prize in chemistry for this major contribution to genetic research.

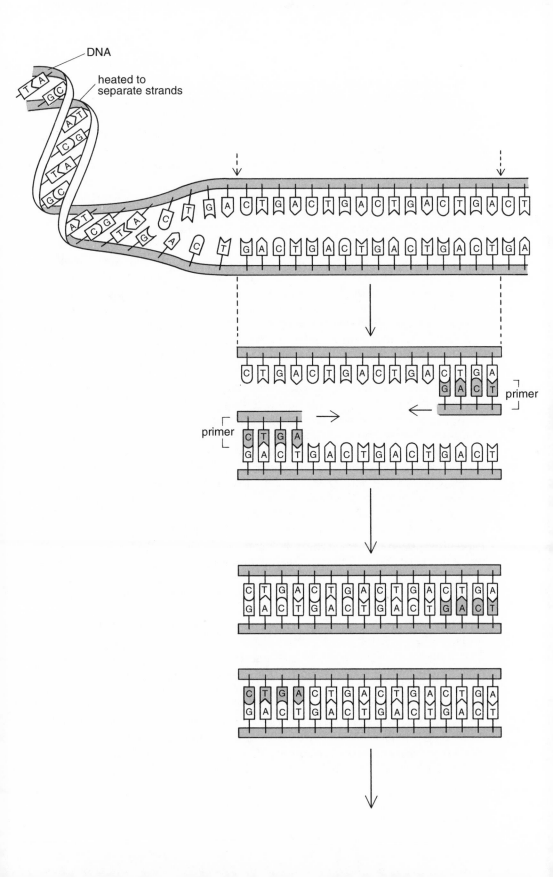

used a conventional DNA polymerase that had to be replaced with each cycle, making the process more complex and more expensive.[59] By mid-1986, the Cetus group was using a DNA polymerase enzyme isolated from a bacterium, *Thermus aquaticus,* found in hot springs, where bacteria routinely copied DNA at temperatures near the boiling point, as used in the PCR process.[66; 67] This sped the process, obviated the need to replace enzyme, and had other technical advantages, such as facilitating automation.

PCR involved no esoteric instruments, only a reliable means to heat and cool the reaction mix and the use of an enzyme and chemical reagents. PCR was wondrously flexible. It made DNA analysis simple enough that laboratories formerly hesitant to do work at the DNA level could do so. It was used to trace human origins to Africa (although this proved more controversial than initially imagined), to detect infections rapidly and with exquisite sensitivity, to diagnose genetic disease, to study the complex evolution of the immune system, and for a plethora of other applications.[57; 68] As PCR was being discovered, for example, the AIDS epidemic was just becoming known. During these same years, the virus causing AIDS was discovered. The problem of detecting infection was an urgent scientific problem, and detection of AIDS became one of the early applications for PCR. The first publication using PCR involved the detection of the sickle-cell gene,[58] demonstrating its usefulness in diagnosis of genetic disease. It was quickly applied to the diagnosis of infections,[69] to the study of immune function, and to the study of cancer. The generality and simplicity of the method created an entirely new market for reagents, heating-and-cooling instruments (called thermal cyclers), primers, primer synthesizers, and enzymes.[70] It spawned a mini-industry.

PCR became the subject of a patent battle when chemical giant Du Pont dug up a series of 1971–1974 papers from the laboratory of H. Gobind Khorana, an MIT Nobel laureate. Du Pont began marketing products for PCR amplification, challenging the Cetus patent and claiming the Khorana papers had placed the idea in the public domain. Cetus took Du Pont to court. In the first legal skirmish, Cetus emerged the winner as the U.S. District Court in San Francisco sustained Cetus's patent claims.[57; 71–75] Chiron, a nearby biotechnology company, announced it intended to buy Cetus a few months later. PCR rights were sold to the Swiss pharmaceutical conglomerate Hoffmann–La Roche as part of the deal. Roche trumpeted its acquisition of PCR rights on December 11, 1991, and formed a new unit, Roche Molecular Systems. The deal gave Roche research and diagnostic rights, while Perkin-Elmer retained rights for reagents, instruments, and nondiagnostic applications.[76] The sale provoked *Nature* to query, "Is Cetus Selling the Family Silver?"[77] Cetus had licensed research and diagnostic uses of PCR to Perkin-Elmer and Roche several years earlier, so *Nature*'s question was too late and misdirected. Cetus licensed PCR to Hoffmann–La Roche because it simply did not have sufficient marketing and distribution capacity for medical diagnostics, and it was seeking a company that would more aggressively pursue PCR applications than the previous licensee (Kodak). In February 1992, Roche announced a relaxation of licensing arrangements, eliminating up-front fees for academic and non-profit institutions and setting a maximum royalty rate of 9 percent for them.[78] While the top brass at Cetus may have neglected PCR for too long, many down in the research and development trenches were well aware of its enormous potential and eager to move toward applications.

Mullis's insight came at a time when molecular genetic techniques had advanced sufficiently to make use of it. Techniques to make the short stretches of DNA used as primers artificially were laborious until automated instruments were developed early in the 1980s. To make useful primers, at least some sequence information on the target DNA was usually needed, so the practical application of PCR awaited facile sequencing techniques. Once it was described in 1985, PCR exploded through the molecular biology community. A bibliography compiled by Perkin Elmer Cetus listed three publications based on the technique in 1985, twenty in 1986, seventy-five in 1987, 280 in 1988, and 860 in 1989.[79] In 1990, Cetus stopped publishing the bibliography because it was growing too fast; by 1992, there were five hundred articles a month published using PCR. The point was proved.

Technological developments surged forward in genetic linkage mapping, physical mapping, and DNA sequencing in the period 1980–1986. These technical developments set the stage for a science policy debate that culminated in ideas for a concerted genome project. Before the project could emerge from the primordial technological soup, energetic people with vision and persistence had to champion new ideas and create institutions to sustain them. The technological groundwork was in place, but the Human Genome Project also required that the new technology be harnessed to a scientific project by securing a budget and establishing a bureaucratic structure. Several individuals independently brought forth their ideas for an audacious new biological enterprise in 1985.

PART TWO

Origins of the Genome Project

5

Putting Santa Cruz on the Map

THE FIRST MEETING focused specifically on sequencing the human genome was convened in 1985 by Robert Sinsheimer of the University of California at Santa Cruz. While the genome project did not grow out of the meeting, or even emerge as a topic of discussion, the 1985 Santa Cruz gathering did plant the seed.

Planning for Sinsheimer's May 1985 meeting at Santa Cruz began the previous October, when Sinsheimer called several faculty biologists—Robert Edgar, Harry Noller, and Robert Ludwig—into his office. Sinsheimer was then chancellor at UCSC. As such, he had been a participant in several major science planning efforts. These included relations with the three national laboratories managed for the Department of Energy (DOE) by the University of California (Los Alamos, Lawrence Berkeley, and Lawrence Livermore national laboratories), discussions of the California state proposal to house the Superconducting Super Collider, and, most directly, the Lick Observatory. The UCSC faculty in astronomy had an international reputation. As a biologist, Sinsheimer wanted biology to achieve similar stature. He wanted, he said, to "put Santa Cruz on the map."[1]

Others had previously conceived of large, concerted mapping projects and technology development, but these did not grow into the genome project.

The European Molecular Biology Laboratory had in 1980 seriously contemplated sequencing the entire 4,700,000-base-pair genome of the bacterium *Escherichia coli,*[2; 3] but that project was judged technically premature. Norman Anderson, who had worked at several DOE-funded national laboratories during two decades, had a track record of devising instruments for molecular biology, including high-pressure liquid chromatography, two-dimensional protein electrophoresis, and zonal centrifugation.[4] He and his son Leigh lobbied during the late 1970s for a national effort to catalog genes and blood

Robert Sinsheimer, as chancellor of the University of California at Santa Cruz, convened the first meeting on sequencing the human genome in May 1985. Although the institute for human genome sequencing that he envisioned for the UC Santa Cruz campus never materialized, the impetus for such a project remained. *Don Fukuda photo, courtesy University of California, Santa Cruz*

proteins,[5] and Senator Alan Cranston pushed for a dedicated $350 million program in the early 1980s. Even then, there was talk of the need to collect DNA sequence data.[3] Father and son continued to urge adoption of their program in the national laboratory system and at DOE. Their efforts were known by DOE administrators, and may indeed have helped set the stage for the genome project, but they had not crystallized into a dedicated science program.

The inspiration for Sinsheimer's DNA sequencing proposal was a telescope.[6; 7] A group of University of California astronomers wanted to build the biggest telescope in the world. The venture was ultimately successful, producing the Keck Telescope on Mauna Kea in Hawaii, which saw first light on November 24, 1990. This success came only after clearing several high hurdles.

In 1984, the costs of enlarging the giant telescope on Mount Palomar or constructing a facility of similar size were estimated in the range of $500 million, a large fraction of the expense associated with manufacturing an enormous mirror. Jerry Nelson of the Lawrence Berkeley Laboratory hit upon the idea of using thirty-six hexagonal mirrors to replace a single large one, reducing cost estimates eightfold. By computer adjustments of the hexagonal array, the complex of smaller and cheaper mirrors could provide the same resolving power. A piece about the telescope appeared in the *San Jose Mercury*. Soon after the article appeared, a Mr. Kane called the laboratory.[8] He thought he might know a donor interested in funding the telescope, the widow of Max Hoffman. Hoffman had made a fortune as the U.S. importer of Volkswagen and BMW automobiles, and had left an estate of several tens of millions of dollars, the Hoffman Foundation, whose trustees were his widow and two others. Mrs. Hoffman signed most of the papers for a $36 million donation for the Hoffman Telescope project the day before she died. It was the largest single gift in the history of the University of California, but it had to be returned. That $36 million return was the event that stimulated the DNA-sequencing idea.

The $36 million donation, generous as it was, fell $30 to $40 million short of what was needed to build the telescope. Further donors were needed, and the University of California was having trouble finding them. Since the telescope was already named for Max Hoffman, it was more difficult to entice further large donations. The University of California finally sought help from Caltech, a private university. The University of California got more than it bargained for. After finding several smaller donations, Caltech got an agreement from the Keck Foundation, built with Superior Oil money, to fund the entire telescope if the name was changed to the Keck Telescope. The Hoffman Foundation, having lost the glory of being the major donor and having lost its most interested trustee, was not interested in helping build a smaller sister telescope or in using its funds for other suggested alternatives.

Sinsheimer wondered if an attractive proposal in biology could recapture the interest of the Hoffman Foundation. He pondered whether there were opportunities missed in biology because of biologists' proclivity to think small, in contrast to their colleagues in astronomy and high-energy physics. Sinsheimer's laboratory had purified, characterized, and genetically mapped a bacterial virus, phi-X-174.[7] Its 5,386-base-pair genome was the first of any organism's to be sequenced, by Frederick Sanger in 1978.[9] Sinsheimer followed the progression of DNA sequencing to larger and larger organisms. As he thought about targets for a large biology project, Sinsheimer struck upon sequencing the human genome, fully a million times larger than the viral genome and ten thousand times larger than the biggest sequencing project to date. He sought counsel of his colleagues at UCSC about establishing an institute to sequence the human genome, and in October 1984, he called the meeting with Noller, Edgar, and Ludwig.[10]

Edgar, Ludwig, and Noller were at first stunned by Sinsheimer's audacity, but as they began to think through the scientific approach that would lead to sequencing the entire genome, they decided that it would be a useful goal and would generate equally useful results along the way. In particular, the process of sequencing would entail physical mapping, a valuable enterprise in its own right. Edgar and Noller prepared a position paper for Sinsheimer on Halloween 1984, which became the basis for Sinsheimer's letter to University of California president David Gardner on November 19.[11] The Santa Cruz scientists proposed that the DNA sequencing institute could be

a noble and inspiring enterprise. It some respects, like the journeys to the moon, it is simply a "tour de force"; it is not at all clear that knowledge of the nucleotide sequence of the human genome will, initially, provide deep insights into the physical nature of man. Nevertheless, we are confident that this project will provide an integrating focus for all efforts to use DNA cloning techniques in the study of human genetics. The ordered library of cloned DNA that must be produced to allow the genome to be sequenced will itself be of great value to all human genetics researchers. The project will also provide an impetus for improvements in techniques . . . that have already revolutionized the nature of biological research. . . .[12]

Sinsheimer urged Gardner to approach the Hoffman trustees with his new idea, asserting:

It is a an opportunity to play a major role in a historically unique event—the sequencing of the human genome. . . . It can be done. We would need a building in which to house the Institute formed to carry out the project (cost of approximately $25 million), and we would need an operating budget of some $5 million per year (in current dollars). Not at all extraordinary. . . . It will be done, once and for all time, providing a permanent and priceless addition to our knowledge.[11]

Sinsheimer also discussed the idea with James Wyngaarden, director of the National Institutes of Health, in March 1985. Sinsheimer noted that Wyngaarden was "attracted by the idea," and he urged Sinsheimer to approach the National Institute of General Medical Sciences if the May meeting reached consensus on the project's feasibility.[13] Sinsheimer concluded he would have to find a source of funds. To do so, he would need the blessing of some internationally recognized scientists to lend the project credence.

The next phase was to call a meeting of experts from around the world. Noller wrote to Sanger, with whom he had worked several years earlier. Sanger's reply was encouraging: "It seems to me to be the ultimate in sequencing and will probably need to be done eventually, so why not start on it now? It's difficult to be certain, but I think the time is ripe."[14] Edgar, Noller, and Robert Ludwig convened the meeting on May 24 and 25, 1985, bringing together an eclectic mix of DNA experts. Bart Barrell was Sanger's successor as head of large-scale sequencing at the MRC Cambridge laboratory. Walter Gilbert represented the Maxam-Gilbert approach to DNA sequencing. Lee Hood and George Church were Americans pushing sequencing technology, Hood through automation and Church (who had done his graduate work with Gilbert) through clever ways to extract more sequence data from each experiment. Those familiar with genetic linkage mapping were also invited, including David Botstein, Ronald Davis, and Helen Donis-Keller. John Sulston and Robert Waterston were invited to report on their efforts toward constructing a physical map of *C. elegans*. Leonard Lerman was a technologically oriented biologist from Boston, and David Schwartz had pioneered the techniques for handling and separating DNA fragments millions of base pairs in length. Finally, Michael Waterman of the University of Southern California was brought for his expertise in mathematics, DNA sequence analysis, and databases.

Over the course of an evening and a day, the group decided that it made sense systematically to develop a genetic linkage map, a physical map of ordered clones, and the capacity for large-scale DNA sequencing.[7] The first sequencing efforts should focus on automation and development of faster and cheaper techniques.[15] The workshop concluded, significantly, that a complete genome sequence was not feasible, as such an undertaking would require large leaps in technology. "In the meantime, one should concentrate on the sequencing of regions of expected interest (polymorphisms, functional genes, etc.). The first few percent should be of great interest."[15]

The idea of sequencing the human genome was out in the open. A later account of the meeting captured its modest aspirations as "Genesis, the Sequel."[6] Sinsheimer sent letters and a summary of the meeting to several potential funding sources, including the Howard Hughes Medical Institute (HHMI) and the Arnold and Mabel Beckman Foundation, but there were no takers.[16–18] Contacts with the Hoffman Foundation, while the initial impetus for the meeting, were not permissible. The University of California president's office now handled the foundation, so the Santa Cruz campus could make no direct approach. The NIH route was blocked by the need to ask for a facility in which to do the work and the large budget required. A major construction effort entailed approval from the UC system. NIH might be approached to fund the project, but not the facility in which to do the work, and not until the facility was built. These were formidable obstacles. Sinsheimer concluded the only solution was to find a private donor for the building first, but his access to large sources of private money also had to go through the UC president's office. The Hoffman funds were never recouped by the University of California.

Sinsheimer later reflected:

I was certain of the value of the proposal. The human genome surely would someday be sequenced, once and for all time. The achievement would be a landmark in human history and the knowledge would be the basis for all human biology and medicine of the future. Why not now?[7]

Sinsheimer contemplated going directly to Congress. He discussed the institute idea with Leon Panetta, his congressman. Panetta was supportive, but indicated his awareness that proposals of such magnitude would have to go through the UC president's office.[19] Sinsheimer was frustrated in his attempts to cultivate interest in Gardner's office. As Sinsheimer neared retirement, prospects for a human genome sequencing institute at UC Santa Cruz quietly died. While he did not get his institute, the Sinsheimer Laboratory for biology was dedicated by UC president Gardner, Senator Mello of California, and Assemblyman Farr, with a public lecture by Charles Cantor, in February 1990.[20] The idea of sequencing the human genome moved on to other pastures, having acquired a life of its own.

6

Gilbert and the Holy Grail

SINSHEIMER HANDED THE TORCH TO Walter Gilbert—Nobel laureate, erstwhile executive, and molecular biologist of legendary prowess. Gilbert began his career in science as a theoretical physicist. As an assistant professor in physics at Harvard, he wanted to learn about the new molecular biology. In 1960, he joined the laboratory of James Watson, who with François Gros was then hot on the track of messenger RNA. In a videotape taken of a meeting to celebrate Watson's sixtieth birthday in 1988, Gilbert described how he was given six papers to read when he first joined Watson and Gros, in contrast to the hundreds a new postdoctoral or graduate student would be handed today.[1] ("Things were different then.") Watson would hold a stopwatch while Gros sloshed a large flask of bacteria and Gilbert poured in ten to twenty millicuries of radioactive phosphate, to label the RNA in the bacteria. Messenger RNA was then a hypothetical entity, postulated to exist by some, but not yet a known commodity. Messenger RNA was, of course, eventually found to exist, and the group at Harvard joined those at the Pasteur Institute in Paris and the MRC Cambridge laboratory in the front ranks of molecular biology.

RNA is copied from stretches of DNA, then spliced, and finally transported out of the cell nucleus to serve as the code to assemble amino acids into proteins. Gilbert's career in molecular biology started with an extremely important problem. His reputation built even more on work that began in 1965 to find the repressor protein, an on-off switch for the gene that produced a bacterial protein. This was one of the most hotly contested races of its day in molecular biology. Gilbert commented on this phase of his work: "By the time the repressors were actually isolated, which was late in 1966, they had become a—Holy Grail?"[2] The mythic theme would return two decades later, by which time Gilbert was among the most respected thinkers in molecular biology.

Gilbert searched for the repressor protein with Benno Muller-Hill of Germany. The *lac* genes, involved in digesting sugars, were turned on and off in response to the presence or absence of sugars in the growth medium surrounding bacterial cells. The simplicity of the *lac* operon system made it a central target of molecular biology. Gilbert and Muller-Hill found the *repressor* protein

Walter Gilbert jots down his estimate of the cost and time it would take to sequence the entire human genome at a rump session of a symposium on the molecular biology of *Homo sapiens,* held at the Cold Spring Harbor Laboratory in June 1986. Gilbert left Harvard University in 1982 to become chief executive officer of the biotechnology firm Biogen; he returned to Harvard two years later and has been there ever since. *Victor McKusick photo, courtesy Cold Spring Harbor Laboratory Library*

that flipped this genetic switch in 1966.[3] It was a period of intense rivalry and cooperation with Mark Ptashne, who worked on a similar problem in a laboratory just down the hall.[4] Ptashne had come to Harvard to work under Watson and was trying to find a different repressor protein, one that turned genes on and off in the bacteriophage, or bacterial virus, named phage lambda.[5] Gilbert and Muller-Hill found their repressor just a few months before Ptashne found his. The next step was to study how the switch was thrown.

In the late 1960s and early 1970s, Gilbert isolated the DNA region that controlled the *lac* genes, called the operon, or genetic-switch region. This was the first segment of DNA isolated.[6] He chose to study the dynamics of the system by analyzing the structure of DNA in the region. This was the work that led to DNA sequencing, described in Chapter 4. Gilbert was thus a part of several landmark developments in molecular biology: the discovery of messenger RNA, the isolation of the *lac* repressor, and the technical miracle of DNA sequencing. It was not the end.

Gilbert joined a three-way race to isolate, study, and express the gene for insulin, one of the most studied proteins in all biology. Since its discovery in the 1920s, insulin had been used in treatment of diabetes. It was the first protein sequenced (by Sanger), and because of its therapeutic use, it was an obvious candidate protein to make using recombinant DNA technology as soon those methods were discovered in the mid-1970s. Gilbert threw his hat into the insulin ring in 1976. This and his past work took him on a short digression into commerce. Gilbert was among the founders of the Swiss-American biotechnology firm Biogen, created in 1978 while Gilbert's laboratory was working to clone insulin. Gilbert was enticed into involvement by a venture capital group hoping to establish the new company. At the scientific end, Gilbert's group at Harvard was the first to trick bacteria into producing the insulin protein, only the second mammalian protein ever so produced.[8]

Gilbert's star rose higher in 1980, when he shared the Nobel Prize for chemistry with Paul Berg of Stanford and Sanger. This was a special year for the Nobel, as these three scientists have a reputation as truly exceptional molecular biologists, even compared to other Nobel laureates. Each has not only left a significant personal legacy of science, but also left a trail of scientists trained in their laboratories and likely to travel to Stockholm themselves someday.

In 1982, Gilbert became chief executive officer at Biogen. Harvard forced him to choose between keeping his professorship and running a biotechnology company. He shook the academic world when he left his American Cancer Society chair at Harvard to direct Biogen. Biogen, however, did not fare well; it lost $11.6 million in 1983 and $13 million in 1984.[8] Gilbert resigned as CEO in December 1984 and returned to Harvard, where he became chairman of the department of biology. (Biogen continued to lose money after Gilbert left the helm.) In 1988, Gilbert was named Loeb University Professor at Harvard.

After leaving Biogen, Gilbert traveled to the South Pacific. The group organizing the Santa Cruz meeting sought him out, failing to locate him for many weeks. Robert Edgar finally reached Gilbert with a letter in March,[9] and Gilbert agreed to come. His addition was significant. After attending the Santa Cruz meeting, Gilbert became the principal spokesman for the Human Genome Project for the better part of a critical year.

Gilbert proved an articulate visionary, transmitting excitement to other molecular biologists and to the general public. He translated the ideas at Santa Cruz into specific operating plans in a memo back to Edgar two days after the workshop. In it he offered a strategy for Sinsheimer's institute, although privately he was not convinced that it should be located in Santa Cruz:

. . . In the early years the institute may want to be a sequencing resource—taking genes and probes from outside and returning sequences, cosmids [clones], and probes to the outside. . . . I expect that the most rewarding information scientifically will be in the first 1 percent of total sequence, if the work is focused, that most of the information, in the sense of interesting differences, will be in the next 10 percent, and the last 90 percent—of intron and intergenic regions—will be the least informative, but the increase in speed of sequencing should make each of these three phases take roughly equal times—or possibly make the last faster than the first.[10]

In this letter, he returned to a familiar motif, noting, "The total human sequence is the grail of human genetics—all possible information about the human structure is revealed (but not understood). It would be an incomparable tool for the investigation of every aspect of human function." Gilbert's Holy Grail proved an enduring rhetorical contribution to the genome debate. Indeed, it captured more than perhaps he intended. The Grail myth conjured up an apt image; each of the Knights of the Round Table set off in quest of an object whose shape was indeterminate, whose history was obscure, and whose function was controversial—except that it related somehow to restoring health and virility to the Fisher King, and hence to his kingdom. Each knight took a different path and found a different adventure.

Gilbert carried the ideas from Santa Cruz into the mainstream of molecular biology. He gave informal presentations on sequencing the genome at a Gordon Conference in the summer of 1985, and at the first international conference on genes and computers in August 1986.[11] Gilbert was extremely well connected, and he infected several of his colleagues with enthusiasm, including James Watson.

Gilbert gave the genome project much greater notice than it would otherwise have achieved. His role was featured in the *U.S. News & World Report, Newsweek, Boston* magazine, *Business Week, Insight,* and the *New York Times Magazine.*[12–17] He joined Watson, Hood, Bodmer, and others as the star of video documentaries on the genome project.[18] Gilbert and Hood wrote supporting articles for a special section in *Issues in Science and Technology* published by the National Academy of Sciences.[19; 20] Gilbert and Bodmer promoted the genome project in editorials for *The Scientist.*[21; 22] Gilbert thus stoked the genome engine, preserving the spirit of Santa Cruz.

Gilbert provoked a major controversy, however, when he decided to try to take the genome project private. He began thinking about establishing a genome institute himself in 1986. In January 1987, Michael Witunski, president of the James S. McDonnell Foundation, approached Gilbert with the idea of foundation support to help create such an institute. This idea died when the foundation funded a study to assess the genome project at the National Research Council of the National Academy of Sciences. Gilbert participated in a spate of meetings convened to debate the genome project during late 1986, and he became a member of the NRC committee. In spring 1987, he decided to take the commercial plunge. He resigned from the NRC committee and announced plans to form Genome Corporation.

Gilbert's idea for Genome Corp. was to construct a physical map, do systematic sequencing, and establish a database.[6] The business objectives included selling clones from the map, serving as a sequencing service, and charging user fees for access to the database. The market would be academic laboratories and industrial firms, such as pharmaceutical companies, that would purchase materials and services from Genome Corp. The purpose was not so much to do things that others could not do at all, but rather to do them more efficiently, so that outside laboratories could purchase services more economically than they could perform the services themselves. In Gilbert's words, "Twenty years ago, every graduate student working on DNA had to learn to purify restriction enzymes. By 1976 no graduate student knew how to purify restriction enzymes; they purchased them. Historically, if you were a chemist, you blew your own glassware. Today, people simply buy plastic."[23] Genome Corp. could free biologists to focus on biology instead of wasting time making the things used in their experiments. These precedents fueled Gilbert's quest for funding from venture capitalists over the course of 1987 and into 1988. By late 1987, however, Wall Street's enthusiasm for biotechnology had turned to skepticism, and the stock market crash in October made capitalizing Genome Corp. all but impossible. The highly publicized efforts to start a genome project in the federal government made prospective investors leery of competing with the public domain. Genome Corp. could succeed only if Gilbert stayed so far ahead of academic competition that others would come to him for services, rather than waiting for the information and materials to be made freely available.

Gilbert was unabashed after the demise of Genome Corp. He remained a highly visible spokesman for a vigorous and aggressive genome project. He was consistently at the high end when making projections of what could be done in the way of mapping and sequencing. He was a technological optimist. Younger scientists balked at his enthusiasm for targeted, production-mode work and feared that he was publicly proclaiming goals too ambitious to attain. They loathed his almost monomaniacal focus on production-style DNA sequencing and bristled at his image of genome research as factory work. They complained bitterly that they would be held accountable for achieving impossible objectives set by policymakers listening to Gilbert; they felt they were being asked to climb Mount Everest after having only strolled a few miles along the Appalachian Trail.

If Gilbert was to blame for setting the sights too high, however, he would at least be there on the firing line with the rest of genome researchers. Gilbert did not indulge in mere rhetoric, but committed his laboratory to be among the pioneers of large-scale DNA sequencing. In 1990, he proposed to sequence the genome of the smallest free-living organism, *Mycoplasma capricolum,* a small bacterium of goats.[24] This project was among the handful of sequencing projects intended to move sequencing from a theoretical possibility to a new way of understanding life. The genetics of the organism were not nearly so thoroughly studied as those of many other bacteria. Gilbert proposed to determine the DNA sequence of the bacterium's 800,000 base pairs, thought to contain five hundred or so genes. He hoped to reconstruct the biology of the organism by starting from its DNA sequence. The idea was not that sequencing would address all the questions of biological interest, but that starting from sequence would answer them faster.

Gilbert's project on *M. capricolum* joined other pilot sequencing projects on model organisms. These were among the grants given out in the first year's operation of the National Center for Human Genome Research at NIH.[24] A European consortium began a multicenter sequencing effort directed at yeast chromosomes. Botstein and Davis also proposed to start sequencing the yeast genome at Stanford (working from the physical map of yeast made by Maynard Olson). The groups working on the nematode *C. elegans* began systematic large-scale sequencing, in a transatlantic collaboration between John Sulston and Alan Coulson in England and Robert Waterston at Washington University in St. Louis.

Gilbert was not content to contribute only to the sequencing effort. His natural talents tended toward more theoretical generalizations. He was among the first to postulate an explanation of why genes were broken into different regions of DNA—with islands of base sequence to be translated into protein separated by long stretches of other sequences. In an article titled "Why Genes in Pieces?" he suggested that the role of fragmentation was to promote the shuffling of useful protein modules throughout the genome, enabling them to be used in different contexts.[25] Indeed, it was his terminology for DNA regions—"exons" for the parts that coded for protein and "introns" for the segments that separated exons—that eventually caught hold. Gilbert and, independently, Russell Doolittle postulated that the exon modules in DNA encoded protein substructures; these could be mixed and matched to serve similar functions in different proteins. They could be moved about in the genome over many generations, and the long intron sequences between the exons made this more feasible physically. Gilbert pushed the idea further a decade later, asserting in a controversial paper that nature had in fact settled on a relatively small set of structures to play with, several thousand or so, and built up the full complexity of existing organisms from a small fraction of the possible permutations.[26]

Gilbert also conveyed an ever enlarging vision of the role of molecular genetics in biology, and the genome project in particular. He foresaw what science historian Thomas Kuhn had termed a "paradigm shift" in biology, with the science becoming driven more by theory. Molecular biologists would do experiments to test ideas first arising from the analysis of masses of information stored in computers. The cloning and sequencing that preoccupied the time of so many graduate students and postdoctoral fellows would be relegated to robots or specialized commercial services. "To use this flood of knowledge, which will pour across the computer networks of the world, biologists not only must become computer-literate, but also change their approach to the problem of understanding life. . . . The view that the genome project is breaking the rice bowl of the individual biologist confuses the pattern of experiments done today with the essential questions of the science. Many of those who complain about the genome project are really manifesting fears of technological unemployment."[27]

A genome program robust enough to sustain such a vision required a bureaucratic structure. The process of erecting this structure was at least as arduous as the science itself. At the beginning of 1987, as Gilbert formulated plans for Genome Corp., there was no center to support these and similar efforts in genome mapping and sequencing. Genome Corp. died, or rather was stillborn. While Gilbert despaired of federal leadership for the genome project, it was eventually two federal agencies that defined it. By the end of 1990, both the Department of Energy and the National Institutes of Health had genome programs with budgets totaling almost $84 million, and there were dedicated genome programs in the United Kingdom, Italy, the Soviet Union, Japan, France, and the European Communities. This remarkable bureaucratic transformation began late in 1985.

7
Genes and the Bomb

B Y PROPOSING A Human Genome Initiative in the Department of Energy in 1985, Charles DeLisi thrust the Human Genome Project onto the public policy agenda. In so doing, he forced the ponderous bureaucracies at the Department of Energy (DOE) and the National Institutes of Health (NIH) into action. Several roots of DeLisi's genome research program can be traced back to the Manhattan District Project to build an atomic bomb. Some led through studies of the biological effects of dropping the bombs at Hiroshima and Nagasaki. Others led through the mathematicians who helped create the initial atomic bomb and, after World War II was over, the hydrogen fusion bomb.

In spring 1985, DeLisi became director of the Office of Health and Environmental Research (OHER) at DOE, the division responsible for funding the bulk of life sciences and environmental research for the department. The Nobel laureate physicist Arthur Holly Compton started the first biology project related to nuclear fission in 1942, at the University of Chicago, site of the first nuclear chain reaction.[1] He was aware of the dangers of radiation to workers, based on early experiences with X-rays and radium. Compton became one of the most important advisers to the federal government in the postwar period, chairing the Committee on the Military Value of Atomic Energy.[2]

Over the years, the mandate of the biological research program broadened considerably to include many biological effects of energy production, in addition to radiation biology. The bureaucracy underwent several reorganizations, from the Manhattan Project to the postwar Atomic Energy Commission (Public Law 79-585) to the Energy Research and Development Administration (Public Law 93-438). Jimmy Carter made a promise to create a Department of Energy in his 1976 campaign for President. The promise was made good in 1977 (Public Law 95-91), carrying with it the biology program that DeLisi later inherited.

In the period immediately after World War II, the Atomic Energy Commission (AEC) was a major supporter of genetics research. The AEC had a relatively large research budget at a time when the National Science Foundation was just coming into existence and the National Institutes of Health were quite small. Even the small fraction of the AEC budget devoted to genetics dwarfed other genetics programs, and the national laboratories funded by AEC grew into centers on the forefront of research. This picture changed as the NIH budget increased steadily for three decades, leaving DOE in the dust. The National Institute of General Medical Sciences (NIGMS) became the principal funding source for basic genetics. Molecular biologists trained in the 1970s and 1980s were accustomed to thinking of NIGMS as the wellspring of genetics; older geneticists who might remember the AEC's role were smaller in number and generally separate from those who founded molecular biology.

Charles Delisi, as director of the Office of Health and Environmental Research in the Department of Energy, set aside the first funding for human genome research at DOE in 1985, in effect putting the genome project on the public policy agenda for the first time. *Courtesy Boston University*

DeLisi's idea for a DOE genome project spun off from an effort to study changes in DNA wrought in the cells of the atomic bomb survivors known in Japanese as the *hibakusha* ("those affected by the bomb"). They had been exposed to one of the most cataclysmic events of all time, but it was just the beginning of their collective nightmare.

The history of the genome project is linked to an attempt to determine if there would be a final, genetic wave of effects from bomb exposure. Specifically, investigators wanted to assess the frequency of inherited mutations caused by exposure to the atomic bombings. Those exposed to the bombings suffered through many phases of radiation effects. Many people were vaporized, burned to death, or otherwise killed immediately by the bomb blast. Among those who survived the first hours, many died of radiation sickness that killed off cells in the immune system, skin, and intestinal lining. Fetuses *in utero* at the time of the bombing had an increased risk of microcephaly (small head and brain associated with mental retardation). Among burn victims, large deforming keloid scars formed in the months after exposure. A few years later, a wave of leukemias passed through the *hibakusha*. After a decade, they began to show somewhat increased rates of cancer in the breast, thyroid, gastrointestinal tract, bone marrow, and other tissues.

The *hibakusha* were severely stigmatized in the postwar period.[3; 4] They were intensively monitored for decades with exhaustive medical follow-up of their health status, in one of the largest, most complex, and longest epidemiological studies ever attempted. In 1947, the U.S. National Academy of Sciences established the Atomic Bomb Casualty Commission (ABCC), with funding from the Atomic Energy Commission, to study the effects of the Hiroshima and Nagasaki bombs. The ABCC used legions of researchers to interview the *hibakusha,* eliciting details related to radiation exposure and health effects. The purpose of the ABCC was to gather information—not to provide treatment, a fact that aroused considerable resentment among the *hibakusha.*[3–5] Eventually, the Japanese government set up special health programs.

In 1975, the ABCC became the Radiation Effects Research Foundation (RERF), based in Hiroshima and Nagasaki, with joint funding from the governments of the United States and Japan. RERF continued the epidemiological investigations and conducted other related research. Most notably, a major reassessment of the nature and amount of radiation exposure was published in 1987, substantially changing dose estimates of those exposed at Hiroshima.[6]

One of the sources of stigma was a belief that the *hibakusha* carried mutations caused by the radiation they experienced. *Hibakusha* women reported they were rejected as mates because they would have deformed children or would pass on mutations and genetic disease. In the early postwar period, the extent of mutational damage to atomic bomb survivors was indeed a hot topic of controversy. H. J. Muller, fresh from receiving a Nobel Prize for his discovery that radiation could induce mutations, used his new fame to sound the alarms. Speaking of the *hibakusha,* he observed that "if they could foresee the results 1,000 years from now . . . they might consider themselves more fortunate if the bomb had killed them."[7] Alfred Sturtevant was even more apocalyptic about radiation exposure: in a letter to *Science,* he warned that atomic bombs already exploded "will ultimately result in the production of numerous defective individuals—if the human species itself survives for many generations."[8]

Such dire predictions were made by some of the most expert geneticists of the day. They fed a growing public fear of radiation that long predated atomic bombs, but was greatly intensified by the mystery surrounding the Manhattan Project and its awesomely powerful products.[9] Nonetheless, the fears were products more of speculation than of observation. The speculations were not purely fabricated; they were based on animal studies, but in this case projections from other organisms proved errant, with distressing effects on the *hibakusha* and their children. The findings from extensive monitoring for three decades were contradictory: according to one expert, "the overwhelming impression that one gains from the analyses of the genetic data . . . is that there is not compelling evidence of genetic change in the offspring of exposed parents."[10] The children failed to show significantly higher rates of cancer or other disease, including birth defects and genetic disorders. If bomb exposure to their parents had produced inherited mutations, they were subtle and hard to detect among the DNA changes that normally occur between generations.[11]

The data were too sparse to drive choices among policies. While radiation clearly increased mutations, no one could say how many or what were the consequences in humans. Historian Susan Lindee concluded that "flexibility in the quantitative side of the argument contributed to flexibility in the 'acceptable' parameter."[5] One group of scientists noted that the species was unlikely to go extinct as a consequence of radioactive fallout, but this was small consolation to a public more interested in intermediate endpoints—the generations destined to live in the meantime.

An enormous range of interpretations was compatible with limited data. The question of whether the *hibakusha* suffered from heritable mutations continued to nag human geneticists. The ABCC studies were expected to produce negative results all along, an odd instance of a major commitment to a project fully expected to be inconclusive.[12]

James V. Neel and others devoted their careers to careful study of the effects of radiation on the genes of the *hibakusha* and their children. Neel founded the first department of human genetics in the United States, at the University of Michigan, based in part on funds to study the genetic effects of radiation. In the mid-1980s, a group sought to apply the emerging techniques of molecular genetics to the quantitative measurement of heritable mutations in humans. Taking the analysis down to the level of DNA sequence was merely an incremental extension of decades of work.

RERF convened a genetics study conference on March 4 and 5, 1984, in Hiroshima. Conferees recommended that cell lines be created from the *hibakusha*, and that "methods for direct examination of DNA should be introduced with all deliberate speed."[13] This recommendation could be interpreted any number of ways, and the International Commission for Protection Against Environmental Mutagens and Carcinogens elected to hold a meeting focused specifically on new DNA techniques. The Department of Energy funded the meeting. Mortimer Mendelsohn of Lawrence Livermore National Laboratory asked Ray White to organize the meeting.

White selected Alta, Utah, as the meeting site. At the same venue where

Botstein and Davis struck upon the idea of systematic RFLP mapping six years before, the masters of technology convened to discuss direct analysis of DNA. White invited an extraordinary mix of molecular and human geneticists to the meeting. The meeting, which lasted from December 9 to 13, 1984, took place in a blizzard. The skiing was memorable; the science was even better.

The 1984 Alta meeting planted the seeds for George Church's embellishments of the Maxam-Gilbert sequencing methods. Many of the young molecular biologists had never met Neel; indeed, some had never heard of him. Maynard Olson, destined to figure prominently in the genome story, was deeply impressed by Neel's commitment.[14] Olson was just beginning to get results on his physical mapping project of yeast. Charles Cantor presented some of the first data using the method he and David Schwartz described for separating million-base-pair fragments of DNA for mapping. The genetic linkage mappers, White foremost among them, had already found their first few RFLPs. Most of the participants had never met one another; as discussion heated up, the meeting became a boiling cauldron of ideas. The roiling broth within contrasted with the blizzard outside, isolating the participants from the world and lending intensity to the discussion.[15]

The conclusion of the meeting was, ironically, that the methods of direct DNA analysis were inadequate to detect the expected increase in mutation frequency from radiation exposure at Hiroshima and Nagasaki.[15–18] In attaining its specific end, the conference was a disappointment, but it brought together a welter of related ideas that would grow into the DOE genome project. The links were a congressional report and Charles DeLisi, a new face at DOE.

The congressional Office of Technology Assessment (OTA) was then doing a report on technologies to measure heritable mutations in man. Exposure to Agent Orange, environmental toxins, and radiation were coming before congressional committees as public policy problems.[19] Mike Gough, then an OTA project director, was present at the Alta meeting and discussed the various technologies in a draft report sent to the Department of Energy for review. The report was published in 1986 as *Technologies for Detecting Heritable Mutations in Human Beings*.

DeLisi had the idea for a project dedicated to DNA sequencing, structural genetics, and computational biology while reading the October 1985 preliminary draft of the OTA report.[20–23] DeLisi was then the newly appointed head of the Office of Health and Environmental Research at DOE. In a scene typical of Washington, he reflected on programs under his direction by reading about them in a report prepared by outsiders.

Once he had the idea, DeLisi moved quickly. He and David Smith, a scientist-administrator also working at DOE headquarters, barraged one another with notes and memos about how to plan this major new initiative. While most of the offices in and around Washington eased into the Christmas lull, Smith and DeLisi were busy crafting a new science initiative. Smith and DeLisi asked the biology group at Los Alamos National Laboratory for comments on DeLisi's idea. The Los Alamos group replied with a dense, scattered, but wildly enthusiastic five-page memo just before Christmas, prepared by physician Mark Bitensky and others.[24] The memo bubbled over with enthusiasm about the potential technical and human health benefits that a structural approach to genetics would open up. The discussion centered on DNA sequencing and barely mentioned physical or genetic mapping. The Los Alamos group found another appealing argument for a concerted research program, arguing that such a project could become a "DNA-centered mechanism for international cooperation and reduction in tension."[24]

The memo saw the national laboratories emerging from the shadow of the atomic bomb. In Bitensky's words, "[J. Robert] Oppenheimer's statement 'I am become death, the Destroyer of Worlds' gives way to 'the National Laboratories are become the ultimate advocates for the understanding of human life.' "[24] He referred to Oppenheimer's quote from the *Bhagavad Gita,* uttered upon the explosion of the atomic fission bomb test at Alamogordo, New Mexico.[25–27] Los Alamos even checked with Frank Ruddle of Yale, to ensure that he would be willing to testify before Congress if called. With this initial encouragement, Smith and DeLisi began to pull the bureaucratic levers in Washington.

DeLisi outlined the political strategy to garner support from the scientific community, from their superiors at DOE, and from Congress.[28] Smith responded with a note about rumors of previous discussions, at a Gordon Conference and at the University of California the previous summer, but he did not know what had come of these.[29] Smith cautioned that criticisms would plague the DOE proposal for some time to come: it was not science but technical drudgery, directed research was less efficient than letting small groups decide what was important, and efforts should be concentrated on genes of interest rather than global sequencing. DeLisi bounced back: "Regarding the grind, grind, grind argument . . . there will be some grind; what we are discussing is whether the grinding should be spread out over thirty years or compressed into ten." He estimated that "we are talking about $100–150 million per year spread out over somewhat more than a decade," and he asserted that such a project certainly would rate as more important than the lower 1 percent of biology grants that funding of this magnitude would displace. The political effort, he argued, should focus not on whether it would displace other work, but instead on how to gain support for new funding.[30]

In order to reach out to the scientific community, DeLisi and Smith asked Los Alamos to convene a workshop: (1) to find out if there was consensus that the project was feasible and should be started; (2) to delineate medical and scientific benefits and to outline a scientific strategy; and (3) to discuss international cooperation, especially with the Soviet Union. A planning group at Los Alamos got together on January 6 to begin planning the workshop.[31] The meeting was shaped in a series of notes and calls back and forth between DOE headquarters and Los Alamos.[32]

The workshop was held in Santa Fe on March 3 and 4, 1986, with "a rare and impassioned esprit."[33] Frank Ruddle chaired the meeting. Discussion at the Santa Fe workshop added an emphasis on integrating genetic linkage and physical maps and the process of making physical maps.[34; 35] Participants agreed on the importance of the new venture and on part of what it should entail, but opinions failed to converge on how to organize the effort. Nobelist Hamilton O. Smith of Johns Hopkins University found that "perhaps the most impressive feature of the meeting was the unanimous consensus that sequencing the entire human genome is doable . . . [although] how to implement such a heroic and costly undertaking is less clear."[36] Anthony Carrano and Elbert Branscomb of Lawrence Livermore National Laboratory stressed the importance of clone maps and warned that "a program whose announced purpose was simply to 'sequence the human genome' might unnecessarily and incorrectly arouse fears of territorial and financial usurpation in the biomedical research community."[37] Events proved their political acumen; fears of a massive mindless sequencing operation became the major threat to scientists' support of the human genome project.

David Comings, a human geneticist from the City of Hope Medical Center in southern California, was further from the mark when he asserted that the whole physical mapping component might be funded "without any stirring up of any congressmen or other related creatures."[38] Those awful creatures proved altogether too alert and intrusive.

Beyond the first rationale, the study of heritable mutations, DOE had a second reason to mount a genome project. DOE managers wanted to capitalize on the resources of the national laboratories, with their ready access to exotic high technology, the best complex of supercomputers in the world, and multidisciplinary teams of scientists.

The Genome Project also fit naturally within a broader DOE mission, and that is the utilization of the Labs to solve nationally important problems in areas that required their unique capabilities. In the case of Genome, the uniqueness was experience with large multidisciplinary projects, and a history of breakthroughs in applying engineering to the medical sciences (nuclear medicine being the paradigm). To the extent that large portions of the project could not be comfortably accommodated at most universities, this second rationale ultimately became as important as the first.[39]

This justification was liable to seem self-serving, however; the arguments sounded like a typical bureaucracy's merely expressing its proclivity for self-perpetuation. And so it was. David Botstein showed his knack for subtle understatement, calling the DOE genome initiative "DOE's program for unemployed bomb-makers."[40] Lee Hood was more diplomatic, noting:

The argument they had enormous technological resources that could be focused on this problem was utterly irrelevant, unless they had the key individuals that could integrate those in a focused and productive way, to take advantage of biology as well as the technology. So on all of those counts, I think DOE had not convinced the world in 1985 that they had the wherewithal to take on the Human Genome Initiative.[41]

The future of the national laboratories proved crucial to the DOE's genome effort. Mutation detection was the intellectual origin, but it was too weak a foundation on which to build a major new program. A new direction for the national laboratories, to channel their ample intellectual and technological energies, became a much more powerful drive once engaged. The laboratories were a natural political base with a well-developed support structure. Scientists at several of the national laboratories were enthusiastic about the idea and were already doing related research. DeLisi's idea started from a narrow base, mutation detection, but then grew to encompass a much larger political goal, the salvation of the national laboratories.

DeLisi discussed the possibility of a genome project with his immediate superior, Alvin Trivelpiece, who supported it and charged the DOE life sciences advisory committee (the Health and Environmental Research Advisory Committee, or HERAC) to report back to him about it. Trivelpiece and DeLisi had discussed why DOE did not have the same high stature in biology that it had in high-energy physics, and they aspired to lift DOE to the forefront of biology on the wings of a genome project. Trivelpiece, as director of the Office of Energy Research, reported directly to the Secretary of Energy (then John Herrington), who in turn reported directly to the President.

On May 6, 1986, six months after his initial idea, DeLisi produced an internal planning memo to request a new line-item budget. This went to Trivelpiece and up through the DOE bureaucracy. DeLisi argued for a two-phase program. Phase I had three components. The first, physical mapping of the human chromosomes, the central element, would take five or six years. The other two components were development of mapping and sequencing technologies and renewed attention to how computer analysis could assist molecular genetics (especially sequence analysis). As physical mapping progressed, parallel efforts would proceed, to prepare for Phase II, the sequencing of the entire genome. High-speed automated DNA sequencing and enhanced computer analysis of sequence information were both essential to making the transition from Phase I to Phase II. DeLisi's background in computational biology, his previous experience in interpreting DNA sequence information at the National Cancer Institute, came to the fore here. Phase II, contingent on success in all three parts of Phase I, was to sequence the banks of DNA clones that constituted the physical map.

DeLisi spoke of a project analogous to a space program, except that it would entail the efforts of many agencies and a more distributed work structure, with "one agency playing the lead, managerial role. . . . DOE is a natural organization to play the lead."[42] A six-year budget of $5, $10, $19, $22, and $22 million was proposed for fiscal years 1987–1991.[43] Plans survived the internal DOE review, and a series of meetings was scheduled, beginning in July 1986, with Judy Bostock, the DOE life sciences budget officer in the presidential Office of Management and Budget (OMB), and with her boss, Thomas Palmieri.

OMB perches atop the federal bureaucracy, with responsibility to oversee management and prepare the President's budget request to Congress each year. Mention of OMB sends shivers of fear down the spines of most who work for the federal government. OMB is the dank home of malicious obstructionists and ax-toting budget officers. The genome project charged into the dark castle—the New Executive Office Building a block from the White House—to face the naysayers and dream-stealers. As the exception that proves the rule, the genome project got a major boost from OMB.

DeLisi's genome meetings with Bostock were focused on planning for fiscal years 1988 and beyond. Bostock was an erstwhile physicist from MIT, intrigued by prospects of improving the speed and efficiency of biological research, who believed that better instrumentation could improve the quality of biology.[44; 45] She saw molecular biology as an extremely inefficient process with postdoctoral and graduate students doing mindless manual work that would be better done by robots or automated instruments. DeLisi was proposing a program to analyze DNA faster and with less human effort, a laudable goal that capitalized on the resources of national laboratories. Bostock bought DeLisi's plans, clearing a major obstacle from the road to Congress.[46]

DeLisi succeeded in his dealing with the DOE and OMB bureaucracies, but he also needed an endorsement from scientists. The OHER advisory committee, the Health and Environmental Research Advisory Committee (HERAC), endorsed the plan for a DOE genome initiative in a report from its special *ad hoc* subcommittee. The subcommittee was a blue-ribbon scientific group chaired by Ignacio Tinoco, a highly respected chemist from the University of California at Berkeley, then on a sabbatical year at the University of Colorado. The HERAC report urged a budget of $200 million per year and made a case for DOE leadership of the effort. The introduction to the report laid out the rationale:

It may seem audacious to ask DOE to spearhead such a biological revolution, but scientists of many persuasions on the subcommittee and on HERAC agree that DOE alone has the background, structure, and style necessary to coordinate this enormous, highly technical task. When done properly, the effort will be interagency and international in scope; but it must have strong central control, a base akin to the National Laboratories, and flexible ways to access a huge array of university and industrial partners. We believe this can and should be done, and that DOE is the one to do it.[47]

Budget projections made by the committee were not directly coupled to the multiyear DOE-OMB budget agreement. The HERAC report was issued in April 1987, at least seven months after DeLisi began to reprogram funds, and four months after the budget agreement with OMB.[48-51] The process of formulating a budget began with DeLisi's notes to David Smith in December 1985 and continued more broadly at a genome conference hosted by the Los Alamos National Laboratory in Santa Fe, New Mexico, in March 1986. In letters sent to the organizers after that meeting, budget estimates covered a wide range and generally focused on only one or two components. By the second Santa Fe conference in January 1987, planning had become more systematic. Several of the participants met over lunch at that conference to discuss what the budget should be. David Padwa, who had previously been involved with founding an agricultural biotechnology company, Agrigenetics, noted some political constraints on the budget. It had to be large enough to command congressional attention, so it would have to be at least $50 million to $100 million per year, but it could not be so large it threatened other research interests. The discussion continued at a meeting of the HERAC subcommittee at the Denver Stouffer's Hotel, February 5 and 6, 1987, a month before their report was to be considered by the full HERAC. Generating cost estimates was delegated to Lee Hood. The second day's meeting started at nine in the morning, and Hood's plane was delayed, so the group began to discuss what could be done within the range of budgets thought to be reasonable for OHER to request. There was discussion of how much physical mapping and sequencing could be done with $20 to $40 million, the maximum thought politically feasible.

Hood entered the meeting at ten o'clock, armed with some handwritten notes, including a menu of technologies and attendant costs. The proposal included technology development, physical mapping, mapping and sequencing of model organisms (yeast and bacteria), and regional sequencing of interesting chromosomal regions (e.g., those packed with genes). His estimates were $200 to $300 million per year for a full program. Someone asked if that was at all possible, since it was a full order of magnitude higher than earlier discussions. Hood did not wait for an answer, and asked passionately whether the budget would drive the vision or the vision would drive the budget. With this, the group deliberated over some technical details of how to make the projections and settled on a figure of $200 million. This brought the budget projection into the range judged politically attractive over the course of previous discussions.

The HERAC subcommittee did not discuss which agency should lead the Human Genome Project at its final meeting to draft its report. This was pointed out to HERAC when it met to consider the subcommittee report in March 1987. By April, when the report was released, Tinoco as subcommittee chairman and Mort Mendelsohn, a member of the subcommittee and chairman of HERAC, had canvassed the members. They wrote the language favoring DOE leadership. Later interviews with members of that subcommittee revealed that at least seven of the fourteen had reservations about giving DOE a blank check, but agreed to the suggested language because they feared inaction on the part of NIH; it was more important to them that the project proceed than that NIH direct it.

Despite the go-ahead from his superiors at DOE, from OMB, and from the scientific community as represented by HERAC, DeLisi's job was still not complete. There was a two-step process in each house of Congress. Before a federal agency can fully implement a major new initiative, Congress has to authorize it and separately appropriate funds for it. These twin processes are interdependent but distinct.

Appropriations committees in the two houses are parallel. They allocate funds according to the executive department expending the funds and follow a relatively stable annual routine. The President's budget proposal is prepared, first by each department and then by OMB. In January the President's budget goes to Congress, where it is referred to the appropriations committees. Except in unusual circumstances (as occurred once during the Reagan years, violating the spirit, and probably also the letter, of the Constitution), the House takes action first, and the Senate works from the House figures. The appropriations committees cannot authorize new programs, but can only fund activities authorized by other committees. The interpretation of these distinctions can be tight or loose, depending on the circumstances. (One of the nation's first large science agencies, the U.S. Geological Survey, for example, was created and operated for years under a rider to an appropriations bill, without an authorization statute.)[52; 53]

To get the genome program started, DeLisi took $5.5 million in funds from the preexisting fiscal year 1987 budget and reallocated them to the genome effort. Such limited "reprogramming" was standard fare, permitted by the appropriation and authorization committees within reasonable limits. For 1988 and later budgets, however, DOE needed support from its authorization committees. DeLisi noted the need for congressional action in his first personal note to David Smith,[28] and he began to hold meetings with congressional staff in 1986. This was unfamiliar territory for DeLisi, who was given to shyness and new to defending a program on Capitol Hill. There was little problem in the Senate, as DOE could in all likelihood count on strong support from Senator Pete Domenici and tacit approval of Senator Wendell Ford, the key figures on the authorization committee. Domenici also sat on the appropriations and budget committees. The problem was in the House.

Staff of the relevant DOE authorization subcommittee in the House were getting mixed signals about the DOE genome initiative. Congressman James Scheuer chaired the subcommittee with jurisdiction over DeLisi's program. Scheuer's staff read the generally negative response to DOE's plans in *Science* magazine; phone calls to biologists elicited both support and opposition. Eileen Lee, the biologist on staff, was uncertain what tack to take. She called on OTA staff, including me, to help plan a hearing, in hopes of penetrating the network of scientists concerned with the genome project.

DeLisi's problem was complicated by the politics of his other programs. Scheuer's staff was generally supportive of DOE staff initiatives, but DeLisi had problematic relations with Eric Erdheim, staff for Claudine Schneider, the ranking minority (Republican) member on the subcommittee. It was unclear to Scheuer's staff whether they should expend the political capital to defend DeLisi against Erdheim on the genome project. Claudine Schneider was generally suspicious of DOE's record on research into environmental health hazards, although she eventually decided DeLisi's program was good. As the hearing approached, the genome project became the battleground for a skirmish between Democrats and Republicans on the subcommittee staff.

About a week before the hearing, I was invited to meet with subcommittee staff from both parties. I could sense the tension in the room, but was blithely unaware of its origin,despite the fact that my wife, Kathryn, worked in Claudine Schneider's office at the time. As we were drifting apart after the meeting, Eileen Lee whispered to me that she thought Erdheim had asked James Watson to testify against the DOE genome program. A few minutes later, as I was preparing to leave the subcommittee's rabbit warren of offices, Erdheim took me aside to tell me he was thinking about calling the Delegation for Basic Biomedical Research to seek testimony from Watson or David Baltimore. Erdheim "had problems with what DeLisi was doing in his programs,"[54] and he was skeptical of the genome proposal. What did I think of that? I suggested that he had better find out what Watson or anyone else from the delegation would say before he invited him.

Eileen Lee arranged for Leroy Hood to testify before the committee. Hood agreed, oblivious to the political maelstrom swirling around him. At the March 19 hearing, he delivered an impassioned plea for the genome project.[55] Hood asserted a role should be found for DOE, NIH, and NSF. He thus deftly if unwittingly ducked the troublesome question of which agency should hold the reins. Scheuer's staff had agonized about the possibility of a Hood-versus-Watson contretemps, but Watson did not show. (Watson later said he was never asked to testify.)[56]

Rep. Schneider's latent distrust broke the surface in a series of questions about forthcoming DOE reports on health effects of radiation among submarine workers, radiation effects among the *hibakusha,* health effects in nuclear plant workers, and "least cost" energy. (DeLisi later noted that Schneider praised these reports when she got them.)[39] Despite the dramatic warning signals, the genome program coasted through the hearings unscathed. The DOE program was probably more vulnerable at this hearing than at any other point in its evolution. DeLisi, unaware of the backroom shenanigans, had cleared his highest hurdle.

The appropriations process was less troublesome than authorization and presented no major obstacles once the genome project had OMB approval. The DOE budget process for fiscal years 1988 and 1989 held true to the initial agreement with OMB—seeking $12 million and $18 million, respectively. It began to exceed the initial agreement only in 1990, when it sought $28 million instead of the original $22 million.

After the March authorization hearing before Scheuer's subcommittee, I escorted Hood (who did not know me then) to the elevators and out to catch a taxi, through the labyrinthine Rayburn House Office Building. He asked, "Is that it?" I asked what he meant. He replied, "Do we get the money?" I was struck, not for the first time, by how much of the process that went into federal research funding was unknown to even the most sophisticated of its recipients. I said something about this being just the first of many steps toward DOE's budget. It was far from a done deal. Hood dashed into a cab and headed for National Airport. He was a long way from home.

DeLisi's ideas found fertile soil in the U.S. Senate, but for reasons different from his own. Senator Pete Domenici was a staunch supporter of the national laboratories in his home state of New Mexico, although he believed that they produced far less long-term benefit for the local economy than they should. He convened a panel of influential policymakers to discuss the future of the national laboratories one Saturday morning, May 2, 1987, in the U.S. Capitol. The meeting featured Barber Conable, a former New York congressman and head of the World Bank; Donald Fredrickson, former director of the National Institutes of Health; Ed Zschau, former California congressman and successful entrepreneur; Jack McConnell, director of advanced technologies for Johnson & Johnson; and the directors of several national laboratories.

In the midst of the meeting, Domenici asked, "What happens if peace breaks out?"[57-60] The bulk of the work supported at the two laboratories in New Mexico was focused on nuclear-weapon production and defense-related research and development. Domenici wanted to know how the immense research resources of the national laboratories could be better integrated into the national economy.[61] He also sought a new mission for national laboratories that did not depend on Cold War rhetoric and that might move them into the growth areas of science, including biology. Domenici knew that sooner or later the Reagan defense spending juggernaut would lose steam.

Donald Fredrickson, then president of the Howard Hughes Medical Institute, asked if the national laboratories might play a role in the human genome project. After the meeting, Jack McConnell helped draft legislation that resulted in Senate bill 1480. By that time, Los Alamos was already beginning its genome program, a year and a half after DeLisi's initial idea. This show of strong support from the Senate nonetheless helped secure the DOE program's future at a time of potential vulnerability.

DeLisi and Smith anticipated many of the arguments that would be made for and against the genome project. But what was missing from their thoughts proved just as important—competition with NIH and acceptance among molecular biologists and human geneticists proved even more important than they might have thought. DeLisi remarked later that "moving unilaterally was not my preference, nor did I consider it optimal."[62] He had a strong potential ally in Vincent DeVita, director of the National Cancer Institute, where DeLisi had worked before. DeVita's power was waning, however, and he was soon to leave the NCI directorship. NIGMS was the NIH institute responsible for funding most basic genetics, but DeLisi's relations with NIGMS were more distant and there was a much greater difference in styles.

DeLisi saw a hole, put his head down, and ran. He put the genome project on the public agenda, but it was not a clean run for the end zone.

The well-known NIGMS response was that if it were to be done, they should do it, but it should not be done. . . . One of my choices was to use the NIH style of cautious consensus building. At times, perhaps most of the time, that is the best procedure; but in my judgment, this was not such a time. I made a deliberate decision to move vigorously forward with the best scientific advice we could muster (HERAC). I am quite willing to take the criticism, rational or not, that such movement provokes. . . . I would have been far more timid about subjecting myself to . . . criticisms . . . if I saw my future career path confined to government.[62]

DeLisi decided to risk attack and push forward. His relations with NIGMS director Ruth Kirschstein, director of the most relevant scientific program at NIH, were intermittent and distant. Those of us observing the process could readily see that the two principal figures in genome politics at DOE and NIH were ill at ease with each other. DeLisi and Kirschstein were both, however, consummate professionals. They avoided direct conflict while encouraging staff exchanges and cooperation. Both later glossed over this period during which their objectives were at cross purposes and their roles inherently cast them in opposition, attributing the perception of conflict to science reporters covering genome politics. The reporters were telling the truth. The tension between DOE and NIGMS was real. The amazing feature of the genome project is that the conflict was contained. It never broke into destructive distrust or resulted in NIH and DOE taking positions that would force them into direct confrontation before Congress. Staff members on Capitol Hill were well aware of the potential for open conflict between NIH and DOE. Some even eagerly awaited the public theater it would provide. Had the battle lines been drawn, the genome project as a whole would almost certainly have been delayed or destroyed.

Several technical elements are remarkable by their absence from early consideration at DOE. There was very little discussion of genetic linkage mapping—the first and arguably the most important step toward making the project useful to the research community—and scant attention to the study of nonhuman organisms as either pilot projects or even scientifically important subjects to study. DeLisi explained these gaps as resulting from a *presumption* that RFLP mapping and work in other organisms would proceed apace, and that the genome program would merely augment the ongoing efforts in these related but distinct areas.[63] A memo from George Cahill corroborates that

DeLisi stressed the importance of comparative genome mapping in man, mouse, and other organisms at the initial meeting of the HERAC subcommittee.[51]

Genetic linkage maps and work on other organisms were, however, clearly subsidiary to the main goals of the initial DOE program: DNA sequencing technology, computation, and physical mapping. By 1990, the genome project was redefined so that genetic linkage maps and physical maps of model organisms and humans were accorded first priority, with sequencing to follow when (and if) it became affordable and sufficiently rapid. In the reoriented genome project, DNA sequencing was subtly removed from the top spot and subordinated to other goals.

The seeming neglect of genetic linkage mapping and nonhuman genetics drove a wedge between DOE and much of the biomedical research community. The enthusiasm driving the DOE human genome proposal proved sufficient to keep it going, but it was a rough ride.

8

Early Skirmishes

IN A COMMENTARY introducing the March 7, 1986, issue of *Science,* Renato Dulbecco, a Nobel laureate and president of the Salk Institute, made the startling assertion that progress in the War on Cancer would be speedier if geneticists were to sequence the human genome.[1] For most biologists, Dulbecco's *Science* article was their first encounter with the idea of sequencing the human genome, and it provoked discussions in the laboratories of universities and research centers throughout the world. Dulbecco was not known as a crusader or self-promoter—quite the opposite— and so his proposal attained credence it would have lacked coming from a less esteemed source.

Like Sinsheimer, Dulbecco came to the idea from a penchant for thinking big. His first public airing of the idea came at a gala Kennedy Center event, a meeting organized by the Italian embassy in Washington, D.C., on Columbus Day, 1985.[2] The meeting included a section on U.S.–Italian cooperation in science, and Dulbecco was invited to give a presentation as one of the most eminent Italian biologists, familiar with science in both the United States and Italy. He was preparing a review paper on the genetic approach to cancer, and he decided that the occasion called for grand ideas. In thinking through the recent past and future directions of cancer research, he decided it could be greatly enriched by a single bold stroke—sequencing the human genome. This Washington meeting marked the beginning of the Italian genome program.[3]

Dulbecco later made the sequencing idea a centerpiece for his September 5 speech to dedicate the Sambrook Laboratory at Cold Spring Harbor on Long Island, New York.[4; 5] Dulbecco sensed a transition in cancer biology: "It seems we are at a turning point in the study of tumor virology and oncogenes."[4] The well-known fact that cancers of certain cell types behaved quite differently in different species meant that "if the primary objective of our endeavor is to understand human cancer, we must study it in human cells."[4]

Dulbecco argued that the early emphasis in cancer was on exogenous factors—viruses, chemical mutagens, and their mechanisms of action. Cancer research had to change strategies, shifting its focus inward: "If we wish to learn more about cancer, we must now concentrate on the cellular genome."[1]

The article as published was considerably shortened from a draft that expanded on how sequence information might tease apart factors explaining the heterogeneity among breast cancer genes.[4] Understanding cancer came from focusing on animal models of cancer, especially tumor viruses. Studying viruses dramatically reduced the number of genes under study and permitted the isolation of individual cancer-associated genes (oncogenes) that would have been forever obscured by studying spontaneous cancers of humans. Molecular biology triumphed by studying the much smaller and more tractable set of genes contained in viruses causing cancer in animals. The study of cloned oncogenes in viruses permitted a reductionist dissection of individual genes contributing to cancer.

Studying oncogenes and tumor viruses could not, however, fully explain the "progression" of tumors—the multiple steps along the road from normal cell maturation to proliferation to cancer. Changes in genes were obviously taking place on this journey, but they could not be easily followed for lack of a road map. The point was not that experiments were impossible, but that they entailed making *ad hoc* maps; much less work would be necessary if there were good global maps of the genome. Dulbecco argued that cancer progression could only be understood once a map was prepared. The DNA sequence was such a map at its ultimate resolution.

While cancer was clearly not a purely genetic disease, in the sense that it was not inherited as a Mendelian trait except in rare families, it was equally clear that the steps leading to uncontrolled cellular growth involved changes in DNA. Changes *were* inherited by groups of cells within the body, even if such changes were not passed on to a person's progeny (since they took place in cells other than those giving rise to eggs and sperm). DNA mutations were thus inherited at the level of the cell, as cells from different organs continually gave rise to new ones. Dulbecco saw the DNA reference sequence as a standard against which to measure genetic changes taking place in cancer. He argued that some such reference was needed, because there was not then and never would be another standard. Human genetic variation was too great, and interbreeding to study specific mutations was unethical. In the mouse, 150 well-characterized, genetically homogeneous strains could be deliberately bred and studied. This well-controlled genetic environment was a vain hope in humans, however, and always would be. Dulbecco saw the sequence information as itself generating new biological hypotheses to be tested by experiment.[6]

Dulbecco envisioned DNA sequence as the lead actor in tomorrow's drama of cancer research. This vision issued from Dulbecco's intuition, more as an inchoate sense of the most productive research strategies for the future than as a concrete step-by-step argument. Indeed, he apologized for "hand-waving," but he did not apologize for his main conclusion, that DNA sequence data would be fundamental to understanding the central problems of biology— cancer, chronic disease, evolution, and how organs and tissues develop.[6] Dulbecco noted the need for biology to encompass some collective enterprises of use to all, in addition to its extremely successful agenda of mounting small, narrowly focused inquiries.

In the *Science* commentary, these arguments for a standard genetic reference genome were given short shrift.[3] Many scientists were puzzled about the scientific rationale behind Dulbecco's proposal, but the *Science* article nonetheless became a catalyst for broader discussion. Sinsheimer convened the first meeting dedicated to discussing whether or not to sequence the human ge-

Renato Dulbecco independently promoted the idea of a massive project to determine the sequence of nucleotides in the DNA of human chromosomes in 1985. Dulbecco, who was awarded the Nobel Prize in physiology or medicine in 1975, is president of the Salk Institute. *Courtesy Salk Institute*

nome, and DeLisi laid the first stones in its bureaucratic foundation, but Dulbecco was the first to publish the idea in a large-circulation journal aimed at the entire scientific community.

By the summer of 1986, the rumor networks of molecular biology were buzzing with talk of the DOE human genome proposal. Dulbecco's proposal helped build the wave. News of the Santa Fe workshop was disseminated by those who attended it; those in the mainstream of molecular biology were beginning to discuss the idea of sequencing the human genome in their phone conversations and at scientific meetings. As is so often the case in molecular biology, Cold Spring Harbor Laboratory on Long Island, New York, became the focal point.

A landmark symposium modestly titled "The Molecular Biology of *Homo sapiens*" took place at Cold Spring Harbor in June 1986, bringing together the giants of human genetics and molecular biology. More than one hundred speakers addressed an audience of 311, reviewing the astonishing progress in two decades of human genetics.[7] The various proposals to sequence the genome were by then hot topics, and they took center stage.

Walter Bodmer, a British human geneticist of broad view, familiar with both molecular methods and mathematical analysis, was the keynote speaker. He emphasized the importance of gene maps and the advantages of having a DNA reference dictionary. He concluded his talk by urging a commitment to systematic mapping and sequencing, as "a revolutionary step forward." Bodmer argued that the project was "enormously worthwhile, has no defense implications, and generates no case for competition between laboratories and nations." Moreover, it was better than big science in physics or space because "it is no good getting a man a third or a quarter of the way to Mars. . . . However, a quarter or a third . . . of the total human genome sequence . . . could already provide a most valuable yield of applications."[8]

Victor McKusick, dean of human genetics and keeper of *Mendelian Inheritance in Man,* the immense compendium of human genetic disease, was next at bat. He summarized the status of the gene map and finished his talk by urging a dedicated effort to genomic mapping and sequencing.[9] He argued that "complete mapping of the human genome and complete sequencing are one and the same thing," because of the intricate interdependence of genetic linkage maps, physical maps, and DNA sequence data. To find a disease gene and understand its function, one would need all three kinds of maps. He urged the audience to get on with the work, and pointed to the future importance of managing the massive flood of data to come from human genetics. Lee Hood enthused about successful early experiments with automated DNA sequencing.[10] The Cold Spring Harbor meeting was also the first exposure many young biologists had to the polymerase chain reaction and to the mix of systematic approaches to mapping and sequencing that were slowly becoming integrated into the Human Genome Project. The synthesis, however, was still a dialectic in transition.

Debate on the genome project came to a head at an evening session not originally on the program. Paul Berg, another Nobel laureate, was unaware of discussions at Santa Cruz and Santa Fe (or within DOE). He read Dulbecco's article and suggested to Watson that it might be useful to have an informal discussion of a genome sequencing effort.[11]

Watson, always well informed through an extensive network of contacts, was aware of the Santa Cruz and Santa Fe meetings. He had talked with Dulbecco and with Walter Gilbert. He called Gilbert at Harvard, asking him to co-chair a genome project discussion with Berg.[12] Berg arrived at Cold Spring Harbor to find himself co-chair of a June 3 rump session intended to ventilate the proposals for a genome project.

Berg led off by trying to channel discussion into the scientific merits of mapping and sequencing, and what technical approaches might make the effort feasible. Gilbert briefly described the Santa Cruz and Santa Fe meetings and then went to the essentials of his post–Santa Cruz missives. He noted that DNA sequence was accumulating at only two million base pairs per year. At that rate, there would be no reference sequence of the human genome for over one thousand years. He thought that could be reduced to one hundred years with no special effort, but that a dedicated effort involving thirty thousand person-years, on the scale of the Space Shuttle project, would produce a dramatic acceleration with enormous benefits. His conclusion from the Santa Cruz meeting was that sequencing the genome "might be doable in a reasonable time," and "it would be inadvisable to do the project in a way which competed with R01 grants [investigator-initiated projects]. . . . the only way in which one could see doing the project was to do it with some structured funding."[13]

Berg took the floor for a short time, and raised the question "Is it worth the cost?" Gilbert had written down numbers—large numbers—large enough to unleash the pent-up fears of younger scientists in the audience. At $1 per base pair, there could be a reference sequence of the human genome for about $3 billion. The audience was stunned. Gilbert's cost projections provoked an uproar. Gilbert seemed to be urging a commitment to a $3 billion project. Sensing a loss of control, Berg called for discussion about whether it would be worthwhile to have the DNA sequence of the human genome, setting aside the cost issue. Berg's white flag was ignored, as the fusillades became too intense to restrain.

David Botstein rose to the podium when he could no longer contain his volcanic energy. Botstein stated that "there are two components to this. One is political, and we shouldn't forget about the political, because we hope to get something, right? And another is scientific, because we hope to learn something. And the question is: How much is it going to cost?" Catching his stride, he moved for the kill, "if it means changing the structure of science in such a way as to indenture all of us, especially the young people, to this enormous thing like the Space Shuttle, instead of what you feel like doing . . . and we should be very careful." He cautioned that "we should not go forward under the flag of Asilomar, okay, because we are amateur politicians and we're about to be dealing with professionals." This was a swipe at Berg, who played a prominent role in the recombinant DNA debate, including a famous meeting at Asilomar on the California coast. Botstein derided the notion of genome sequencing, noting that if Lewis and Clark had followed a similar approach to mapping the American West, a millimeter at a time, they would still be somewhere in North Dakota. Botstein closed by pleading that molecular biologists "maybe accept the goal, but not give away our ability to decide what is important because we have decided on the Space Shuttle."[14]

This broke the dam, and applause resonated through the audience. Gilbert responded that Botstein was basically right, and that the initial efforts should concentrate first on the 1 percent of the genome containing biologically known function, then do the next 10 percent, and only then finish the job, devoting

equal resources to each phase. A Gilbert-Botstein-Berg exchange then went on for several more minutes, reaching consensus on an important point when Gilbert cautioned that "we shouldn't confuse, let's say, sequencing the human genome with a total knowledge of all science." And Botstein responded, "That's what I hope will not happen." Gilbert then stated the main goal of the project: "essentially the total speeding up of all the things that laboratories [now have to do one gene at a time]."

The exchange went on until Maxine Singer of the Carnegie Corporation (a nonprofit foundation) broke it off by focusing on the notorious failure of science to predict its future. Several speakers followed, including many prominent scientists who reiterated Botstein's sentiments. Others supported the notion of a sexy proposal that could attract public support but were ambivalent about its impact on science. David Smith from DOE spoke on the focus of the DOE proposal, which did not embody a commitment to DNA sequencing per se, but only developing the technologies and infrastructure necessary to a future commitment, but he was clearly on the defensive, ceding in response to one question that perhaps DOE should not lead such an effort. His comments were largely swept away as the dam broke, although he noted that many people in the audience later came forward privately to indicate their support. Berg struggled intermittently and unsuccessfully to contain the flood, pleading for a discussion of the technical and scientific aspects. The emotional torrent was simply too strong, however. Molecular biologists were not enthused by the DOE Human Genome Initiative, perceiving it as a misguided bureaucratic initiative and, more important, as a direct threat to their own research funding.

At the time, the Cold Spring Harbor symposium seemed to stall the momentum toward a massive DNA-sequencing effort. DOE's effort, in particular, was under heavy fire. The symposium, however, proved merely a short losing battle in a longer war from which the genome idea emerged triumphant. That hardly seemed the likely outcome in June 1986, however. The symposium proved to be the opening event in an international tour culminating in a restructured genome project that commanded worldwide consensus. It marked a transition from emphasizing the sequencing of the human genome to a broader plan for genetic linkage mapping, physical mapping, and the study of nonhuman organisms.

Prospects for the Human Genome Project reached a nadir in June 1986, at a special rump session of a Cold Spring Harbor Laboratory symposium on the molecular biology of *Homo sapiens*. In the upper photo on the facing page, David Botstein, a geneticist from Stanford University, attacks the notion of mindlessly sequencing the entire human genome, as Walter Gilbert listens. In the lower photo, Paul Berg, who chaired the session, attempts to quell the unruly crowd following Botstein's remarks. Berg, also from Stanford, shared the 1980 Nobel Prize for chemistry with Sanger and Gilbert. *Victor McKusick photos, courtesy Cold Spring Harbor Laboratory Library*

The goals of the Sinsheimer, Dulbecco, and Gilbert formulations were simple and clear: a complete reference DNA sequence of the twenty-four human chromosomes (X, Y, and the twenty-two nonsex chromosomes). DeLisi's program was justified primarily as the first step toward that goal. In the genome project that began to emerge in the wake of the Cold Spring Harbor meeting, however, the goal was a useful set of chromosomal maps, not only of humans but also of some other organisms. DNA sequencing—particularly the technology to make it faster, cheaper, and more accurate—was still important but no longer dominant. Sequencing dropped from being the primary or only goal to a goal subsidiary to these more general objectives. In the redefined genome project, the goal of the project was to bring the new techniques of molecular genetics to bear on a massive scale, to enable approaches to human genetics analogous to those long employed to study yeast, nematodes, fruit flies, and other organisms.

The DNA-sequencing goal continued as a source of controversy, with many equating the project to the initial sequencing goals. Sequencing the genome became the butt of jokes. A letter to *Nature* suggested that "sequencing the genome would be about as useful as translating the complete works of Shakespeare into cuneiform, but not quite as feasible or as easy to interpret."[15] Robert Weinberg of the Whitehead Institute was surprised that "consenting adults have been caught in public talking about it. . . . it makes no sense"[16] and worried that geneticists would be "wading through a sea of drivel to merge dry-shod on a few tiny islands of information."[17; 18]

Joseph Gall of the Carnegie Institute noted that DNA sequencing might be an inefficient way to study genetics, since complex organisms like nematodes and fruit flies could get by with only 3 to 6 percent as much DNA as humans, while salamanders and many plants had ten times as much.[19] More DNA did not necessarily imply greater complexity, and deciphering the information content of DNA was more than simply reading off the order of base pairs. Gall suggested a two-pronged attack, expanding on the work on nematodes to construct physical maps on one front, and sequencing of individual genes of interest on the other front. The sequencing part might be expedited by a large-scale project to catalog and sequence those parts of DNA directly coding for proteins. A letter to *Nature* pointed out that the pace of gene cloning and sequencing could not continue to explode without displacing all other biology, but noted tongue-in-cheek that "Man's feeling of self-importance will probably not be satisfied until the last bit of his genome has been sequenced and filed somewhere."[20]

The first reports of the *C. elegans*– and yeast-sequencing projects began to make sequencing look like a more efficient way to ferret out and study the function of large numbers of unknown genes, but the results took five years of scientific effort. By then, it began to seem that more traditional projects directed at individual genes and genome-scale sequencing were not interchangeable strategies. Gene analysis and "the sequencing of entire genomes are not

alternatives, but are rather complementary approaches. . . . [Studying expressed genes] cannot by definition throw much light on regulatory processes, on the reasons why some genes [are interrupted] and others are not . . . and on how the genome got like that, anyway."[21] But securing the future of the genome project required a broadening of the political and scientific base, building bridges to both genetic linkage mapping and more traditional genetics.

The originators of the genome idea differed in their assessments of whether the redefinition of the genome project resulted more from political pressures than scientific ones. Gilbert stuck to his guns at many public forums, stating his views that sequencing was still the ultimate goal, and the faster the better.[22;23] He viewed the redefinition as a step backward, while committing a part of his laboratory to original project of large-scale sequencing. Dulbecco believed he presented the "most extreme case" in 1985 and was merely watching a normal scientific reformulation of the sequencing idea as it met with the need for realistic goal-setting.[6] Sinsheimer believed that genetic and physical mapping remained only stepping stones to the true objective, added to the agenda of the project mainly to gain political support.[24] DeLisi felt the program was unfolding more or less as he anticipated.[25]

The initiators thus viewed the new project as only slightly changed, in that the physical mapping, genetic linkage mapping, and study of other organisms were always part of the process of moving toward genome-scale sequencing. The explicit redefinition, however, enabled those who did not view genome-scale sequencing as the end goal to fall in line to support a genome project with broader goals. This was a subtle but important transition. If the originators were correct that DNA sequence would in the end be the most important objective, then the project could rededicate itself to that goal in the future, but the redefined project made room for them to be wrong and nonetheless produce something quite useful. Sequencing moved from the primary to a subsidiary goal. The broader definition of mapping extended the political support base within science, enhanced the scientific integrity of the project by increasing the likelihood of attaining at least some of its goals, and hedged bets on exactly which kinds of genetic maps would ultimately prove most important. Without such support, the funding to make the massive sequencing projects possible, and the biology to make them meaningful, would not have been in place.

The disputes at the June 1986 Cold Spring Harbor symposium were covered by Roger Lewin for *Science,* the first signals of the debate to come for many in science and in government.[26;27] These articles highlighted Gilbert's quest for the "Holy Grail" in a call-out quote and introduced the history of the idea for a DNA sequencing project: "During the past twelve months there have been half a dozen separately organized small gatherings scattered across the country, each one discussing the prospect of obtaining a complete nucleotide sequence of the human genome."[27] Lewin chronicled the shift in objectives, quoting Ray White—"Humans deserve a genetic linkage map. It is part of the description of *Homo sapiens*"—and elaborating on the utility of physical maps. But Lewin captured the confused mix of issues and supposed goals of a genome project, ending his piece on an ambivalent note by quoting Nobelist David Baltimore—"The idea is gathering momentum. . . . I shiver at the thought."[27]

At the June 3 Cold Spring Harbor session, Carnegie Institute biologist Maxine Singer observed, "Of course we are interested in having the sequence, but the important question is the route we take in getting it."[27] A consensus was forming, but it had not jelled and could not yet be articulated. The process of building a consensus reconstructed the genome project and resulted in a dedicated program of map-making with new organizational bases in the federal government. As the goals shifted, the debate moved from the scientific Mecca, Cold Spring Harbor, to the political Gomorrah, Washington, D.C.

PART THREE

The Support Structure

9

The Odd Legacy of Howard Hughes

T HE HOWARD HUGHES MEDICAL INSTITUTE (HHMI) was founded on December 17, 1953, six months after Watson and Crick published their double-helical structure of DNA. In an irony of American capitalism, the largest biomedical research philanthropy in the United States was founded as a tax dodge for a defense contractor.

Howard Robard Hughes, Jr., set the institute up a week before his forty-eighth birthday. He apparently had ideas about establishing a medical research institution of some kind as early as 1926, according to George Thorn, who was the institute's scientific director from 1956 to 1978.[1; 2] Planning got more serious in 1946, when Hughes commenced discussions about a medical institute with his personal physician Verne Mason. Hughes was then recovering from injuries sustained in the crash of a prototype XF-11 photoreconnaissance plane he was test-piloting. He was, perhaps, feeling especially appreciative of modern medical technology and its scientific underpinnings and may have been thinking about what sort of legacy to leave behind.[3] In 1950, Hughes gave a total of $100,000 directly to the first three Howard R. Hughes Fellows; he funded another four in 1951.[2]

The institute was created as part of a legal package that carved the Hughes Aircraft Company out of the Electronics Division of Summa Corporation.

Summa was part of the original Hughes Tool Company, an oil-drilling-equip-
ment company which Howard Hughes inherited from his father at age nine-
teen. In late 1953, Hughes was under intense pressure from his chief client,
the U.S. Air Force, to make management improvements or risk losing future
business with the government.[4] The reorganization was the result. The medical institute was given ownership of the Hughes Aircraft Company, a part of
the larger Hughes conglomerate. Hughes himself was sole trustee of the institute from its founding until his death on April 5, 1976. During those twenty-
three years, the institute spent $63 million for medical research.[2]

The charter of the new institute stated:

The primary purpose and objective of the Howard Hughes Medical Institute shall be
the promotion of human knowledge within the field of the basic sciences (principally
the field of medical research and medical education) and the effective application thereof
for the benefit of mankind.[3]

The Internal Revenue Service begged to differ. In November 1955, the
IRS concluded that the institute was "merely a device for siphoning off otherwise taxable income to an exempt organization and accumulating that income."[4] The IRS denied tax-exempt status to the institute until March 1, 1957,
when it was granted. (In the meantime, Hughes underwrote a loan to Donald
Nixon, brother of the Vice President and future President. The loan was never
repaid, although no connection to the IRS approval of tax-free status was
established.)[4]

The institute dramatically increased both the number of investigators and
its financial commitment to them, beginning in 1976, in response to a report
from its medical advisory board. Hughes, the sole trustee, died that year.
Before he died, the institute had already chosen to focus on three fields believed
fundamental to understanding human disease: genetics, immunology, and the
study of metabolic-endocrine disorders.[3] In 1976, the institute's spending
increased dramatically to $4.7 million, and by 1980 it reached $25.8 million.[5]
An important new element was support of research itself, in addition to payment of the investigators' salaries. Until then, at Hughes's direction, only
salaries had been paid by the institute.[6]

In 1984, the Delaware Court of Chancery removed management of the
institute from its executive committee, which had run it during the eight years
of litigation following Hughes's death. The reins were handed to an eight-
member board of trustees appointed by the court. The trustees elected a ninth
member later that year. Donald S. Fredrickson, who had directed the National
Institutes of Health from 1975 to 1981 and joined HHMI as vice president in
1983, was appointed president.

Early in 1985, the trustees decided to sell Hughes Aircraft. The sale was
prompted by the tax dispute with IRS that had hounded the institute for more
than two decades. Fredrickson explained, "We could not settle this controversy
with the IRS without knowing the exact worth of our endowment, and the

only way to do that was to sell the company."[7] Hughes Aircraft was put on the auction block, and General Motors bid highest. GM paid HHMI $2.7 billion in cash and created 100 million shares of a special stock valued at roughly the same amount. The GM settlement was itself a source of friction until a February 1989 agreement between GM and Hughes.[8; 9] The sale of Hughes Aircraft made HHMI the largest private philanthropy in the nation, with assets then valued at $5.2 billion.

From 1985 through 1987, as the genome debate was intensifying, HHMI was in the throes of managing its explosive growth. Its research budget more than doubled from 1985 to 1986, reaching $214 million, with another $17 million for administration.[5; 10] On March 2, 1987, the institute reached a settlement with the IRS, ending the tax dispute.[11] HHMI agreed to spend 3.5 percent of its endowment in support of research each year, to pay the U.S. government $35 million to forgive any tax obligations it might not have met (although not admitting to any), and to give out at least an additional $500 million for special projects related to the HHMI mission over the next decade (ending August 1997).

The tempest of controversy was not quite spent. Donald Fredrickson took a leave of absence in April 1987 and resigned as president on June 2. His highly publicized forced resignation was viewed as "both a personal tragedy and a public loss."[12] He left amid allegations that his wife had improperly wielded authority over renovations of an HHMI property located on the NIH campus. The HHMI trustees had voted to bar Mrs. Fredrickson from HHMI meetings in December 1986 and hired a firm to investigate HHMI financial dealings. They got back a report of more than three hundred pages that did not find conclusive proof of wrongdoing, but convinced the trustees that decisive action was necessary.[12; 13]

The institute continued to consider action on the genome project amid all the turbulence caused by selling Hughes Aircraft, planning what to do with its newfound wealth, negotiating a settlement with the IRS, and losing a president. The science went on, and HHMI played a pivotal role in the genome debate.

The HHMI interest can be followed along several paths. HHMI staff credit Ray Gesteland and Charles Scriver as the people principally responsible for getting HHMI interested in gene mapping. Gesteland was a student in the Watson laboratory in the mid-1960s and later became an independently supported Howard Hughes Investigator at the University of Utah. Soon after publication of the RFLP mapping paper in 1980, Gesteland suggested to George Cahill, then HHMI's director of research, that HHMI might support systematic RFLP mapping along the lines proposed by David Botstein at MIT. Botstein had independently raised the idea of RFLP mapping with HHMI trustee George Thorn.[14] Ray White, still at the University of Massachusetts at Worcester, had by then contacted David Botstein about RFLP mapping.

Cahill recruited White to go to Utah.[15] In the background was a desire among some HHMI trustees to strengthen ties to Salt Lake City, because of Howard Hughes's Mormon connections. White was attracted in part because of the incredibly rich and detailed Mormon pedigrees kept by the university that might be useful for clinical genetic studies. The large and well-documented families were a unique resource. They would be invaluable not only in the search for RFLP variants, but also for disease-gene mapping once the markers were in place. White commenced work to construct a genetic linkage map when he moved to Salt Lake City in November 1980, building on his recent success in finding the first RFLP marker with Arlene Wyman.

Scriver was a Canadian geneticist of international reputation serving on the HHMI medical advisory board from the late 1970s into the mid-1980s, a period during which the institute's annual funding for biomedical research increased more than twenty-fold. Scriver was fascinated by the prospect of a Human Genome Project, thinking primarily of the immense impact systematic mapping could have on medical genetics. He was concerned about science, but also about the patients he saw every working day in his Montreal genetics clinic. He called the decision to fund genetic linkage mapping, for which he became a champion on the medical advisory board, "a close thing" on the part of HHMI. Several years later, the trustees and scientific advisers to HHMI considered this among their most productive investments.[16]

Scriver became convinced that support of genetics databases was an essential next step, mainly through conversations with Frank Ruddle of Yale, and later with Ray White. Scriver worked to persuade the other members of the medical advisory board. He spoke three times at board meetings. The first two times, he sparked little enthusiasm; by the third meeting, however, George Cahill had warmed to the idea and supported an HHMI commitment. Scriver then made a presentation to the trustees in December 1985. He caught trustee Hanna Gray's attention by referring to how genetic maps would introduce a neo-Vesalian era into medicine. (The Flemish anatomist Andreas Vesalius prepared diagrams based on actual dissections in the mid-1500s. His methods were not universally applauded—indeed, they were regarded as sacrilegious and macabre by many at the time. His anatomy paved the way for the functional studies of William Harvey in England a century later, and the Italian Giovanni Morgagni's studies of how disease affected the body in the mid-1700s.) The trustees agreed to support genetic database efforts for five years, from 1986 to 1991, subject to annual review.

Having secured trustee approval, Scriver, Cahill, and Fredrickson convened a special meeting to discuss the HHMI human genetic resources at HHMI headquarters in Coconut Grove, Florida, on February 15, 1986. (Headquarters moved to Bethesda, Maryland, in 1987.) Fredrickson opened the meeting, followed by Scriver. Next were presentations from the Utah human genetics team and the New Haven group that ran the Human Gene

Mapping Library. The focus was on how to manage the massive increase in information about genetic marker maps, locations determined by somatic cell genetics, and new DNA probes. There was also much discussion of the emerging broad outlines of the genome project, particularly its underlying technologies.

The Coconut Grove meeting took place two weeks before the DOE meeting in Santa Fe on sequencing the genome, but the two meetings were in separate orbits. Ruddle was central to both efforts, but his was one of the few areas of overlap, and the emphases of the two discussions were quite different. DOE concentrated on sequencing, viewing physical mapping and genetic linkage maps as steps along the way. HHMI had a major commitment to genetic linkage mapping, supporting the largest group in the world, and was moving toward support of databases. The initiatives converged only later under the umbrella of the Human Genome Project. In March 1986, the emerging HHMI interests and DOE's nascent genome initiative were worlds apart. In June, attention shifted to the contretemps at Cold Spring Harbor. In July, the HHMI and DOE programs were brought face to face at an HHMI public forum, Washington's first major public event dedicated solely to the genome project.

After the Coconut Grove confab, HHMI wanted a much larger and more public forum on genome research. The July meeting was scheduled to prepare Fredrickson for another HHMI trustee meeting on August 5. Maya Pines, a professional science writer, was commissioned to describe gene mapping and sequencing to the HHMI trustees. Genetics was HHMI's strongest research area, constituting roughly a third of its research. HHMI polled its contacts about what its role should be. James Watson met with Fredrickson on April 1 to indicate his strong support for an HHMI presence in genome research.[23; 24] While DOE came to human genetics mainly form technology—laser-activated sorting of chromosomes, DNA sequence database experience, and projects to make clone libraries of the different human chromosomes—HHMI was smack in the middle of mainstream human genetics, supporting many of the most prestigious groups in the world. As a consequence, HHMI drew on an international constellation of stars for its July forum.

In the wake of Cold Spring Harbor, there was a palpable tension surrounding the HHMI forum. The stage was set; science journalists and others interested in science policy flocked to watch the sparks fly. It was a gala event held on the NIH campus, July 23, 1986. Sinsheimer, Watson, Gilbert, and Lloyd H. (Holly) Smith (chairman of the HHMI medical advisory board) sat next to one another cribbing notes in the Nobel laureates' corner, next to the dean of human genetics, Victor McKusick. Donald Fredrickson, former director of NIH and then president of HHMI, introduced and closed the meeting, which was chaired by Walter Bodmer from the Imperial Cancer Research Fund laboratories in London.

The HHMI forum turned into a love fest for a redefined genome project. There were several brief presentations about the technologies and what was going on in U.S. agencies and in other parts of the world. But mainly, it was a show of power—a battleship summit for molecular biology. The HHMI forum was a turning point, but the new direction was not entirely clear at the time. *Science* reported that "the drive to initiate a Big Science project to sequence the entire human genome is running out of steam."[17] Leroy Hood asserted that massive sequencing was premature, and that the focus should instead be on improving the technologies.

Under the surface, however, a new consensus was emerging. The meeting rechanneled the genome project, rather than rejecting it outright. Those attending the meeting agreed that the time was ripe to mount a special initiative in gene mapping and technology development, to redress deficiencies in the infrastructure undergirding genetics. This agreement was obscured by more conspicuous disagreements about priorities and the proper style of leadership. At one critical juncture, chairman Bodmer could not contain himself when David Smith presented an outline of the DOE genome initiative. Bodmer interjected that the DOE proposal did not acknowledge the importance of genetic mapping. While Smith continued, a bit shaken, Sydney Brenner, seated at the meeting table, conspicuously passed a note to Gilbert and Watson that was read by those around them: "This is a retreat."[18; 19] DOE was on hostile turf, in the homeland of NIH and HHMI.

A two-phased strategy emerged from the HHMI forum. The first phase would concentrate on genetic linkage mapping, physical mapping, and development of technologies for DNA sequencing and for analyzing genetic information with computers. The second phase, contingent on reassessment as Phase I progressed, would concentrate on DNA sequencing. This could aim to sequence the entire human genome or not, but it would clearly start first with genes and regions of interest. Leroy Hood urged that the emphasis fall on developing new technologies during the early years, so that later efforts would be faster and more efficient.

James Watson allowed that while he was strongly in favor of a genome project, everyone else he talked to at Cold Spring Harbor Laboratory was against it.[17] He reflected that young scientists feared a massive sequencing project might subtract from the pot of funds available for their work,[17] but he pointed out that the mapping work deserved special attention or it would not get done. "We know how useful this Phase I work is. There should be more of a sense of urgency about it. Are we going about it as if we were in a war?"[20]

Walter Gilbert added a Phase III, understanding the function of the genes uncovered in Phases I and II. In Gilbert's mind, Phase I would last five years or so, Phase II a decade, and Phase II would be much of biology in the twenty-first century. Walter Bodmer summed up his sense of the meeting by focusing on the primary goal of the genome project: understanding human disease. There were subsidiary goals related to understanding evolution and variations among different populations, but the central thrust of the project should be to expedite all biomedical research.[17; 20; 21]

After the Coconut Grove meeting and the July forum, HHMI moved on several fronts. Maya Pines completed her briefing paper for the trustees, *Shall We Grasp the Opportunity to Map and Sequence All Human Genes and Create a "Human Gene Dictionary"?*[20; 22] The answer was clear; "the monumental project of mapping the entire human genome has moved from a pipe dream to a realistic goal which is arousing increasing enthusiasm."[20] HHMI proceeded to fund several international human gene mapping workshops, where geneticists from around the globe assembled to hammer out the best maps of various genes and RFLP markers. HHMI also directly funded the on-line version of McKusick's *Mendelian Inheritance in Man* at Johns Hopkins, carrying forward work previously supported by the National Library of Medicine. HHMI picked up funding of the Human Gene Mapping Library in New Haven from NIH from 1985 to 1989. During 1988 and 1989, HHMI convened a special review panel to chart the future of genome databases. The committee advised HHMI to phase out the Human Gene Mapping Library in favor of a new Genome Database, to be developed by Peter Pearson, Richard Lucier, and staff of the Welch Medical Library at Johns Hopkins. HHMI also supported a parallel effort to computerize a genetic database for mouse genetics at Jackson Laboratories in Maine. HHMI gave funds to the Centre d'Etude du Polymorphisme Humain (CEPH) in Paris to support the international collaboration unifying genetic linkage mapping efforts around the world.

HHMI support of databases and international collaborations knitted together a disparate and widely dispersed genetics community, weaving an informal network of electronic and personal ties. Finally, HHMI commissioned Maya Pines to write a public document describing the genome project. This booklet was released in December 1987, as Congress was preparing for the 1989 budget, and as prospects for the future of the project in Congress were still unclear.[25]

Hughes also lent support as a new scientific organization, the Human Genome Organization, formed to mediate international collaborations. Agreement to form HUGO, as it was soon nicknamed, came from an impromptu meeting called by McKusick at the first annual Genome Mapping and Sequencing Symposium at Cold Spring Harbor on April 29, 1988. HHMI paid for the thirty-two founding members from eleven nations to meet in Montreux, Switzerland, in September. Early in 1990, HHMI agreed to give HUGO a $1 million startup grant for four years. Diane Hinton of the HHMI staff was appointed administrator of the Americas office for HUGO. HHMI housed HUGO in the United States for several years, and in November 1990 donated a Bethesda, Maryland, office condominium owned by HHMI to HUGO for its use.

HHMI moved decisively to support genome research at several points. It established the first and largest RFLP mapping effort soon after the idea was published. It moved to support several computer databases for human and mouse genetic information. And it helped to establish HUGO as the international coordinating body. HHMI also played a more informal role in the interagency jockeying that took place as the genome project unfolded. George Cahill and Diane Hinton emerged as central mediating figures. After Cahill retired, Max Cowan assumed his position as a major coordinator and funder of genome research internationally. HHMI was presumed to be a neutral party in most disputes, a philanthropy with international reach and a commitment to shared informational resources in human genetics. HHMI was smaller than the federal agencies, but large enough to sustain significant commitments of money and personnel. It responded rapidly when federal agencies did not, filling niches left vacant by the NIH behemoth. Howard Hughes had his name, however inadvertently, on the flight of the genome project as it lifted off the runway a decade after his death.

10

The NAS Redefines the Project

THE JULY 1986 HHMI forum began a several-year period during which the National Institutes of Health and the Department of Energy jousted over the genome project. Indeed, which agency would prevail became the dominant topic of discussion about the genome project until well into 1988. Only then did an emerging consensus about the importance of a concerted research effort begin to displace divisive debates over who would get to run it.

At the July 1986 Howard Hughes Medical Institute forum, the question imperceptibly shifted from the central focus of the Cold Spring Harbor meeting. Whether to start a genome project gave way to what it encompassed, how best to do it, and who should lead it. By the end of 1986, it was already clear that DOE was committed to mounting a genome project and HHMI would play a small but significant part. The big question was what would happen at NIH.

The consensus was not yet apparent, however, and having an endorsement from a single meeting convened by HHMI was not necessarily politically persuasive. HHMI was already committed to genome research, and the most prominent skeptics—such as Maxine Singer and David Botstein—were not at the HHMI forum. A more formal scientific review was necessary before the genome project could be said to command a scientific consensus. The natural body to perform this function was the National Research Council of the National Academy of Sciences.

President Lincoln signed legislation creating the National Academy of Sciences on March 3, 1863, during the Civil War. The Academy was the brainchild of Alexander Dallas Bache, who in 1851 first proposed a national scientific body to rival the French Academy and the British Royal Society. Language for the bill came from a Saturday-night dinner meeting on February 19, 1863, that included four members of an informal scientific group called the Lazzaroni. Louis Agassiz, an eminent natural historian from Harvard, interested Senator Henry Wilson of Massachusetts in sponsoring a bill, which was introduced on February 21 and passed well after seven in the evening of the last day of the legislative session, without a complete reading and with no debate. The legislation listed the initial fifty scientist members, designated by those at the February 19 meeting. Bache became the first president, but his health soon declined and with it the Academy's. The Academy was rescued from lassitude by the eminent physicist (and head of the Smithsonian Institution) Joseph Henry, behind whose back it had been created. Despite being excluded from the negotiations to create the Academy, Henry stepped in to direct it. He put it on firm footing by getting Congress to permit an expansion of membership and by focusing its energies on basic science.[1] The enacting legislation decreed:

The Academy shall, whenever called upon by any department of the Government, investigate, examine, experiment, and report upon any subject of science or art, the actual expense of such investigations, examinations, experiments, and reports to be paid from appropriations which may be made for the purpose, but the Academy shall receive no compensation whatever for any service to the Government of the United States. [An Act to Incorporate the National Academy of Sciences, 37th Congress, Session 3, Chapter 111, March 3, 1863]

In 1916, the astronomer George Ellery Hale and a committee he chaired wrote to President Wilson, offering the services of the National Academy of Sciences and urging the establishment of a National Research Council (NRC) to provide advice to the government. American involvement in World War I was looking more and more likely, and scientists saw a great opportunity to further the war effort if a broader scientific body were to become more actively involved with providing advice than the Academy itself. The National Research Council could involve scientists who were not members of the Academy, greatly expanding the range of available expertise. The president accepted the Academy's invitation, and the NRC had its organizational meeting in September 1916. Hostilities with Germany broke out in 1917, and the NRC busied itself, under the direction of physicist Robert Millikan, with fundraising and organization to help the war effort. After the war, an executive order of May 11, 1918, made the NRC permanent, and it began to dispense funds from the large private foundations. In this period, the NRC "so nearly lost touch with the federal government that it was neither a coordinating center for science in the bureaus nor an active adviser," according to one historian.[1]

Relations between the National Academy of Sciences, its NRC advisory arm, and the federal government waxed and waned. In general, during times of war, science came closer to government, but then it drifted away in times of peace, until after World War II. In the postwar period, the permanent federal commitment to science grew substantially as the links among science, technology, military power, and economic growth became widely held cultural beliefs. It was not always so. A Colorado congressman commenting on the Academy in the late 1870s quipped:

This Academy has never published but one work, and that was a very thin volume of memoirs of its departed members. And if they are to continue to engage in practical legislation, it would have been well for the country if that volume had been much thicker.[1]

Such sentiments ebbed to a small minority as science and technology become more central to national life. The National Academy Complex, which to the Academy proper and the NRC added a National Academy of Engineering in 1964 and the Institute of Medicine in 1970, had an annual budget of roughly $160 million by the end of the 1980s. In any given year, a thousand committees worked on various projects, with the federal government supplying more than three-fourths of the total funding.[2; 3]

In 1986 and 1987, debate about the wisdom of mounting a genome project was largely confined to the scientific community and was of particular interest to geneticists and molecular biologists. The pro and con arguments began to converge on what should be done and how to do it, questions well suited to the report-writing process employed by the NRC. The NRC process ensured a systematic assessment often absent from open-ended debate. A report from the NRC also carried special weight in Congress and in executive agencies.

Convening a panel on the genome project was a considerable risk for genome proponents at the time, because sentiments were largely against the DOE proposal, which had dominated discussion to that point. An NRC report that equivocated or came out against a genome project would likely kill the idea in any agency, for several years at least. A positive report would not guarantee its success, particularly if it asked for extra funding, but a negative report would be an almost insurmountable obstacle.

Plans to involve the Academy coalesced soon after the Cold Spring Harbor meeting in June 1986. On July 3, John Burris, executive director of the Board on Basic Biology at the Academy, wrote a short proposal to fund a small group meeting to discuss the genome project as an add-on to an August 5 meeting in Woods Hole, Massachusetts.[4] This was a new addition to the agenda, not anticipated at the March 3 meeting of the board.[5] The project was quickly approved, and on July 10, Burris sent out a tentative agenda and background materials, including clips from the *New York Times* and a summary of the DOE meeting in Santa Fe.[6-9]

Francisco Ayala of the University of California, Irvine, chaired the August 5 afternoon session. David Smith was there for Charles DeLisi, who declined an invitation to go himself. Ruth Kirschstein, director of the National Institute of General Medical Sciences, represented NIH. The group included a high concentration of powerful figures in science.[10] The board noted its support for physical mapping and expressly withheld its support from a massive sequencing program. The board met again the morning of August 6 and agreed that there should be a series of technical workshops and that the NRC staff should develop a proposal for a study. The Commission on Life Sciences, parent body to the board, met the next day and also encouraged development of an NRC study. Meeting minutes suggested that "the project should be designed primarily to look at mapping of the genome, with consideration given to the international cooperation in this effort."[11] A study proposal was written and approved by the Academy's governing board on September 23, as proposed by the Commission on Life Sciences and the Board on Basic Biology.[12] At Watson's suggestion, Burris forwarded the proposal to Michael Witunski of the James S. McDonnell Foundation. The McDonnell Foundation responded

Bruce Alberts chaired the National Academy of Sciences / National Research Council committee that crafted the scientific strategy for the Human Genome Project. Alberts, who was then a professor of biochemistry at the University of California at San Francisco, is now president of the NAS. *Courtesy National Academy of Sciences*

positively to the proposal and had a check to the Academy to fund the study in less than a month.[13]

The NRC appointed Bruce Alberts chairman. Alberts, a molecular biologist at the University of California at San Francisco, was a brilliant choice. He had written an editorial in *Cell* the previous year that argued against Big Science in biology.[14] He had taken no position on the genome project, but would be seen as neutral or inclined to oppose it. Furthermore, his experience in writing a major textbook *(Molecular Biology of the Cell)* confirmed his talents in managing large writing projects. The original hope was to complete the NRC study in six months, or at least by midsummer 1987.

Several others identified as skeptics were appointed to the panel, notably Botstein and Shirley Tilghman. The committee was peppered with Nobel laureates, including Gilbert, Watson, and Nathans. Sydney Brenner was invited to represent the views of British mappers and sequencers, and John Tooze from the European Molecular Biology Organization in Heidelberg was a second well-connected European. Russell Doolittle from the University of California at San Diego was a pioneer in analyzing data about protein sequences and managing databases. Charles Cantor was a physical mapper, Hood was the expert in instrumentation, and Ruddle was a major force in somatic cell genetics and databases for gene mapping. McKusick, Leon Rosenberg (dean of Yale Medical School), and Stuart Orkin (whose Harvard Medical School laboratory had done seminal work on chronic granulomatous disease and several other diseases) represented human genetics.

Alberts and Burris hatched a strategy intended to slowly build consensus, if that proved possible. The first meeting on December 5, 1986, was intended to give the committee a general sense of the lay of the land, with presentations from the U.S. organizations with special genome-related activities—NIH, DOE, the National Science Foundation, HHMI, and the congressional Office of Technology Assessment (OTA). A survey of activities in Europe followed. The remainder of the day was devoted to what the report should cover and what further information the committee needed to make policy recommendations.

Burris and Alberts elected to focus early meetings almost exclusively on technical background and to postpone discussion of policy options and funding until the technical stakes were clear. The committee opted to bring in those with "hands-on" experience in the technologies under discussion, prudently divining that subsequent policy debates would be less acrimonious as the facts themselves settled many points. At a January 19, 1987, meeting, Maynard Olson from Washington University and others discussed physical mapping, genetic linkage mapping, and somatic-cell genetics.[15] Scientists running major sequencing projects were asked to ground the sequencing discussion in technical realities.[16] Several investigators with direct experience in mathematics, computation, and database management covered informatics and mathematical analysis. Alberts became convinced that he needed someone like Olson, a person deeply involved in actually doing the molecular biology, to balance more senior technological optimists on the committee.[17] After that meeting, the dynamics of the committee took an interesting turn.

In February 1987, Walter Gilbert announced plans to form the Genome Corporation, to map and sequence the genome in a private company. He resigned from the NRC committee to avoid a conflict of interest. Gilbert had consistently proselytized for a fast-track genome project and had always been optimistic to the extreme in projecting timetables and budgets. He despaired of the government ever acting decisively to mount a genome program. Several committee members felt Gilbert was such a strong champion that he impeded consensus; his assertiveness provoked a backlash. His resignation paradoxically made it possible for those skeptical of the project to participate in redefining it.

Helen Donis-Keller, James Gusella, and Ray White, the leading figures of genetic linkage mapping, opened the next NRC committee meeting in March 1987. The afternoon was a snapshot of the political landscape, with presentations from OTA and Wyngaarden about prospects for securing funds from Congress and a discussion of NIH-DOE politics.

Maynard Olson was appointed to the committee just before Gilbert resigned. Alberts later called the addition of Olson his "major contribution to the NRC committee."[18] Olson's work in physical mapping and large DNA fragment cloning was at the heart of the science under discussion, and he

Maynard Olson served with Alberts on the NRC committee on the Human Genome Project. A specialist on the physical mapping of yeast chromosomes, Olson and his colleagues at Washington University in St. Louis developed the yeast artificial chromosome, or YAC, which made it possible for very long DNA sequences to be replicated. He is now at the University of Washington in Seattle. *Courtesy Victor McKusick*

brought quiet but occasionally biting insights to the discussion. Because he had direct experience with the methods, he could argue convincingly that the seductive theoretical schemes glibly promulgated in the press and on the committee were unlikely to work as advertised.[17] Moreover, Olson's philosophical approach and dry humor were well suited to illuminate conceptual muddles and to forge consensus on a technical base.

It was Olson who noted the importance of having sufficient genetic linkage markers to help orient a physical map, thus cementing the union of genetic linkage and physical mapping. Olson also concisely articulated what might distinguish genome research from other genetics. Olson's guiding philosophy was an extension of Frederick Sanger's tradition at the MRC laboratory in

Cambridge, England—to illuminate function by analyzing structure in projects of increasing scale. (By fostering projects regarded as just barely possible,
the boundaries of technique could be extended, and used to elucidate biological function.) Olson argued that projects should be considered genome research only if they promised to increase scale factors by threefold to tenfold
(size of DNA region to be handled or mapped, degree of map resolution,
speed, cost, accuracy, or other factors.) By the end of the March meeting, it
was clear that the skeptics had been converted by redefinition of the genome
project's goals.

The consensus emerged most clearly in a discussion of budget recommendations. A subgroup was delegated the task of producing budget options.
Botstein spearheaded this effort, and presented three options: $50 million,
$100 million, and $200 million, with completion dates sooner for the higher
figures (the year 2000 for $200 million versus 2025 for $50 million). The
estimates were based on technical presentations at previous meetings, but were
adjusted to reflect how many people in how many laboratories could be funded
at the different budget levels. Watson objected to the process, noting that it
would naturally incline the committee members to seem reasonable by choosing the middle option. He therefore suggested an option of $500 million per
year. Since Botstein had already dubbed the $200 million annual budget the
"crash program," Watson's became the "crash crash." Comments started from
Botstein's right and went counterclockwise around the table.

One by one, members supported a dedicated genome effort, although
there was no discussion of which budget figure to choose until a second round.
Having demonstrated unanimous acceptance of the importance of a genome
program, the committee in the second round achieved general acceptance of
something near the $200 million figure. Botstein was made responsible for
reviewing the figures again after the discussion. The committee ultimately
projected a need for $200 million a year for fifteen years: $60 million for ten
centers, $60 million for grants and technology development, $55 million per
year in early years for construction and capital costs, and $25 million per year
for administration, quality control, and review.[19]

While the NRC process was in midstride, the Council of the American
Society for Biochemistry and Molecular Biology issued a policy statement on
the genome project. Bruce Alberts was also involved in this effort, which
recommended physical and genetic linkage mapping. The council sharply con-
demned a genome project of the type promoted by Gilbert and was also clearly
concerned the project might involve only a few national laboratories under the
thumb of DOE. The statement concluded:

A large-scale, massive effort to ascertain the sequence of the entire genome cannot be
adequately justified at the present time. . . . The Council wants to state in the clearest
possible terms our opposition to any current proposal that envisions the establishment
of one or a few large centers that are designed to map and / or sequence the human
genome. . . . It is of the utmost important that traditions of peer-reviewed research, of

the sort currently funded by the National Institutes of Health, not be adversely affected by efforts to map or sequence the human genome.[20]

The statement was intended to thwart a DOE end run that might carry off the genome project, and also to cut short a Gilbert-style monolithic project. Genome politics by now engaged the entire biological community, the national laboratories, NIH, and DOE. The NRC committee was a microcosm of these politics. Its deliberations for the first time systematically assessed the arguments for and against a dedicated genome project, among a panel deliberately selected for balance. Producing a full-length report, through the standard NRC process, required the committee to justify its recommendations at some length.

The NRC committee surveyed the various technical components necessary to produce a coherent genome project and merged them into a scientific strategy. Alberts called the NRC committee "the most fun of any committee that I have worked on" because of "the talented people on it, the rapid learning process it entailed, the uncertainty of its outcome, and its direct impact on policy."[21] The NRC report succeeded to a remarkable degree in setting scientific agenda. This was the critical missing element from 1986 to early 1988. The NRC committee experience changed the course of Alberts's own career, as he subsequently became chair of the Academy's Commission on Life Sciences, and in July 1993, president of the entire NAS.

The NRC report stated that "acquiring a map, a sequence, and an increased understanding of the human genome merits a special effort that should be organized and funded specifically for this purpose."[19] It then outlined goals for genetic linkage mapping and physical mapping. Regarding the vexing question of sequencing, the committee said that "the ultimate goal would be to obtain the complete nucleotide sequence of the human genome, starting from the materials in the ordered DNA clone collection. Attaining this goal would require major (but achievable) advances in DNA handling and sequencing technologies."[19] Olson's scaling idea became a major bullet. Toward this end, the committee recommended pilot sequencing projects and a program to improve sequencing technology. Research on animals and selected lower organisms was marked as essential to make the human maps intelligible.

Agencies were urged to start first with a peer-reviewed grant program and progress toward larger and more targeted projects only as technologies matured. The direct impact of the NRC committee was seen in the March 1988 hearings for NIH's 1989 budget, when Representative William Natcher, chairman of the House appropriations subcommittee that funded NIH, referred directly to the report. It had been released several weeks earlier, and Natcher referred specifically to its budget projections.[33]

The Academy report had one critical weakness—its recommendations about how the project should be organized. The scientists on the committee made little attempt to survey what the agencies were doing in any detail. Their interest and experience were not in science administration in the federal government, but in science. Yet NIH, DOE, and Congress were percolating ideas about genome projects vigorously. The NRC committee members had informal contacts, principally with NIH, but there was no systematic attempt to gather information about the bureaucratic elements. The federal bureaucracies were highly complex, and the political process of their interactions with one another and with Congress was unpredictable. Having an impact on policy required knowledge about the workings of large bureaucracies, jurisdictional boundaries in Congress, and the histories of pivotal figures making decisions. One former science agency director was sent the penultimate draft of the NRC report as a reviewer. He was "appalled" by the organizational options and conveyed his dismay to the committee and NRC staff, provoking a rewrite of the section on administration.[22] Other reviewers had similar, although less pointed, concerns about the organizational options. Subsequent interviews with committee members indicated that the committee did not have enough data on which to base a recommendation, but felt it had to do so to execute its responsibility.[23] There had not been a meeting to discuss project organization and administration, and last-minute phone calls did not crystallize a solid consensus.

The report was released recommending that there be one lead agency, but "in a move that may leave those in Congress scratching their heads, the committee declines to specify whether it should be NIH or DOE."[24] The report failed to address what would happen if Congress tried to choose between the agencies. If plans had been drawn from scratch, the NRC's recommendation would clearly have been the preferable organizational structure. By early 1988, however, each agency had a multimillion-dollar budget, advisory committees, planning documents, and just as important, expectant constituencies and congressional patrons. If the committee intended that one agency should have a formal mandate to complete the genome project, with funding coming from several pots, then it would have been politically feasible, but ineffective. How could NIH decide how DOE should spend its funds, or vice versa? If a lead agency controlled all the funding from one pot, then either the NIH or the DOE program would have to be dismantled. Creating a program was considerably easier than burying one; the NRC recommendation proposed a politically hopeless task and invited open warfare between NIH and DOE, a war that might well kill the very project NRC intended to promote. In the end, Congress ignored the organizational recommendation and correctly read the bottom line of the NRC report to be a powerful endorsement of the genome project's scientific content.

The report was released to great fanfare on February 11, 1988, in time to be discussed at an annual meeting of the American Association for the Advancement of Science later that week. *Science* covered the release, noting that "the committee, like much of the biological community, was divided . . . when it began, but after a year of deliberation, it came out resoundingly in favor of the project."[24] Other scientific journals and daily newspapers reported on the Academy's approval of "the genome project"; many of the headlines featured the budget recommendation.[25-32] The NRC conception of the genome project promptly displaced the original proposals focused on sequencing. NIH, with some backroom politicking by committee members, took the NRC report as an invitation to active participation.

James Wyngaarden, director of NIH, used the NRC report as the organizing focus for an *ad hoc* planning meeting held two weeks after the NRC report was released. This organizational meeting involved many NRC committee members and set the agenda for the growing NIH commitment. The course charted by the NRC committee gave Wyngaarden enough rudder to break the inertia of an unwieldy NIH ocean liner, and to steer it in a new direction.

11

The NIH Steps Forward

JAMES WYNGAARDEN played the central role in securing NIH's genome budget. President Reagan nominated him to become NIH director in spring 1982. Wyngaarden came from Duke University, where he was chairman of the department of medicine for fifteen years. He was highly respected as a clinician and human geneticist. He accepted the job with some reluctance, and said so openly in confirmation hearings before the Senate: "I did not actively seek the post. . . . my acceptance of that honor is out of a sense of obligation based on an awareness of the vital role of NIH in biomedical research."[1] He accepted the position because of considerable worry about what might happen to NIH if a caretaker was nominated instead of a person thoroughly familiar with biomedical research.[2]

NIH's preeminent role in biomedical research drew on its hundred-year history. NIH began with $300 in laboratory equipment in an attic of the Marine Hospital on Staten Island, New York. Joseph Kinyoun, who directed the small Staten Island laboratory, had been trained by one of the most prominent bacteriologists of the day, Robert Koch. The laboratory was rescued by Surgeon General John Hamiliton from a congressional assault in 1888, and Kinyoun remained its director until 1899. It moved to Capitol Hill in Washington, D.C., as the Hygienic Laboratory in 1891. Senator Joseph Ransdell introduced a bill to increase appropriations for the Hygienic Laboratory and to create a separate National Institute of Health in 1926. When he combined the two organizations into one, he won passage in 1930. The Hygienic Laboratory was thus transformed into the National Institute of Health. The Wilson family of Bethesda, M.D., donated part of their estate to the government in the 1930s, and construction of facilities began in October 1938.

Biomedical research flourished in large part because of increasing federal funds. The national institute became plural with the establishment of the National Cancer Institute in 1937.[3] The NCI was made a part of the NIH when President Franklin D. Roosevelt signed the Public Health Service Act of 1944 (Public Law 78-410).[4] NIH moved to Bethesda in 1940. During the war, biomedical research joined many other parts of science in the war effort.

The Office of Scientific Research and Development guided science policy during the war. Much of the biomedical research effort focused on keeping soldiers healthy; researchers studied venereal diseases, yellow fever and malaria, and mental health. As the war neared an end, polio, cancer, and motor-vehicle accidents emerged as major health problems. As the wartime Office of Scientific Research and Development closed down its efforts, it transferred many programs to existing agencies. The set of contracts for biomedical research was quietly turned over to the National Institutes of Health, with no public notice.[5] This thwarted a more centralized approach to science that would have placed all research under a single agency. Vannevar Bush, principal architect of the plan to sustain science after World War II, favored a single science foundation that would encompass biomedical research as well as all other fields. Senator Claude Pepper, chairman of the Senate Committee on Health and Education, and several others favored a separate biomedical research organization. Pepper and his congressional colleagues held sway. The NIH emerged as the dominant funder for biomedical research during the several-year post-war delay in creating the National Science Foundation.

NIH's emergence from the shadows was not a carefully orchestrated strategy. Rather, NIH took shape through diligence and serendipity. In 1945, Cassius J. Van Slyke had a heart attack. He was running a venereal disease service for the Public Health Service, and his friends wanted to find something less strenuous for him to do. Ernest Allen had joined him in 1943, amid the war effort, at a facility in Augusta, Georgia. Allen was invited to administer sixty-six federal contracts slated to terminate on June 30, 1946, an "incidental, part-time, lower-left-hand-drawer-of-the-desk sort of activity."[4] Congress appropriated $8 million for 1947, of which $4 million was for extramural grants. A big part of the budget was for antibiotics. When the cost of making penicillin dropped precipitously, existing contracts could be smaller, thus freeing up funds for other purposes. Van Slyke and Allen cast about for ideas about what to do with the residual funds.

Van Slyke and Allen innocently sent a letter to medical school deans: "We have limited funds available for research purposes. If you have investigators who need these funds, let us know by return mail." They provoked a landslide. Allen later called the missive "the most naive letter ever to emanate from the national government in Washington."[4; 6] Within a year, they received more than a thousand responses. Between January 1946 (when the OSRD contracts were transferred to NIH) and the end of August 1947, NIH doled out $10 million to nongovernment institutions.[4] Even so, there were more proposals for research than funds to support them, and NIH sought an increased research budget the next year, the first of what would thereafter become an annual ritual. Thus began the largest extramural grants program in the world. So much for Dr. Van Slyke's part-time job.

NIH grew explosively and with aggressive lobbying efforts expanded its mandate and funding. The pivotal event was a coalition of media supporters, advertising executives, members of Congress, White House contacts, and a lobbyist hired specifically to bolster biomedical research funding.[7; 8] Mary Lasker and Florence Mahoney were leaders in the outside effort, working closely with Senator Pepper and other members of Congress. These champions convinced Congress to view biomedical research as a public investment and focused on diseases of high prevalence and severity, particularly cancer, heart disease, and mental illness. Biomedical science administrators were at first reluctant to expand their vision so grandly, and leaders of science (including the president of the National Academy of Sciences) were suspicious of a larger federal presence. The prospect of a biomedical research establishment outside their control, however, induced leaders of the Public Health Service, including the NIH, to join in the battle. The expansion of NIH's mission and money attracted a second generation of able administrators with great vision.

James Shannon, who became NIH director in 1955, regarded his first few years as critical. The NIH budget tripled from $98 million in 1956 to $294 million in 1959.[5] Before he became director, Shannon developed close working relationships over the years with Representative John Fogarty and Senator Lister Hill, chairmen of the appropriations subcommittees with jurisdiction over NIH. Shannon also cultivated relationships with their staff directors. When he became NIH director, he chose his political strategy. "I was less concerned with the budget per se than the development of a conceptual base for the mission of NIH in broad terms. . . . My salable item was opportunity, not need. . . . We did not ask for specific things in terms of specific budget increases we would like. Rather, we discussed long-range goals and the strategy we would use in their attainment, given adequate funding."[5] This powerful triumvirate of Fogarty, Shannon, and Hill fell apart in the late 1960s. Fogarty died in January 1967, Shannon left the NIH directorship in 1968, and Hill retired from the Senate six months later.[5]

Shannon was the first of many NIH directors to find himself in a delicate position, with divided loyalties between his superiors in the executive branch and Congress. Executive branch administrators generally looked at NIH budget increases with some alarm, while congressional patrons consistently fostered them. Part of the tension grew from the bureaucratic locus in which NIH found itself. The NIH director represented but one of several agencies within the Department of Health, Education, and Welfare (later, Health and Human Services). The director sat several layers down in the bureaucracy. Shannon's budget expansion was so fast that it made Health Secretary Marion Folsom uncomfortable. Folsom sought outside advice, but the committee he appointed endorsed Shannon's strategy.[5]

The expansion of NIH got a major boost when the War on Cancer became a presidential issue. Mary Lasker and a legion of biomedical research advocates, grown sophisticated over a decade and a half of experience, were key forces behind the initiative. In his January 1971 State of the Union speech, President Nixon announced:

The time has come in America when the same kind of concentrated effort that split the atom and took man to the moon should be turned toward conquering this dread disease [cancer]. Let us make a total national commitment to achieve this goal.[9]

After a circuitous political route, the President signed the National Cancer Act on December 23, 1971.[9] Congressional largess also extended to the rest of biomedical research. As the NIH grew for several decades, the NIH budget, while remaining a very small part of the total disbursements of the department, accounted for a larger fraction of "discretionary" funds.[10] Most of the funds in the department went to entitlement programs—Social Security, Medicare, and Medicaid—over which administrators had relatively little control. NIH programs, in contrast, were subject to annual appropriations, more subject to administrative control. Congress, however, treated NIH with special care. The NIH director often found it easier to find support for NIH programs in Congress than among his superiors in government.

The NIH director thus walked a tightrope between serving his bosses under the President, answering to a rambunctious constituency of biomedical researchers almost monomaniacally obsessed with more research funds, needing to establish research priorities to respond to a changing world, and attending to the wishes of congressional patrons. Congressional support was often the most reliable element, but entailed its own difficulties as individual representatives and senators desired specific programs from NIH.

The job of NIH director could be awkward. Bernadine Healy wrote about the NIH three years before becoming the first woman to direct it:

The NIH is a jewel—loved by all, especially Congress. But the budget process, by precedent and design, keeps the NIH under constant scrutiny, subject to much micromanagement. . . . NIH is subject to staggering political manipulations, disease- and constituent-directed earmarking, and intense programmatic review—more extensive than that experienced by virtually any of the other research and development agencies. The pressure is from both the Congress and the Office of Management and Budget . . . [but] the bureaucratic infrastructure keeps NIH and institute directors sufficiently low on the totem pole that they do not have the ready political access their exposure and scrutiny would suggest they deserve.[11]

NIH was a remarkable institutional innovation of the postwar era, the preeminent biomedical research organization in the world. It was nonetheless amazing that it ever got anything done.

NIH remained the largest single funding source for biomedical research until 1982, when private industry, led by the pharmaceutical industry, exceeded NIH's contribution.[12] This was the same year that Wyngaarden became director. Robert Marston, who directed NIH from 1968 to 1973, believed that the tacit understanding of NIH in the upper reaches of the executive branch dissolved during the 1980s, replaced by fiscal and political concerns.[13] This was the climate that prevailed as James Wyngaarden contemplated the human genome.

Wyngaarden first focused attention on the genome project in mid-1985. Robert Sinsheimer spoke to him about the first ideas for a Santa Cruz sequencing institute in March 1985, but Wyngaarden barely remembered this meeting several years later.[2; 14] Wyngaarden was present when Renato Dulbecco unveiled his idea for genome sequencing at the Italian embassy in mid-1985, but Dulbecco remembered this, not Wyngaarden.[2] The idea hit home only when Wyngaarden heard about DOE genome plans in London, at a meeting of the European Medical Research Council, June 4–7, 1986. Someone, he does not recall who, asked him what he thought about a DOE plan to spend $3.5 billion on sequencing the genome. He was shocked; the idea seemed to him "like the National Bureau of Standards proposing to build the B-2 bomber."[2] At roughly the same time, Ruth Kirschstein, director of the National Institute of General Medical Sciences (the main patron of basic genetics among the NIH institutes), began to get feedback from DOE's March 1986 workshop in Santa Fe.

Charles DeLisi had invited an NIH representative to the Santa Fe meeting,

James Wyngaarden, as director of the NIH from 1982 to 1989, was responsible for securing the first NIH budget for human genome research. The NIH quickly overcame the DOE headstart to become the lead agency for the U.S. genome effort. *Courtesy National Institutes of Health*

but the invitation got lost in the deluge of mail that pours into the NIH director's office. DeLisi sent materials about the meeting afterward, as preparation for a meeting with Wyngaarden and Norman Anderson, but this got little attention until Wyngaarden returned from London. Upon his return from London, Wyngaarden asked Kirschstein to convene a group to decide how NIH should respond to DOE's overture. Kirschstein summarized the

June 27 meeting of that group in a memo to Wyngaarden, noting that "first and foremost, while it is clear that the Department of Energy has taken, and will continue to have, the lead role in this endeavor, the NIH must and should play an important part."[15] The bottom line was profound ambivalence, the tale of pushmi-pullyu translated into NIH argot.

The NIH group recommended that Wyngaarden focus the October Director's Advisory Committee meeting on the genome project because "the debate started at Cold Spring Harbor and reported in the *Science* article should be extended in order to determine how the international scientific community truly views this project."[15] Between October and the end of the year, NIH could incorporate any recommendations into the fiscal year 1988 budget. They also noted the need for increased support of GenBank, the database funded by NIGMS to archive and disseminate DNA sequence information.[15] Kirschstein wrote to DeLisi—and Wyngaarden to DeLisi's boss Alvin Trivelpiece, establishing the bureaucratic pecking order—agreeing that the NIH and DOE should talk at the July informational forum organized by the Howard Hughes Medical Institute (HHMI).[16; 17]

The October 16–17, 1986, meeting of the NIH Director's Advisory Committee followed on the heels of the HHMI forum and featured another all-star cast. The aura of Nobel laureates and aspirants suffused Conference Room 10 in Building 31, the same site used for the HHMI forum. David Botstein opened the technical discussion with an overview of the role of genetics in understanding biology. Charles Cantor reviewed physical mapping. Nobelist David Baltimore, from the Whitehead Institute in Cambridge, Massachusetts, spoke to the science policy issues. He asserted that a proposal to sequence the genome would be useful, but would merit at best a mediocre score if he were voting in a scientific review group. Lack of sequence data was generally not a limiting feature, although lack of a global genetic map was. He urged strongly that the genome project not become a "megaproject" that would become a political creature of its own right, handing control from scientists to bureaucrats and politicians. He echoed these sentiments in a subsequent article.[18] Dulbecco challenged Baltimore by saying that the aggregate cost of doing many small projects would drastically exceed the cost of an organized program. Botstein countered that the costs would be higher, but the amount of information beyond mere sequence data would also be significantly greater.[19; 20]

The meeting then systematically surveyed NIH, DOE, HHMI, and NRC genome efforts, and then, briefly and incompletely, efforts in Japan and Europe. Allan Maxam presented a group of guiding principles that would make the program most useful. It should be (1) promoted as the Human Genome Project, and have human genetics as its ultimate goal; (2) designed to include comparative genetics with other organisms; (3) linked to protein and RNA catalogs and research efforts; (4) made relevant to related fields such as anthropology, paleontology, and evolutionary biology; and (5) planned so that new databases were compatible with existing genetics databases and others in related fields. Russell Doolittle noted that if the purpose was mapping, then databases were essential, analogous to the star map or the coastal surveys conducted with federal funds. Yet existing databases were already collapsing under the weight of new data. GenBank, for example, was behind in data entry and was precluded from developing algorithms and software under its NIH contract. Donald Lindberg, director of the National Library of Medicine (NLM), spoke about the manageable magnitude of the database task, but also the need to dedicate resources to it. Ronald Davis, from Stanford, reviewed work in yeast that presaged the yeast artificial chromosome technique that enabled cloning of megabases of DNA, and Charles Cantor from Columbia University, in his second talk of the day, reviewed techniques for analyzing such large DNA fragments.

The format was more structured than that of the HHMI forum three months before, and the policy issues were more salient. The main conclusions were that (1) NIH should eschew Big Science or a crash program, (2) the study of nonhuman organisms was important to make map and sequence data useful, (3) it was too soon to start sequencing, but pilot projects might make sense, and (4) information-handling was already a problem.[20] *Nature* observed, "The initial polarization of opinions has given way to a more constructive consensus that some concerted effort can begin without rending the fabric of biological science."[21]

An NIH Working Group was appointed after this meeting. Wyngaarden chaired the working group, which also included the directors of several NIH institutes, centers, and divisions: Kirschstein, Duane Alexander (director of the National Institute of Child Health and Human Development), Betty Pickett (director of the Division of Research Resources), Lindberg (NLM), and Jay Moskowitz (Program Planning and Evaluation). George Palade (Nobel laureate from Yale) was the lone outsider. Rachel Levinson was named executive secretary of the working group, the staff person most closely tracking genome activities. The group met on November 6 and December 16. The November meeting focused on improving communication among the managers of various NIH-supported databases and information resources, to be followed by linkage to non-NIH American ones and finally international nodes.[22]

In early December, Wyngaarden briefed staff from several House legislative committees and from OTA about NIH genome activities, and Senate staff requested a similar briefing. At the December 16 meeting at NIH, Wyngaarden indicated that he wanted to signal Congress that NIH's support of genome research should be channeled through routine grant mechanisms. The working group agreed to work toward program announcements over the next few months.[23] These were indications of interest from NIH, but did not carry setaside funds or entail special review procedures. The program announcements were published May 29, 1987.[24; 25]

The NIH Working Group also discussed how to respond to an OTA request for information about what NIH was already doing in the way of genomic research. At a December 4 meeting with Kirschstein and Levinson, Patricia Hoben and I (on behalf of OTA) explained our need to gather information about NIH activities for Congress. Kirschstein saw no easy way to answer our questions, because the categories of interest did not fit those used to sort grant applications. There was no way to pull out computer records by a simple search. The NIH Working Group decided to make some uniform definitions and ask the various institutes, bureaus, and divisions of NIH to cull through their portfolios.[23] This proved an elaborate and time-consuming task, and ultimately proved a politically divisive exercise.

Wyngaarden's early concern was to ensure that NIH had a major role in any large genome program, but he did not want to make any premature long-term commitments. He was in favor of the concept of the genome project "from the very start," but did not want to get too far in front of his biomedical research constituency when there was so much dissension among them. He likened his position on the genome project to Lincoln's waiting for success at Antietam before announcing the Emancipation Proclamation, so as not to jeopardize Union support in Europe. His second analogy was Roosevelt's delay in pushing the pre–World War II Lend-Lease Act until public sentiment supported the course he had already chosen.

Wyngaarden did support the genome project where it counted the most— in the appropriations process. This was a process he had come to understand well. The NIH had an unusually complex and protracted budget process. Each individual institute, division, or center submitted a budget request two years in advance. This was integrated in the NIH director's office into a set of budget requests, although technically there were eighteen separate budgets. The NIH aggregate requests went to the assistant secretary for health, responsible for all Public Health Service (PHS) agencies, including NIH. This budget went then to the department secretary's office and was made to fit with other arms of the department. Finally, the departmental budget was forwarded to the Office of Management and Budget, and then to the President.[26] Each institute's budget, therefore, had six levels of negotiation before it was presented to Congress (1, institute; 2, NIH director; 3, PHS; 4, department; 5, OMB; 6, President).

Article 1 of the U.S. Constitution gave Congress sole authority to tax and spend for the federal government. The President's budget request, from the congressional perspective, was merely the first step in the real budget process. Within Congress, the budget was handled first by an appropriations subcommittee and then a full committee in the House, passed by the full House, and then forwarded for action by the subcommittee and committee in the Senate. The subcommittee chairmen, ranking minority members, other subcommittee members in both houses, and staff (not necessarily in that order) all had power over the NIH budget. Wyngaarden, like all other NIH directors, had to master this process, with its immense parliamentary complexity and ample opportunity for blundering into some appendage attached to one of the players. Wyngaarden took his handling of the budget as a point of pride. He opened an article on the centennial of NIH by stating that "the most satisfying aspect of my first five years as director of the NIH has been the sustained growth of NIH funding."[26] Wyngaarden knew that most NIH policy was determined by the appropriations process, and regarding the genome project, he focused on this objective.

In his summary statement to the House and Senate appropriation committees for fiscal year 1988 (in February and March 1987), he cast gene mapping in high profile. The description of the extra dollars requested for gene mapping did not match the scientific strategy then being outlined by the NRC committee, hinting only at extensions of ongoing gene hunts, but it was still a new line item in the NIH budget. In his statement, Wyngaarden mentioned NIH's centennial, the urgency of AIDS research, and then the genome project. A straight reading of his text suggested gene mapping was second in priority only to AIDS.[27]

The NIH appropriation for genome research did not require a special authorization, as it clearly fell within the bounds of NIH's biomedical research mission. Unless someone in Congress objected, much could be done through appropriations alone. Fiscal year 1988 was one of the years when the NIH budget dance ignored the beat of the administration request, as Congressman David Obey made explicit in his comments. Because the NIH director was part of the administration, however, Wyngaarden had to toe the administration line, defending the official administration requests before Congress. Testimony before legislative or appropriations committees was reviewed by officials in the Department of Health and Human Services and in the Office of Management and Budget. The ponderous bureaucracies had notoriously thin skins and brooked little deviation from settled policy; in the absence of an explicit policy directive from above, new initiatives were viewed with skepticism.

Wyngaarden noted the delicate balancing act by quoting an order from Harry S. Truman to his agency chiefs, directing them not to request more funds from Congress than he as President requested. Congress had leeway, however, to ask "factual questions" to which agency heads could respond.[26] The bureaucracy could not interfere with Congress's authority to ask whatever questions it liked, and interfering with honest answers was a violation of federal whistleblowing laws. Over the years, appropriations committees had devised hypothetical questions as a way to tease apart NIH's true priorities from administration malarkey. Each year, they asked the NIH director what he would do with sums of money in addition to those requested, in $100 million increments.

In his replies to the House Appropriations Committee for fiscal year 1988, Wyngaarden asked for $30 million in genome research funds as part of the fifth $100 million increment, and another $15 million in the eleventh increment (of twelve).[27] Michael Stephens, on the staff of the House Appropriations Committee, recalled making some minor modifications in the final NIH budget, adding a few special projects, and stopping somewhere between increments five and ten in NIH budget additions that year.[28] The starting point, however, was Wyngaarden's blueprint, based on his personal judgment, input from the Director's Advisory Committee, and the NIH genome Working Group.

After Wyngaarden testified in early spring 1987, Nobelists David Baltimore and James Watson briefed members and staff of the House and Senate appropriations committees. They were invited to speak informally as part of a series of meetings occasionally arranged by Bradie Metheny of the Delegation for Basic Biomedical Research (affectionately known around NIH as the Nobel Delegation). Baltimore and Watson met briefly before the session on May 1 to go over their remarks. The meeting included Congressmen William Natcher and Silvio Conte, chairman and ranking Republican of the NIH appropriations subcommittee, and also Rep. Joseph Early, a subcommittee member and staunch NIH supporter of many years. Senator Lowell Weicker, who had been chairman of the Senate appropriations subcommittee for NIH until late 1986, was also present.[29–31] Just as important, the senior staff from the appropriations subcommittees of both houses were present. The principal aim of the meeting was to promote funding for AIDS research, but Watson availed himself of the opportunity to seek another $30 million for genome research.[32]

The House responded to Wyngaarden by appropriating $30 million for genome research, the amount in the fifth increment and the same as the amount requested by Watson. The Senate was less enthusiastic. Maureen Byrnes, staff to Senator Weicker, recalled that he was not as enthusiastic about the genome project as the House delegation; other senators, such as Tom Harkin, were more enthusiastic but less senior, and did not get to set the mark.[33] Michael Hall, staff director for the new chairman of the Senate subcommittee, Lawton Chiles, got no clear signal of strong support and put $6 million in the Senate bill. The House and Senate bills went to a conference committee for resolution of differences. The arithmetic mean of $18 million emerged from the House-Senate conference. The bill passed and became law. A Gramm-Rudman-Hollings recision reduced the final appropriation to $17.2 million for genome research at NIGMS that year. In private conversations, NIGMS staff estimated that $5 million of this was diverted from existing programs and the rest was "new" money.[34; 35]

An additional $3.85 million funded a new National Center for Biotechnology Information at the National Library of Medicine (NLM). The NLM regents, principal overseers of the library, identified molecular biology as an important area. An outside support organization, the Friends of the National Library of Medicine, took up the cause. The result was a bill, largely drafted by NLM and the Friends of NLM, prepared for Congressman Claude Pepper, the same man who as a senator more than three decades earlier had been instrumental in establishing NIH as the dominant biomedical research agency.

The NLM bill was to establish an information management center to support the biomedical research and biotechnology efforts in the United States. It added a new NLM budget authorization rising to $10 million annually. Pepper held a moving hearing on the bill on March 6, 1987, at which the victims of genetic diseases testified. Victor McKusick recounted his positive experiences working with NLM to computerize *Mendelian Inheritance in Man* and underscored the urgency of managing the explosion of information about human genetics. Other prominent scientists, NLM director Donald Lindberg, and HHMI President Donald Fredrickson also voiced their support.

The bill got caught in legislative complexities, however. The hearing was held in Pepper's subcommittee of the Select Committee on Aging, which had no legislative authority. Lindberg was present when Pepper called Henry Waxman, chairman of the subcommittee with legislative jurisdiction. Pepper indicated his wish to hold the hearing, and Waxman apparently agreed. The news did not reach staff members, who ultimately were delegated authority for action in the full committee, however. They learned of the hearing through OTA and NIH.[36; 37] NIH, unlike many other agencies, is authorized for three-year intervals as a rule, and 1987 was not one of the years when such a bill was in Congress. There was thus no logical vehicle to which the NLM bill could be attached, and so it stood alone. The NLM provisions were finally folded into the NIH authorization bill that passed more than a year later. The Appropriations Committee acted before then, however, by appropriating $3.85 million for fiscal year 1988, with the understanding that it was to be spent toward the purposes specified in the languishing Pepper bill. Pepper himself testified in favor of this action by the Appropriations Committee.

NIH appropriations for fiscal year 1989 were more or less routine. NIGMS requested $28 million for genome research. This was the final year of the Reagan administration. Congress and the President had agreed on a two-year budget plan the previous fall, in the wake of the October 17, 1987, stock market crash, and the President's budget request held to this agreement. 1989 was the one year under Reagan when the NIH request was taken seriously by the appropriations committees, and the requested amount was granted. The House appropriations subcommittee again asked Wyngaarden what he would spend with additional funds, and the genome project was listed as a potential beneficiary, but this year the incremental funding game was not the basis for appropriations.

There was one sidelight in the 1989 appropriation hearings, in that the NRC report was newly available.[38] Rep. Natcher led off a series of human genome questions by asking Wyngaarden how the $28 million genome budget request from NIH fit with the $200 million recommended by the NRC committee.[39] This gave Wyngaarden an opening to explain that there would be higher budget requests in future years. Wyngaarden ended his answer to this question by noting that "NIH considers the human genome initiative a *very* high priority."[40]

NIH's appropriations for 1990 involved several complications. NIH sent its budget request to the department and thence to OMB, with a final request of $62 million. When the President's budget request came back from OMB, it sought $100 million for genome research at NIH. The $62 million was apparently increased to $100 million by divvying up monies freed up by removal of other programs during OMB review.[41] The increase surprised NIH and signaled support for the NIH genome project high in OMB or elsewhere in the White House. Confusion surrounded the process, as this was a time of transition from Reagan to Bush, and it was not clear whether support for genome research was a carryover from the Reagan administration or indicated fresh support to be expected under President Bush. In the end it did not matter, as appropriations subcommittee staff in the House used the initial request level, known from NIH planning documents, as the basis for their deliberations. The 1990 appropriation was $59.5 million after some final adjustments.

During negotiations on the 1990 budget, Wyngaarden discussed the need to create a separate administrative center for the genome project, as the genome budget had become sufficiently large. He got agreement from the House to allocate the 1990 budget request to a new center that the Department of Health and Human Services would create by administrative fiat. The Senate agreed to roughly the same budget figure, but left the funds in NIGMS. In conference, the report followed the House, creating a new budget center—the National Center for Human Genome Research. The center was administratively created by Secretary Louis Sullivan, Jr., in October 1990, a new baby in the NIH family.

The baby had to fight for its patrimony even as it was born. The 1990 NIH genome budget was subject to last-minute negotiations in a Senate looking for ways to fund new initiatives elsewhere in the Department of Health and Human Services. One eleventh-hour Senate proposal, put forward and then withdrawn by Senators Pete Domenici and Ted Kennedy, had a genome budget reduction from $62 million to $50 million. The $12 million from the genome office, along with funds taken from elsewhere in NIH, would go to to programs for the homeless.

This illustrated the twofold vulnerability of the genome project as a new program at NIH. First, activities that showed a rapid growth were highlighted by their percentage budget increases and tracked closely by appropriations staff. Second, NIH's large share of the discretionary budget in the Department of Health and Human Services made its budget attractive as a possible benefactor for new initiatives. NIH's $8 billion budget was a plump fruit to be squeezed for new programs in health and social services. Any new sprout on such a bounteous tree was subject to pruning.

Wyngaarden made his greatest contribution to the genome project by securing its budget. He also established the Office of Human Genome Research to coordinate efforts during 1988 and 1989, and convinced his departmental superiors to create the National Center for Human Genome Research.

This gave the genome project a more permanent home in the NIH bureaucracy, and "center" status assured independence by conferring the power directly to disburse funds. Without special attention from him, it is unlikely NIH would have moved with as much dispatch. Wyngaarden did indeed react to the DOE initiative, rather than generate the idea for a program, but his task was no smaller. In an interview just before he left NIH as director, Wyngaarden noted the decline in autonomy of the NIH director since the 1960s. In particular, he noted the "more and more tortuous process for documents of all sorts through the Department and OMB. . . . It's not unusual now for a new policy document to be a year in transit through those various offices."[42]

A 1984 Institute of Medicine report on the structure of NIH noted an "absence of the trappings of bureaucratic authority; hence the director manages largely on the basis of persuasion, consensus, and knowledge."[43] The genome project showed that the NIH director could exercise considerable authority, but not quickly. The NIH process entailed an elaborate advisory mechanism and a slow consensus-building effort, both within NIH and with its outside constituency of investigators.

From the 1970s through the 1980s, NIH grew into a ponderous bureaucracy, with little control exercised by its director. The NIH budget doubled while Wyngaarden was director. The genome project was established in large part through his direct efforts, and it got disproportionate attention. Wyngaarden characterized his job by paraphrasing Yogi Berra; directing NIH was "90 percent damage control, 50 percent budget, and the rest was fun."[2] The genome project counted as fun.

Wyngaarden left NIH in July 1989, several months into the new administration. He served briefly in the White House Office of Science and Technology Policy and then became foreign secretary for the National Academy of Sciences and the Institute of Medicine. In July 1990, the Council of the Human Genome Organization (HUGO) appointed him director. This made Wyngaarden the central figure in HUGO, the last best hope of smooth international collaboration for the genome project. He held this position into 1992, when HUGO took off in a new direction.

The fiscal year 1988 through 1990 budget decisions in Congress, along with the creation of NIH's National Center for Human Genome Research, set up independent projects in both NIH and DOE. They would continue to be funded by separate appropriations bills. In this sense, Congress had spoken, granting both NIH and DOE their wishes for a genome program. The size of the budgets ($58.5 million at NIH and $26 million at DOE) was an implicit measure of relative power. The way the programs would coordinate the work, however, was far from clear. As the genome programs became established, the major topic of discussion moved from whether NIH and DOE should have genome projects at all to which agency should lead the effort. Congress might be the place where that decision was made as well.

12

Tribes on the Hill

T HE HONORABLE GEORGE E. BROWN was the first member of Congress to take note of the Human Genome Project. He noticed an article in the *Washington Post* reporting on the July 1986 HHMI forum[1] and offered his encouragement to build toward a truly international effort.[2] Brown was one of a handful of members interested in science, but he attained a position from which he could take action only five years later, when in early 1991 he assumed chairmanship of the House Science, Space, and Technology Committee. While the genome project was forming, he had to content himself mainly with watching developments.

The organization of Congress is inscrutable to newcomers. J. McIver Weatherford, an anthropologist, spent a year on Capitol Hill as a fellow in the program run by the American Association for the Advancement of Sciences. (This was the same program that first brought me to Washington, as well as Eileen Lee, Lesley Russell, and several others who figure in this story.) He wrote a book analyzing the similarities between Congress and tribal organizations.[3] The book, while somewhat flippant, captures some of the central forces that keep an otherwise chaotic institution from completely disintegrating. Vestiges of tribal organization prevail not only in Congress, but also in executive agencies. The decentralized bureaucracy at the National Institutes of Health was a particularly good example, with the director of each institute, center, or division a chief of his or her domain. The two large tribes on two hills—on Capitol Hill and at the NIH campus at the foot of Pook's Hill in Bethesda—organized into multitudinous clans. The bureaucratic organization of the genome project emerged from a Byzantine process of negotiation and politics.

When appropriation committees funded both NIH and DOE genome programs, the question of how the two programs would coordinate their efforts began to dominate discussion. The pressure mounted as budgets grew from $4.5 million DOE-only in 1987 to almost $30 million (both NIH and DOE) in 1988 and to $46 million in 1989. Management of such a rapidly growing enterprise lagged behind efforts to convince congressional patrons of the value of the science. The coordination question first came up in the appropriations process. In the spring 1987 House appropriations hearing for the 1988 budget, Congressman Obey asked several questions about genome research, stimulated by another article in the *Washington Post*.[4] Obey wanted to know why DOE was proposing to lead such a project, to which Wyngaarden replied that DOE had legitimate interests in detecting mutations, but NIH was outspending DOE by a hundred to one in the relevant fields, "and so NIH should . . ." Wyngaarden was about to finish his policy recommendation when Obey interrupted, asking for further clarification of DOE's interest. Wyngaarden later told *Nature* that he thought it was presumptuous of DOE to claim leadership when it was spending less than $10 million a year in the area,[5] but he was not pressed by Obey or others on what NIH should do about it. *Science* opened its "Research News" section with a depiction of interagency squabbling[6] and captured the confused positions of scientists and administrators during this formative period as they jockeyed for position at NIH and DOE.

A year later, Wyngaarden was ready with an answer. In questions about 1989 appropriations, subcommittee chairman Natcher again asked what agency should take the lead. Wyngaarden was unequivocal and direct: "I think NIH is the appropriate agency."[7] This flew in the face of what DOE preferred. It disagreed sharply with the recommendations of eight months earlier, from the advisory panel set up to advise DOE.[8] Each agency, not surprisingly, wanted to manage the genome project itself. Asking DOE and NIH which agency should lead was unlikely to provide a consensus. If the agencies reached agreement between themselves, so much the better, but that did not appear to be happening. The President presided over both agencies, providing a theoretical point of coordination, but because the efforts were in two separate departments, there was no lower authority than the White House where negotiations could take place.

The White House had other matters to attend to, and the genome project, while its coordination did reside in the cabinet for a short while, under the Domestic Policy Council, was ill suited for such a scientific and technical task. The upper reaches of the Reagan administration, even more than most, were almost oblivious to biomedical research. The way was left open for Congress to have a decisive voice in how the project unfolded. Congress had already commissioned a report on the genome project, at its Office of Technology Assessment (OTA), and that report focused heavily on just this question of interagency coordination.

The first interest in an OTA report surfaced in late spring 1986. OTA's Gary Ellis and Kathi Hanna invited Victor McKusick to present arguments for mapping the human genome at an OTA biotechnology meeting just before the Cold Spring Harbor symposium. Bernadine Healy chaired the biotechnology panel, and was intrigued. Others at OTA took notice, although no direct action ensued. When news from the Cold Spring Harbor meeting was reported in *Science* and *Nature,* it attracted the attention of several congressional staff members. Lesley Russell, science assistant to chairman Dingell of the Energy and Commerce Committee, and I (then working at OTA) instigated an OTA assessment of the genome project. Interagency conflict was already apparent at that early date, and it seemed likely that Congress would become involved in more than just appropriations. In a memo, I argued for a "simple and direct" project, which would be easiest with a single requesting committee in each house.[9] The proposal that followed focused on several issues:

The expertise to perform the sequencing resides in several different executive agencies, primarily DOE and NIH. Funding and coordination would thus be complex. Second, this could be among the first "big science" projects in biology, requiring substantial resources over a sustained period. Third, the technologies to do the sequencing and gene mapping would have significant clinical applications, scientific consequences, and industrial spinoff for biotechnology. Fourth, an international effort to map the human genome would have to contend with conflicts between free exchange of data and technology, on one hand, and proprietary and nationalistic interests, on the other.[10]

By pure happenstance, the OTA and NRC projects were approved within an hour of each other on September 24, 1986. I directed a team of exceptionally capable staff for the OTA project. Patricia Hoben, trained in molecular biology at Yale and fresh from a postdoctoral stint at the University of California, San Francisco, kept abreast of technical developments and wrote scientific background chapters. Jacqueline Courteau had obtained her B.A. in the history of science at Radcliffe and was just completing her master's degree from the science writing program at Johns Hopkins. She focused on databases and repositories and on international genome efforts,[11] ultimately compiling the first systematic data on international efforts.

The scientific consensus laid out by the NRC committee was a necessary precondition for the political decision about whether and to what degree a program should be funded. OTA added little to the scientific rationale, but instead focused on the bureaucratic structures and processes. The NRC and OTA reports thus complemented each other. NRC performed the first and most important function by articulating a scientific program that captured the need for collective resources and focused efforts. OTA gathered information about bureaucratic moves and political choices more systematically and acted as a well-informed but neutral observer, expert in science policy rather than scientific strategies. When the NRC committee recommended organization under a lead agency but neglected to say which one, the mess was left for OTA to clean up. There was no avoiding the issue, since the interagency rivalries were well publicized and known throughout Congress.

In the end, no bill on interagency coordination became law, but the very real threat of legislation induced DOE and NIH to make peace. At the level of daily operations, there was never war; indeed, there was less conflict than often prevails between different institutes within NIH. The fact that *Science, Nature,*

and the *Washington Post* were following genome organization carefully undoubtedly made administrators think twice before acting upon their territorial
urges. In the upper reaches of NIH and DOE, and among congressional
patrons, however, talk of domination and potential conflict abounded. The
process of achieving a suitable working arrangement between NIH and DOE
took several years.

The battle was fought over two related Senate bills. On July 10, 1987,
Senator Pete Domenici introduced S. 1480, the bill crafted by Jack McConnell
and Domenici's staff to promote technology transfer from DOE-funded national laboratories.[12] In the section covering the genome project, Domenici's
bill gave the Secretary of Energy a mandate to map the human genome. The
Energy Secretary was to direct a research consortium dedicated to this purpose
by chairing a National Policy Board on the Human Genome that included the
NIH director, the NSF director, the Secretary of Agriculture, and other officials. Domenici considered adding the bill as an amendment to the trade bill
then in the Senate. His staff called other Senate and House committees with
jurisdiction. Senator Chiles, chairman of the NIH appropriations subcommittee and of the full Budget Committee, and Senator Kennedy, chairman of the
NIH authorization committee, were in critical positions; their support was
essential to pass the bill.[13–19] The Domenici bill was a surprise to most on the
Hill, and even more of a surprise to the scientists who heard about it. After an
initial blush of enthusiasm, Senator Domenici's office dropped the idea of
adding it to the trade bill. The main reason was a roadblock thrown up by
Senator Chiles. After a conversation with staff member Rand Snell en route
from the Senate floor to another meeting, he said "it just didn't feel right" that
DOE would lead the effort and NIH would not.[13; 14; 20]

Chiles was generally accepting of NIH initiatives in biotechnology. Through
the tortuous connections of congressional politics, the genome project was
linked to orange groves in Florida. Chiles's interest in biotechnology stemmed
from a 1982 or 1983 meeting with a Florida constituent, Francis Aloysius
Wood, dean of the School of Agriculture at the University of Florida.[13] Wood
caught the senator's attention by describing how gene manipulation could
move the frost belt sixty miles north. This meant more land could be devoted
to cultivating a large crop of immense importance to Florida. Wood had found
a graphic way to explain how deletion of genes that cause ice crystals to form
on fruit, by recombinant DNA manipulations to create "ice-minus" bacteria,
might lower the temperature necessary to cause fruit damage. If oranges could
be grown in climes just a few degrees colder, it would reduce the annual
worries of Florida's orange growers and would significantly expand the territory acceptable for planting. It was good for Florida; that was enough for the
senator. The lingering message was the power of the new biology.

When the Senate majority reverted after six years to the Democrats in the
1986 election, Chiles ascended to chair of the appropriations subcommittee

that funded NIH. His main interest at NIH was biotechnology policy. The genome project became linked to biotechnology through Domenici's bill and some of the commercial rationale promoted by DOE, national laboratories, and NIH-supported investigators. When Domenici's office contacted Chiles's office about the DOE genome amendment to the trade bill, Rand Snell began to probe the network of scientific contacts around Chiles. Patricia Hoben from OTA happened to meet with Snell on the competitiveness of U.S. biotechnology. When she heard about the Domenici genome bill, Hoben mentioned there was already an OTA project on the genome project, of which she was also part, and asked whether there had been outside consultation with university researchers. Hoben suggested that Snell call Bruce Alberts in particular, as he was chairman of the NRC committee. Alberts was noncommittal, but did indicate that there was ambivalence about DOE leadership and a strong feeling among some on the NRC committee that NIH should be the lead agency.[20]

The *Washington Post* was one main reason the trade bill amendment died. Larry Thompson from the *Post* reported on the Domenici gambit on July 21, noting the year-long debate about NIH and DOE leadership. DOE "already has been aggressively pursuing the project," while Wyngaarden was "personally very interested in the genome project," yet "NIH's leaders have been criticized for waiting for the money to fall into their laps." Charles DeLisi, leader of the DOE program, shunted the issue aside, saying, "We are eliminating the debate by simply doing it."[4] The coverage provoked a storm of protest calls into Domenici's office, with spill-over into the Kennedy and Chiles bailiwicks.[13–16; 21; 22]

Domenici did not give up. He held a genome workshop in Santa Fe, August 31, 1987. It was Charles DeLisi's last day on the job at DOE.[23] Domenici pronounced his strong support for a DOE role in genome research in stentorian tones. At the meeting, Norman Anderson pulled out all the stops in a moment of zeal:

I think so far as the man in the street is concerned . . . to say that here is the possibility at one shot of finding the cause of some 2,500 human diseases is really stunning. . . . A century from now, as history books are written, the big projects that were important in this century are the genome project, and after it possibly space and then the atomic bomb (the order of those, I don't know). But the man who first proposes to do the genome project in the United States Congress is in history.[24]

It was a good way to get the senator's attention.

Domenici and Wyngaarden came to loggerheads several months later. At hearings on Domenici's bill on September 17, 1987, Wyngaarden articulated his desire for what might be paraphrased as "the mission and the money, but not the management." This came during an exchange with Domenici following Wyngaarden's testimony:

Domenici: If you were assured that it was not the intention of the legislation to in any way denigrate or detract from your ongoing activities, would you recommend that the United States of America have a policy of mapping the human genome as expeditiously as possible?"

Wyngaarden: Yes, sir. Unequivocally, yes.

Domenici (several exchanges later): If Congress wants to do it, how do we do it? Just give the NIH more money under their existing program and give DOE some more money . . .

Wyngaarden: I think that is a very good way to do it.

Domenici: And would it get done?

Wyngaarden: Yes.

Domenici: Without any changes in the law?

Wyngaarden: I think so.

[James Decker, representing DOE, concurred with Wyngaarden.]

Domenici: I love you both and I think you are great. But I absolutely do not believe you. I believe it would get done. But I am quite sure that it would not get done in the most expeditious manner, because I do not think you would be charged with doing that. I do not think you would send up any requests of a priority nature with reference to it, because you do not have enough money to do what you are doing. And if you tried to send up the request, it would be thrown in the waste basket at OMB. . . .[25]

Wyngaarden and Domenici locked horns for several minutes more, over definitions of what the other had meant, but it was clear that the basic issue was one of mutual distrust between the legislative and executive branches of government. Congress, in the person of Domenici, did not trust the agencies to act quickly, and the agencies, principally in the person of Wyngaarden as supported by Decker, did not want to have Congress intruding into matters regarded as technical and scientific. Neither side could win decisively, and the policy process unfolded over many months of thrusts and parries.

The OTA report was released on April 27, 1988, at a hearing before Rep. John Dingell, chairman of the House Energy and Commerce Committee. Dingell was a powerful figure, with a reputation for hardball politics, and was in control of the committee engaged in much of the substantive legislation in the House. The focus of the hearing was how genome programs in two agencies could be integrated into a single project. In my testimony, I likened the situation of contending with genome projects to the problems facing Congress a century earlier, when trying to integrate the various surveys of land and coastal regions in anticipation of the second opening of the American West. We at OTA favored letting each agency contribute, but forging some interagency task force to coordinate efforts. If the project had started from scratch, it would indeed have been better to have a single lead agency with the vast bulk of budget control. But while NIH waited for a green light, DOE had mounted a substantial effort. The process of killing either effort would be politically costly and would serve mainly to undercut support for the resources needed to do the job overall. An interagency effort was suboptimal, but at least promised pluralistic funding and respite from internecine warfare. *Science* and

Nature coverage of the OTA report centered on interagency politics.[26–28] Well before the OTA report was released, Congress began to move on the issue.

When Chiles blocked Domenici's bill, Domenici was furious, but unable to maneuver around Chiles. As chairman and ranking minority member of the Budget Committee (Senators Chiles and Domenici chaired the committee under Democratic and Republican control, respectively), they had worked together on many issues. They decided to overcome their immediate feelings about Domenici's genome amendment and work toward an accommodation. Domenici shuttled the semiconductor, high-temperature superconductor, and several other provisions of his original bill (not related to the genome project) into the Defense authorization bill. Senators Domenici and Chiles worked with Senator Kennedy to patch together a separate genome bill. This became the Biotechnology Competitiveness Act of 1987, which included several biotechnology provisions in addition to the genome project. The Chiles-Kennedy-Domenici bill, S. 1966, included a genome project provision modified from Domenici's, most notably giving NIH and DOE joint leadership. The genome provisions focused squarely on interagency coordination and management.

Kennedy's position was critical as chairman of the NIH authorization committee. Kennedy was an opinion leader in the Senate on health and biomedical research, matters far less partisan than most others in the same committee (which also had jurisdiction over labor-related issues). Mona Sarfaty and Stephen Keith worked on several parts of the bill for Kennedy. Supported by the powerful triumvirate of Chiles, Domenici, and Kennedy, the bill sailed through the Senate, passing 88–1 (Senator William Proxmire was the lone dissenter).[14; 29] Lisa Raines, working with staff for Kennedy and Chiles, polled the Industrial Biotechnology Association, a trade association for the larger biotechnology companies. The survey showed a strong consensus in favor of funding a genome project, but only under the aegis of NIH.[30] Snell noted, "This is a consensus bill. It's going to be difficult to do anything that sidesteps this." He neglected the always-simple option of killing it and the small matter of the House of Representatives.

After the bill cleared the Senate, it was sent to the House, where it was referred to the Energy and Commerce Committee and the Science, Space, and Technology Committee. The latter committee held hearings on the bill on July 14, 1988. It was a time of intense politicking between DOE and NIH, and the House committee was generally disposed in favor of the bill. At that hearing, I testified that the interagency committee specified in the Senate bill was one viable option, but it might be preferable to let NIH and DOE forge a workable coordination mechanism themselves, rather than impose a permanent and inflexible structure by legislation.[31–33] I thought the House might want to pass the bill to put pressure on the agencies, but then remove the interagency coordination structure from the bill in conference committee if NIH and DOE had by then reached some agreement. The bill was reported out of the Science, Space, and Technology Committee favorably, and awaited action in Dingell's Energy and Commerce Committee.

NIH and DOE could have preempted genome initiatives in Congress by negotiating an agreement between themselves. Yet until the release of the NRC and OTA reports, and indeed for months after, staff from both NIH and DOE appeared to believe that Congress would somehow designate their agency the genome leader. The myopia of those working for one agency or the other was reinforced by internal conversations with different constituencies that did not overlap significantly. DOE was fostered by the national laboratories, while NIH had its university-based investigators. The bureaucracies heard only the messages passing through their separate constituencies. In conversations with congressional staff, including me, both DOE and NIH representatives voiced disappointment that NRC and OTA ducked a tough call (that is, choosing their agency to lead); yet there was no call to make. The existence of twin genome programs was set as soon as DOE got its first authorization and appropriation through Congress and as long as NIH decided not to abdicate its role. There were forces in both agencies moving toward accommodation, but they did not quite find their way to successful completion until faced with the prospect of legislation.

DOE and NIH vied for advantage from the summer of 1986 until well into 1988. After NIH missed the first Santa Fe meeting in March 1986, NIH and DOE consistently sent representatives to each other's meetings through 1987 and 1988. This was no small commitment. Genome meetings proliferated as more and more players sought a voice, and DOE and NIH were both represented at each meeting—NRC committee meetings, OTA panel meetings and workshops, and a myriad of others. The profusion of meetings brought staff at the program level together frequently, and they maintained cordial relations. Irene Eckstrand of NIGMS sounded a conciliatory note in a memo reporting on the HERAC subcommittee meeting in February 1987. She observed that "the subcommittee and OHER staff are clearly committed to this project and seem committed to working with other agencies. They are anxious to learn NIH's plans and stated their view that NIH should be more actively involved."[34]

At the level of agency heads, David Kingsbury of NSF emerged as a mediator for a time, attempting to channel the conflict, first through the Biotechnology Science Coordinating Committee in the White House (formed principally to deal with interagency disagreements over the release of genetically altered organisms into the environment) and then through the Domestic Policy Council (a cabinet-level group).[6; 35; 36] Kingsbury's mediating role meant that NSF had to stay out of the direct competition. NSF's policy position was quite clear for several years—it had no genome program *per se,* although NSF support for instrumentation and nonhuman biology was directly relevant. It was a position crafted in the bureaucratic netherworld where incompatible concepts reside comfortably in the same paragraph. In Washington, this was not only acceptable, it was occasionally compulsory.

Kingsbury's political base eroded quickly when he was implicated in a financial conflict of interest. The Department of Justice began an investigation related to his financial connections with Porton, Inc.,[37] a biotechnology and instrumentation company. NSF was thus taken out of the loop for several years, and it reentered only in 1989 with its instrumentation centers and plans for a plant genome research focused on *Arabidopsis thaliana,* a plant with a conveniently small genome and short generation time.[38] NSF thus entered the genome sweepstakes late, at least as a declared contestant, but on a strong base in plant science and instrumentation.

At NIH, Wyngaarden had to make a clear choice between creating a special genome effort or maintaining the *status quo ante,* but with a larger budget for genetics. Wyngaarden became chairman of the Domestic Policy Council genome working group, which at its May 5, 1987, meeting decided to survey its members about their spending for genome research.[39] He was thus at once at the nexus of a dispute about genome project administration within the NIH and also an arbiter of NIH's role relative to DOE. Incompatible visions of what the project entailed kept NIH and DOE apart. The seed for this disagreement was the genome research funding information so meticulously gathered by Ruth Kirschstein.

Kirschstein was keen to protect basic genetics research from a political juggernaut. NIGMS was the largest source of funding for basic genetics in the world. She led the effort to retain NIH leadership under a regime that would expand funding for genetics research. Her response to genome enthusiasm was to enlarge the pie while retaining the existing NIH bureaucratic structure. The NIH funding figures were her main tool. She noted that NIH funded $313 million in fiscal year 1987 on projects that involved mapping or sequencing, of which just over $90 million was for work in humans. The NIGMS program announcements demonstrated a willingness to support genome research, but did not formally set aside funds for this purpose. She argued that the announcements were "not exactly business as usual, but not highly targeted either." Rachel Levinson, staff to Wyngaarden, whose job included genome policies, corroborated this, arguing there was no need "for a concerted effort because it is not new. Every institute has work related to mapping and sequencing."[6]

Kirschstein's data-gathering showed the robust science base NIH was laying in genetics, but her interpretation of the data dismayed those who wanted a genome mapping and sequencing project, as opposed to a series of individual gene hunts. Several members of the NRC committee, among them Watson, wanted to recommend NIH leadership outright. Before they could take such a position, however, they needed some indication that the project was understood and supported at NIH. They wanted NIH to push vigorously for complete genetic and physical maps and development of new technologies. David

Botstein, another member of the NRC committee, felt strongly that NIH had neglected RFLP mapping, and he wanted a clear signal of support. In the eyes of several committee members, NIH did not need to take any position about the DOE initiative, but merely to make a commitment to a concerted research program. NIH would then emerge as *de facto* leader by virtue of its scientific preeminence and overwhelming size advantage.

Kirschstein's expand-the-pie-but-don't-fire-the-cook position was intended to assuage fears that genome efforts would threaten investigator-initiated research. Her constituency included those pursuing individual research projects in genetics and basic research. The constituency favoring the genome project was, in contrast, many senior and powerful opinion leaders in molecular biology. Kirschstein's positions were cold comfort for them. The position that NIH was already committing hundreds of millions on genome research and the implied failure to distinguish individual gene hunting from global genome mapping undermined support for NIH leadership. The concern was based in part on NIH's historical failure to support genetic linkage maps in the

Ruth Kirschstein, as director of the National Institute of General Medical Sciences at NIH, presided over the science that spawned the genome project. Her vision of how to conduct the project ultimately lost out to that of James Watson and the NRC committee, and the new NIH genome center was, in effect, carved out of her institute. *Courtesy National Institute of General Medical Sciences*

early 1980s and in part on the NIH's own justification documents. In the same memo that presented the figures on NIH genome research, for example, three projects were listed to indicate the kind of work that might be undertaken with further genome funding. All three projects were searches for individual genes: the retinoblastoma gene, the gene for Alzheimer's disease, and a gene implicated in autoimmune disease.[40] These were important targets, and indeed all were hotly pursued over the next year by NIH researchers, but they were not what the committee had in mind when it asked for a comprehensive genome map. Small project grants of themselves were not enough; the problem was not what NIH was doing, but what it had so far failed to do.

Kirschstein's position was a delicate one. Her institute based its reputation on supporting the best basic research throughout the country. NIGMS had no in-house research program on the NIH campus, unlike most other institutes. It was the very success of the NIGMS grants program that kindled the idea of the genome project in the first place. Kirschstein's position was aimed at preserving a strong base for undirected research. This put her in conflict with those, such as Watson, who argued that genetics now demanded a more deliberately planned and coordinated approach, with attention to technology development and completion of maps. At root was a disagreement about the best process for investing public funds in the future of genetics. Kirschstein believed that a large genome program could well cut into the core of undirected research; such a program was inimical to the style of managing grants in her institute.

Contentiousness was apparent at an August 1987 OTA workshop. Dave Guston, working with us at OTA as a summer fellow, arranged the meeting (originally suggested by Rachel Levinson of NIH) to project the costs of genome research. Nobel laureate Paul Berg chaired the meeting. After a long day trotting through the various component items necessary for a genome program, the group addressed administrative costs. Watson recounted that he had urged Wyngaarden to set aside genome research funds in the director's office. Kirschstein responded that the first genome moneys, then still awaiting final action in Congress, were initially considered as an add-on to the director's office, but Wyngaarden had subsequently assigned them to NIGMS. She noted there was an internal NIH working group and that there had been several program announcements.

The tension between Watson and Kirschstein was palpable. Kirschstein noted that "we sent them [program announcements] out to people who don't bother to read them when we send them out to them . . . people don't seem to realize that we want to do this. I don't know how to get the word across."[41] Minutes later, Watson came back to program administration, asking: "Are we going to have a large program run by a committee, or is there going to be one person who is in charge of it? I instinctively believe one person should be in charge of it who understands the scientific issues and who is not chosen purely to be an administrator."[41] Walter Gilbert supported Watson's notion that there had to be much more than a grant program, or the map and sequence data would never coalesce into useful maps and databases. The conversation drifted to other topics, but again Watson snapped back to his foremost concern. He aimed a question at John Sulston, directing the physical mapping effort on *C.*

elegans: "Doesn't one person really have to finish up that last 10 percent and live or die for the thing?"

Sulston, finding himself unexpectedly invited into the crossfire, temporized until he could gather his wits. "Yes, I expect so . . . you can have your political division into chromosomes . . . I don't see how you could ever have a number of labs generating raw data and fitting it all together." He then wondered whether sequence data from separate laboratories might indeed be pooled in a way impossible for physical mapping data.

Watson was talking about something altogether different, however. "You're going to get bad years, and the program will be under attack, and there's got to be someone who sees all the components and fights for it."

Sulston then worried about one person "administering this huge empire. Do you really want that?"

Watson didn't miss a beat. "I think someone has got to do it."

Sulston shot back, "You would like to do it."

Watson ducked. In a manner befitting a political figure, he offered up a non sequitur: "It's a question whether it's an active scientist or a retired scientist."

Sulston then elaborated how it was important to cultivate "individual dreams . . . surely that is the way to do it."

Watson then interjected a humorous deflection. One of Watson's heroes was Franklin D. Roosevelt, a master of deliberate inscrutability. Watson's next remark would have done Roosevelt proud. "Well, I couldn't think of a job I'd like less. I'm very relieved." The exchange broke up in laughter. Leroy Hood suggested that instead of a scientific director, perhaps the chairman of an advisory panel could serve this purpose. Kirschstein then stepped in, noting an outside scientist was already on the panel (referring to George Palade, a Nobel laureate from Yale). Watson shot back, testily: "That sounds good, but he has no qualifications for the job at all." Kirschstein bristled: "Well, he thinks pretty well."[41]

Paul Berg drew a halt to the swordplay before someone lost an arm. He asked me to summarize congressional interests. I noted that Congress was working on legislation directly concerned with how the project would be coordinated between NIH and DOE, referring to the Chile-Kennedy-Domenici efforts then beginning to take shape. Berg adjourned the workshop, noting that debate about genome research had shifted in the year since he and Gilbert had chaired the Cold Spring Harbor meeting. The question had moved from whether to how.

In later conversations, both Kirschstein and Watson recalled the meeting with residual anger. They had incompatible visions of how the genome project should proceed. It was a clash between two of the most powerful figures in molecular biology. It was science policy formation at its passionate best; each champion cared about and fought for a persuasive vision. But only one could prevail. John Sulston had sliced to the heart of the matter. Watson would be king.

In late 1987, after the Wyngaarden-Domenici and Watson-Kirschstein contretemps, NIH vigorously opposed the Domenici bill. In a letter drafted for the Secretary of HHS, attached to a memo drafted by Kirschstein, NIH made clear its opposition to the bill because it gave leadership to DOE and did "not fully recognize the importance of research in genetics to human health and the dimensions of the historical and continuing commitment NIH has made to such research."[42]

As the NRC and OTA reports came out in 1988, NIH and DOE were faced with two options: conspicuous cooperation or a strong likelihood of legislation that embodied Congress's preferred framework. In December, James Decker, acting director of the Office of Energy Research at DOE, wrote to Wyngaarden seeking some agreement between the agencies.[43] There is no reply letter in the NIH files, and DOE officials reported never having received an answer. After a December 1987 meeting between Watson and Wyngaarden, while the NRC report was nearing completion, momentum seemed to favor NIH. NIH thus had little to gain by such an agreement at that time, and Congress might yet declare NIH the leader. If there were to be a single lead agency, it had to be NIH, and the OTA report said this quite plainly. Events did not conform to NIH's highest hopes, however.

NIH and DOE had to settle into a permanent peace. Judith Greenberg, new director of the genetics program at NIGMS and a chief in the Kirschstein tribe, proffered an olive branch in the *Washington Post,* noting "there's certainly room for more than one agency. . . . The National Science Foundation, for example, has supported a lot of research that is closely related to the research that NIGMS supports. In fact the two have a way of complementing each other, and I don't think there should be any difference with the DOE-NIH cooperation."[44]

Greenberg's sanguine views began to dominate as 1988 drew to a close. NIH's program would be significantly larger, but DOE would also have a substantial budget. In light of the standoff, the agencies opted to sign a memorandum of understanding, in hopes of staving off House action on the bill. They reached a tacit agreement with Lesley Russell from the Energy and Commerce Committee, and the committee never reported the bill for a floor vote. The content of the memorandum was less important than agreement on a process for joint planning, forcing NIH and DOE to face the political reality that the other would also have a genome program.

There was one more way that Wyngaarden could assert leadership, however, and he exercised this option with a masterful stroke. Wyngaarden established a bureaucratic center for genome research, and he secured the all-important budget. Wyngaarden was best remembered, however, for a single personnel decision, the appointment of James D. Watson to direct NIH's genome effort.

13

Honest Jim and the Genome

URING 1988, Jim Watson emerged from the back room to
tower over the genome project. Watson seized the moment,
noting, "I would only once have the opportunity to let my scientific life encompass a path from double helix to the three billion steps of the human genome."[1]
As presaged at the 1987 OTA workshop on costs of the genome project,
Watson became head of the NIH genome office. By this stroke, NIH director
James Wyngaarden assured NIH a dominant voice in genome politics. It was
a strong voice, but not necessarily in close harmony with the others. Bob Davis
of the *Wall Street Journal* remarked on Watson's distinctive tactics: "Taunt the
Germans for their fears about science, bash the Japanese for 'freeloading,' and
dish out money to your opponents to knock the program."[2]

Watson was among the world's most famous scientists, surely the best-
known biologist of the day. Upon his appointment, attention naturally gravitated toward him. His every word was taken as an NIH policy statement.
Although his direct authority extended only over the NIH program, the scope
of Watson's informal influence was far vaster. As Norton Zinder of Rockefeller
University expressed it, Watson was "standing like a colossus over the whole
program."[2] By appointing Watson, Wyngaarden made NIH the center of
power in genome politics and harnessed one of the dominant talents in molecular biology.

Watson's authority was rooted in history, as codiscoverer of the double-
helical structure of DNA, one of the most significant revelations in twentieth-
century biology. His influence also stemmed from a reputation for having an
extraordinary nose for the important questions of biology. Watson made a big
impact as an impresario of molecular biology. At Harvard in the 1960s, he
pulled together one of the "hot" laboratories in molecular biology, one of the
"big three," standing beside the MRC laboratory in Cambridge and the Pasteur Institute in Paris. He became the director of Cold Spring Harbor Laboratory in 1968. The laboratory had been rescued from financial oblivion by his
predecessor John Cairns. Watson took a quiet institution and made it a robust
world center of molecular biology.[3] Watson's personal energy and focused
commitment were the ultimate sources of power.

Watson's scientific and administrative careers were built on an independent sense of priorities. "Just do good, and don't care if it doesn't seem good to others."[4] His highly intuitive manner focused on character judgments and results, and scientific results were the ones that counted most. Good science was elevated to a guiding moral principle. Watson's credo was that "the essence of most good science is very deep curiosity with some way of knowing the comparative importance of things you are curious about."[5] His ability to ferret

James D. Watson set in motion the whole chain of events that led to the Human Genome Project when he and Francis Crick discovered the double-helical structure of DNA in 1953. The longtime director of the Cold Spring Harbor Laboratory, he served as the first head of the NIH genome program, from October 1988 to April 1992. *Courtesy Cold Spring Harbor Laboratory Library*

out the central problems of molecular biology was legendary; it ranked as a salient quality along with an almost reckless openness and a brutal honesty.[5–7] Watson moved institutions to stay atop the rapidly shifting sands of science. "The one word that comes to mind on Jim as an administrator is bold," noted Ray Gesteland, Watson's former student.[5]

In October 1988, Watson accepted an appointment as associate director of NIH and head of the Office of Human Genome Research, housed in the

NIH director's office and answering to NIH director James Wyngaarden. Watson's boldness rose to the surface immediately. At the press conference to announce his appointment, Watson declared that the ethical and social implications of genome research warranted a special effort and should be funded directly by NIH.[1; 8] In making this path-breaking commitment, he took a leaf from the National Research Council and Office of Technology Assessment reports. This was nonetheless an astonishing event. While NRC and OTA had indeed suggested that analysis of social, ethical, and legal implications of rapidly advancing human genetics should go forward in parallel with genome research, it was still surprising to be taken seriously, particularly by Watson, whose public image as an *enfant terrible* did not meld well with support for careful deliberation and expenditure outside science. Remaining in character, he made the public commitment before conferring with Wyngaarden or anyone else at NIH.[9]

Watson first got interested in genetics late in his undergraduate years as one of Robert Maynard Hutchins's "Whiz Kids" at the University of Chicago.[10] He was fortunate to choose the University of Indiana for his graduate work, when Harvard and the California Institute of Technology rejected him.[7] At Indiana, he found Salvador Luria, who became his mentor. In April 1948, Max Delbrück came to Bloomington, and he deeply impressed Watson. Delbrück was the philosopher-king of the "phage" group, which studied viruses that infected bacteria, among the smallest of living things, chosen because their biology was amenable to approach through precise experimental formulations. The phage group included Delbrück, Luria, and Alfred Hershey. The phage group's central thesis pursued the line laid out by physicist Erwin Schrödinger in his book *What Is Life?*[11; 12] Life could be explained by chemical mechanisms.

Delbrück was an erstwhile physicist turned biologist. His guiding principle was to study systems simple enough to explain by molecular mechanism, and to explain function by elucidating structure.[13] The ideal experiment was one that had a simple yes or no answer and shed light on molecular mechanism. Molecules, cells, and life were elements in a reductionist dreamscape.

In the summer of 1948, Watson and another student, Renato Dulbecco, went with Luria to Cold Spring Harbor Laboratory, summer home to the phage group.[10] During his graduate work, Watson became convinced that the structure of DNA held the key to understanding important questions of molecular biology. This was not yet a widely shared belief, but it was becoming a central premise of the group surrounding Watson.[14]

Watson was fortunate to happen upon Francis Crick, who shared his passion for DNA. Crick later reflected: "It's true that by blundering about we stumbled on gold, but the fact remains that we were looking for gold. Both of us had decided, quite independently of each other, that the central problem in molecular biology was the chemical structure of the gene."[15] About Watson, he commented, "He just wanted the answer, and whether he got it by sound methods or flashy ones did not bother him one bit. All he wanted was to get it *as quickly as possible.*"[15] Watson's taste for important problems was established early. His signature was also quickly established: choose a scientific goal and push toward it relentlessly. He adopted any method that looked promising, regardless of its disciplinary origin or the politics of its genesis.

The double helix became the central icon of molecular biology. The discovery was important of itself, but Watson thrust it into the public eye by writing *The Double Helix,* a young man's chronicle of the process. The book, written in the years after Watson, Crick, and Maurice Wilkins received the Nobel Prize in 1962, became a best-seller because of its lively and personal tone. The literati loved the book; those who wanted to preserve the image of selfless scientists pursuing knowledge for its own sake hated the book and vilified Watson for writing it.[14] Watson self-consciously cast himself as the antihero, a compulsively competitive brash cynic obsessed not only with DNA but also young women's chests and backsides. The inglorious image he projected in *The Double Helix* haunted him for life. It was the persona that others expected to encounter, and projected on him. He explained away this unflattering self-portrait as driven by his desire to tell a racy tale.[16] If he made himself a hero, his story would be undermined. To preserve the liberty to be nasty to others, he was nastier to himself. David Schlessinger, a former student, noted that "critics who ordinarily suspect any autobiography of distortion have swallowed the novel whole, apparently because the author has been so 'honest' about himself."[17] Schlessinger called the book a novel; Crick likewise reflected that Watson did not truly resemble the monomaniacal Nobel aspirant portrayed in *The Double Helix.*[15; 18]

Watson started *The Double Helix* by recounting how Willy Seeds snubbed him on a walk in the Swiss Alps. Seeds asked, "How's Honest Jim?" and walked on by. *Honest Jim* was the working title of Watson's book, cadged from Kingsley Amis's then-current novel *Lucky Jim. Honest Jim* remained a nickname of sorts for Watson thereafter, for many years mainly as a term of derision. Those who worked closely with Watson recognized a kernel of truth in the epithet. It captured one of Watson's dominant qualities, his unremitting candor, both admirable and at times vexing. Watson exercised it vigorously, often exploding at students and telling his administrative superiors precisely what he thought. He could be petulant, at times almost vindictive. Watson long remembered swearing at the president of Harvard when he was passed over for tenure.[4; 19] He used his fame as the raging bull of molecular biology to feed his power and did not hesitate to tramp onto politically muddy turf. He dubbed Nixon's War on Cancer "lunacy" in the early 1970s, for example.[20] But quips and tirades were not Watson's main contribution. Such quirks were cultivated eccentricities, freedoms earned by making a great discovery at age twenty-five. The hard thing was to prove it was not a fluke.

Prove it he did. First at Harvard and then at Cold Spring Harbor, the scientific results flowing out of his laboratory showed that he was more than a person who happened to be in the room when Francis Crick was about. More important, he trained a cadre of researchers who subsequently proved their worth, and who now reflect on him warmly.[21] He created a laboratory pedigree as impressive as any in molecular biology. During the 1960s, he slowly grew out from under Crick's shadow in his own mind.[4; 10] Others also took note. In September 1962, Watson and David Rogers were the only biologists on *Life* magazine's roster of the hundred most important men and women in the United States.[22] In 1990, Jonas Salk and Watson were the only biologists to earn a place among *Life*'s hundred most important Americans of the century.[23]

While Watson's public persona was defined by his self-portrait in *The Double Helix,* his impact on molecular biology came through another book, *The Molecular Biology of the Gene.*[24] Almost to a person, molecular biologists over the age of forty-five anywhere in the world can recall where they were when they first read the book. It was an unusual text, a book strongly written by a single author who was pushing a line of argument. It was authoritative, but willing to call on intuition and to piece together plausible, if not fully demonstrated, results to tag the most important and central scientific questions facing molecular biology. This book earned him the admiration of a generation of scientists, the first to grow up in thrall to molecular biology.

Watson had an aptitude for teaching himself new skills. He learned how to write well in the popular voice as he worked on *The Double Helix.* He made another career as textbook author, moving *The Molecular Biology of the Gene* through three editions that all sold well,[25] and helping to edit several other scientific texts. He mastered an entirely different set of skills to raise millions each year for Cold Spring Harbor Laboratory. His talents were highly plastic. With the Human Genome Project, he turned his talents to national politics—science policy Washington-style.

As 1987 gave way to 1988, Wyngaarden was beset by disagreement about the proper strategy to pursue for genome research. Ruth Kirschstein and Jim Watson favored incompatible administrative options, exemplifying an ideological rift within the biomedical research community. Wyngaarden had to choose. Watson and Baltimore met with Wyngaarden on December 17, 1987 to discuss AIDS research and the human genome. Watson forcefully expounded his view that NIH had missed the boat on the genome project; he railed against Kirschstein's stay-the-course approach. The next day, Wyngaarden met with NIH staff to convene a planning meeting in Reston, Virginia. That meeting became the pliers to extract an NIH commitment.

There was extensive overlap between those invited to the Reston planning meeting and the membership of OTA and NRC committees. The NRC report was released with a great ballyhoo in February, and many of its members participated in a session at the AAAS annual meeting in Washington. After that session, a group got together and decided they were not interested in coming to NIH again to urge a major commitment, with no assurance of a response. Charles Cantor, who had been sent the draft agenda for the upcoming Reston meeting, believed a tentative tone would appear in disastrous contrast to the enthusiasm so pervasive in the wake of the NRC report. He called Rachel Levinson, the staff person most responsible for planning the Reston meeting, to sound the alarm.[26; 27] Victor McKusick was testifying at the same hearing as Wyngaarden the following week. Wyngaarden called McKusick to discuss what they would say. McKusick suggested a model of funding similar to that used for sustaining research infrastructure, through NIH's Division of Research Resources. Wyngaarden followed up with calls to several others from the NRC committee who had been at the AAAS meeting. He got readings consistent with McKusick's.

Rachel Levinson completely revised the format of the meeting, less than a week before it was to take place.[26] The purpose shifted from seeking consensus about whether NIH should proceed to planning how best to do so. The Reston meeting took place from February 29 to March 1, 1988, and refined the scientific framework laid out in the NRC report. With the backing of an NRC report, the kernel of a coherent scientific plan, and congressional signals auguring well for substantially increased appropriations in future years, Wyngaarden chose the high road.

A group met with Wyngaarden during a lunch break at the Reston meeting. They waited for Watson to leave before starting the discussion. They indicated that Watson was the only person who could credibly lead the NIH effort.[26; 28-30] Watson's involvement would enshrine the project as first-class science. He would bring his prodigious organizational talents to NIH, and his status as the discoverer of DNA would carry considerable weight on Capitol Hill. Wyngaarden had already been thinking about Watson before the meeting. Following that meeting in Reston, Wyngaarden was widely known to be considering Watson as the head of genome efforts at NIH.[31] At Reston, Watson agitated strongly for the same goal he stated at the OTA workshop seven months earlier—for NIH to leave its genome effort in the hands of an active scientist. He later noted, "I did not then realize that I could be perceived as arguing for my own subsequent appointment."[1] One could be forgiven for thinking Watson was not so naive; John Sulston had months before pushed him publicly on precisely this point.

Expected or not, it was uncharacteristic of Watson to care what others thought when an important goal was at stake. "Just do good, and don't care if it doesn't seem good to others." He wanted the job done right. In October 1988, he agreed to start directing NIH's genome program, which took shape as the NIH Office of Human Genome Research.[32-34] The *New York Times* commented that "the appointment seemingly completes his metamorphosis into a senior statesman of science."[35]

Watson hired two NIH stalwarts, Elke Jordan and Mark Guyer, from the National Institute of General Medical Sciences (NIGMS). Jordan and Guyer carried out most of the work in the new NIH genome office as its first full-time staff. Jordan had directed the genetics program at NIGMS. Watson knew her from the 1960s, when she was a molecular biologist working in Matthew Meselson's laboratory, down the hall from Watson at Harvard. Guyer, an erstwhile bacterial geneticist trained at Berkeley, had worked at Genex Corp, before joining NIGMS.

NIH and DOE signed a memorandum of understanding in the fall of 1988, to avoid the threatened Chiles-Kennedy-Domenici bill. The memorandum ratified an existing informal arrangement, but grew into substantially more, as bona fide joint planning came to seem advantageous to both agencies. Throughout 1988 and 1989, staff at NIH and DOE met to discuss how to carry out the terms of the memorandum. They appointed a joint NIH-DOE advisory group, composed of members taken from the advisory panels for each agency.

Randy Snell, from Chiles's personal staff, and Michael Hall, staff director of the NIH appropriations subcommittee, inserted language into the conference report for the fiscal year 1989 appropriation, a document that accompanied the law to explain congressional intent. The conference report expressed concern about interagency coordination and stipulated that NIH and DOE report back to Congress "the optimal strategy for mapping and sequencing the human genome."[36] Watson insisted on responding with a "serious" planning document, rather than a coordination plan that worked only on paper.

For many months, an informal coordinating committee met monthly to make the logistical arrangements among the various agencies and organizations. Mark Guyer, Diane Hinton of HHMI, Irene Eckstrand of NIGMS, John Wooley of NSF, and Ben Barnhart of DOE formed the core, and others attended occasionally. Their plans began to jell when combined with the powerhouses of genome research in a retreat at the Banbury Center, Cold Spring Harbor Laboratory, August 28 to 30, 1989. The advisory committees and staff for both DOE and NIH, and a few additional experts in genome research, began to plan in a truly cooperative mode.

Norton Zinder, chairman of NIH's genome advisory committee, organized the discussion into task areas. Working groups were asked to specify goals and means of achieving them. Much of the meeting focused on how to construct physical maps. Maynard Olson and others concentrated on the idea of using short stretches of DNA sequence as unique "tags" that would serve as landmarks on the chromosomes. Laboratories using different methods could thus compare results directly. Cassandra Smith raised the idea of using sequence information as an index earlier that year at the Cold Spring Harbor symposium. Olson and Botstein hammered the notion home at Banbury, stressing its importance as a common reporting language to unify physical and genetic linkage mapping. Unique short DNA segments of known sequence could bridge the two kinds of maps and also link them to DNA sequencing as mapping moved toward its ultimate level of resolution. The idea was seized upon and given the name Sequence-Tagged Sites (STS). A group agreed to prepare a paper for *Science*.[37; 38]

NIH and DOE staff hoped to prepare a five-year plan going into the meeting, but uncertainties about the proper strategy held their aspirations in abeyance. The genome project could yet become an amorphous grand idea of imprecise scope and indefinite goals. Barnhart and Zinder thought that no specific planning draft would emerge from the retreat,[39] but they proved themselves wrong. The plan took shape after the retreat, and staff adopted the goal-directed format to focus the report.[40; 41]

The NIH genome office had planning capacity, but still lacked direct budget authority. Watson, Jordan, and Guyer put together a genome advisory body, linked NIH to DOE, and began to plan how NIH might best contribute to genome research. Statutory authority to spend federal funds remained with the NIGMS council. The difference was largely symbolic, as NIGMS staff and council cooperated closely with their former colleagues. The character of the genome project was drifting away from its traditional NIGMS moorings, however, and budgetary independence was an important step in this process.

A year after the NIH genome office was created, as the budget grew to $59.5 million for fiscal year 1990, the Office of Human Genome Research became the National Center for Human Genome Research (NCHGR), with Watson as director. Louis Sullivan, Jr., Secretary of Health and Human Services, conferred center status on the genome project.[42] NCHGR thus gained direct control over its budget, similar to the authority wielded by other NIH institutes, centers, and divisions. Mark Guyer announced at the 1990 NCHGR Christmas party that staff had grown from two to twenty-two or twenty-three in a year (depending on whether I counted as an outside consultant).

Watson had made another career shift to direct a part of the federal bureaucracy. He was an unusual director. Watson's stature as a scientist and public personality was his power base; his NIH duties were almost incidental by comparison. Unlike other NIH institute and center directors, he did not come to power by dint of position, but tapped directly into the community supported by NIH. He used his status as the "father of DNA" to secure resources from Congress and to guide the scientific strategy of genome research. Stephen Hall noted in a *Smithsonian* profile that "it is precisely Watson's candor and integrity, and his willingness to take the heat," that earned support among his colleagues.[6] He now used the same tactics in Washington.

Watson declared *ex cathedra* that the genome project would officially start in October 1990, at the beginning of fiscal year 1991. He argued that the first few years were taken up by getting organized. Strong crosswinds hit the genome project as it lifted off the runway. A battle over the 1991 budget severely tested support for the project.

The early stages of the budget year looked promising. NCHGR requested $108 million (up from $59.5 million), and the genome office at DOE sought

$46 million (up from $26 million), getting within range of the aggregate $200 million plateau foreseen in the NRC plan. If both agencies got their requested budgets, the budget target would likely be reached in 1992. At that point, the high-growth phase could stop, leaving the programs on a firm footing and less conspicuous as targets for budget cuts. Initial budget requests survived departmental and OMB review; indeed, they drew strong support, but a storm was brewing outside.

The problem first surfaced at House appropriations hearings.[43] Congressman Obey took Watson to task for failing to find a way to prevent genetic information from being abused by insurance companies and employers. Obey asked: "What can you point to besides your personal hope" that abuses would not occur? He continued "wondering if whether we would really be doing any great long-term damage if we were to go to $100 million of money to go to R01s [individual investigator-initiated grants]" and warned, "I think you should not underestimate the weakness of the political system in terms of its ability to deliver the kind of protections that need to be afforded this information."[43] Throughout the hearings on various NIH institutes, it was clear that the committee was looking for loose change, and would give programs with either large budget increases or heavy reliance on research centers careful scrutiny, in strong preference for support of R01 grants.

The genome budget was growing very rapidly as a percent of its 1990 base, and it was set to establish centers as a foundation for its future work. The 1990 budget already included some funds for genome research centers, but 1991 was the year the center program was to expand considerably. Rumors that the budget was in trouble reached NCHGR a week before the House subcommittee was to consider the final budget marks. Watson and Jordan went to Capitol Hill to meet with Michael Stephens, chairman Natcher's aide, who handled the NIH budget in the House of Representatives. Watson was extremely discouraged, and he threatened to resign in a stormy meeting. Stephens and Natcher were unimpressed.

The genome budget's growth phase coincided with a drop in the proportion of grants funded during each review cycle. This stemmed from several policy decisions taken by NIH beginning in 1986. That year, NIH began to respond to a chorus of voices from the scientific community by agreeing to extend the length of the average research grant. The idea was to free investigators from perpetual fund-raising, giving them a year or two of working budget without the distraction of filing new grant applications. In the standard three-year cycle, an investigator would work for a year, then begin to write grants during the second year, so that there would be funding when the third year ended. New proposals thus began with only a year's new data; investigators spent inordinate energies as supplicants for money, and the system encouraged a proliferation of proposals to hedge bets against uncertainty. If the grant period was extended to five years, the argument went, then investigators should be able to work relatively worry-free for three years, and would have to apply only 60 percent as often. This would increase productivity and reduce instability.

Lengthening the grant period meant, however, that the carryover commitments for old grants would mount each year until a new equilibrium was reached. This necessarily cut into the amount of new funding available each year. Carry-over commitments increased from 67 percent of the NIH budget in 1985 to 76 percent in 1990.[44] In the long run, fewer applications would be expected, as the stability of the five-year cycle caught hold. In the interim, however, the number of new grants would necessarily fall. The effect of longer grant commitments was exacerbated by an increase in the average yearly award, because of inflation of research costs. From 1982 to 1990, the average length of award increased 23 percent, from 3.3 to 4.3 years, and the average commitment for each grant increased from $107,000 to $208,000, a 94 percent increase.[45] Most of the increase was due to rising personnel costs.[44] Thus because of both the greater length and higher annual costs of grants, the number of new grants dropped from a third to a fourth of total applications. Moreover, the expected drop in new applications from grant term extension, which should have been felt by 1989 as the 1986 policies took effect, never materialized. The number of new applications stayed the same, even as the number of new grants that could be funded dropped.[44]

New scientific vistas were opening in almost every field, but confronted worsening funding prospects. This translated into immense frustration. Some investigators went hunting for a scapegoat. There was some sniping at AIDS, which accounted for roughly 10 percent of the NIH budget, but invective was also directed at the genome project. The genome project was vulnerable, as it lacked a disease constituency and its bureaucratic base was as yet small and feeble. Its indomitable director, however, was not.

University-based research centers, as opposed to individual project grants, were under attack throughout NIH, as scientists supported by small independent grants felt squeezed. The House Appropriations Committee responded by setting a cap for all centers at NIH and calling for review of the proposed genome centers. It gave NCHGR $71 million for fiscal year 1991, enough to carry forward its previous commitments, but with few funds for new grants or centers. It also left an additional $18 million available for genome research, but gave authority to spend it only to a permanent NIH director.[46; 47] There was no permanent NIH director, nor would there be one soon. Wyngaarden had resigned in July 1989 to join the White House Office of Science and Technology Policy, and the NIH director's position remained vacant until Bernadine Healy was confirmed by the Senate on 9 April 1991.[48] The budget set-aside was a shot across Watson's bow. It signaled congressional ire that the administration had failed to appoint an NIH director, and clipped Watson's wings. A fourth of his budget could be held hostage to a political process well beyond his control. His threat to resign was not appreciated. The tactic that worked well at Harvard and Cold Spring Harbor, where losing Watson was a serious threat to the institution's prestige, did not work nearly so well in Washington.

Meanwhile, opposition mounted among investigators, particularly young ones, within the research community. This opposition was independent of the House Appropriations Committee's concerns, and centered on a different, although related, set of issues. Watson's former mentor Salvador Luria wrote a blistering letter to *Science* asserting that "the program has been promoted without public discussion by a small coterie of power-seeking enthusiasts."[49] One could only wonder why he couched his judgment in the plural, as the target was so clearly Watson. The problem with Luria's statement was not its accuracy. There were indeed enthusiasts, and they were seeking power. The sting of his judgment was, however, that they were using the genome project to seek power, rather than using power to get the genome project. This was a motivational judgment Luria was poorly situated to make.

Opposition took the form of letter-writing campaigns. In January 1990, Leslie Kozak, from the Jackson Laboratory in Maine, sent a letter to Senator William Cohen stating that the genome project "threatened the quality and conduct of our nation's health-related research effort."[50] In February, Martin Rechsteiner of the University of Utah wrote letters to acting NIH director William Raub, presidential science adviser D. Allan Bromley, and Senators Al Gore and Ted Kennedy. The letter began: "The human genome project is mediocre science and terrible science policy."[51] Rechsteiner's letter questioned the origins of the project in DOE, warned of sequencing drudgery, challenged the value of sequencing the human genome, and urged that the project be curtailed to reduce divisiveness within biology. Thus began a string of letters, some of which included copies of the Rechsteiner letter, indicative of a concerted campaign against the genome project. *Science* and other periodicals got wind of the campaign and reported that Rechsteiner had sent his letter to five hundred scientists.[52; 53] Rechsteiner cited Bruce Alberts's *Cell* editorial warning against Big Science in biology, apparently unaware that Alberts had chaired the NRC panel that crafted the genome strategy.

A series of other letters began to pass through the electronic mail networks in biology. One such letter by Michael Syvanen (University of California, Davis) and his colleagues urged that scientists write to their own congressional representatives to kill the genome project.[47; 52; 54] Robert Martin, an intramural NIH researcher, expressed concern that the genome project was overly concentrating research funds and would misdirect biology.[55] Even in peer review meetings for genome grants and on-site visits to prospective genome research centers, dissension was palpable among the reviewers. The opposition was critiqued by Dan Davison at Stanford, who observed that the central issue was not the genome project *per se,* but the paucity of investigator-initiated project grants. The genome project was indeed going to be more targeted, and should be more open about it, but the genome budget was but a drop in the bucket, and killing it would do little to ameliorate the funding squeeze.[56] In June, the *New York Times* juxtaposed features on mounting opposition to the genome project and on the wrenching dilemmas faced by young investigators.[57; 58] The dilemma was real, although the genome budget was not the root cause.

In mid-July 1990, the four offices targeted by Rechsteiner indicated they had received thirty or forty letters on the genome project, running four or five to one against it. (By comparison, the Superconducting Super Collider generated about ten times more mail at the White House Office of Science and Technology Policy, in more or less the same ratio.) When genome supporters got wind of the campaign, they reacted with letters of their own, so that by mid-August the odds were evening up.[59-62] Each of these four offices read the opposition primarily as a reflection of self-interested group politics. Congress and executive agencies were hearing strong support from the clinical genetics community, which wanted rapid progress on human genetic disease, and from industry.

The Industrial Biotechnology Association (IBA) did a survey of its member companies in fall 1987 and found strong support as long as the project was conducted by NIH.[63] At the IBA annual meeting in May 1988, Patrick Gage (then of Hoffmann–La Roche) waxed rhapsodic about the genome project in a talk titled "Why We Should Do It—Now!"[64] Two other respected leaders of pharmaceutical research and development teams, George Poste (Smith, Kline & French) and Ralph E. Christoffersen (then at Upjohn), were equally upbeat at an October 1987 meeting of the Pharmaceutical Manufacturers' Association.[65] When the University of California at Los Angeles did a survey, principally among industrial groups interested in the genome project, it found that government and industry figures were much more likely than academic scientists to consider the genome project "a worthwhile use of taxpayers' money and scientific resources." The survey found 62 percent of industrial respondents, 70 percent of the financial community, and 88 percent of those from government agreed, but only a slim majority among academic researchers (thirty-five for to thirty-one against).[66] When the Industrial Research Institute surveyed its member companies about five major science projects—the Superconducting Super Collider, the Hypersonic Airplane, the Strategic Defense Initiative, the Space Station, and the Human Genome Project—the genome project came out on top by a wide margin (with more than twice the votes of the next-nearest project).[67] Industry and government administrators clearly thought the project had promise, at least by comparison to other large science projects. One reason the academic backlash failed to make inroads was its confinement to academic circles.

Opposition within science peaked with the publication of a commentary from Bernard Davis and his colleagues in the microbiology department at Harvard Medical School, in the July 27, 1990, issue of *Science*.[68] This short article made clear that competition for research funds drove the opposition. Davis saluted the redefinition of the genome project by the NRC, but argued that "it is doubtful that they [genome projects] could generate the strong political appeal of the original proposal." The letter urged that any sequencing projects be targeted at "units within the chromosomes that have functions." Finally, it hit the center of contention, by asserting that work on model organisms lacked "obvious justification for insulation from competition with other kinds of research," signaling that the real basis for suspicion was that genome research was protected from peer review, or escaped comparison to other research priorities. Given that all genome grants were being channeled through standard NIH peer review, contention centered on whether genome research deserved its own bureaucratic center and how large its pot of gold should be.

Davis agreed that some elements such as databases needed centralized management, but the genome office was getting too big a slice of the biomedical research pie. The letter asked for a reevaluation of the project and questioned whether it deserved funding "at a level equivalent to over 20 percent of all other biomedical research," although the origin of that estimate was not specified. If the source of concern was the 1991 budget, between $60 and $70 million of the $108 million request was uncommitted, amounting to 4 or 5 percent of the funds slated for new and competing grants at NIH that year. Five percent was the figure Davis himself used later in congressional testimony.[69] The Davis letter made clear that 20 percent (or 5 percent) was too high, but 0 was too low. How much was the right amount?

The House appropriations subcommittee was attempting to preserve all it could for investigator-initiated grants, and it cut the 1991 genome budget, but this was not due to the letter-writing campaigns. Subcommittee staff learned of the campaigns only when Watson mentioned them, and when Leslie Roberts of *Science* called to ask if the Rechsteiner and Syvanen campaigns were related to the cuts.[47] The campaigns failed to target members of the appropriations subcommittees in either House, thus violating the first principles of interest group politics. The principle of preserving funding for small grants, even at the expense of other programs, was the motivating force behind the budget cuts, and did the work that those hoping to lobby failed to do.

As the genome project neared its official starting date of October 1, 1990, the curtain rose on a public drama. On July 11, the Senate held a hearing before its Committee on Energy and Natural Resources, chaired by Senator Wendell Ford, with Senator Domenici as the ranking minority member and star prosecutor. The Subcommittee on Energy Research was holding its second set of hearings on the genome, two years after the first. Matthew Murray, a former student in Leroy Hood's laboratory at Caltech, now working at the Lawrence Berkeley Laboratory, was doing a short internship in Domenici's office. The idea for a hearing began as a survey of the DOE genome program. The hearing was one of the first highly public acts for David Galas, the new director of DOE's Office of Health and Environmental Research—successor to Charles DeLisi, who first conceived a federal genome research program.

Galas announced that Lawrence Livermore would become the third designated national laboratory genome center, joining Lawrence Berkeley and Los
Alamos in the DOE constellation.[70] He also announced a new DOE plan to
map and sequence complementary DNAs, regions known to code for protein.[71] As plans for the hearing progressed, Ben Cooper, staff director for the
subcommittee, wanted to let genome critics have a voice. He believed their
opposition was largely due to misunderstanding of the budget dynamics, and
he wanted to give them a chance to speak and be questioned. Cooper invited
Martin Rechsteiner and Bernard Davis to testify.

Watson and Galas opened the hearing, followed by the heads of the three
DOE genome centers—Robert Moyzis of Los Alamos, Anthony Carrano of
Lawrence Livermore, and Charles Cantor of Lawrence Berkeley. Kirk Raab,
CEO of Genentech, spoke of industrial support for the project. Leroy Hood
dazzled Domenici with reverential references to technology transfer, speaking
from his own experience in developing new instruments and methods.

The hearing began as a showcase for the DOE program, but ended with
Davis's summarizing the *Science* letter that questioned the urgency of the
genome project and asking for a reevaluation of its funding levels.[68; 69] Domenici slammed into Davis with a zeal hearkening back to his past as a lawyer. "As
someone who is supposed to know all about the federal budget, I am rarely in
a position where I can look at a program and say that it is exciting enough to
keep somebody like myself energized while we are trying to reduce the deficit,
but I have found one here."[72] There was a carry-over debate in *The Scientist,*
with Davis arguing that the genome project was prone to become too bureaucratic, and again attacking the notion of sequencing the entire genome.[73]
Leroy Hood retorted that the tasks of molecular biology were inherently
repetitive, and that the idea of the genome project was to provide the information, methods, and instruments that would liberate molecular biologists
from some of the tedium.[74] Martin Rechsteiner missed the opportunity to
greet Domenici's wrath.

Domenici was an important Senate figure in the budget summit meetings
to trim the federal deficit. He was among the members of Congress working
with President Bush, OMB director Richard Darman, and other senior administration officials to hammer out a two-year budget agreement. A budget
summit meeting was scheduled in conflict with the genome hearing time, and
the hearing had to be rescheduled. Staff notified the witnesses, but could only
leave a message for Rechsteiner, who had already left Utah. Rechsteiner entered the hearing room at noon, two hours before the time he thought it
would start, only to find it had just adjourned. I greeted him and ushered him
to the front, introducing him to Domenici, who was giving post-hearing press
interviews. For Rechsteiner, it was a long trip from Utah; it must have been
an even longer trip home.

As the House appropriations subcommittee prepared to mark up its bill
for the full committee, a critical step in the budget process, Watson exhorted

other titans of molecular biology to write support letters to committee members. I and other staff members feverishly called our contacts to notify them of the genome project's plight. Watson spent days in phone-to-phone combat, turning his considerable energies to shim the sagging fate of his program. The same ardor that assembled the double helix from cardboard and wire models in 1953 constructed the political structures to shore up the NIH genome project in 1990.

In the end, NIH salvaged a livable budget. Representative Obey slashed $36 million from the President's $108 million request in the House,[75; 76] but Senator Harkin restored the full request and even added a couple hundred thousand dollars.[77] The final budget was the arithmetic mean, less a few more whiskers shaved off all NIH programs and totaling $89,731,000.[78] The House contingency budget, sequestering genome funds in the NIH director's coffers, was excised. Funding was sufficient to launch six genome centers by February 1991[79; 80] and another three by the end of the fiscal year. The centers anchored the strategy sketched out by the NRC committee. Much of the budget remained in investigator-initiated grants, but Watson and his advisers were convinced that data and technologies would come together with sufficient force only if NIH cultivated teams large enough to mount significant interdisciplinary efforts. Letting a thousand flowers bloom would generate marvelous science, but a field of wildflowers could not feed the army of researchers who needed systematic maps and databases.

Those who argued in 1987 and 1988, as many did, that the genome project should proceed only with "new" money came a cropper in 1990. Opponents shattered a fragile argument. If the project was so important, why should it proceed only with "new" money? Was this not just a way of saying it could not compete on its own, and had to be insulated from budget competition? The rhetoric of "new" funding, and the disingenuous argument that genome research would never displace funding from other work, distracted from the central question. Elke Jordan and others argued that the genome project would enlarge the pie by presenting a new objective that Congress could readily support.[53]

The central question remained how much of the NIH budget should go to collective organized efforts, to establish an infrastructure for future genetics, and how much to undirected research and other worthy ends. Initial NIH genome funding in 1988 was just under 2 percent of a budget request increment (or 0.2 percent of the overall NIH budget.). Was Wyngaarden right in his decision to dedicate the funds to genome research? Those who contended that the genome project's budget would displace other science generally argued from an abstract perspective, but there was some evidence to support their position. While the genome project, because of its small size relative to all NIH, had a negligible impact on NIH overall, it did cause some transient "collateral damage" (to borrow a military term) to the basic genetics program at NIGMS.[81–84]

Truth lay on both sides of the "new money" argument. The funding available for investigator-initiated basic genetics grants at NIGMS dropped from $187.8 million in 1989 to $159 million in 1990, before rebounding in 1991 and thereafter. Those same years, the genome budget jumped from $28.2 million (1989) to $59.5 million (1990), the year of maximum adverse impact at NIGMS. The increase in genome budget clearly gave genetics investigators a new place to apply for funds, but the nature of the science was not the same, as it was principally aimed at map construction and technology development rather than undirected analysis of basic genetic mechanisms.

If the genetics program at NIGMS, birthplace of the genome program, had grown at the same rate as other parts of NIH from 1987 through 1992, its budget would have risen to roughly $255 million by 1992. Between the genome research budget and the NIGMS budget, the total was instead $334.4 million, of which $229.5 was at NIGMS and $104.9 was at the genome center. A reasonable conclusion is that the genome project attracted an additional $80 million per year of funding that would not otherwise have been anywhere in the NIH budget while pulling away $25 million that would otherwise have remained at NIGMS. Three-quarters of the genome budget was "new" money appropriated by Congress in response to a new idea, while somewhat less than a fourth was carved out of NIGMS.[85]

The choice facing policymakers was analogous to deciding when a new territory was crowded enough to build roads or make rules about land and water use. When John Wesley Powell surveyed the American West, he realized the central importance of water. He urged a communitarian political solution, one quickly killed by the politics of the day. He lost his position as head of the U.S. Geological Survey, arguably the most powerful scientific position of the time. Senator Walter Stewart, a lawyer from Nevada, deposed Powell through his position on the Appropriations Committee. An ideology of false abundance, including a denial that water was a scarce and controlling factor in the arid regions, doomed the American West to cycles of boom and bust linked to drought and plentiful rain.[86] Powell was a messenger who spoke too soon, before the politics could catch up.

The scarce resource for biomedical research was public dollars, which would also wax and wane. The genome was also largely virgin territory, but molecular biologists had begun to stake claims. When was the time to plan resources for the common good? When should NIH build roads as well as fund more explorations of the vast genetic terrain? The answer hinged on whether the genome project filled an unmet need that if not addressed would exacerbate the scarcity of research dollars.

The NRC committee and leaders of the biomedical research community identified a weakness in the pattern of NIH funding—a neglect of genome-scale mapping efforts, inattention to development of new technologies, and insufficient funding of databases and shared resources. The most senior officials at NIH agreed with that appraisal. How much was it worth to fix the problems? The strongest argument for the genome project was that in the long run, it would make finding genes faster and cheaper. Finding genes was emerging as a strategic bottleneck in many disparate areas of biomedical research. The genome project was worth doing, and need not have rested on the perilous rhetorical perch of "new" money. An unfortunate choice of semantics failed to acknowledge that the genome project was a response to a policy failure; genetics research had outstripped its support structure. Insisting that genome funding *never* come at the expense of other initiatives belittled its central importance to the future of biology, despite the truth that the bulk of its funds did not come at the expense of research grants. The genome project did cause a several-year downward bump in the NIGMS basic genetics program, although it would be hard to argue that it had a substantial impact elsewhere in NIH.

By the end of 1990, the frustration directed at the genome project had found new targets. The debate continued, but its rancor diminished. The wrath of investigators shifted to the Institute of Medicine (IOM) in late 1990, after it released a report on biomedical research funding. The IOM report concluded that "the allocation policies of the past decade have focused too heavily on short-term problems and solutions and have neglected the long-term integrity of the research enterprise."[44] The committee did three alternative funding scenarios, favoring two growth scenarios: 2 percent or 4 percent per year over inflation. Responding to its charge, however, the committee also analyzed a no-growth scenario. Under that scenario, the committee recommended boosting training and construction funds even if it cut into other programs. This recommendation precipitated a firestorm of criticism among those who feared any incursions into extramural grant funding. (Two years later, however, the strategic planning process at NIH itself had reached many of the same conclusions.)

Animosity about genome research funding also dwindled as more scientists began to appreciate that the goals had been broadened to encompass more than just human genetics, and that it would focus for several years on genetic linkage mapping, physical mapping, and technology development—all consensus goals. Bernard Davis, for one, was mollified.

A rearguard action was fought at the October 1990 Genome II meeting in San Diego. Davis held out an olive branch, telling *Science*, "I don't want to say I have been converted, but there is much less disagreement than there was a year and a half ago."[87] Not to mention four months ago. Michael Syvanen claimed that opposition had not diminished, but his claim rang hollow.[88] Donald Brown from the Carnegie Institution fought against a retreat. He argued that the genome project was "overtargeted, overbudgeted, overprioritized, overadministered, and has to be micromanaged."[87; 89] Brown remained at one pole, unconvinced that the genome project should displace funds from project grants. He cited the fact that oncogenes and other major discoveries of the 1980s had flowed from research throughout NIH,[89] much as farmers and ranchers pointed out in the 1860s that *they* were the producers in the American West.[86] Who was John Wesley Powell to suggest that the West would benefit from planning, especially regarding water use? Communitarian values were invidious, and collective actions antithetical to the prevailing ideology.

In his opening talk at the San Diego meeting, Watson anticipated criticism of DNA sequencing efforts: "Saying that you support mapping without sequencing is like saying I'll marry you but there will be no sex."[87] Vintage Watson. Watson later got into a shouting match with Brown. Walter Gilbert closed the meeting with a call to arms. Biology was undergoing a fundamental change, and genome researchers were the shock troops. "The paradigm of molecular biology that Don Brown and Bernie Davis spoke from was that biology is a purely experimental science. In my mind that paradigm is shifting."[87]

Planned research efforts were new to molecular genetics. Shifting from a set of completely independent project grants to a coherent program with definite goals gave the research community some bumps and bruises. One early controversy about achieving goals erupted over refinement of the genetic linkage map. The NRC report had set the goal of a one-centimorgan map, so that the density of ordered markers along the chromosomes would be spaced an average of one million bases apart—close enough to help orient physical mapping efforts. With a map at this degree of resolution, a gene running in a family would be located with enough precision to go directly to analysis of DNA from the region. At the December 1989 meeting of the NIH program advisory committee, Maynard Olson commented: "There is a zero probability that we will develop a one-centimorgan map unless there is a major change of policy. Is this a goal or not?"[90] David Botstein seconded Olson's concern, and the issue was covered prominently by *Science* and *The Scientist*.[90; 91] Five months later, *Science* reported the map was "back on track" following a meeting of genetic linkage mappers just before that year's Cold Spring Harbor genome conference.[92]

The goals of genetic linkage mapping were scaled back to a map two to four times less dense with ordered markers, in large part because physical mapping techniques and PCR had made it seem likely that fewer markers would be needed to assist physical mapping efforts. It was also beginning to seem plausible that physical mapping and regional sequencing might well assist construction of a genetic linkage map as much as the reverse. This possibility was not apparent when the NRC report was issued early in 1988. The genome project had made a course correction.

A second controversy about targeting research broke out over DNA sequencing projects. Even as the budget battle for fiscal year 1991 quieted down, the question of large-scale sequencing remained an active issue. Passionate disagreements centered on whether, when, and how to begin systematically determining the DNA sequence of large stretches of the human genome. The two policy documents had parted company on whether a complete human genome sequence was an explicit goal. OTA stopped short of endorsing the idea of sequencing the entire genome, at least directly,[93] believing it was still an open scientific question whether it would be desirable to sequence every last region. The long arm of the Y chromosome, for example, seemed likely to contain large expanses bereft of genes. The NRC report was bolder. The committee reviewed the argumentes for and against a dedicated sequencing project, finally judging that "the ultimate goal would be to determine the complete nucleotide sequence of the human genome."[94]

The origins of the genome project were indeed Sinsheimer's, Dulbecco's, and DeLisi's visions of a fully sequenced genome. The Human Genome Project, however, evolved into a program with considerably broader goals. There was little disagreement that there would be much more DNA sequencing in the future, and that methods to perform it faster, cheaper, and more accurately were essential. There was further agreement that sequencing large expanses of model organisms was laudable and would be highly useful—sequencing genomes of the bacterium *Escherichia coli,* the nematode *Caenorhabditis elegans,* and baker's yeast, *Saccharomyces cerevisiae,* began to move forward. In these organisms, genes were tightly packed, investigators could take advantage of a vast array of genetic manipulations to test gene function, and the scale of the effort was a mere order of magnitude or two beyond demonstrated technical capacities. Sequencing projects in these organisms, funded by the NIH program and a massive multicenter collaboration organized under the European Community, pushed forward the technical frontiers and began to produce impressive results by 1992.[95; 96]

The wisdom of the National Academy committee's recommendation to fund animal model work also became abundantly clear in a mouse genome project. The mouse project was not directed at chromosomal sequencing, but rather at constructing a useful map of markers for genetic linkage mapping. There were hundreds of inbred strains of mice, and thousands of mutations defined by their effects on development, behavior, and physiological function. A genetic linkage map would greatly expedite the search for genes underlying the traits and would also enable rapid detection and targeting of new mutations. A genome center organized by Eric Lander of the Whitehead Institute in Cambridge, Massachusetts (with collaborators at Rockefeller University, Princeton, and MIT), produced a map that covered almost the entire genome with highly variable markers.[97] A few gaps remained, but the automated technique promised to quickly fill the gaps with new markers.

The project applied a concerted and systematic search for markers to an entire genome and produced an extremely useful map in a matter of a few years. The collaboration was a model of things to come, combining the resources of an NIH-funded genome center, several investigator-initiated grants from NIH and the National Science Foundation, and private funding from the Markey Charitable Trust.[97] While this was not a sequencing project, its technology was based on the polymerase chain reaction and many steps entailing large amounts of sequencing. While the purpose of the project was more "biological," in that it produced a tool for analysis along the lines of classical genetics, it was highly automated and dramatically expanded the scale of analysis to at least the same degree as other genome projects. It was not exactly what Sinsheimer, Dulbecco, and DeLisi had in mind initially, but it just might prove even better.

When it came to the human genome, however, there was a fixed chasm between those who wanted to sequence regions of known interest and those who argued that it would be more difficult to find out what was important than to sequence the genome and then pick out the juicy bits. Walter Gilbert long espoused the view that sifting through the genome and deciding what to sequence would be less efficient than sequencing it and using the information to guide biology. Many others argued that sequencing should be restricted to genes and regions of known interest, at least to regions with many genes and densely packed information, until the technology made sequencing considerably less costly.

The sequencing task would be formidable, regardless of which DNA fragments were first selected. Bart Barrell, who had managed several of the world's largest sequencing projects to date at the MRC Laboratory of Molecular Biology in Cambridge, injected a sober note, testing the reality of 1990 sequencing rates against those projected in 1985 by the Santa Cruz position paper. The Santa Cruz group had speculated that sequencing rates of fifty thousand base pairs per week might be feasible by 1988. Barrell shattered this optimism; he estimated that in 1990 the average rate in most laboratories was no more than fifty thousand base pairs *per year*. Barrell pleaded for "a prime goal . . . to make the existing technology more efficient both by making the methodology simpler and more automated and by better strategies that narrow the gap between the theoretical sequencing rate and the practical sequencing rate."[98]

The genome offices at NIH and DOE basically adopted Barrell's commonsense suggestion, which temporized, delaying a commitment to sequencing the entire genome until better technologies were in hand. NIH and DOE sponsored grants to develop entirely new sequencing methods, to sequence the genomes of organisms with densely packed genes and well-developed genetics, and to sequence relatively small regions (one or two million base pairs) of the human genome known to be of intense interest. The sensible position was to support different approaches and see which avenue proved most productive. This did not entirely rein in the controversy, as many in the community remained convinced that the genome project harbored an implicit commitment to sequence the entire genome no matter what.

The joint NIH-DOE five-year plan followed the NRC in stating "determination of the complete sequence of human DNA and of the DNA of selected model organisms" as its third major objective. The specific five-year goals, however, were to mount pilot projects on model organisms, to reduce sequencing costs to 50 cents per base pair (including preparation and analysis costs), and to sequence ten million bases of contiguous DNA (0.3 percent of the genome).[82] These were hardly utopian goals, or even a Manhattan Project for sequencing.

How to perform large-scale sequencing projects remained controversial. One group favored continued reliance on manual sequencing methods, automating some laboratory steps, but retaining a legion of human DNA sequence readers. A University of Wisconsin project to sequence the E. coli genome, directed by Fred Blattner, was the largest effort along these lines. He proposed to sequence the five million bases in that genome by using robots and a small army of undergraduate and graduate sequence readers. George Church at Harvard Medical School, Ray Gesteland at the University of Utah, David Botstein and Ron Davis at Stanford, and others experimented with an embellishment of the Maxam-Gilbert sequencing method, "multiplex sequencing," that enabled dozens of sequence stretches to be read without running new electrophoretic gels. In essence, this was a way to run twenty to fifty sequencing analyses (of three hundred or so base pairs) at a time in parallel. The multiplex method could also be automated, but the automation strategy focused on different components. The complicating factor here was how accurate the sequence determinations would be, and how to meld sequence data from the thousands of experiments into DNA sequence information for long contiguous stretches of chromosomal DNA. One major problem was how to reduce the amount of time humans had to spend fitting the data together. This approach confronted a daunting task of automated image analysis, and another task in detecting DNA sequence matches.

Other groups were committed to pushing the emerging automated DNA-sequencing machines to their limits. Groups at Caltech, Baylor, the National Institute of Neurological Disorders and Stroke (NINDS), and elsewhere mounted large-scale sequencing efforts that relied primarily on DNA sequencing instruments rather than manual sequencing methods. A group at Baylor, directed by C. Thomas Caskey, sequenced a part of the X chromosome containing the *hprt* gene, whose mutant form caused a terrible disease of self-mutilation among boys called Lesch-Nyhan syndrome. The ALF sequenator pioneered by the European Molecular Biology Laboratory, and modified for commercial sale by LKB-Pharmacia, played the central role in this effort.[99] The Caltech group concentrated on sequencing regions of the genome involved in regulating immune functions in both mouse and man. The group at NINDS was madly sequencing regions surrounding the neurofibromatosis region of chromosome 17, the tip of chromosome 4 (in search of the Huntington's disease locus), and regions that contained receptor genes for neurotransmitter receptors—proteins involved in nerve cell and muscle communication.[100; 101] This group then turned to sequencing short stretches of

DNA known to code for proteins,[102] and ultimately a split from NIH under private corporate sponsorship.[103; 104]

The Caltech and NINDS groups employed the Applied Biosystems sequencing instrument as their main sequencing tool. Other groups used these machines to generate the first wave of data and supplemented them with the ALF machine for those sequencing runs that started from specified sequences.[96; 101] By running a series of machines in parallel, massive amounts of sequence data could be generated. The reagents were expensive, and these techniques also required detecting sequence matches and reassembling thousands of short stretches of DNA sequence information into long contiguous sequences.

Walter Gilbert suggested a novel means of extending his and George Church's methods to directly determine DNA sequence data from chromosomal DNA, and he proposed to test the idea on the smallest free-living organism, *Mycoplasma capricolum,* a bacterium that infected goats. William Studier of Brookhaven National Laboratory proposed to use yet another approach to DNA sequencing that relied heavily on both automated DNA sequencers and also automated instruments to generate short stretches of synthetic DNA. Bruce Roe of Brookhaven had early success testing this idea out.

Entirely new approaches to sequencing also began to surface. Several groups—one surrounding Lloyd Smith at the University of Wisconsin, another at a startup firm named Genomyx in the San Francisco Bay area, and yet another involving a collaboration between the University of Utah and the University of Alberta—focused on using small capillary tubes and highly sensitive detection methods. If successful, these would dramatically reduce the amount of DNA needed for analysis and at the same time increase the speed of sequencing a hundredfold or more. Exotic methods were tested as well, in the various national laboratories and small pilot projects supported by DOE and NIH. One method hoped to apply direct analysis of individual DNA molecules through scanning-tunneling electron microscopy; another chipped one nucleotide at a time off a DNA molecule and would require new technology to detect a single labeled molecule.

These contesting methods were hotly debated among sequencing wizards. J. Craig Venter, C. Thomas Caskey, and Jack McConnell convened the first DNA Sequencing Conference at Wolf Trap, in the Virginia suburbs surrounding Washington, D.C. This conference took place in October 1989 and included a gala reception at the Phillips Collection near DuPont Circle in the District of Columbia. Sequencing enthusiasts saw this conference as the one that put large-scale sequencing on the map. Sydney Brenner closed the session, noting that "most individuals involved in the genome project had effectively written off the initiation of large-scale DNA sequencing for at least five or more years. This conference has clearly moved DNA sequencing to the forefront of the genome effort."[105] Those committed to use of automated DNA sequenators took the meeting as a vindication, a turning point in the genome project.[106]

Craig Venter and C. Thomas Caskey organized the second DNA Sequencing Conference in Hilton Head, South Carolina, a year later.[101] It began on October 1, 1990, Watson's decreed starting date for the genome project as a whole. Watson affirmed the importance of sequencing and directed his plea to the sequencing enthusiasts. He forcibly noted that rhetoric had sustained the DNA sequencing promoters for several years. It was time to produce mountains of data, and let the usefulness of DNA sequence information prove itself to the skeptics. He believed sequencing data from some projects—*Escherichia coli* and *Saccharomyces cerevesiae*—were so important that they should be pursued as crash projects, even if there was no agreement on one "best" method. Watson asserted that DNA sequencing was the focus of those who contended the genome project was bad science and that Congress had made a mistake in funding it. The earliest proposals for DNA sequencing were "shouted down" by the sequencing experts themselves in review committees. The community was highly fractious and opinionated, to the detriment of the field. Those interested in demonstrating the effectiveness of DNA sequencing as a scientific strategy would do better to convince skeptics by overwhelming them with important results rather than continuing to talk vaguely about the promise of sequencing.

Watson also noted the need for a common policy on the release of sequence data. He personally favored release of data as soon as investigators were confident about accuracy. The criteria for accuracy were unclear, however, as was the relevance of sequence data to proprietary uses (e.g., for commercial diagnostic tests or as data important to secure patents). Those doing large-scale DNA sequencing had to discuss criteria for data release as a first step toward consistent policies. Watson sought answers about these questions because he felt certain Congress would want to know who would have access to sequence data and when data would be deposited in public databases.

Watson's impact was felt far outside the Washington Beltway, the ring that marks off that part of the nation whose main preoccupation is the federal budget process. He had perhaps his greatest impact establishing the long-term scientific strategies of the genome project in its critical opening phase. One measure of his success was the quality of the investigators he brought into the fold. Genome centers and large genome grants established in the first wave of NIH funding were directed by scientists of supremely high caliber. Many were candidates for Nobel Prizes or thought likely to achieve that stature in the coming decade: David Botstein and Ronald Davis (Stanford), C. Thomas Caskey (Baylor), Francis Collins (Michigan), Glen Evans (Salk), Eric Lander (Whitehead Institute), Rick Myers and David Cox (University of California, San Francisco), Ray White and Ray Gesteland (Utah), and Walter Gilbert (Harvard).

Another administrator might have secured the same budget, but none could similarly create the ambience of "hot" science. If molecular biologists were sharks cruising the seas in hunt of tasty morsels, Watson was a great white. Norton Zinder, chairman of the NIH program's advisory committee, commented that "the spring of 1988 saw a quantum leap in the program's credibility" when Watson agreed to serve.[107] Watson took away one of the strongest arguments offered by genome critics—with Watson at the helm, it was difficult to argue that the science was mediocre. Watson had the trust of those who were initially skeptical of the genome project. Botstein commented on Watson's strong sense of priorities in the genome project: "We need to test its progress, regulate its growth, and slap it down if it becomes a monster. Jim Watson understands the dangers as well as any of us."[108]

Watson trudged through the muddy battlefield of genome politics in 1990. In early 1991, the genome project achieved its first direct presidential endorsement, along with a request for $110.5 million in the presidential budget. (DOE had requested $59 million.) In his remarks to the National Academy of Science in April 1990, and again at the American Association for the Advancement of Science in February 1991, President Bush saluted small science, but also gave explicit commitments to the genome project, high-speed computing, and a global change science program.[109; 110]

Watson shepherded the genome project at significant personal cost. The project needed him far more than he needed it. In this regard, Luria's biting comment, aimed at Watson in a thinly veiled reference to "power-seeking enthusiasts," was especially wide of the mark.[49] Watson did gain power, but he lost any semblance of normal life. He had long been famous, but the genome project dramatically increased press requests and solicitations for articles. There were more profiles to add to his already well-stuffed files. The travel schedule was brutal for a man in his mid-sixties, and much of the attention was intrusive and unpleasant. Robert Wright pilloried Watson in *The New Republic*. Watson's photo graced the front cover under a question in large, boldface type: "Mad Scientist?"[111] The public criticism stung, however little Watson cared to admit it. He said he could take the heat, but this was hardly fun.

Watson did not direct the NIH genome program for fame. It was the attraction of power, but not so much power over people as power over the future of science. The project seemed important to the future of biology, and Watson wanted it done right. That sense of duty extracted a commitment from him. It was clear the pressure and attention wore Watson down; at times he seemed almost resentful. Had he sensed there was another person able to carry the ball so far and so fast, he would have handed it off gladly.

Every public statement he made was taken as a policy proclamation. This was a particularly onerous adjustment for Watson, who was accustomed to a smaller and more local power base, where his outbursts washed away quickly and his personality was simply accepted as part of the Cold Spring Harbor firmament. Watson had to learn to restrain his characteristic strong-minded statements about what should be done, because his incomplete thoughts and trial balloons became instantly associated with NIH policy. He could restrain his impulses, but not obliterate them. He still made a much more colorful director than most. Like senators, members of Congress, and other public figures, Watson learned the confinement of power. He did not relish his temporal power over others, and he resented its intrusion into his personal life. He was, therefore, a great if reluctant leader.

Like Powell before him, Watson fell afoul of policy decisions by those more powerful. He resigned in April 1992, after a long-standing disagreement with his boss, NIH director Bernadine Healy. His stint as director of the NIH genome center was already drawing to an end, and he said as much in January 1992 at a genome advisory committee meeting. His exit was accelerated by a controversy over the patenting of DNA sequences that began in the summer of 1991. Before we turn to the Watson exodus, however, the international and domestic political stage must be set.

PART FOUR

Genome Gone Global

14

First Stirrings Abroad

THE GENOME DEBATE that began in the United States quickly spilled across national borders. The international ethos of science had little regard for political boundaries. As scientists in many nations approached their governments to seek funds for new genome research programs, many met with success. The first success was Charles DeLisi's program in the U.S. Department of Energy. The next was in Italy.

Italy's genome project began as a pilot project in 1987, under the Italian National Research Council, less than a year after the first DOE reprogramming began in the United States. The Italian genome program traced its origins to Renato Dulbecco's 1985 Columbus Day lecture in Washington, D.C., in which he first unveiled his idea of sequencing the human genome as the next major step in cancer research.[1] Consensus formed around Dulbecco's subsequent *Science* editorial,[2] and a program was quickly formulated and ratified by the Italian government. It was announced in May 1987 by the National Research Council.[3] Italy saw genome research as a road to world stature in molecular genetics. Dulbecco was appointed the project coordinator, with Paolo Vezzoni as the deputy, and the project grew from fifteen participating groups initially to twenty-nine by 1989.[4] The budget for the Italian program was $1.25 million for each of the first three years.

Dulbecco explained that "from the beginning, the project was organized on the concept that it would be carried out by many units scattered throughout

the country, because none of the units had all the necessary skill and equipment. To give unity to the project, a common objective was selected: [the end
of the long arm of] the X chromosome (Xq 28-Xq ter). Representatives of the
various units [met] two or three times a year. This approach led to active
collaboration among units. Collaborations also developed with various laboratories in Europe and the United States."[5]

The next national program emerged in the United Kingdom. The program
there was deeply rooted in the history of molecular biology. British science
was intimately woven into the fabric of molecular genetics. Indeed, the need
for a specific genome research program was less acute in the UK, since research
very much along the lines of the genome project was philosophically in line
with long traditions of British science. The Medical Research Council (MRC)
Laboratory of Molecular Biology in Cambridge honed the cutting edge of
DNA sequencing and physical mapping technologies and remained one of the
world centers for molecular biology, with or without the genome project. On
the other hand, the laboratory's traditions fostered the rapid emergence of the
genome project, and so while it may not have been needed to the same degree
as elsewhere, the genome project was a natural extension.

Representatives from the MRC laboratory in Cambridge attended the
major genome meetings in the United States, beginning with John Sulston's
presence at the first genome-specific meeting in Santa Cruz. British scientists'
views were almost automatically solicited because molecular genetics there was
a major part of the science, and there was only one world science. Wherever
one looked at molecular biology, scientists from the UK were engaged in
audacious projects to push the limits of structural biology. Two figures loomed
especially large in the UK.

Walter Bodmer and Sydney Brenner were immediately in the fray as the
genome debate began. Bodmer directed the the Imperial Cancer Research
Fund (ICRF) in London, a privately funded research center with an international reputation, particularly in genetic aspects of cancer and other diseases.
He was known throughout the world for his work on genetic variations in the
immune system and in cancer, and on human population genetics,[6–12] and he
was chosen by Watson to keynote the Cold Spring Harbor meeting on human
molecular biology in June 1986.[8] He also chaired the HHMI meeting in
Bethesda two months later. Sydney Brenner from the MRC lab attended
several early meetings and was later invited onto the National Academy of
Sciences committee. Brenner and Bodmer were positioned close to the centers
of power in science in the UK and served as British ambassadors to the larger
world of molecular biology. Brenner was highly positioned in the MRC,
holding several different posts while the genome debate was underway.

The British genome debate began in 1986, when Brenner suggested to the
MRC that he start a Molecular Genetics Unit that would include genome
research. Brenner jump-started the UK genome program with funds from a

private £300,000 ($525,000) award he received from the Louis Jeantet Foundation.[13] He proposed to apply the physical mapping methods developed for *Caenorhabditis elegans,* pioneered by John Sulston and Alan Coulson of the MRC laboratory in Cambridge, to the human genome.[14]

Brenner thus drew MRC directly into genome planning. At Brenner's request, MRC established a scientific advisory board, chaired jointly Sir James Cowan, as secretary of the MRC, and Bodmer, as director of the ICRF. Membership reflected the interests of other research councils and private charities. ICRF and MRC were expected to contribute roughly equal funding, and coordination was via a joint scientific advisory committee. In February 1989, the secretary of state for education and science announced that £11 million would be provided to the MRC over three years.[15; 16] The UK genome program officially began in April 1989 as a three-year project, but expected to attain a stable annual budget of £4.5 million beginning in 1992.[15; 17]

The scientific strategy was a two-pronged approach. One prong was to coordinate ongoing work, including databases and material exchange centers; the second prong was intended to goad basic genome research with additional funds and ambitious technical aims. The approach to international coordination was, in essence, "Let's get started and then we'll talk." Brenner noted that once "we have established a center in the UK which has already been of value to our research community, then we will be well placed to play an active role in international efforts."[16]

During its first year, the UK program focused on automation, new techniques, and mapping regions of special interest. It then shifted to focus on cloning, mapping, and a stronger emphasis on protein-coding regions.[18; 19] The genome program included work on mice, especially a mapping effort based on cell lines taken from back-crosses between two species of mice that enabled straightforward physical mapping. Work on *C. elegans,* of course, remained a prominent feature of British genome research. The UK effort was subsequently focused even more on protein-coding regions of the human genome and informatics. The program imported the yeast artificial chromosome library developed at Washington University in St. Louis, to make the clone set available in Europe. It collected a set of DNA probes at ICRF and increased access to human cell lines stored in a repository at Porton.[20] Access to the Genome Database at Johns Hopkins was acquired in 1991, making the UK the first node outside Baltimore.

Progress was recorded in a newsletter that disseminated information about resources, ongoing projects, program plans, new techniques, and summaries of major meetings. Carrying on a long tradition of British humor, the first three issues were named the *G-String*.[21–23] (This name, of course, referred to a DNA sequence of guanines.) The title was then changed to *G-Nome News*[24–27] Early issues focused on the UK, but later editions broadened to cover other countries in Europe. Tony Vickers was appointed human genome mapping project manager in 1990,[24] and the Human Genome Resource Center was established at Northwick Park, Harrow, later that year.[25] Vickers remained director until late 1992, when Keith Gibson took the reins.

One novel element of the UK effort was a transatlantic bridge built on the nematode work. The *C. elegans* physical mapping effort, one of the prototype projects for physical mapping, grew into a U.S.–UK collaboration to sequence the genome.[28] The nematode was once again to serve as the pioneer for a new technological feat. Brenner's chosen organism was yet again to push back the frontiers of biology. The *C. elegans* collaboration was so successful that it became the backbone of a major expansion of genome research efforts in the UK during 1993, which included John Sulston becoming the director of genome research at a newly founded Sanger Centre funded by the MRC and the Wellcome Trust, at facilities in Hinxton Park, south of Cambridge (see Chapter 20).

During the first two years of the genome project, attention in the UK turned to hosting the eleventh Human Gene Mapping Workshop (HGM 11) in London, with the ICRF as host. UK scientists came to feature prominently in international genome politics, far beyond the budgets supporting their work. Bodmer replaced Victor McKusick as president of HUGO in December 1989. He gave a presentation to the Parliamentary and Scientific Committee, a science policy forum, to bolster support for genome research.[29] Malcolm Ferguson-Smith of Cambridge became chairman of the European Community working party to formulate the EC genome analysis program.[19] Brenner, one of the Nobel committee's more conspicuous oversights, remained a major scientific force. Work on *C. elegans,* pioneered in the UK, continued to serve as a prototype for human genome research.

Even as plans hatched in Italy and the United Kingdom, a genome project of a different character was beginning in the USSR. The project there took root in scientifically rich but politically unpromising soil, and by the time it reached full flower, the USSR was the former Soviet Union and science was struggling for its very life in hard economic times.

The instigator of the Soviet program was Alexander Alexandrovich Bayev, who survived the Gulag from 1937 to 1954 and in old age weathered the many government transitions of the late 1980s. In the far north, he was known for having built a hospital for children in Norilsk. He was sent there in exile, having survived a term in one of the most brutal prisons in the Gulag. Bayev met his second wife in Norilsk (his first wife divorced him after he went to prison). The same perseverance and vision that created a pediatric hospital during the 1940s led in 1988 to the creation of the world's third national genome program.

Bayev's irregular life story is but one of millions—the legacy of Stalin's destructive repression. Bayev trained as a biologist and had obvious talent as a young man, but his career was interrupted during its most productive phase, from age thirty-three to fifty, by imprisonment and exile. Bayev was born in 1904 in Chita, east of Lake Baikal and north of Mongolia. He studied mathematics and physics and later transferred to medicine, graduating in 1927. He decided to pursue a research career. In 1930, he became a graduate student under Vladimir Alexandrovich Englehardt in Moscow.[30] Bayev's trouble began in 1937, in purges against those who might possibly be supporters of Stalin's rivals Bukharin and Trotsky.[31] These purges directly caused the death of millions, rent the social fabric of the Soviet Union, and hurled Bayev to the outer reaches of Russia.

Bayev was sent to Solovetskiy Special Prison, an infamous detention center on an island in the White Sea. He was then exiled to Norilsk, where he survived by reverting to his role as physician.[30] In the frozen north, Bayev was connected to the outside world only through his mother, who soon died. He remained entirely isolated for several years, not wanting to endanger friends by writing to them. In a brief political thaw during 1944, Englehardt got a letter to Bayev inviting him to Moscow. Bayev ventured to Moscow, working out of Englehardt's apartment because he was unable to visit libraries. Bayev returned after a few months to Norilsk, leaving behind a dissertation on molecular biology that culminated in "candidate of science" status. He was briefly freed in 1947, but arrested again and sentenced to exile "forever." Forever lasted until 1954, less than a year after Stalin's death. Englehardt managed to retrieve Bayev when a period of enlightenment somehow held the Soviet bureaucracy at bay for a year or two. Bayev resumed his scientific career in Moscow, first at the Institute of Biochemistry and later at the Institute of Molecular Biology.[30]

Bayev missed out on the beginning of the revolution in biology taking place in the 1950s. He was in exile when Watson and Crick published their structure of DNA. The USSR contained few scientists able to appreciate the achievements of molecular biology. T. D. Lysenko exercised a vigorous ideological aversion to genetics and made sure others shared his enormous blind spot. Lysenko killed the field by repressing its practitioners, turning Soviet genetics into a wasteland populated only by those few brave souls willing to bet that times would change, or too tough to care. For two decades, genetics and molecular biology were systematically suppressed.[32; 33] After Stalin's death, Lysenko slowly began to lose his lock on biology and agriculture. Bayev's mentor Englehardt was in the vanguard opposing Lysenko; as a consequence, Englehardt temporarily lost his position in the section of biology of the USSR Academy of Sciences in 1958. Lysenko was almost deposed in 1959, as an Academy of Sciences committee was poised to censure him. The committee was thwarted the evening before its report was to be made official, when Nikita Khrushchev rescued Lysenko.[33; 34]

Englehardt persevered. In 1962, he presented a paper to the Academy of Science asserting, against the Lysenko ideology, that the recent accomplishments of molecular biology were not flukes, but the first fruits of a scientific revolution.[33] In June 1964, Englehardt worked to block appointment of a Lysenkoist to the Academy of Sciences. He was supported in his efforts by Andrei Sakharov.[33] Sakharov's prestige, derived from physics and thus independent of Lysenko's power base, was essential to the effort. Indeed, it was Sakharov's struggle against Lysenko that first gave Sakharov fame as a political actor, with an impact well beyond his scientific field.[35]

Khrushchev reacted sharply to Englehardt's brashness, calling for an investigation of the Institute for Physical Chemistry and Radiation Biology, where Englehardt worked, but the academy stood behind Englehardt, failing to admit Lysenko's colleague. Englehardt and others ultimately prevailed; Lysenko was finally debunked publicly later that year.

Englehardt became director of the Institute of Molecular Biology and brought Bayev to work there. A young student working for Bayev, Andrei Mirzabekov, was attaining recognition as a rising star. Mirzabekov was born the year Bayev was arrested, in 1937. His family moved to Moscow in 1943, and he became interested in biology. In 1971, Mirzabekov was permitted to go to the West. Mirzabekov went to the MRC laboratory in Cambridge, where he worked to crystallize transfer RNAs for analysis by X-ray crystallography, a technique used to study the three-dimensional structure of DNA and proteins. He managed to extend his stay for six months through the machinations of Aaron Klug, Francis Crick, Frederick Sanger, and Max Perutz (all Nobel laureates).

Mirzabekov was a link between the USSR and world molecular biology in the mid-1970s. He felt the full force of molecular biology from its epicenter, at the MRC laboratory in Cambridge. Mirzabekov returned to the West in time to participate in several major events. He attended the 1975 Asilomar meeting about the safety of recombinant DNA. While in the United States early in 1975, Mirzabekov had a famous lunch with Walter Gilbert, Allan Maxam, and Jay Gralla at Harvard, related to what later became a technique for DNA sequencing. Mirzabekov discovered that DNA could be destabilized at specific base residues by dimethyl sulfoxide, adding methyl groups to guanine and adenine, and causing the DNA to fragment at positions containing those bases. Gilbert and Maxam used Mirzabekov's chemical modification methods to study the binding of protein to DNA[36] and extended the method into the Maxam-Gilbert DNA sequencing method.[37] (See Chapter 4.)

Through the 1970s and 1980s, Mirzabekov and Bayev continued to work at the Institute of Molecular Biology in Moscow, now called the Englehardt Institute. (By tradition, institutes of the USSR Academy of Sciences were named for their founders several years after their death.) They and others in Moscow and Leningrad brought new approaches to Soviet biology, adopting recombinant DNA research and the other new techniques. In 1986, at age eighty-two, Bayev declined to become director of the Englehardt Institute. He recommended Mirzabekov for the job, and Mirzabekov became director that year.

Mirzabekov remained active in science, although his time was increasingly devoted to administrative duties and politics necessary to preserve the health of the Englehardt Institute. Several Soviet scientists were particularly interested in developing techniques for DNA sequencing that might be less demanding of reagents and robotics, commodities hard to find in the Soviet Union. Less reliance on high technology and Western reagents made the likelihood of a Soviet contribution to the sequencing effort more feasible. His laboratory continued to analyze protein-DNA binding. Hans Lehrach of the Imperial Cancer Research Fund in London suggested a DNA sequencing technique based on binding very short segments of known sequence, in the range of eight or more base pairs in length, and determining whether these bound specifically to a given DNA fragment. If they bound, then that sequence was present on the DNA fragment. By binding a large number of such short segments and identifying which short sequence stretches were present, the sequence of the fragment could theoretically be determined. There were several difficult technical obstacles for such a method. A positive score, when bases of the short segment exactly matched a sequence on the fragment to be sequenced, for example, had to be reliably distinguished from a negative, if the match was inexact. Moreover, for the method to be practical, thousands, perhaps even hundreds of thousands, of the short DNA sequences had to be tested and scored at once, despite subtle differences in the strength of DNA binding for each short segment. The potential advantages in speed and simplicity once a system was set up, however, made this sequencing scheme tantalizing. Mirzabekov's group worked to make it a sequencing method, as did a Yugoslav group under R. Drmanac and R. Crkvenjakov[38; 39] and groups in London and the United States.

Bayev and Mirzabekov became the champions of the USSR genome program. Bayev learned about the genome project from Walter Gilbert and James Watson, during a visit to the United States in 1986. Mirzabekov attended the 1986 symposium "The Molecular Biology of Homo sapiens" at Cold Spring Harbor that same June. Once back in the USSR, Bayev and Mirzabekov worked to build support for a Soviet genome program.[40–47] Their timing was propitious, capturing the initiative at a time of great change under Mikhail Gorbachev.

The genome project benefited from the new policies of *glasnost* and *perestroika*. *Glasnost,* openness, made it possible to acknowledge, at long last, the damage Lysenko had wrought on genetics and molecular biology, and to begin repairs. *Perestroika,* restructuring, as applied to biology, sought to link science and biotechnology to national economic goals. A first step was to bring Soviet molecular biology up to world standards. The Soviets renewed attention to peer review and other aspects of science funding and science administration.[48] Bayev and Mirzabekov wrote to their colleagues to build support for a USSR genome program, intended to match the movement gathering force in the United States. Bayev argued that "there are times in the history of science when far-reaching decisions must be made, and in the field of molecular science, one such moment is upon us."[49] The genome project met opposition, principally based on its importance relative to that of other areas of science, in 1987 and again in February 1988, when Bayev made a presentation to the general assembly of the USSR Academy of Sciences. Bayev and Mirzabekov persisted, however, and eventually brought their colleagues and political patrons around. A program was presented to the Council of Ministers in 1988. It was approved, and in 1989 the State Committee for Science and Technology listed it as one of fourteen priority areas in science.[44–47; 49–51] Genetics was becoming central to several other initiatives in biology and agriculture as the shadow of Lysenko inexorably shrank in the bright sunlight of modern biology.

The Soviet genome program thus became an instance of *perestroika*. Funds under the project were distributed partially through traditional mechanisms, controlled by the directors of national institutes and laboratory directors within the institutes. Another, more innovative part of the budget was modeled on the NIH project-specific grant system. Mirzabekov, as director of the Soviet genome program, eagerly sought information about peer review and grant administration from his Western colleagues, hoping to reinvigorate Soviet molecular biology by applying Western methods of science administration. The NIH was seen as the most successful agency in cultivating and sustaining molecular biology in the world, and Mirzabekov wanted to bring its peer-review methods, with appropriate modifications, into the USSR.

The budget for the Soviet genome program grew despite hard times for the Soviet economy. The 1988 budget of 25 million rubles was increased to 32 million in 1989. The fax line from the Englehardt Institute to the West exemplified the high priority of genome research, requiring a direct international phone line and a machine purchased with scarce foreign currency. Yet continuing economic turmoil within the Soviet economy imperiled the genome program. Science was caught in the crossfire, in a debate over decentralized planning, and in the tumultuous transition from a centralized communist economy toward a capitalist base. Nevertheless, in 1991, a 40-million-ruble genome program was approved in the national USSR budget.

Many of the institutes of the USSR Academy of Sciences, funded through the central USSR government, were thrown into chaos in the transition from a centralized economy, as the national republics began to wield more political and economic power. The tumult following the failed coup against Mikhail Gorbachev in August 1991 was a period of great uncertainty. With the dissolution of the USSR, the Russian Republic picked up the scientific institutes housing most genome research, but budgets were extremely tight. Science was a relative luxury in an economy reeling out of control after years of central management. Getting food onto tables, building houses, and cultivating other elements of a consumer culture were more important than science.

The genome program was relatively spared during this period of confusion.[52] Indeed, even as science disintegrated through 1992, the genome program was given a line item budget under the Russian Academy of Sciences.[53] This may have been because Bayev and the Englehardt Institute were never closely associated with the political power under the Brezhnev "period of stagnation" and were long associated with reform. Because the genome project was spared, it became the chief vehicle supporting all of molecular biology in the former Soviet Union. Bayev's time in the Gulag and his integrity after returning from it were credentials of great value in the new era.

While attending a scientific meeting in the USSR in June 1989, I visited both the Englehardt Institute in Moscow and the Institute of Cytology in Leningrad (now St. Petersburg). The science at both institutes was seriously constrained by limited infrastructure, but the minds were keen, and ideological constraints were conspicuously absent. The Leningrad institute, in particular, was well populated with erstwhile renegades from the Baltic republics. In hotel rooms where only months earlier the talk would have been hushed for fear of KGB interlopers, there was bold discussion of national politics. Evening discussions became a delightful mix of science and new-wave politics, with predictions of how the central government would dissolve and courageous talk about how that process might be expedited.

Bayev and Mirzabekov presented the genome project as linked to economic development from the beginning. Bayev promised growth through biotechnology in his advocacy for the project. As Russia and the other republics threw off the old order, economic revitalization was the order of the day. The genome project was one of the programs already in place.

Scientists attached to the genome project were all too aware of the daunting task ahead. They were ambitious and bright, but hampered by a legacy of repression. Their work in experimental areas was impeded by limited access to instruments, materials, and technologies from domestic suppliers, and extremely tight budgets for increasingly expensive foreign goods. Regina Eisner, a young molecular biologist from the Englehardt Institute, summarized the prospects for Soviet participation in the genome project: "Soviet science is very good when it does not depend on technology. We have brains and courage. If there are things that need only those, then we can participate."[54]

Bayev embodied the courage and endurance that ran so deep in his culture. He and his fellow scientists had crafted their genome project with an eye to the future. The shape of that future was completely uncertain, but the genome project appeared likely to survive into the new era, a brick in the new edifice.

As genome programs sprang up in Italy, the UK, and the USSR (and its successor political units), parallel developments were taking root elsewhere on the European continent. French genome research began from several centers already deeply involved in human genetics. In 1988, Prime Minister Jacques Chirac announced a FFr 8 million ($1.4 million) program to bolster genetic linkage mapping, to cultivate DNA sequencing, and to foster informatics. Jean

Dausset, director of the Center for the Study of Human Polymorphism (CEPH), chaired a scientific advisory committee to oversee the program.[55] CEPH had long been the collaborative core of genetic linkage mapping in humans. It subsequently expanded its efforts into physical mapping and technology development for sequencing and mapping. Through CEPH and several prominent research groups, France played an important role in the initial genome mapping efforts.

France initially supported selected specific grants in genome mapping, and also to EC programs and the Labimap project for automation (a joint project involving the UK, France, CEPH, and the British company Amersham Ltd.).[20] The early phase of individual grants from French science agencies evolved into a significantly larger and more directed program between 1988 and 1990. Prime Minister Chirac flagged genome research as a national research priority, and by May 1990, the government announced a FFr 8 million ($1.4 million) budget for it, distributed through a committee chaired by Dausset.[56] It was a beginning. The Ministry of Research expressed its intention to mount a more centralized genome research program in June 1990.[57] The ministry charged the National Institute for Health and Medical Research (INSERM) to plan a research program to be formalized later in the year. Philippe Lazar, director of INSERM, delegated the task of formulating plans to Philippe Kourilsky of the Pasteur Institute, who drafted the necessary language. Hubert Curien, the minister of research and technology, formally announced the program on October 17, 1990. The French program was budgeted for FFr 50 million ($8.75 million) in 1991 and FFr 100 million ($17.5 million) in 1992,[58] but 1991 funding actually fell far short of this projection.[59]

A private effort took off faster and produced impressive results quickly. The French muscular dystrophy organization (Association Française contre les Myopathies, or AFM) raised money through telethons and poured the funds into a high-technology approach to human genetics, pursued in conjunction with CEPH. Daniel Cohen, the CEPH director, had worked with Dausset since 1978, and fourteen years later he found himself a leader of French molecular biology.[60] He wowed the crowd of researchers attending the annual genome meeting at Cold Spring Harbor in May 1992, unveiling results on a physical map of chromosome 21 far more advanced than most groups had expected.[61; 62]

A collaborating center, the Généthon facility in Evry near Paris, aspired to become the most advanced technological center for human genetics in the world, and seemed likely to achieve that goal, at least for awhile. Between them, CEPH and Généthon employed a staff of 250; the AFM monies provided about 70 percent of the CEPH-Généthon genome budget. Généthon purchased a large number of Apple computers as tools for public education, and a dozen Applied Biosystems automated DNA sequenators.[61; 62] Scientists with CEPH produced an impressive stock of yeast artificial clones with great speed and expanded the size of the DNA fragments contained in them through technical innovations, improving on the other clone collections and thus expediting the direct study of DNA regions in the human genome. When James Watson was asked for his assessment of the best national genome effort outside the United States at the 1993 budget hearings, he responded that "through Généthon, the French have moved to super-production first," and when pressed about which effort was the "number two country," he again replied, "France. Also important are the UK and Japan."[63]

The French genome efforts grew out of a strong tradition of molecular biology. In 1958, when President Charles de Gaulle appointed a committee to look into reorganizing French science, molecular biology emerged as the top priority. Jacques Monod, who shared a Nobel Prize in 1965 with fellow Frenchman François Jacob, chaired a subcommittee that urged the science ministries to foster small problem-oriented units rather than major thematic centers.[64; 65]

Jean Dausset's involvement in genome research began with his work on the cellular systems involved in determining tissue compatibility and immune function. An enormously complex family of genes made up the histocompatibility complex. Teasing apart the component genes and proteins took decades. Dausset and his colleagues at the Hôpital St. Louis in Paris were constantly in the fray. Dausset, as leader of the French team, shared a Nobel Prize in 1980 with two U.S. scientists (Baruj Benacerraf and George Snell). An important element of Dausset's work centered on genetic differences among individuals, an essential feature of the histocompatibility complex, and led naturally to an interest in genetic linkage mapping. Dausset took a seed grant in 1983 and brought together the two large groups assembling genetic linkage maps—Ray White's groups in Utah and Helen Donis-Keller's group at Collaborative Research, Inc., near Boston. An art dealer's bequest and partnership with the AFM put the effort on a firm financial foundation,[60] and the science took off. These and other, smaller groups from around the world formed the CEPH collaboration. Paris became a coordination center for producing a human genetic linkage map.

The French genome program grew from several years of discussion, involving INSERM (the National Institute for Health and Medical Research), the National Center for Scientific Research (CNRS), scientists at the Pasteur Institute, CEPH, and several genetics research centers throughout France. Scientists at the Pasteur were less enthused about a massive assault on the human genome.[66] While the private efforts raced ahead, the government program worked its way through the Ministry of Science and Technology.

Genome research, like other research, labored for several years against the rigidities of the French national research system.[58; 59] The 1958 commission had not succeeded in freeing molecular biology from the traditional French university system, despite success in nurturing selected groups within it. The private funding through CEPH and Généthon was not so encumbered, and it progressed rapidly. The new government program likewise attempted to re-

move some of the shackles from genome research. It had three major goals—
to isolate and sequence protein-coding regions of DNA, to support the complete sequencing of small genomes such as that of *Bacillus subtilis,* and to
encourage the development of analytical software. The effort was organized
under a quasi-public organization, Groupement d'Intérêt Public (GIP), that
enabled the participation of private French firms.[58] The GIP was headed by
Jacques Hanoune; François Gros was president, and a scientific advisory committee was to help coordinate efforts and plan strategies.[59] Well into 1992,
however, the dedicated genome research program remained a shell without a
core of fiscal support. Starved of funds, it teetered. The private Généthon
funds, in contrast, were a stable base from which France raced ahead of other
nations.

Although Germany had Europe's largest economy, its contributions to
human genetics lagged behind those of the United Kingdom and France. It
ran a distant third in the number of articles on human genetic mapping, barely
edging out Italy and the Netherlands in a bibliometric assessment for 1990.[67]
This sustained the pattern that prevailed over the previous decade.[3] Part of the
laggardly pace of German genetics was explained by the long shadow of eugenics and racial hygiene in German culture.

The contributions of German scientists to the ideological foundations of
the Nazi racial hygiene programs before and during World War II began to be
openly discussed in Germany just as the genome project was gaining momentum in other nations. Benno Muller-Hill, a molecular biologist who had worked
with Walter Gilbert in the 1960s, wrote an angry book about such "murderous
science."[68] His book was merely the first in a a long list of German books about
scientists' complicity in Nazi ideology. This ended a long and conspicuous
silence. Many of the most forthright racial hygienists from the Nazi era had
taken academic jobs in human genetics after the war, and the role of science in
Nazi ideology had remained taboo for an entire generation.[69–71] Decades of
silence regarding the Nazi activities of researchers lent credence to public
suspicions of the academic elite. Many books on the history of eugenics and
racial hygiene were also published in English,[70; 72–76] but the cultural sensitivi-
ties in Germany were more combustible. What was a subject of interest mainly
to historians elsewhere was inflammatory in Germany.

I encountered the difference firsthand in 1989 at a bioethics conference at
the Ruhr University, Bochum. I was one of several speakers at a conference on
ethics and human genetics at the city outside Düsseldorf. The meeting was
almost halted because local students threatened to demonstrate against it. The
conference was held, but students were selling booklets alleging that the conference organizer, Hans Martin Sass, was a closet apologist for racial hygiene.
I was spared any personal attacks, in part because I was obscure and in part
because I was introduced as having worked at the Office of Technology As-

sessment, whose 1983 report on genetic testing in the workplace was lauded several times during the discussion.[77] Although I had little to do with that report (except helping explain RFLP mapping to one of its authors), it nonetheless served to protect me by association. My conversations—with students concerned about the implications of genetics, with clinical geneticists involved in genetic counseling, and with scientists interested in human genetics—made it clear to me that German science would pay a penalty for its long silence. Why would young and able scientists or physicians choose to enter a field so inherently suspect, so widely perceived as tainted in their culture?

The Green movement in Germany was another obstacle to genome research.[14] The Greens had strong suspicions of biotechnology in general and genetic engineering in particular. During the late 1980s, while genome research was first being debated, the Greens were a growing force, and they remained so until caught flatfooted during the 1990 elections, unprepared to deal with the initial enthusiasm for reunification with East Germany. The Greens were concerned at how the results of human genetic research might harm individuals, particularly the use of genetic tests by employers and private insurers. One countervailing force was the AIDS epidemic and the demand it evoked to use molecular genetics to combat a major public health threat. Opposition to genetics had to be tempered by appreciation of its potency in thwarting at least some diseases.

Despite the relative paucity of human genetic research in Germany, scientists there were eager to join in the worldwide genome research effort. Many learned molecular genetics abroad, where it was not subject to the same degree of stigma. They hoped to build a science in Germany that would be seen as a boon to society, rather than a threat. This called for putting genetics on a new moral footing and directly contending with the legacy of racial hygiene.

The German Research Council (DFG) commenced a program centered on human genetics in 1986. In September 1987, representatives of the DFG rejected a position paper prepared by a group of scientists to mount a concerted genome project.[3] Grant funding for individual projects continued, however, under a program named Analysis of the Human Genome by Molecular Biological Methods, which included data analysis and data handling, technology development, basic genetics, and support of European Community programs.[20] This budget was renewed in 1990 for six years at DM 5 million ($2.2 million) per year.[20]

Scientists' other proposals to mount genome programs were rebuffed. A position paper prepared by scientists for the German parliament (Bundestag) died in the Ministry of Research and Technology.[20] A June 1988 meeting in Frankfurt precipitated a consensus that German efforts might concentrate on informatic aspects of genome research, under funding from the Commission of European Communities. This effort led to a three-part program under the German Cancer Institute, for a genome database network node at Heidelberg,

development of a genomic database integrated with the Genome Database at
Johns Hopkins, and an initiative to identify open reading frames in DNA
sequence data.[20]

The 1990 unification of Germany merged two very different scientific
structures. Human genetics in the former German Democratic Republic, or
East Germany, had focused by necessity on clinical applications. A program
started there in 1986 began to introduce molecular techniques to the diagnosis
of the three most common genetic diseases: cystic fibrosis, Duchenne muscular
dystrophy, and phenylketonuria. Because of restricted access to Western technologies so necessary to molecular genetic research, scientists in East Germany
had little to contribute aside from access to family resources with excellent
clinical profiles.[78–80] When the eastern republic joined the western one, it
brought a social structure that supported a much higher proportion of scientists. East German scientists were starved but numerous; they had previously
been hampered by limited funds to conduct research and limited access to
reagents and instruments. Many were now faced with the prospect of unemployment. As 1990 moved into 1991 and the genome program gathered force,
the euphoria of reunion gave way to recognition that the two Germanies had
indeed drifted far apart in four decades of separation. True unification would
be a long process attended with uncomfortable discontinuities on both sides
in the early phases. One happy product of this situation was the new Max
Delbrück Institute for Molecular Biology in Berlin, at an institute previously
part of the East German scientific establishment. One of the founders of molecular biology was thus honored posthumously in the country he fled five
decades before, a fitting signal of new directions.

Human genetics in Denmark had a long and distinguished history. Danish
medical offices had for many years maintained scrupulous clinical records, and
Denmark established repositories containing thousands of cell lines for human
genetic research. A special effort had produced a large collection of well-
characterized, apparently normal families (i.e., lacking known genetic diseases).[3] Most families were small, although one was large enough to be part of
the CEPH family set. Attention to normal families was complemented by a
strong capability in clinical genetics. The bulk of genetic illness was referred to
a single hospital, the Rigshospitalet in Copenhagen, dramatically simplifying
the process of building a genetic registry. While small families were less useful
for making a genetic linkage map, the thorough documentation and consistency of clinical assessment were major advantages for hunting down specific
disease genes. Danish genome efforts therefore continued the traditional emphasis on clinical genetics.

According to one observer, both the government and the public in Denmark were "more interested in genome research being concerned with disease-
related problems than mapping *per se*. Both were content for the United States
and Japan to undertake the latter."[20] A genome research center was one of ten

recommended by an international committee of experts in 1989, in evaluating fourteen ongoing biotechnology centers. The Human Genome Research Center at Aarhus University was a reincarnation of the former Bioregulation Research Center that operated from 1987 to 1990. The Genome Research Center commenced work on January 1, 1991, with a mandate to do genetic linkage mapping and physical mapping, to characterize mutations causing human genetic diseases, and to study various functional properties of genes.[20] Its annual budget of 10 to 15 million kronas ($1.8 million) was contingent on government funds from the Medical Research Council being supplemented by the university and other sources.

The Commission of European Communities (EC) hoped to knit genome research in the various EC member states into a coherent whole. The EC program began to emerge early in the genome debate. It grew from a convergence of interests among member states and a desire not to be left in the dust.[19; 20; 81] Sydney Brenner alerted officers at the commission with a short proposal received February 10, 1986.[82] Further discussions elicited support for projects on *S. cerevesiae, B. subtilis, Drosophila,* and *Arabidopsis thaliana,* and multinational efforts commenced in 1988.[20] These projects were sponsored by various biotechnology programs of the EC. A program on the pig genome was added in 1991.[20] The sequencing of yeast chromosome 3, organized by the EC, was one of the first major triumphs of genome research anywhere in the world.[83] Despite skepticism that a collaboration involving so many laboratories could produce results, the yeast sequencing project nonetheless produced the longest continuous DNA sequence achieved to date. Its progress was undoubtedly slower than it would have been had it been done at a single center, as indicated by the very small amount of sequence data derived from automated methods, and there was ample criticism of delayed access to the data as it was being assembled, but in the end it reached its goal.[84] There was political wisdom behind the choice of a logistically complex collaboration. The widely distributed collaboration avoided a divisive debate over which country would get the political plum, and the support for the project produced by its broad base helped to make it a major success.

Europe promoted several efforts to automate DNA sequencing and to develop other instruments for DNA analysis. The European Molecular Biology Laboratory (EMBL) in Heidelberg received funds from a variety of European governments under a multilateral agreement. In addition to ongoing work in genetics, it also maintained the European node of the DNA sequence database, shared initially with GenBank in the United States. (In 1987, the DNA Database of Japan was also brought in.) EMBL was also the center of an effort to develop a fluorescence-based automated DNA sequencing instrument. EMBL scientists developed a prototype that was later modified and marketed by the Swedish firm LKB-Pharmacia as ALF. Several EC programs focused on biotechnology instrumentation, including the Labimap project and a joint effort between the University of Manchester and the University of Konstanz. The EC quickly found agreement that informatics and computer analysis of genetic data were important targets not only for genome research, but for biotechnology more generally. The need for data exchange across borders was readily apparent, and agreement on the importance of informatics was readily achieved.

The EC program in human genetics provoked more controversy than studies of other organisms, delaying approval of a human genome program. German research minister Heinz Riesenhuber was a major force promoting EC involvement in biotechnology, including genome research. His interests stemmed from wishing to see cooperative European efforts in biology, but also from the difficulties that research programs in human genetics encountered within Germany. The EC provided a lever to secure support from the German national government for multinational European programs. The EC funds were also an independent pot of funds for which German scientists might apply, entirely avoiding the problems of domestic funding.

Peter Pearson of the Sylvius Laboratories in the Netherlands chaired the working party charged with formulating genome research plans, until he moved to Johns Hopkins University in 1989. Malcolm Ferguson-Smith of Cambridge University then became chairman. The name of the proposal to support an EC human genome program was changed from "Predictive Medicine" to "Human Genome Analysis,"[85; 86] signaling a recognition of social concerns.[19] The original title had offended German sensibilities, particularly those of Benedikt Härlin, a German Green Party member. Härlin was a member of the European Parliament who served as "reader" for the genome research proposal in its science and technology committee. He sought to ensure that a program to examine the social implications of the research progressed in parallel with the scientific effort.[87] Explicit inclusion of a program to consider the ethical, social, and legal aspects (ESLA) of genome research cleared the way for approval.[88] While the proposal was under consideration, an ESLA working party was appointed, chaired by Martinus F. Niermeijer of Erasmus University, a well-known human geneticist. The ESLA program was allocated 7 percent of the budget, and with the understanding that the genome research program would implement confidentiality protections and would exclude germ line genetic manipulations, it was approved by the council on June 29, 1990.[89]

European efforts to keep abreast of U.S. science spawned several reports in late 1990 and early 1991. The Medical Research Council of the UK was commissioned by Academia Europaea, an organization of academic specialists from a wide variety of disciplines, and the European Science Foundation to survey genome research throughout the world. Diane McLaren, executive secretary of the UK human genome mapping program, did the most exhaustive world survey of genome research to date.[20] This survey fueled conclusions from the ESF and Academia Europaea reports, which agreed in their strategic conclusions, suggesting that Europe should coordinate its efforts to become a major player in the international arena.

The Academia Europaea report bluntly warned that "there is a need to scale up the contribution of European scientists to human genome research."[90] The ESF report noted that the EC program lacked a single figurehead comparable to James Watson, and concluded, "European efforts appear fragmented, and command individually, fairly insignificant levels of support. . . . There is a danger that the European contribution to genome research may thus be dismissed as insignificant, that European researchers are ignored in the context of international meetings, and that the major players seem to speak for the entire genome community."[67] The reports concurred in their central strategic aims, but differed over tactics.

The ESF report called for stronger central direction and systematic peer review, specifically in the EC program,[67] while the Academia Europaea committee believed "funding for human genome research should remain primarily a national objective."[90] The academic scientists on the Academia Europaea committee recommended a decentralized approach with formation of a new coordinating body (Eurogene) analogous to the task force favored by OTA. ESF favored a bolstering of the EC and ESF multinational institutions to sustain a more coordinated approach. The ESF thus favored cultivation of the existing national research efforts rather than a more tightly coordinated effort. The ESF report got right to the point, arguing that HUGO had failed to articulate its role and pointing to inadequacies in EC program administration.

As 1991 progressed, Bodmer had his hands full organizing the eleventh Human Gene Mapping Workshop and attending to increased financial pressures at the ICRF, which forced him to lay off personnel. The genome project was an opportunity to demonstrate the unity of European science, but the struggle for control revealed the parochial interests of scientists and politicians in the various member states. Lennart Philipson commented candidly in *Nature* on "the animosity and struggle for power within and between the different European organizations involved in funding biological research."[91] Philipson's fervent desire for a coherent but ecumenical planning process for research was widely felt, but the mechanism to achieve it was elusive. It was far easier to specify the end than to devise the means.

In Canada, the genome debate recapitulated debates in Europe and the United States. Canadian genetics was highly esteemed, among the most internationally conspicuous contributions of Canadian science. Canada's genetic services were the envy of their U.S. counterparts, with particularly strong networks in British Columbia, around Toronto, and in Quebec. Charles Scriver of McGill University, who helped involve the Howard Hughes Medical Institute in genome research, tried to work the same magic in Canada. He was not alone. Ronald Worton from Toronto was on HUGO's council and was well known for his participation in the successful search for the Duchenne muscular dystrophy gene. Canadian geneticists angled for genome funds in the cool waters of distal North America.

In the spring of 1989, interested scientists gathered in Toronto to discuss the possibility of a Canadian genome project. The four meeting organizers (Ford Doolittle, James Friesen, Michael Smith, and Ronald Worton) produced a White Paper, including a long list of supporting scientists.[92] In October, the White Paper was sent to the government and Canada's three main granting councils. The response from Canada's minister for science was swift and positive, but given an austere budget climate, he wanted the granting councils to support the new venture with existing funds. The National Sciences and Engineering Research Council (NSERC) developed a model in which the project would be defined in advance and submitted a large application for funding. In June 1990, the White Paper's authors rejected this monolithic model, favoring an open-ended project like that pursued in the United States. They proposed funding from the Ministry of Science.[93] The Medical Research Council (MRC) agreed to champion this alternative, and the NSERC and the Social Sciences and Humanities Research Council formed an Inter-Council Human Genome Advisory Committee, chaired by Charles Scriver.

In early 1991, the committee recommended "the immediate creation of a genome program in Canada."[94] "Immediate" proved to be a relative term. As the genome project entered its third year in the United States and Italy, Canada's scientists became concerned about their ability to contribute to an international genome effort. As Norton Zinder from Rockefeller University observed, the genome project was "a really exciting global initiative in which Canada is noticeably absent."[95] A summary document prepared for policymakers by Scriver argued:

Without a Program, in one form or another, Canada: (i) will not long be competitive in medicine, agriculture, the pharmaceutical industry, or biotechnology, etc.; (ii) will not attract or keep the best workers in their fields; (iii) will be marginalized in all life sciences (biology) within the decade. With a Program there will be a sea-change in the way we do life sciences in Canada.[96]

The delayed response from government was only partially bureaucratic. Other factors also contributed, including a severe economic recession and a less elaborate set of connections between science and government. In the end, however, the government flagged genome research as a priority and gave it a fiscal boost.

On June 2, 1992, William Winegard, the minister of science and technology, announced the Canadian genome program at the International Biorecognition Conference. Ronald Worton was named director of a four-year program with $12 million of new funding, a $5 million commitment from the National Cancer Institute, and $5 million from the Medical Research Council.[97] Its goal was to "comprise a coherent, collaborative activity in mapping and sequencing of genomes, both human and nonhuman; the collection and distribution of data; the training of human resources; the development and transfer of associated technologies; and the evaluation of associated ethical, legal, and social issues." Like its U.S. and European counterparts (except that of the UK), the Canadian program earmarked a fraction of its budget, 7.5 percent, to look at social, legal, and ethical issues. The Pharmaceutical Manufacturers of Canada were expected to supplement this funding, but had made no final decision when the program was announced.[98] The hope was to match the $22 million in government funds, for a total of $42 million over the four years, or just over $10 million per year.[94]

The Canadian genome program thus emerged as a joint effort of three granting councils and the National Cancer Institute of Canada. It was a new independent effort with a management committee chaired by a scientist (Worton) and representation from all four agencies. Peer review committees reported to the management committee, which had the ultimate funding authority. This autonomous program was a departure from the way research was normally funded in Canada, an institutional innovation responding to the need for multidisciplinary research.[93]

As the genome debate become highly public in 1986 and 1987, and as more nations began to hop on the bandwagon, the need for international collaboration became apparent. One of the first responses was to hold international conferences, a natural reflex in the scientific community. The first major international conference on human genome research was organized by Santiago Grisolía of the Institute of Cytology in Valencia, Spain. Grisolía was a biochemist, but he was intrigued by the notion of genome research and fascinated by the cast of characters participating in the debate. In the summer of 1987, he began to organize a lavish conference in Valencia. The idea was initially to invite fifty or so scientists from around the world to discuss mechanisms to promote international collaboration. The conference soon took on a life of its own, as influential scientists from more and more countries, who could not easily be turned away, expressed interest.

The Workshop on International Cooperation for the Human Genome Project took place October 24–26, 1988. The participants were regaled with Spanish high life, including special "genome wine," conference T-shirts, and city buses displaying the conference logo. The workshop proved to be a reality check on what could honestly be expected from mapping and, especially, sequencing efforts.[99] It also revealed a flurry of simultaneous activity in many nations moving toward genome research efforts. Most participants first learned of the Japanese and Soviet genome efforts at this conference. The modest initial French effort and the multicenter yeast mapping and sequencing efforts under the EC were just getting under way. The meeting produced a one-page "Valencia Declaration" encouraging international cooperation, although the precise mechanism provoked a minor controversy.[100; 101] An early draft of the declaration urged involvement of both the Human Genome Organization (HUGO) and the United Nations Educational, Scientific, and Cultural Organization (UNESCO). Victor McKusick and James Wyngaarden chaired the final plenary session where the declaration was discussed, and the final document emerged with reference only to HUGO.

UNESCO had a new director-general, Federico Mayor, a Spanish biochemist. Mayor wished to renew UNESCO's commitment to science, which had lagged under his predecessor, Amadou-Mahtar M'Bow of Senegal. The United States and United Kingdom left UNESCO in 1984 under M'Bow's reign, strongly objecting to proposed press restraints under a proposed New World Information and Communication Order, but also alleging that UNESCO was an inefficient, expensive, and bloated bureaucracy. Mayor wanted to woo back both countries and saw genome research and other scientific efforts as good opportunities to do so. Scientific efforts were likely to prove less divisive and less purely political than many other programs within UNESCO's purview.

Congress held hearings in April 1989 about whether the United States should rejoin UNESCO. The genome project was mentioned prominently as an opportunity for UNESCO involvement and was listed as the first candidate under life sciences.[102] Mayor visited Washington a year later, in the wake of a State Department statement reiterating opposition to U.S. funding for UNESCO, hoping U.S. policy would change.[103] UNESCO continued to get positive reviews of its reforms, but difficult economic times and emerging isolationist sentiment in the United States undermined support to rejoin UNESCO.[104]

Following a February 1989 meeting of genome advisers from Europe, the United States, Japan, the USSR, and Australia, Mayor appointed a Scientific Coordinating Committee to steer UNESCO's genome program.[105] The UNESCO program began to take shape at meetings in February (Paris) and June 1990 (Moscow). The UNESCO program, budgeted for $260,000 over two years, emphasized training of scientists from countries that would otherwise be unable to participate (from the Third World and Eastern Europe, for example), helping Third World countries to participate directly in genome research, and exploring ethical issues through multicultural exchanges.[106–108] UNESCO contributed funding to several international meetings in 1990 and 1991, notably a high-profile meeting of genome luminaries in Paris, February 1991, and a second conference in Valencia, November 1990, which centered on ethical issues. The centerpiece of the UNESCO program was a short-term fellowship program cosponsored with the Third World Academy of Sciences. This program provided travel funds and stipends for young scientists from the Third World and Eastern Europe to seek training for several months in laboratories in Asia, Europe, and North America. Applications were reviewed just after the November 1990 meeting in Valencia; sixteen fellowships were awarded in the first year of what was to become an annual program.[109] UNESCO also contracted with the Third World Academy of Sciences to produce a directory of centers interested in or engaged in genome research.[110]

Jorge Allende, an energetic biochemist from Chile, maintained myriad collaborations with scientists in Europe and North America. He shepherded a resolution promoting human genome research through a meeting of the Latin American Network of Biological Sciences in Quito, Ecuador, June 29 to July 1, 1988. The resolution called for developed countries to ensure that the genome project enabled the participation of developing countries, such as those in Latin America. It also urged Latin American governments and scientists to assess local resources and organize into a regional network. The conference participants asked Allende to carry forward his plans for a June 1990 regional workshop on genome research in Santiago, Chile, and asked for partial funding from UNESCO.[111; 112] A June 1990 workshop, "Human Molecular Genetics and the Human Genome: Perspectives for Latin America," brought together scientists from twelve countries of Latin America and drew upon scientists from North America. It officially launched the Latin American Human Genome Program.

Allende also edited a special issue of *FASEB Journal* devoted to genome research throughout the world, and described the Latin American efforts to promote training and international collaboration with genome efforts in the technologically advanced countries.[113] It promised to organize into a mechanism for North-South cooperation; the workshop produced another resolution of similar tone and sought continued UNESCO support.[114] UNESCO hoped to see similar regional networks established in Africa, Southeast Asia, and the Middle East.

Populations residing in the Third World were centrally important to understanding human origins and genetic diversity. Consanguinity rates of over 20 percent were not unusual in some regions, particularly where traditional patterns of marriage prevailed under Islam, making recessive genetic diseases more common.[115] Several other religions and local customs encouraged consanguineous marriage. Not only were recessive genes more likely to be detected, but knowledge of consanguinity also presented an opportunity to map genes. The technique relied on the availability of genetic linkage maps to compare chromosome regions from distantly related relatives. If patients with a disease consistently inherited the same chromosomal region, the gene causing the disease was likely to be located there. A gene could thus theoretically be mapped with only a handful of patients, far fewer than needed for more traditional family studies.[116; 117]

Large families to enable gene hunting studies were, moreover, often found in the Third World simply because so many more people lived there. The search for the gene causing Huntington's disease was immeasurably expedited by one enormous family in Venezuela; hemoglobin disorders were studied primarily among those who lived in the malaria belt (the Mediterranean basin, Southeast Asia, and parts of the Middle East) or whose ancestry could be traced there.

The call for Third World involvement was thus more than an empty gesture. Genetic disease was a serious problem in several regions, ranking among the most serious health concerns in the Mediterranean basin and parts of southeast Asia. Hemoglobin diseases were among the major killing diseases over large expanses of Africa and southern Europe. Diagnostic methods derived from genome research would be quite useful in the developing world, but only if they were inexpensive and reliable enough. Efforts to study diseases that primarily affected Third World populations would likely be neglected, and technologies to make tests cheap and simple might well languish without help from the technologically advanced countries.

HUGO was the great hope for finding a mechanism sufficiently durable to sustain vigorous international collaboration but flexible enough to avoid bureaucratic encrustation. It was a brainchild of the genome elite, founded on April 29, 1988, at the first annual Cold Spring Harbor meeting on genome mapping and sequencing. Victor McKusick circulated among the conferees, describing Sydney Brenner's notion of a new international genome organization. An impromptu session was scheduled at five in the afternoon. McKusick urged the thirty or forty individuals gathered in Grace Auditorium to form an organization modeled on the European Molecular Biology Organization (EMBO). Watson rose to reminisce about the early years of EMBO, which had been modeled on the European Center for Nuclear Research (CERN). Lee Hood endorsed the idea of a new organization to foster international cooperation and argued for an open membership structure and a strict focus on science. Sydney Brenner suggested the name HUGO, for Human Genome Organization (although he said he personally preferred THUG). Brenner urged that membership be by election, rather than open to all, and nominated McKusick as president of the new organization. McKusick was elected by those present.

McKusick followed up on May 3 with a memo summarizing the discussion, which he sent to a core group.[118] Those on the list were all senior biologists and constituted a presumptive founding council.[119] Bodmer and Matsubara soon pledged financial support, and the group planned a September meeting to begin the next steps.[120] The Howard Hughes Medical Institute funded HUGO's initial meeting to organize more formally in Montreux, Switzerland, on September 6–7, 1988.[121] The council had since expanded to forty-two members from seventeen countries. The Montreux group decided to focus on databases, physical mapping and sequencing, nonhuman species, ethical issues, and human disease mapping. The council drafted a brief organizational plan and elected McKusick president. Vice presidents were elected from Europe (Bodmer and Dausset) and Japan (Matsubara). The Montreux meeting was occupied in part with deciding between open and elective membership options. Those advocating elective membership won the day, although HUGO progressively diminished restrictions and eased the process of election between

1988 and 1990. HUGO's next efforts aimed to secure a financial base. Walter Gilbert was elected treasurer, and his first job was to find operating funds.[122]

HUGO weathered 1989 with great difficulty. Cash was scarce, and none of the organization's goals could be accomplished without it. In December 1989 there was only $25,000 in the bank. At a politically delicate meeting in Bethesda, Bodmer was elected president.[123] McKusick was named founding president, and the vice-presidential posts were designated by region. Matsubara stayed as vice president for Japan; Charles Cantor became vice president for the United States and Mirzabekov for Eastern Europe. Bronwen Loder performed most staff work in London, and ICRF footed the bill. Diane Hinton staffed the Americas office, on loan from HHMI. Loder and Hinton prepared a series of funding proposals to seek private funding so that HUGO could begin to operate.

The funding picture brightened in 1990, when the Wellcome Trust in the UK and HHMI in the United States both announced substantial multiyear grants to HUGO. Michael Morgan, from the Wellcome Trust, announced the award for 1990 "in the order of £200,000" ($350,000)[124] and said there would be further support over the next two years. HHMI announced a $1 million, four-year award to HUGO on May 3.[125] A week later, Cantor published a letter in *Nature* that described HUGO plans to coordinate physical mapping efforts chromosome by chromosome.[126] He proposed building on the existing Human Gene Mapping Workshops, which had committees for each chromosome and met at one large conference every other year.

HUGO hoped to go beyond biennial meetings to more frequent meetings focused on putting together maps of regions or chromosomes.[126] In July 1990, Wyngaarden was appointed executive director of HUGO, becoming the first permanent staff member. In August, the HUGO council agreed to Cantor's proposal to focus on chromosome-specific workshops. The HUGO Americas office would apply for funding from government agencies and private sources to hold the meetings, standardize the reporting format, and ensure speedy publication of individual workshop reports. HUGO would also search for a single facility that could be consistently used, enabling computer links to databases and possibly accumulation of a library and other resources.[127] As 1990 drew to a close, HUGO struggled to consolidate a financial base, to hire a staff in three offices (Bethesda, London, and Osaka), to track international developments, and to broker agreements on sharing data and materials.

As the genome project officially began in late 1990, it was clearer that HUGO needed to succeed than that it would actually do so. It proved difficult for an international organization to attract government funds. Most governments limited grants to domestic organizations, requiring HUGO to incorporate in each country or to seek a formal intergovernmental agreement. This was a slow and costly process. Many private funders also emphasized domestic interests. Another problem with private funding was its predominance in the United States, with one nation home to most foundations able to contemplate grants of a size commensurate with HUGO's task (the Wellcome Trust in the UK and the French muscular dystrophy funds were notable exceptions). It was also extremely difficult to devise a staffing pattern that would retain substantial autonomy in each regional office (and hence attract good staff) but also enable coordination among offices on three continents. Most international precedents for international scientific projects were supported from the start by multilateral agreements or existing international organizations such as the UN or international scientific unions. These usually built from national government science agencies, coordinating science administration rather than scientists themselves.

HUGO attempted to reverse this strategy by starting from a private funding base and then cultivating government funding by applying for specific projects. The structure of HUGO was yet another new precedent that the Human Genome Project hoped to set. HUGO aspired to coordinate various national governments' programs through an organization established and directed by scientists. HUGO was based on the premise that the balance of power had shifted, so that scientists could exercise power over an international research program by creating their own organization. In a generous assessment of HUGO's accomplishments, John Maddox noted that "despite its modest successes so far, HUGO will find it has to keep running hard if it is successfully to play the ambitious role it has set for itself. The long-term objective is to command the respect of the world's genome sequencers individually."[128] HUGO had a long way to go, but then so did the genome project.

By 1993, the genome project was a well-established international effort. Nine countries and the EC each had one or more human genome programs. HUGO was five years old. At least six countries were actively considering whether to start genome programs (the Netherlands, Australia, Chile, Sweden, Korea, and New Zealand).[20; 67]

In a 1989 column, political columnist George Will urged readers to "pay at least as much attention to science news as to political news. Political choices are made in contexts that politicians cannot choose, and the contexts are increasingly shaped by science."[129] Governments might not succeed in capturing the benefits of genome research for their domestic economies, but they could certainly try.

Management of the genome project was only partially within the reach of national governments, and yet government funding was, in most countries, its principal sustenance. The conflict between international scientific aspirations, to use the human genome as a vehicle for international cooperation, flew in the face of intense nationalist fervor premised on economic competition. The rhetoric of economic nationalism pervaded arguments across the Atlantic, but took a backseat to stronger and more established scientific norms of collaboration for the most part. Japan was a special case that brought trade tensions into sharp profile.

The genome project developed in a period when Western Europe sought unity and the former barriers to Eastern Europe were being dismantled. The early European genome efforts took origin almost simultaneously with the U.S. initiative. The Italian program, for example, can be traced to Renato Dulbecco's 1985 Columbus Day speech in Washington. The British and Russian programs developed in parallel with the NIH effort, formulated partially in response to DOE plans. The European Community program began soon after the first genome debates took place in the United States. The early European efforts thus trace their roots to the same sources as the U.S. effort.

A second wave of genome efforts was formulated in part to respond to the U.S. effort. In the context of a drive toward European economic and political unification, the rhetoric of keeping up with American programs crept into justifications for the genome program. This provided political justifications—preservation of national prestige and maintenance of a position in a field related to biotechnology—in addition to the original scientific rationale. Commitment to a genome research program became not only and end in itself, but a necessary investment to thwart American domination of an important frontier. The Canadian program shared this political justification with many of its European counterparts.

In the USSR, and then Russia, the genome project was relatively spared (although still seriously affected) by the turmoil that halted much science. This was because the originators of the program were associated with the reform movement from the start, the economic rationale behind the genome project was used from the beginning, molecular biology was widely regarded as a central filed for any future biotechnology, and the genome project carried a substantial fraction of all molecular biology. Molecular biology in the USSR had not attained the size and scope of its Western counterparts, in part because it had never fully recovered from the ravages of Lysenko. Only one full generation had passed. As the USSR dissolved, the Russian components of the Academy of Sciences attempted to sustain a genome program.

The importance of private initiatives proved an important feature of developments in Europe. The privately organized CEPH consortium greatly facilitated genetic linkage mapping. CEPH became even more powerful when it forged an alliance with the private muscular dystrophy association AFM. The resulting Généthon became Europe's most notable innovation in genome research, contributing to both genetic linkage and physical mapping. Its high-tech approach and heavy emphasis on automation created a prototype for similar centers set up later in the United States. In the UK, the private Imperial Cancer Research Fund became an equal partner with the government Medical Research Council. It also quickly adopted approaches intended to foster automation. The Wellcome Trust was an early supporter of the Human Genome Organization and came in to rescue an underfunded transatlantic collaboration. In 1992 and 1993, it stepped up its commitment to genome research, soon dwarfing the government contribution, and bringing the UK effort to rough equivalence (per capita or relative to the size of the economy) with that of the United States, the only country that could claim such parity. The private support for genome research in the UK built upon a long-standing tradition of excellence in genetics and molecular biology greatly in excess of the nation's economy or even its biology budget.

These early contributions from nonprofit organizations in Europe paralleled the early involvement of the Howard Hughes Medical Institute in the United States, but their impact was relatively greater. This was, in part, because their relative financial contribution was greater. The government contributions to genome research in Europe were, by and large, imposed on relatively inflexible research bureaucracies, and the infusions of new funding were small by comparison to those in the United States, even after adjusting for the relative size of the national economies. The private funding sources were more flexible and their financial contributions relatively greater.

The hope for a European science may have driven some interest in genome research, but the structures to unify science were relatively weak. Most planning took place within the structure of individual government science and technology ministries. The European Community programs were notable exceptions. The EC successfully coordinated a highly complex collaboration that sequenced yeast chromosome 3, a major accomplishment, and previous EC biotechnology instrumentation programs figured in the creation of Généthon. The human genome component of the EC program was slow to start, however, because of concerns about its social implications, and the EC's genome research budget was no larger than the national programs. The EC efforts were thus important, but hardly sufficient to coordinate European programs to the same degree as NIH-DOE joint planning.

The rhetoric of European unity failed to translate into a carefully orchestrated genome research program, but it did succeed in garnering funds from various national governments. It also extracted commitments from the EC and various organizations supporting biological research throughout Europe, such as the European Molecular Biology Organization and the European Molecular Biology Laboratory. The desire for a more coherent program also led to cooperation among European governments. The European Science Foundation and Academia Europaea both commissioned reports on how to proceed, but even without those reports, the number of joint meetings and the degree of collaboration were unusual. The genome project thus served as an example of progress toward unification of science, but also an illustration of how far there was still to go.

15

Japan: A Special Case

I N JANUARY 1872, thirteen-year-old Chokichi Kikkawa disembarked from the magnificent steamship *America* onto American soil. The young man looked for the first time at San Francisco Harbor, at one of the most beautiful cities on the continent that would be his home for the next eleven years. Kikkawa thus began his education in the ways of the West. He was one among one hundred Japanese in the Iwakura entourage, among the first delegations to venture out from Japan four years after the February 1868 enthronement of Emperor Meiji.[1] It was a revolutionary period, and the Iwakura entourage was among the first of many human connections to America established after Commodore Matthew Perry forced open Japan's doors.

Takayoshi Kido, a former samurai, was also on board the *America*. Kido was one of the three principal leaders who restored power to the emperor in the tumultuous 1860s.[2] Kido was sent to the United States to renegotiate the terms of treaties signed in 1854 and 1858, documents that granted the United States access to Japan. The *America*'s voyage began with great fanfare at 1:00 P.M. on December 23, 1871, when it left from Yokohama to the echoes of a nineteen-gun salute.[3] Kido did not succeed in renegotiating the treaties, but the trip produced a wealth of knowledge about the outside world for a Japan craving just such information.

Commodore Perry, a hero of the Mexican War, ended Japan's two centuries of nearly complete insulation from foreign influence that began with a 1638 decree from the Tokugawa shogunate. When the American "black ships" sailed into Naha Harbor in May 1853, the threat of foreign military power disrupted Japan's internal order. Superior arms and foreign technology shattered the crusty feudal regime. Perry's fleet included steamships, which were quickly replacing the famed American clippers, and he brought novel technologies—pistols and rifles, a telegraph, and even a locomotive engine complete with one car and several hundred yards of track.[4] Perry and his new technology thus cracked the wall that had separated Japan from the rest of the world. He ended the Tokugawa era—two and a half centuries of rule under the shoguns.

The technology of sea travel overcame the geographic barriers that kept Japan separate; advanced foreign military technology made Japan vulnerable.

The importance of technology left a strong imprint on the newly exposed Japan. After a decade and a half of chaos and intrigue, Emperor Meiji gained power in 1868. Japan's leaders recognized the degree to which the nation had fallen technologically behind the West and aggressively promoted policies to catch up. The emperor asserted that "knowledge and learning shall be sought after throughout the whole world, in order that the status of the Empire of Japan may be raised higher and higher."[5] Thus began a trend that persisted throughout the twentieth century. Those who went abroad were sources of knowledge useful in modernizing Japan. Kido and Kikkawa were among the pioneers, human bridges from Japan to the West.

Takayoshi Kido returned to Japan in 1873, maintaining his role as imperial councillor and helping to dismantle feudalism in Japan. He became a leader in reform movements and helped establish a constitution modeled on Germany's, doing much of the writing himself. Kido was regarded as "the most liberal and humane member of the government, even as his power waned."[2] Among the "Meiji triumvirate" who engineered the emperor's resurgence, Kido alone died of natural causes. The second committed *seppuku* (ritual suicide by stabbing the viscera) later in 1877, and the third was assassinated by disgruntled samurai the following year.[2] Kido spent his last years criticizing government policies that impoverished his former samurai brethren and the peasantry. The teenage Chokichi Kikkawa had a less treacherous if more meandering road home from America.

Young Kikkawa traveled by train from San Francisco to Boston, where he stayed from March to August 1872 with the Rev. Charles Nathaniel Folsome, who ran his household according to rigid puritanical precepts. Kikkawa learned to revere strict personal habits and dedication to objective truth. He then spent a year at the Rice Grammar School and another four years at Chauncy Hall School. He graduated *summa cum laude,* with an award for English composition. In June 1879, Kikkawa passed his entrance examination for Harvard, and matriculated there in October, living in 23 Matthews Hall. He was among Harvard's first Japanese graduates, perhaps its first. He graduated in 1883 and set sail for Europe.[1]

Kikkawa returned to Japan in December 1883, twelve years after the voyage on the *America* began. He was persuaded to join the Foreign Office by the foreign minister, serving in Tokyo from 1883 to 1886 and then in Germany for four years. He returned to Japan in 1890, joined the House of Peers, and was married in 1892.

Kido's and Kikkawa's families reunited to form another human bridge to the United States on June 28, 1929, with the birth of Akiyoshi Wada—great-grandson to Kido and grandson to Kikkawa, from a different branch of the family tree. His father, Koroku Wada, was president of the Tokyo Institute of Technology, dedicated to advancing Japanese technologies in space and aeronautics. Technology ran in the family. Akiyoshi Wada studied at the point of intersection between physics and chemistry and ultimately worked in biophysics. His contact with the United States was solidified when he spent 1954 to 1956 working on protein structure with Paul Doty of Harvard, where he learned "the basic spirit of science, which is to serve mankind."[6]

Wada became a critical figure in DNA sequencing before the genome project was conceived as a special program. In 1981, Wada was appointed chairman of a project named "Extraction, Analysis, and Synthesis of DNA." This was supported by a special fund from the Science and Technology Council of Japan, funded through the Science and Technology Agency (STA). The project had two aims: to reduce the tedium of biological research and to engage the interest of companies from outside biology, including firms whose technology base was robotics, electronics, computers, and materials science.[7] Wada's strategy was to automate existing protocols used in molecular biology rather than to invent entirely new approaches.

The project focused on DNA sequencing because it was clear it would become increasingly important. Japan was greatly interested in robotics and automation, and it was thought relatively straightforward to automate laboratory processes involved in DNA sequencing.[7-9] Wada enticed Seiko, Fuji Photo, Toyo Soda, Hitachi, and Mitsui Knowledge Industries to join the project team.[7] This first phase, 1981–1983, was funded at ¥910 million ($3.7 million at the then-current 240¥ / $). It produced a microchemical robot made by Seiko and a standardized electrophoresis gel system made by Fuji Photo. In 1984, the project was funded again under another branch of STA, now titled "Generic Basic Technologies to Support Cancer Research" and funded at ¥450 million (or $2.05 million at the then prevailing rate of 220¥ / $). Seiko developed a DNA purification system and another microchemical robot, Fuji began to mass-produce its gel, and Hitachi developed a prototype DNA sequencing machine.[10]

The base of operations was moved as a "Research Project on Gene Composition" to the RIKEN Institute in Tsukuba Science City (officially, the Institute of Physical and Chemical Research, or RIKagaku KENkyusho). The RIKEN Institute was established in 1917, during the Meiji era, with support from the imperial household, government, and private sources. Just before and during World War II, RIKEN was the home of Japan's efforts to develop an atomic bomb,[11] providing yet another historical link between genome research and bomb projects. The Tokyo laboratories were largely destroyed in the war, and the institute was reestablished in 1958.[12]

In 1985, Wada's DNA sequencing project was swept into the debate about a human genome project. The connection was Charles DeLisi, who explained:

In 1985 when I was director of the Department of Energy's Health and Environmental Research programs, I was impressed by the need for a more efficient and cost-effective approach to DNA sequencing. When we started developing the Human Genome Project . . . it was picked up by the American press as a new and bold initiative. In fact, it was not at all new in Japan. I received a note from Minoru Kanehisa [who had

worked at the GenBank database at Los Alamos National Laboratory before moving
to Kyoto University] indicating that a similar project had been initiated there five years
earlier. . . . My old colleague, Professor Wada, whom I had known through his distinguished contributions to a somewhat different area, turned out to be the person to
speak to. . . . It became obvious that he had already done what we were just beginning
to think about. . . . Just this one initiative alone would have been sufficient to rank
Professor Wada as a major figure in world science, and a hero of Japanese science and
technology.[13]

Akiyoshi Wada, who headed a Japanese project to
automate the biochemical processes involved in determining DNA sequences in the early 1980s, later
led attempts to organize an international genome
project. Wada's efforts were particularly influential in
stimulating the U.S. program. *Courtesy Akiyoshi Wada*

In late 1986, Wada came to the United States, seeking support for an
international DNA sequencing effort, hoping to consolidate an international
base of support for his project. He met with those involved in the nascent
human genome debate in Washington, at the National Institutes of Health,
the Department of Energy, and the Office of Technology Assessment. He also
visited several research centers, including Los Alamos National Laboratory,
and returned to Los Alamos to speak at a workshop on robotics in January
1987. Wada's vision was a series of international centers dedicated to rapid,
inexpensive DNA sequencing. He thought that Japan would be an early leader
in automated DNA sequencing, and it would be logical for Japan's efforts to
concentrate on that strong suit. He intended large-scale DNA sequencing to
be a unifying force, bringing the United States and Japan closer together
through collaboration. The politics of the day would not have it so.

Wada's interest in visiting the United States was not only to form new
collaborations with American scientists, but also to generate political support
for his project in order to bring foreign pressure on Ministry of Finance
bureaucrats back in Japan. Wada was following an established strategy, *gaiatsu,*
using foreign presence to pry funds loose from the Japanese government,
notoriously stingy in its support for basic biology. Press reports from abroad

could be used as ammunition in the battle to capture funding, essentially embarrassing the Ministry of Finance into loosening the purse strings.

Wada obtained continued funding for the RIKEN project, although not at the levels he desired. He had hoped for a major commitment to a large sequencing center; what he got was a continuation of the RIKEN research effort. Indeed, his colleagues in Japan faulted him for raising hopes too high and for exacerbating tensions between the United States and Japan by scaring Americans with a high-profile project.[14; 15] Critics also pointed out that industrial partners were abandoning the project, that Seiko was not marketing its machines, and that Hitachi was selling its sequencer only in Japan. Hitachi left the project, and Fuji was about to do so. The Fuji ready-made gels for DNA sequencing were test-marketed, but then quietly withdrawn. Wada became preoccupied with other projects, as he became dean of the faculty of sciences at the University of Tokyo and worked toward reforming science in Japan.

In 1988, the reins of the automated sequencing project were turned over to Koji Ikawa and Eichi Soeda at RIKEN. They reassessed the technical objectives and concluded that initial cost and speed estimates were too optimistic.[16] The goals were scaled back to a sequencing capacity of 100,000 DNA base pairs per day, down from Wada's million. By 1989, the automated DNA sequencing project had been under way for eight years, yielding a set of machines capable of sequencing roughly 10,000 bases per day.[16] A new series of projects, to automate cloning and other processes in molecular biology, was commenced under Isao Endo of RIKEN. Endo's project bore fruit in June 1991, when he reported it had attained a potential raw output of 108,000 base pairs per day.[17; 18] The project involved ¥ 600 million ($4.5 million) from STA over the decade, and an unknown total of in-kind expenses from Hitachi, Seiko, Cosmic, Mitsui Knowledge Industries, Tosoh, and Fuji Film. The sequencing part of the system, humorously named Human Genome Analyzer or HUGA, was a fluorescence-based DNA sequenator made by Hitachi.

STA also began to fund selected projects in universities, including chromosome mapping on chromosomes 21 and 22 under Nobuyoshi Shimizu at Keio University. The annual STA genome budget was approximately ¥ 200 million ($1.3 million) in 1989 and 1990[16; 19–25] and rose to ¥ 1.2 billion ($8.6 million) in 1991.[17] The 1991 STA program commenced a formally approved extension of the previous pilot projects, with specific component projects led by Eichi Soeda (RIKEN), Masaaki Hori (National Institute of Radiological Sciences), Isao Endo (RIKEN), Joh-E Ikeda (Tokai University), and Hiroto Okayama (Osaka University).[26]

The RIKEN DNA sequencing project was pulled into the vortex of American debates about genome research. It was not used to promote cooperation, however, but rather to goad the U.S. government into funding the American genome project. The Japanese sequencing project was held up to Congress as evidence that Japan had a five-year lead in a crucially important technology. This surfaced in the first, critical congressional hearing on the DOE project.[27]

Congressman David Obey noted in hearings associated with NIH's first genome budget that "given the competitiveness issue which we have in this
country, and the trade issue . . . it sounds to me like this argument is about to
be couched in terms of them versus us."[28] Indeed it was.

Senator Pete Domenici commented on the Japanese genome project in his
opening statement for a field hearing on the genome project in Santa Fe, New
Mexico, in August 1987: "It came to me very quickly during the debate on so-
called trade and competitiveness that an issue such as the mapping and sequencing of the human genome . . . while terribly important in terms of our
understanding diseases and being able to cure them, that it was becoming a
very, very significant competitive situation, vis-à-vis at least the Japanese, but
not limited to them."[29] Staff in Congress (not Domenici's) discussed adding
riders to NIH appropriations forbidding purchase of Japanese instruments
under federal grants, or restricting NIH funding for foreign postdoctoral and
graduate students. These discussions did not bear fruit, but their existence
clearly indicated the dominant congressional concerns.

Wada's bridge had become a wedge, driving the countries apart. In the
United States, the Japanese effort was seen as a technological threat, and
another instance of Japan neglecting basic science in order to promote work
on development of something that could be exported and sold, in this case
DNA sequencing machines. The Japanese genome project got stuck in the
tarbaby of U.S.–Japan trade tensions.

American perceptions of the Japanese genome project as a biotechnologi-
cal Trojan horse—a premeditated assault on one of the remaining bastions of
U.S. preeminence—were grounded more in loose historical analogies with
automobile manufacture and electronics than in direct observation of the policy process. The Japanese genome program as a scientific effort was largely the
result of scientists aspiring to join the international ranks. Industrial partners
were, at least in the opening phase, more reluctant participants than instigators. The genome program was more a dream of what Japanese science could
become than a cornerstone in some grand economic plan.

Ken-ichi Matsubara tried to broaden Japan's genome effort by giving it a
stronger academic grounding. Matsubara was the director of the Institute for
Molecular and Cellular Biology at Osaka University. He got his bachelor's and
Ph.D. degrees from the University of Tokyo, then did postdoctoral fellow-
ships at Harvard and Stanford. Most of his work was at the interface between
molecular biology and biochemistry, looking at the process of cancer formation in the liver and also at how hepatitis viruses infected liver cells. He became
director of the Osaka institute in 1982.

Following press reports of emerging genome projects, Matsubara visited
the United States with a small group in February 1988. He was an adviser to
the STA project, but had hopes that the Ministry of Education, Science, and
Culture (MESC, commonly known as Monbusho) could also be drawn in,

perhaps on a larger scale than STA. Monbusho funded the vast majority of academic science in Japan, principally at the nine major universities and forty or so smaller universities throughout the prefectures. Monbusho's university base contrasted with STA's emphasis on government-funded laboratories and institutes. Matsubara hoped Monbusho would make genome research a priority area, beginning in the 1989 budget.[23]

When Matsubara returned to Japan, he became chairman of a group advising Monbusho. Monbusho did indeed smile favorably upon genome research, giving it ¥300 million per year for 1989 and 1990 ($2 million per year at ¥150 / $). This commitment to a pilot project, quite significant by Japanese standards, did not seem as grand to those outside Japan. The issue came to a head in the fall of 1989.

The dark clouds gathered slowly. The potential for conflict loomed in the background of the first international conference on the Human Genome Project, held in Valencia, Spain (October 1988). As noted earlier, this lavish conference was organized by Santiago Grisolía, a Spanish biochemist greatly intrigued by the genome project. By the time of the Valencia meeting, the DOE project was beginning its third year and NIH its second. Italy, the UK, the USSR, and the European Communities had described their respective genome programs at a session presenting efforts from around the globe. Yoji Ikawa described the STA sequencing project and mentioned the Monbusho project, then in planning phase. The annual budget was comparable to Italy's, but far lower than those of the EC, USSR, and UK programs. There was some comment to this effect after the session, but in muted tones and only outside the formal session.

The next major meeting on international collaboration was held in Moscow, in June 1989. The meeting, cosponsored by UNESCO and the USSR Academy of Sciences, featured James Watson, Walter Gilbert, Victor McKusick, François Gros, Charles Cantor, and several other luminaries. Soviet academicians Andrei Mirzabekov and Alexander Bayev hosted the meeting, considering it an opportunity to showcase the Soviet genome project. By then, the STA project in Japan had secured another two-year funding commitment and the Monbusho project was getting underway. Matsubara had completed a pair of documents outlining the Monbusho strategy,[30] and a similarly favorable report had been completed for STA.[31] Matsubara could not attend the Moscow conference because a colleague died, leaving Ikawa as the lone Japanese representative.

Watson pulled Ikawa aside on the steps of the Hotel Ukrainia, one of Stalin's seven "wedding cake" skyscrapers in Moscow, to indicate his irritation. Watson asserted that Japan was now a great nation with vastly greater resources than any country except the United States, and Ikawa should go back to Japan and tell his government to put more funding into genome research, or there would be problems. Watson hinted that Japan was becoming isolated by its niggardly ways, and if need be, the United States would make access to databases and research materials difficult. Absent a bigger commitment, Japan would be shut out of planning the international enterprise. Honest Jim Watson was apparently carrying on the mission of Commodore Perry.

Watson was in the process of deciding whether to visit Japan later in the year. Matsubara had invited him to visit Japan to meet fellow scientists and government officials, among other things to convince them of the importance of genome research. Watson was ambivalent because of his experience years earlier, when Itaru Watanabe attempted to establish an Asian molecular biology organization. Watson went to Tokyo to support that effort, but it came to naught, devolving into a nasty fight between Watanabe and government officials. Watson recalled a meeting when "they behaved like twelfth-century shoguns." Watson did not want to be similarly used again. He indicated that he would come only in response to a signal that the government was ready to deal.

In July, Watson wrote to Matsubara that he would not be coming to Japan. He went on to urge Japan to ante up by supporting the Human Genome Organization and to bolster basic science funding. Watson testified on October 19, 1989, before the Subcommittee on International Scientific Cooperation of the House of Representatives.[32] Japan inevitably came up. Soon Leslie Roberts, a reporter for *Science,* found out about the Watson letter to Matsubara by talking to American scientists embarrassed about it. Watson declared, "I'm all for peace, but if there is going to be war, I will fight it."[14] After this outburst, other scientists opened up and released copies of Watson's letter. Roberts quoted from Watson's letter and Matsubara's reply in a *Science* news feature, "Watson versus Japan."[14] This made the spate public, and it was echoed elsewhere.[15; 33–35] In the United States it was seen as another Watson temblor; in Japan the quake was larger and left more rubble. In *Asia Technology,* the story was about "The Human Gene War."[15]

John Kendrew spoke for many in the scientific community when he scribbled a note to Watson: "I hope this report is not true, because if it is, you should be ashamed of yourself! All the best for 1990."[36] Watson shot back a crisp reply:

I am not ashamed of myself. The issue now is whether to make the human sequence data available to all before those labs which have generated megabase stretches have a first go at their interpretation. Your LMB [Laboratory of Molecular Biology, Medical Research Council] did not distribute broadly its viral sequences before they were published. . . . The UK seems resigned to becoming economically a Japanese colony, but there are many of us in the States who will fight like hell to prevent a similar situation. With the Cold War gone, at last we have a chance to ask where we are going. See you in Paris (?)[37]

Japanese scientists were particularly incensed at the paternalistic tone of Watson's letter to Matsubara. While allowing that he was correct to say the

Japanese government was giving insufficient funding for basic research, they deeply resented the public humiliation. Many ascribed Watson's remarks to racism or fashionable "Japan-bashing." This underestimated the degree to which Watson captured what irked Americans and Europeans about Japanese science policy. Watson had expressed admiration for Japan, and had urged that the United States imitate its macroeconomic policies at a Harvard University talk a year earlier.[38] If his words were Japan-bashing, they were not of the simple-minded xenophobic variety.

Several Japanese scientists saw a policy advantage for them as a result of the furor. Privately they hoped the publicity would put pressure on the Japanese government to increase research funding. Michio Oishi noted that Watson's strong focus on genome research and his insistence that Japan do its part made Japanese government bureaucrats take notice.[39] This was again a classic instance of *gaiatsu*, bringing foreign pressure to bear on the Japanese government when domestic pressure was ineffectual. If the strategy was to create foreign pressure for government funds, then Watson was the ideal messenger; it certainly made the papers.

A year after his unpleasant encounter in Moscow, Ikawa explicitly acknowledged the impact of Watson's statements in a January 1991 summary of Japanese genome efforts, saying that "Dr. J. D. Watson has criticized the inadequacy of Japanese participation." Ikawa closed his article by noting that Watson had visited his laboratory twelve years before and expressing his hope that "this review will aid him and others outside Japan to understand this country's slow but steady movement forward in this important field."[40] The article was clearly aimed at Watson.

The U.S.–Japan dispute did not become news until late 1989, but it was anticipated in Washington for some time. Aki Yoshikawa, who wrote a commissioned paper on Japan for the Office of Technology Assessment in 1987, concluded the paper with the observation that "efforts to cooperate by well-meaning scientists from the two countries . . . may end up in conflict."[41] The conflict grew out of competing goals—free international exchange of scientific data, on one hand, and economic nationalism regarding biotechnology, on the other. Factions espousing free scientific data flow and others emphasizing economic competition contended for the upper hand in both countries. Yoshikawa noted:

Japanese scientists do not have credibility problems—the quality of Japanese science is well recognized elsewhere in the world. Japanese scientists are eager to conduct research with American colleagues. However, it is the process of Japanese policy formulation—the role of bureaucrats and the close ties between government and business in Japan—that foreign observers have sometimes questioned. . . . Although friendships within the scientific community may be sufficient to bring an informal "small" project to fruition, a "large" science project that requires commitments from more than two governments as well as scientists is a more complicated matter.[41]

A mechanism to match benefits to contributions, with each country paying a proportionate share, was difficult to envision. Governments were pursuing policies to bolster national commercial interests. Industrial competitiveness was the policy buzzword of the day, and it was much heard in Washington.

The genome project, because of its timing and its high-tech, whiz-bang aura, was coupled directly to biotechnology, and thence to industrial competitiveness. In covering U.S.–Japan competition in technology, *Time* gave considerable weight to the genome project, indeed more than its due: "The centerpiece in the U.S. response is the government's mammoth effort, known as the genome project."[42] *Roll Call,* a newspaper widely read on Capitol Hill, ran a special feature on competitiveness in July 1989, centered on the Japanese threat to American dominance in technology. Senator Domenici had the lead article, and cited the genome project as a case where the United States had to remain in front.[43]

It was clear that the United States was preeminent in biotechnology. Analysts differed markedly, however, in assessing the future. Some, including the U.S. Department of Commerce, believed that the major contenders would arise in Europe rather than Japan.[44] Others, including the Office of Technology Assessment, believed Japan would figure more prominently.[45] The debate centered on two factors with opposing trends: the importance of deliberate government interventions to foster technology and the importance of a solid scientific base.[46] In other industries, Japan had successfully promoted economic expansion through targeting specific areas of technology; biotechnology was now targeted for special treatment. The United States was clearly ahead in science, but it was a matter of debate whether science could be confined within national borders.

Some believed that Japanese government policies would have little impact on biotechnology,[47] while others thought Japan's targeting of biotechnology would confer a critical edge to Japanese industry.[48–51] A 1991 OTA report noted that "there are two prerequisites for a nation to fully compete in biotechnology: (1) a strong research base and (2) the industrial capacity to convert the basic research into products."[46] The United States was clearly preeminent in research, but the health of its industrial capacity and policies to foster the translation from science to product were less certain. OTA noted that the U.S. science base remained the world's most robust. Japan had strong industrial policy direction and a strong track record in applying new technologies in other industries, but was relatively weak in the industrial sectors most relevant to biotechnology—pharmaceuticals, agriculture, and environmental remediation.[46] Europe had a strong research base, although it was far more fragmented into national programs, and had strength in the relevant industrial sectors, but the climate of public opinion was turning sour for biotechnology regulation.

One salient feature of U.S. biotechnology was a group of more than four hundred dedicated biotechnology companies. These were relatively small and new firms, most of which sprang up between 1980 and 1984. Other countries had a few such firms, but there were far more in the United States. These companies had grown in parallel with the power of genetics and the recognition that the science would find commercial application. A web of intertwined relationships developed between U.S. university scientists, small and large biotechnology companies, and Japanese firms, in all combinations and in a complex mix of arrangements.[52; 53] Analysis of how these arrangements would play out was confused by the ambiguous status of multinational corporations, whose interests resisted simple classification according to where the headquarters were located.[54-56] The main conclusion, however, had to be that these firms were the engines of innovation in U.S. biotechnology, but their precarious financial position invited investment by foreign firms (and larger U.S. firms). Small biotechnology companies were a national asset, but were a channel of technology transfer abroad, and were vulnerable to purchase or domination by foreign forms.

A 1992 report from the National Research Council studied the flow of ideas between the United States and Japan, principally via agreements between large Japanese firms and small U.S. biotechnology companies. It focused on the main trend—ties between small U.S. dedicated biotechnology companies and large Japanese pharmaceutical, chemical, and agricultural companies. The report concluded:

Despite the strengths of the U.S. biotechnology industry today, the NRC working group is not sanguine about the future and the ability of the U.S. biotechnology industry to compete in the twenty-first century. Significant potential problems were identified that cannot be adequately addressed on an ad hoc basis because active collaboration of government, industry and universities will be needed.[57]

Given the high stakes of the outcome and the vast cultural disparities, the debate about international competition and biotechnology was vigorous and emotional. The genome project was but a small vessel sailing through the rough seas of international trade tensions.

The Watson-Matsubara correspondence embodied the global conflict in microcosm. Matsubara bristled at the fact that the U.S. government created a program and then apparently expected the rest of the world to play by its rules. Watson resented the fact that the world's second most wealthy nation was apparently content to freeload off other countries' research, focusing only on those aspects that promised economic payoff in the form of exportable goods. The genome project caught the United States reeling economically from a decade of mismanagement of its banking and financial sectors, overwhelmed by debts, excessive defense spending, and the inexorable expansion of federal entitlement programs. Japan, in contrast, exuded confidence while the genome project was first formed. The 1980s were stacked atop four decades of consistent economic growth that transformed Japan into the second-largest economic power, one with a growth rate well in excess of that of number one, the United States.

Japan was completing a decade of remarkable economic expansion, although the rate of expansion slowed in 1989 and 1990 and declined even more sharply through 1991 and 1992. Unbridled optimism slowly gave way to a more temperate view. Japan nonetheless boasted seven of the world's ten largest banks and assets valued greater than those of any other nation, in large part because of the extremely high cost of land and housing on the densely populated islands. The future of the Japanese economy was less certain than it had seemed in the era of unrestrained self-confidence, as competition within the Pacific basin intensified, Japanese stocks plunged by more than a third, and real estate values—the main asset of many economic heavyweights—became unstable, having become so outrageously high that government action to control spiraling land inflation seemed inevitable.

The United States, for its part, had considerable doubts about its economic future. In 1991, the nation reasserted its military might in the Persian Gulf, but the underlying economic strength was more questionable. In 1991, U.S. biotechnology firms raised a record $17.7 billion in public offerings, and biotechnology stocks rose,[46] but a protracted recession cast a pall over national policy and intensified doubts about the underlying strength of the U.S. economy. The debate about long-term competitiveness remained a subject of much speculation, particularly relative to Japan. The U.S.–Japan relationship had to adapt to circumstances starkly different from those prevailing during the 1950s and 1960s, when most of the framework for interactions had been constructed.

As Japan in the 1970s and 1980s demonstrated its technological prowess, becoming the world leader in consumer electronics, automobile manufacturing, and other areas, its technological preeminence was oddly wedded to neglect of basic science.[58] Research was much more heavily supported by government in the United States than in Japan. In the United States, government supplied almost half of all research and development dollars, compared to one in five for Japan.[59] Specifically in biotechnology, the U.S. federal government expended an estimated $3.5 billion in 1990, and industry another $2 billion.[60; 61] The U.S. federal contribution was even more dominant in biotechnology than in other parts of research and development, while in Japan, the government funded at most a quarter.[46; 57] The obvious inference was that U.S. funding went predominantly to basic research whose results were published and shared, while private funding in Japan was more focused on specific corporate interests.

The enigma of Japan was not new. In 1877, a British commentator noted the logical and dedicated education given to engineers in Japan, as an early reform of the Meiji era.[62] In 1904, Henry Dyer wrote to *Nature* that "all the latest applications of mechanical, electrical, and chemical science have been freely and intelligently employed. . . . The inventions and improvements which have been made by Japanese officers, engineers, and scientific men disprove the charge, which is very often made, that the Japanese have no originality. Even in the matter of pure science, Japanese investigators have shown that they are able to take their places among those who have extended the borders of knowledge."[5] He could have written the same text nine decades later. Japanese dedication to technology long predated World War II.

The decentralized and feudal organization of the science bureaucracies harked back to the Tokugawa era. A May 1990 Department of State memo on genome efforts in Japan observed: "Government of Japan Ministries continue to conduct human genome research efforts independently of any government-wide coordination. . . . Japanese participation [will emphasize] commercial applications rather than . . . basic science research."[63] In 1990, aggregate funding for the genome effort in Japan was estimated at between $5 and 7 million, depending on how costs were counted—more than tenfold lower than the U.S. government contribution through NIH and DOE in absolute terms, or sixfold lower relative to GNP.[16; 34; 64–68]

There were indications, however, that Japan might be changing. From 1985 to 1987, Japan moved from fourth among nations to second, behind the United States, in number of scientific papers published.[69] Japanese science attained world standards in more and more areas. Japanese papers were numbers two and three in the list of top ten papers cited in biology in the second half of 1989.[70] Japanese groups had international stature in some areas of molecular neuroscience, biophysics, and other fields. The United States consistently contributed 35 to 45 percent of the articles on human gene mapping and sequencing in the decade 1977–1986, more than twice the share of any other nation, and roughly comparable to the total for all Western European nations combined. Japan contributed only 2 percent in 1977, but showed consistent growth to 5 percent by 1986[71] and to 6.1 percent by 1990.[72]

Japanese science was spotty, but the spots were growing in number and size, and the bright spots were luminous indeed. The system of funding was antiquated and inflexible, but complaints were being aired publicly,[68; 73; 74] offering some hope for remedy. In part because genome project planning was new, it incorporated reforms. The Monbusho proposal for 1991–1996 included funding for postdoctoral students, otherwise lacking in Japanese biology, and also followed the worldwide trend to include a program in "ethics."[75] If this trend continued, support for basic science might grow as the linkage between science and technology became apparent, the types of technologies ripe for commercialization increasingly blended with basic research in biology, and as Japan became more conspicuous as a world power responsible for shouldering a burden for providing the world with public knowledge. But tomorrow was not here, and the prevailing ethos—fear of Japanese technological domination—colored international genome politics.

Watson faced intense pressures in Congress not to "give away the family jewels," in the form of public data generated at U.S. taxpayer expense. On a personal level, Watson regretted having written the Matsubara letter as he continued to face hostile questions about threats to withhold scientific data.[76] At a June 1991 international conference on ethical aspects of genome research, Watson and Matsubara shared the stage.[77] Not a word was spoken about restricting access to databases or stock centers. The emphasis was much more on building an international support system to fund worldwide databases and other shared resources. The storm was finally spent between the two principals, although it continued between their governments.

In Japan, biologists knew of their sorry state in comparison with their industrial counterparts and wished for a bigger slice of the national economic pie. Policy was dominated by bureaucrats and industrial interests only slowly learning the connections between science and technology, quite distant from the science base. The difficulty of forging a coherent plan was made clear by the proliferation of bureaus mounting genome projects. For scientists in Japan, the future might be bright, but it seemed a long way off.

The STA and Monbusho projects sought substantial budget increases to begin in 1991.[40; 78] The Monbusho plans crafted by Matsubara and the scientific advisory committee were trimmed by Monbusho officials in the fall of 1990. The Ministry of Finance, responding to the paucity of government funds, cut even more deeply. *Nature* noted the dousing of scientist's ambitions under the headline "Japan's Project Stalls."[79] Monbusho and the Ministry of Health and Welfare eventually got 1991 budgets of ¥400 million ($2.96 million) each.[17; 80] As budget negotiations began for 1992, the Monbusho program appeared likely to plateau far below the aspirations of its university proponents, while the STA genome budget surged ahead with a 50 percent increase.[81]

In addition to the Monbusho and STA programs, the Ministry of Health and Welfare mounted a genome research program that concentrated on disease-associated human genes and new technologies to find them.[82]

The commitment to the study of ethical, social, and legal issues paralleled the U.S. and EC efforts, but the degree of commitment was dramatically less. The Monbusho and Ministry of Health and Welfare programs each had such components, but they constituted a much smaller fraction of a smaller budget than their European, American, and Canadian counterparts. The Monbusho "ethics" program was directed by Norio Fujiki of Fukui Medical School, but its annual budget was $30,000,[83] a paltry sum. Tadami Chizuka, a professor of European history at Tokyo University, had a similar budget of ¥5 million ($37,000) from the Ministry of Health and Welfare, most of which went to translate genome policy documents from abroad.[83] Japan hosted a major international bioethics conference on genetics, which generated the Declaration of Inuyama.[84] Bioethics in Japan, however, was clearly not at the same stage of evolution as in North America or Europe, and was unlikely to grow in the absence of a commitment by academic centers and government funders.

The Ministry of Agriculture, Forestry, and Fisheries announced a ¥621 million ($4.1 million) 1991 budget to map and sequence the rice genome. This was later scaled back to ¥372 million ($2.7 million), and raised questions about the commitment of Japanese industry to the project and the degree to which information would be shared or closely held by participating companies.[85] By 1992, however, the rice genome project had a dramatic resurgence, fueled by the Japanese practice of funneling a quarter of the proceeds from horse-race betting into science and technology projects. When it came to horse-race funds, MITI and the Agriculture Ministry had the inside track on the universities and the Science and Technology Agency.[86] In a reversal of the human genome pattern, it was the American scientists who came from their government empty-handed, the U.S. Department of Agriculture fearing a long line of researchers seeking genome research funds for their particular crop plant.

Even as the science agencies in Japan cried poverty and pointed to the future promise in their domain, and Ministry of Finance bureaucrats trimmed the wings of their nation's best scientists, the Chiba prefectural government and private corporations announced plans to build the Kazusa DNA Research Institute, on the opposite side of Tokyo Bay from Japan's capital.[87] The advisers to this project overlapped extensively with those advising the four government agencies: Monbusho, the Science and Technology Agency (STA), the Ministry of Health and Welfare (Kosei-cho), and the Ministry of International Trade and Industry (MITI). One part of the Chiba project, led by Mitsuru Takanami of Kyoto University and slated to start in 1993, was to serve as a nonprofit DNA sequencing center for Japanese laboratories. Another part was to focus on technology development, research, sequence analysis, and structural biology.

The DNA research institute was to be a centerpiece in the Kazusa Academia Park, a magnet for private industrial research institutes. Itaru Watanabe and recent Nobelist Susumu Tonegawa were influential in convincing Chiba prefectural officials to support the institute, securing ¥5 billion ($37 million), most of which was from the prefecture but some of which was in the form of contributions from Nippon Steel, Tokyo Electric Power Company, Tokyo Gas, Hitachi, Mitsui Toatsu Chemicals, and local banks. By 1991, the pool of funds stood at stood at ¥9 billion ($67 million). The institute was said by prefectural officials to be "unrelated" to genome plans by various agencies in Japan; to outsiders it seemed that the Kazusa institute was as unrelated to the Human Genome Project as a son is to his father. Its power base was indeed different, because the people championing it were different, and so to its Japanese sponsors it may have seemed unrelated. To the outside world, the functions it would perform were clearly related to genome research, however, and so it seemed integral to some larger plan. But there was no such master plan.

This numbing litany of budgets amounted to roughly ¥2 billion ($15 million) for the nonagricultural parts of government-supported genome research,[17] compared to $160 million in the United States. This was a significant increase from 1990, but still small by comparison. The figures were not strictly comparable, however, as the Japanese budgets did not include salaries, which would have boosted their funding to the equivalent of $21 million. Despite the Ministry of Finance cuts, the higher funding levels in 1991 brought Japan to rough and transient parity with the United Kingdom, surpassing genome funding in all other countries except the United States, after adjusting for GNP.[19; 25; 88–92] (In 1993, the U.K. again surged ahead through an infusion from the Wellcome Trust.)

Meanwhile, the powerful Ministry of International Trade and Industry (MITI) was hatching plans of its own. MITI hoped to use excitement about the genome project to galvanize the interests of corporations not hitherto associated with biology. An official within MITI first floated the idea of a genome program in 1987. Sumitomo Electric picked up the signal and responded with enthusiasm.[33] Michio Oishi of Tokyo University was designated the academic contact for planning the MITI foray into genome research, which was intended to welcome companies from electronics, robotics, and other sectors. Project planning in academia was spearheaded by Oishi, while industrial support was organized by the Japan Biotechnology Association, a private organization interposed within the triangle defined by MITI, academia, and industry.

The Japan Biotechnology Association linked the three vertices of this triangle, incubating ideas and staffing the activities necessary to generate consensus behind new initiatives. The idea of the MITI project was to improve research instrumentation, to cultivate interest in basic biology among powerful industrial interests new to the field, and to encourage a long-term commitment by a consortium of companies using a mix of roughly equal portions of government and private funding.[89; 90] MITI conferred its blessing on the Kazusa DNA research institute, lending it stature and credibility, although the ministry committed no funds to its operation.[87] It seemed likely that in the ensuing years, when that institute opened its doors, MITI might have a substantial effort of its own underway.

The genome project thus spawned programs in several ministries, whose planning processes and basic constituencies were for the most part separate. Wataru Mori, a member of the Science and Technology Council that advised the prime minister, attempted to bring some coordination to the disparate programs. He appointed a genome committee with scientists associated with the various agencies' genome programs.

Japan learned well the lesson brought to its shores by Commodore Perry. Technology was power. The genome project, with one foot planted in technology and the other in pure basic science, was seen in radically different ways by American and Japanese cultures. In the United States, it was conceived as a government-funded public good—an informational resource and the source of new research methods. In Japan, there was ambivalence about how to manage such a project. One group saw the genome project as an opportunity to found science on a Western base, with autonomous science institutions pursuing knowledge for the public good. This group dominated the early genome planning effort. Matsubara noted that in planning the Japanese genome program, "we gave top priority to international collaboration."[26] This ultimately carried over into patent policies for the government-supported projects, although of course the policies pursued by university scientists could not bind their private counterparts. This goal of open science contended with those who would couple biotechnology, and the science related to it, directly to the industrial base and adapt it to corporate interests. As the genome project grew, it was not clear which faction would ultimately prevail.

Policies in the United States and Japan were, in many respects, drifting in opposite directions. In the United States, a relatively small number of administrators in science agencies successfully launched the genome project. In Japan, the ministries vied for dominance without the same pressure for cooperation faced by DOE and NIH in the United States. The resulting effort was more atomized. Bureaucrats with little understanding of the underlying science, or its importance, made more decisions in Japan; and far more of them had to be convinced. Genome research programs in Japan were thus likely to be more independent bureaucratically than the joint NIH-DOE effort. Against this bureaucratic independence, however, was a countervailing trend. The proclivity to rely on very senior scientists with international stature meant there was a limited number of advisers, and they communicated with one another and often sat on advisory committees for several agencies.

In the United States, scientists identified an objective for the genome project and quickly persuaded Congress and federal science agencies to use public resources to attain it. In Japan, the scientists appeared to wield far less direct power over the agencies that funded their work. The federal agencies in the United States controlled the lion's share of funding for basic biology research; in Japan, the anemic government support for basic research left more uncertainty about the future character of genome research. Would it be dominated by academics striving to put Japan atop the world of science, by industrial firms hoping to capture the power of the new biology, by local governments whose relative fiscal health opened the door to a new regionalism in Japan, or by the national ministries that aspired to extend their reach by replicating previous industrial policy successes in electronics and automobiles?

Policy on the genome project confronted uncertainty in several layers. The connection between genome research and industry was loose, although real. A vigorous and seemingly irresolvable debate centered on whether direct government promotion conferred advantages for international economic competition. And the objectives of national economic policies in the face of large multinational corporations were ambiguous. Congressional patrons of genome research saw the genome project as a vehicle to maintain a technological advantage over Japan.

As the Human Genome Project officially began in October 1990, by Watson's decree, the world of molecular biology was in transition. The dominance of American science was giving way to uncertainty, with feisty bickering among scientists over increasingly intense grant competition and the prospect of a shrinking or constant science funding pie. In Japan, there was much hand-wringing, and growing consensus about the need to reform science funding. Great confidence in Japan's long-term economic vitality might or might not translate into policies to sustain scientific research. The question was not whether the economic engine was powerful enough, but whether scientists would be allowed into the control room.

PART FIVE

Ethical, Legal, and Social Issues

16

A New Social Contract

NANCY SABIN WEXLER radiates openness, integrity, and commitment; she is a doer with a big heart. Her passion for the science of genetics is grounded in personal experience. Her mother, Leonore Sabin Wexler, died of Huntington's disease, as did her uncles Jesse, Seymour, and Paul, and their father, Abraham Sabin (Nancy's maternal grandfather). As Alzheimer's disease did in the Ross family, Huntington's disease cut a wide swath through the Wexler family. Nancy, her sister, Alice, and her father, Milton, watched as Leonore, a woman of formidable intellect, developed uncontrollable movements and deteriorated mentally.

On May 14, 1978, it was over. Her body was cremated, according to her wish. The funeral was strictly family. We spent the time reading letters she had written in the early days of her marriage. They were cheerful, exuberant, and full of intelligence. They recreated the woman who had been vibrant and alive. Now that it was finally over, we could afford to remember her when she was healthy and allow ourselves to feel the enormity of the loss.[1]

When Milton Wexler first discovered the diagnosis in 1968, he called Nancy and Alice home to Los Angeles and explained the prospects. Nancy was in London at the Hampstead Clinic Child Psychoanalytic Training Institute, having just graduated from Radcliffe. She went on to do graduate work in psychology at the University of Michigan, and wrote her dissertation on how those at risk lived with the threat of Huntington's disease. She then taught for two years at the New School for Social Research in New York City.

Opportunity knocked for Nancy Wexler at age thirty, when she was hired as executive director of the Congressional Commission for the Control of Huntington's Disease and Its Consequences (the Huntington's Commission). The history of the commission is another story of a woman's persistence. After singer Woody Guthrie died of Huntington's disease, his wife, Marjorie, formed the Committee to Combat Huntington's Disease to focus attention on the disease and to lobby for action in Washington. Congress subsequently created the Huntington's Commission, which was housed at the National Institute of Neurological and Communicative Disorders and Stroke (NINCDS) within the National Institutes of Health (NIH). The commission's task was to recommend what Congress should do to combat Huntington's disease. Milton Wexler had formed the Hereditary Disease Foundation in 1968, to focus on the science,[2-4] and Nancy's expertise and family background made her a logical candidate to direct the commission.

The commission published its report in 1978,[5] and Wexler went to work at NINDS to implement its recommendations. She became the impresario of Huntington's research, enticing the best scientists she could find into the field. It became an American success story, balancing the strengths of the public and private sectors. The Hereditary Disease Foundation was the private arm that could move quickly. It convened a series of informal workshops that would have been more difficult to engineer under federal auspices. NINDS had much deeper pockets, and the infrastructure to cultivate the best science through grants.

In 1979, the Hereditary Disease Foundation held a workshop on applying recombinant DNA technology to search for the Huntington's gene. Alan Tobin, a UCLA researcher and the foundation's scientific director, was convinced that direct study of DNA was the fastest route to solving the problem of Huntington's disease.[3] It seemed farfetched to many,[6] but the idea of a genetic linkage map of the human genome had begun to grow among a small group of cognoscenti. A group in Boston that included David Housman (MIT) and Joseph Martin (Massachusetts General Hospital) was thinking seriously about genetic linkage mapping as part of a nascent Huntington's research center.[3] The Botstein et al. paper was not yet published, but the notion was becoming known, particularly in Boston. Arlene Wyman and Ray White were just beginning work to find the first DNA marker in nearby Worcester, in collaboration with Botstein at MIT.[7;8] Botstein came to the workshop, where he scribbled furiously on the board in a persuasive display of intellectual pyrotechnics. The foundation placed a bet on genetic linkage mapping.

P. Michael Conneally at the University of Indiana searched for linkage between Huntington's disease and protein markers, while a team led by James Gusella began to work with the new DNA markers. Gusella was a graduate student with David Housman at MIT and later went to work at the new Huntington's Center Without Walls at Massachusetts General Hospital. The NINCDS-sponsored program grew out of the Huntington's Commission recommendations; its NIH project officer turned out to be Nancy Wexler.

The DNA marker project took several years to get up to speed. By then, the prospects had brightened considerably, although DNA markers had never been used to find disease genes, and many doubted they could be. David Housman and Richard Mulligan chaired a May 1983 workshop convened by the Hereditary Disease Foundation in Cambridge, Massachusetts. The topic was "What Can Be Learned About Huntington's Disease Once the Gene Has Been Located?" According to one report, the meeting ended "on a note of sobriety for the distance to be traveled and genuine offers of assistance some five or ten years hence when a marker would be found."[9] Wrong.

Barely three months later, Gusella's group turned up a promising lead. They found a possible linkage between Huntington's disease and a marker on chromosome 4. This marker, designated G8, was among the first tested.[3; 4; 10] (Thereafter, Gusella was known as "Lucky Jim."[4]) An August workshop, titled "Clinical Impact of Recombinant DNA Research on Neurogenetic Diseases" and once thought premature, was suddenly playing catch-up.[11] That workshop took place "in an atmosphere of elation and stunned disbelief."[9] The results were published in *Science* that November, by which time the linkage was well established.[12] A November workshop focused on issues that might emerge as the marker was used to predict who might develop Huntington's disease.[11]

At a January 1984 workshop, just months after Gusella and others found the approximate chromosomal location of the Huntington's gene, a consortium of laboratories formed spontaneously to search for the gene itself and the DNA alteration that caused the disease. Wexler and the Hereditary Disease Foundation were the spokes supporting the wheel. The consortium held together through the 1980s and into the 1990s. Other groups outside the consortium—such as Rick Myers and David Cox in San Francisco, Michael Hayden and his coworkers in Vancouver British Columbia, and groups in Europe and Asia—continued in generally friendly competition for a decade, until the gene was found.[3]

Once the gene's location was found, many hoped it would be only a few years until the gene were found. The hunt for the gene itself proved much more arduous. It took a decade of dedicated work, but Gusella's group did lead the effort that eventually uncovered the gene and the nature of the mutation causing Huntington's disease.[13] The gene was more elusive than some because it was embedded in a complex and confusing region. In the end, the article announcing the end of the search was authored by the entire Huntington's Disease Research Group, which by then included fifty-eight authors in six groups spanning the Atlantic.

The hunt for the Huntington's disease gene was far more than luck. It involved a large international collaboration and a decade of intensive work

with many false starts. Another critical factor was the discovery of a large family with Huntington's disease living near Lake Maracaibo in Venezuela. This family had been discovered by a Venezuelan physician, Americo Negrette. Wexler flew down to investigate in 1979. It turned out to be an enormous pedigree, containing thousands of living members, with an immense toll of Huntington's disease. Thus began an annual rite of visitation that continues to this day. Wexler became a local fixture, known as La Catira ("the Blonde"), and the Venezuelan families became an extension of Wexler's family—a group with whom she shared an emotional bond deepened by mutual suffering and the fierce struggle against a common enemy.[2] Dr. Negrette described the feeling of working among the Maracaibo families in the company of La Catira:

I arrived in their homes and their shacks, and left feeling destroyed inside because I felt incapable of solving the problems. . . . At times I would distance myself from them— for years. . . . and feel guilty. But now as I grow older I have become more sensitive to the pain of others. So much so that it now no longer feels apart. It is my pain, this pain that they feel. And it is for this that I love La Catira. Because she comes every year, for more than ten years to battle. . . . She brings them medicines and she brings them projects for their social welfare. But she brings them something more precious yet. She brings them an immeasurable love. She pours on them a warm contagious care. I have seen her embracing women and embracing men and kissing children. Without theater, without simulation, without pose. With a tenderness that jumps from her eyes. And her fingers are claws of love mingling with tenderness and passion. . . . I adore La Catira who has as hair a hanging waterfall of gold. Like the love she gives.[14]

The new ability to detect Huntington's disease, particularly in the decade between finding the gene's chromosomal location and discovering the gene itself; brought complex medical, family, and social choices. Who would take the test? Who *should* take the test? Who should offer it, and under what conditions? How could the quality of laboratory work be assured? How much counseling should be required before administering the test? Who other than individuals and their physicians should have access to test results? Should information about Huntington's disease in one individual that was relevant to another be communicated without knowledge of, or over the objection of, the person tested? If so, under what conditions and how? These questions had long been debated, but in the abstract. In 1983, technology called everyone's bluff. The stakes were very high—life itself. The game was, in Nancy Wexler's words, "genetic Russian roulette."[15]

When the test became available, the questions that had been merely rhetor-

---

Nancy Wexler, a leader in the search for the genetic basis of Huntington's disease (and herself a member of a family affected by the disease), was picked to head the NIH group devoted to exploring the ethical, legal, and social implications (ELSI) of human genome research. She is shown here with a young Huntington's patient, a member of a large family with the disease living near Lake Maracaibo in Venezuela. *Peter Ginter photo, courtesy Nancy Wexler*

ical suddenly became urgent and real. Some of the answers were surprising. Nancy had always assumed she would want to know; she would therefore take the test. But Milton Wexler pointed out that it was not entirely an individual decision. The family was being tested, and if both Nancy and Alice took the test, the chances were three in four that one or both of them would turn out to have the Huntington's gene.[2] If either had the gene, all three Wexlers would be crushed. Did they really want to know? Taking the test required more thought. Wexler asked herself:

Would I change my job? No, I love what I'm doing. Would I work any less? No. Would I work any more? I'm not sure I can. Would I be any less frantic and obsessional? Probably not. Would it change personal relationships and friendships? No. There's an awful lot it wouldn't change. . . . I'm already happy, how much happier am I going to be? Part of me realized how happy I am, being part of this whole research process that's going to make a difference in the future.[16]

Even disclosing whether she had taken the test or not was an issue. Wexler was a highly public figure, but why should the public know about her private decision? She wanted to make clear that she might take the test or she might not. She believed it was a matter of personal and family choice, not a matter of public record.[17]

The technology of genetic testing replaced implacable fate with agonizing choice. A majority of those eligible to take the test end up not doing so after counseling. Those who opt in favor of testing face a difficult psychological travail, whether the results show the Huntington's gene to be present,[18] or absent,[19] or prove inconclusive.[20] The first empirical study of the benefit of predictive testing for Huntington's disease suggested that those who discovered they were at decreased risk fared better on psychological measures of distress soon after testing than those whose risk status was unchanged—those who chose not to be tested or for whom the test was inconclusive. After a year, both those who learned of increased risk of Huntington's and those with decreased risk scored better, suggesting that even bad news with increased certainty might be perceived as better than lingering uncertainty.[21] The tests might indeed provide a psychological benefit, but the complexities of family testing nonetheless still demanded care and caution.[22] The control group in the study combined those who deliberately chose not to have the test with those who had it but did not get conclusive results, which would intuitively seem to be very different psychological situations. While the first study was encouraging, therefore, this was not a green light so much as a flashing yellow one.

Producing a diagnostic test was not the purpose of locating the gene; it was a side effect. The ability to use DNA markers to predict Huntington's disease was, in Wexler's words,

a way-station on a more important journey: the isolation and sequencing of the HD gene with the aim of treating the gene or its consequences. . . . If the initial steps on the road to finding treatment can be of clinical use for presymptomatic and prenatal detection for some at risk, so much the better, but the fact that the HD gene now has a chromosomal localization will hopefully speed the day when effective treatment can be offered to all families.[9]

Wexler's charm and warmth enclosed a powerful engine of change. Her passionate devotion to genetics was born of seeing it as the sole salvation for herself, or at least for future individuals facing the same gruesome choices. Knowledge might or might not yield power; but ignorance was certain impotence. Hers was a smooth and almost imperceptible style of exercising great power. Huntington's disease may have stolen Nancy Wexler's mother, but it also gave her life a meaning it might not have found otherwise: "The struggle against hereditary disease has given me purpose and direction."[1] She has shared this wealth.

The decision to commence a program to anticipate the social implications of genome research was made by James D. Watson alone, without conferring with anyone else at NIH. It was one of Watson's first acts on joining NIH. "Some very real dilemmas exist already about the privacy of DNA. The problems are with us now, independent of the genome program, but they will be associated with it. We should devote real money to discussing these issues. People are afraid of genetic knowledge instead of seeing it as an opportunity."[23] Watson thought NIH should set aside 3 percent or so of its genome funds for this purpose.[24] He argued that the genome project was "completely correct" to go after gene maps and DNA sequence data as fast as possible, but it was essential to be completely candid about how such information could be abused and to suggest laws to prevent such abuse, because "we certainly don't want to mislead Congress."[25]

Having made a commitment as his first public act, Watson then had to carry it out. He officially assumed his NIH associate director position on October 1, 1988. Three weeks later, the first major international meeting on genome research took place in Valencia, Spain. Organizer Santiago Grisolía had achieved his goals of a high-profile meeting by attracting Nobelists Christian Anfinsen, Hamilton Smith, Jean Dausset, and Severo Ochoa, as well as Watson. Victor McKusick, James Wyngaarden (still NIH director), and many other prominent scientists joined this star-studded cast of scientific heavyweights in October 1988 at the Hotel Sidi Soler along the Mediterranean coast. The meeting had an unexpected benefit. Watson had just begun his NIH duties and was nearing completion of the list of outside advisers to appoint. Nancy Wexler was at the meeting to discuss medical applications of genome research, which were still only a sideshow in the genome debate, in a period when most discussion centered on cost and scientific strategy.

At one of the sumptuous Valencian meals, Wexler found herself in the company of Wyngaarden, Watson, McKusick, and several others while they discussed who should represent human genetics on the advisory committee.

Wexler was, of course, an ideal candidate—as a psychologist, a person at risk of genetic disease, a fieldworker on pedigree research, and someone intimately familiar with the science. The initial interest in appointing Wexler came less from her interests in ethical and social issues than from her ability to balance the scientific background of the advisers with a broader view of human genetics. When the NIH Program Advisory Committee on Human Genome Research broke into working groups, however, it was obvious that one group had to concentrate on ethics, law, and social policy. Nancy Wexler alone among the advisers had standing to chair such a group. She and McKusick were both appointed, and she was designated chair of a working group on ethical, legal, and social implications (ELSI) of human genome research.

Congress had signaled concerns about ethical issues even earlier, as genome plans first surfaced. In 1986, Edwin Froelich, physician adviser to Senator Orrin Hatch, called Charles DeLisi to his office, soon after having learned of the Department of Energy schemes for a genome project. Senator Hatch was the ranking Republican in the Committee on Labor and Human Resources. The committee had jurisdiction over NIH, but not DOE. Froelich nonetheless expressed grave concern to DeLisi that DOE's genome research should be scrutinized for its broader impact, particularly whether it would lead to more prenatal diagnosis and abortion. Froelich also called Ruth Kirschstein, director of the NIH institute central to genome research planning, when he heard of NIH's emerging interest in 1987. Kirschstein and W. French Anderson went down from Bethesda to Capitol Hill to assure Froelich that NIH was indeed concerned about these matters. Froelich wanted explicit attention to ethical issues, or the human genetics program would be in jeopardy.

When we learned of these concerns at OTA, we relayed them to Chase Peterson, president of the University of Utah and a member of OTA's overall advisory committee. In addition, I called Ray White, whose genetic linkage group at the University of Utah was becoming a world hub of human genetics—one of Utah's most conspicuous intellectual landmarks. Peterson met with Hatch's staff to help clarify the importance of genome research in Hatch's home state.

Meanwhile, John C. Fletcher, chief of the bioethics program in the NIH clinical center, wrote a memo to Kirschstein expressing concern that "the NIH should not appear to be driven by a technological imperative. . . . Are we as concerned about preparing society to find the wisdom to live with a control of this new knowledge as we are with seeking the knowledge? I hope so, but those who work on the proposal need to have a plan to examine the issues."[26]

Independent of these activities, Senator Barbara Mikulski of Maryland met with me on June 15, 1988, to express her concern that "go-go" science would race far in advance of prudent policies. It could become difficult to contain its adverse impact on individuals and society. Enthusiasm for the biology needed to be tempered by a public policy process to anticipate its social impact. As the

NIH authorization bill went through the Labor and Human Resources Committee in the Senate that fall, Senator Mikulski again raised this concern. Senator Edward Kennedy, the committee chair, echoed it, and noted his long support for the National Commission and President's Commission (previous bioethics commissions), and his hopes for the congressional Biomedical Ethics Board, of which he was a member. These meetings about the social, legal, and ethical implications of genome research, however, were for the most part hidden from public sight and only tangentially related to planning at NIH and DOE until late 1989.

The American ELSI program was unfocused as it began. The first announcement of the NIH grant program asked five general questions: "What are the concerns to society and to individuals arising from the Human Genome Project? What specific questions in the broad area of ethics and law need to be addressed? What can we learn from precedents? What are possible policy alternatives and the pros and cons of each? How can we inform and involve the public and stimulate broad discussion?"[27] The response to this somewhat vague solicitation was understandably diffuse and general. Bettie Graham, acting administrator of the ELSI grants program until a permanent staff person could be hired, noted: "We have very little experience in the area and we need a point of reference."[28] Many of the first grant applications were to host conferences that would consider all the issues. Ten such conferences were eventually funded that first year.[29]

Nancy Wexler's ELSI working group met in September 1989, to set forth a series of objectives for the program. It was the first meeting of the group and led directly to a refinement of the program announcement to guide those seeking grants. The five general questions became five pages of background and a series of nine topic areas, ranging from immediate policy questions—fairness in use of genetic-test results in employment and insurance—to philosophical issues—how conceptions of personal identity and responsibility might change in light of new genetic knowledge.[30]

Wexler proposed activity along several fronts. She wanted to hold a series of small workshops with the working group, as well as larger town meetings to solicit broader input. The working group would also continue to help steer the research program of grants and contracts. What more the ELSI group should do was open to debate, particularly whether the group should become a forum for policy deliberations. Questions about how far the ELSI group should go into policy analysis surfaced repeatedly at meetings in February and September 1990 and January 1991. The working group took its cues less from internal debate, however, than from events swirling about the genome program.

Congressman David Obey forced NIH's hand in hearings on the 1991 budget. He had a long exchange with Watson, expressing his view that genome research might best be delayed until prospects for protecting genetic information were better.[31] He followed up by inserting report language that mandated a program to devise policy options to thwart adverse uses of genetic testing, and made explicit the need for a more activist approach.[32] Obey thus tilted the balance in favor of the more activist members, among whom I counted myself.

The ELSI group worked on several issues simultaneously. It was composed of a core group with a long-standing interest in the social uses of genetics. Tom Murray, head of the bioethics program at Case Western Reserve and long associated with genetic-testing issues through work at the Hastings Center in New York and then at the University of Texas, prepared an overview of issues.[33] Jonathan Beckwith of Harvard Medical School had helped to isolate the first bacterial gene and later was involved in the recombinant DNA debate of the mid-1970s. He was a prominent antagonist in a controversy over whether males with an extra Y chromosome were more prone to criminal behavior, and had been a vocal skeptic of claims that IQ was genetic.[34] Tom Murray and Beckwith cochaired the insurance task force under ELSI. Robert Murray, a clinical geneticist from Howard University who had direct experience with the problems of sickle-cell-screening programs of the 1970s, agreed to oversee activities related to the introduction of genetic tests into medical practice. Patricia King was a law professor with extensive policy background. She had served on the recombinant DNA advisory committee and on both the major federal bioethics commissions. Victor McKusick was, of course, the dean of human genetics, and also chairman of the Human Genome Organization's ethics committee until mid-1991. I was the youngster, on the group for its first meeting, off for the second (because I was working as a consultant to NIH), and then back on again after leaving NIH employ. I was chosen for my background on the Hill, where I had written OTA reports on gene therapy and the genome project, and for my experience as acting director of a congressional bioethics commission, the Biomedical Ethics Advisory Committee.

The ELSI program at NIH got a major boost when Elke Jordan hired Eric Juengst to direct it. Eric had a Ph.D. in philosophy from Georgetown University, where he worked on issues related to bioethics. He subsequently worked at two other major bioethics centers before joining NIH—at the University of California, San Francisco, and the Hershey Medical Center in Pennsylvania. Juengst brought a broad background in the history of biology, philosophy, and pragmatic bioethics to NIH, and he had previous experience as an administrator at the National Endowment for the Humanities to boot. Eric had twin responsibilities—to help the ELSI working group formulate policy and also to administer the NIH portion of the ELSI grants program.

Michael Yesley, a lawyer, joined the ELSI working group at its third meeting. During the mid-1970s, Yesley worked at NIH, where he was executive director of the National Commission for the Protection of Human Subjects of Biomedical and Behavioral Research, the nation's first federal bioethics commission. He then went into consulting work and eventually to Los Alamos

National Laboratory. When the DOE program moved into bioethics, he was already working for Los Alamos, and he became a logical point man for DOE interests.

Nancy Wexler decided to structure the effort by keeping the working group small, but supplementing its expertise at a series of workshops on different topics. Outside experts could be brought in at each meeting. Juengst had line authority over the NIH grants program, for which the ELSI working group would serve as a steering committee. Oversight of DOE grants fell to Michael Yesley as a consultant, and DOE line staff in the Germantown headquarters. After the first round of DOE "ethics" grants, Daniel Drell became the principal DOE staff person.

The grant mechanism supported conferences and outreach, with large "town meetings" planned later. By the end of its first year, the genome office was supporting sixteen projects extramurally, through grants and contracts. These ranged from small conference grants, to an Institute of Medicine study of genetic testing in clinical practice, to substantial funding for public education that included a Public Broadcasting Corporation production, *The Future of Medicine*.[35] By September 1991, NCHGR had funded twenty-five extramural grants and ten national conferences.[29] The five vague questions of March 1989 had grown into a ten-page strategic plan for Congress[36] and a growing portfolio of projects supported throughout the nation.

The program attracted some attention from other NIH centers and institutes that had contemplated programs in social and ethical issues before, and now had an experiment to observe. The National Cancer Institute, National Institute of Neurological Disorders and Stroke, and National Institute of Child Health and Human Development watched closely. When Bernadine Healy came in as the new NIH director in mid-1991, she commenced a major strategic planning exercise for the institution. Attention to ethical and social issues became a part of this planning exercise, and Juengst emerged as the NIH staff person with the most direct experience, taking the lead in preparing the strategic planning documents on social, legal, and ethical issues in biomedical research.[37] He thus became one of the principal architects of the NIH-wide proposal to address such issues.

One potentially adverse effect of the ELSI genome program was the concentration of resources in a relatively narrow field of biomedical research. As support for bioethics related to the genome project grew, and with few resources available for other lines of bioethical analysis, many bioethics programs developed modules on genetics. This may have helped achieve the goals of the genome office, but it also skewed concern with bioethics toward the genome research. Where cash went, ethics followed.

Nancy Wexler personally took the lead on efforts to encourage field trials of genetic testing for cystic fibrosis (CF). In late 1989, this was emerging as the most urgent policy problem related to genetics. It began with discovery of the CF gene,[38-40] a great technical triumph. It turned out, however, that CF testing would be far more complex than expected. The CF gene encoded a membrane protein regulating the transit of chloride ions across cell membranes, and the cellular defect could be corrected by inserting the normal gene.[41] The gene's DNA sequence, however, was marred by a staggering array of different mutations in different patients. One common mutation, the loss of three base pairs, accounted for the majority of cases in most populations. Scores of different mutations were also associated with the disease, however, making impractical a simple DNA test to detect them all, at least until new technologies developed.[42] DNA sequencing might disclose the full range of mutations, but sequencing remained for the time being too expensive and too slow for routine clinical testing.

Different population groups varied widely in the relative frequency and diversity of CF mutations. In northern Yugoslavia (still a single country at the time), the three-base-pair-deletion mutation accounted for only 26 percent of CF genes, compared to 88 percent in Denmark. In North American groups, the range went from 3 percent (among those of Eastern European Jewish background) to 84 percent (in a mainly Caucasian group).[43]

Genetic complexity in populations was further confounded by clinical heterogeneity. The disease varied in severity and range of symptoms. Most symptoms stemmed from viscous mucus that plugged duct systems in the lungs and pancreas. Lung plugs sealed off pockets that became breeding grounds for recurrent infections. Clogged pancreatic ducts obstructed the secretion of digestive enzymes into the intestines, so that foodstuffs were poorly digested and absorbed. The life span of CF patients soared in the 1970s and 1980s, with better antibiotic treatments and supplementation with digestive enzymes. CF children in the past had generally died before age twenty, but now most lived well into their twenties and even beyond. With CF, judgment of clinical severity was more uncertain, in contrast to other unequivocally horrid genetic diseases such as Tay-Sachs, in which infants begin to die even as they are born. Everything about CF was more complicated than previous genetic diseases, and yet it was far more prevalent in the American population.

Testing for the most common mutation would pick up, on average, about 70 percent of carriers with a single copy of the CF gene. It would thus miss the 30 percent of potential CF genes caused by other mutations. If two prospective parents were both carriers, each of their children had a one-in-four chance of developing the disease. Having one CF gene did not cause the disease, but if both copies were defective, disease inevitably ensued. The problem was that the test would miss many CF carriers, and so many couples would not be aided by the test.

The unanticipated diversity of mutations immensely complicated the process of testing individual patients and screening populations for CF. A poll in England taken just before the gene was discovered found that 80 percent of those who had heard of CF wanted to know if they were carriers.[44] That desire for a test confronted considerable technical and logistical obstacles. There was fear in the United States that profit incentives for private testing laboratories would combine with fear of malpractice to unleash a massive wave of CF testing. If physicians did not offer the test, they could be sued. Yet interpreting test results for CF was even more complicated than for most other genetic diseases, which were already hard to explain. Genetic counseling and other genetic services were strained even without a massive increase in demand.[45] The existing CF tests would add many new clients, a substantial fraction of whom would require lengthy counseling to understand the disease and the meaning of equivocal test results.[45; 46]

The American Society of Human Genetics adopted a statement at its annual meeting on November 13, 1989, hoping to stave off premature population screening. The society endorsed CF testing for those who had a close relative with CF, but indicated population screening would become practical only when the test was far more sensitive. Genetic testing for CF was a research topic and an adjunct to individual genetic counseling, not a standard of medical practice.[47] The statement was intended to thwart malpractice suits and to apply a moral brake to private laboratories promoting CF testing.

In March 1990, an NIH workshop on population screening concurred that population screening should not be undertaken until the tests detected a much higher fraction of total CF genes and until the medical care system was much better prepared.[48] The Office of Technology Assessment commenced a study of CF testing later that year.[49] The public statements were expressions of consensus, not unanimity, and they referred mainly to the American health care delivery system. A review of the arguments for and against wide use of CF testing,[50] published in the *American Journal of Human Genetics,* found strong arguments on both sides. In the United Kingdom, Canada, and Denmark, testing programs went forward.[42] In the United States, CF testing programs became a focus of controversy, sparked by the disarray of health-care financing and fueled by the vitriolic abortion debate.

Among clinical geneticists, concern shifted quickly from stemming the tide of genetic testing to analyzing how a CF test might best be introduced into practice. The American Society of Human Genetics indicated a need to evaluate pilot testing programs.[47] The NIH statement was even stronger: "Pilot programs investigating research questions in the delivery of population-based screening for cystic fibrosis carriers are urgently needed."[48] The need may have been urgent, but there was no eager patron. The Cystic Fibrosis Foundation, which had sponsored research to find the gene and was now funding work to understand how the gene led to disease, saw its mission as research, not test development and genetic services. The National Institute of Diabetes, Digestive and Kidney Diseases (NIDDK) viewed the problem similarly, at least initially. A focus on biomedical research, narrowly defined to exclude research on health services, became the public rationale for inaction.

Another rationale, voiced privately, was a judgment of political risk. The coalition supporting CF research might fracture over the abortion issue.[51] This would severely hamper the research efforts at both NIH and the CF Foundation. Pilot testing meant mostly testing for carriers. That was fine, but the problem came further down the line. The primary reason to test most carriers was to inform them about reproductive choices. If both parents were carriers, they stood a one-in-four chance of having a child with CF in each pregnancy. They could choose not to have children, to seek artificial insemination, or to take their chances and have children. If a fetus tested positive for CF, some families would clearly carry on the pregnancy, judging the disease insufficiently severe to merit abortion. Other families, however, would choose to abort.

With sustained research the primary objective, divisive debate about CF testing and abortion could only undermine political support. The power of this fear was exemplified in the NIDDK call for Small Business Innovation Research grant applications. NIDDK sought companies to prepare educational materials for those entering screening programs "about the risks, benefits, and limitations of the test and helping people found to be carriers of the cystic fibrosis gene defect interpret and understand the test results." A final caveat made this laudable exercise a charade: "The scope of this topic does not include materials related to reproductive decisions."[52] Companies could take the horse to water, but not let it drink.

One proposal for a CF testing pilot program came to NIH during this period, but it was raked over the coals by incompetent peer review.[53] The proposal made the mistake of technological optimism, asserting that 95 percent of CF mutations would be identified within a year. In the lag between submission of the proposal in January 1990 and peer review in June, the conventional wisdom changed. The optimism that all CF genes would be identified quickly that held sway immediately after the gene was identified gave way to recognition that the task was more formidable. The urgency of pilot programs did not hinge on finding 95 percent of mutations, and it was a mistake to make the claim. This red herring, however, is not what doomed the proposal, as the peer review statement made quite clear.

The peer reviewers believed a more serious weakness of the proposal was that nothing more could be learned from the pilot project. The relevant information was already known. The review sheet opined it was "not clear that it [the pilot project] will uncover new and significant information not already available from previous population studies involving other genetic diseases such as hemoglobinopathies and Tay-Sachs."[54] The study section thus judged that the proposed study would turn up little useful new information.

This errant judgment failed to account for vast differences between those diseases and CF, different technical uncertainties of the genetic test, and changes in genetic services in the intervening decade. CF testing and screening would have to be entirely different from previous programs for sickle-cell anemia, thalassemia, and Tay-Sachs disease. CF was more clinically variable than Tay-Sachs; the size of the population at risk was ten times larger, and the disease affected those of Caucasian descent, raising a completely different set of social issues related to ethnicity, economic status, and educational background. Federal support for genetic services over the previous decade had atrophied, and public health programs in the states were retrenching. The shifts in the delivery of health-care services, the technical basis of the tests, and the complexity of interpreting test results overwhelmed any similarities to previous experience in genetic testing, but the peer reviewers were mainly laboratory geneticists unfamiliar with the wider problems confronting genetic services. A few lonely voices dissented from the majority, but their votes could not raise the priority score to a fundable level or change the consensus against the pilot project.[55] It was as if the study section were to decline to study a gene for prostate cancer because someone had found the one for breast cancer a decade earlier.

Based on this shoddy evaluation, the NIDDK council passed over the only pilot testing project under review in fall 1990. NIH's inaction frustrated clinical geneticists, who began to raise a ruckus.[51] Nancy Wexler courageously steered the ELSI working group straight into the storm.

Wexler's ELSI working group was just catching its stride as the CF controversy hit. The working group became a natural forum because of its composition and, more to the point, because it was the only conspicuous place to discuss the pressing policy issues in the federal government. Virtually all the working members had long been involved in public policy regarding genetic testing. They were keenly aware that CF was at once the single most common recessive single gene defect in the American population and also the prototype for a long list of genetic tests to be developed over the ensuing decade. At a gathering in Williamsburg, Virginia, in February 1990, the working group identified CF testing as a priority item. An NIH workshop on CF screening was held in March, cosponsored by NIDDK and the genome center. The ELSI working group held another, smaller meeting of CF experts from the United States, the United Kingdom, and Denmark in September 1990, and prepared a summary statement that stressed again the urgent need for action by NIH and a sense of growing frustration among medical geneticists.[42]

When stories appeared about disgruntlement in the wake of that year's annual meeting of the American Society of Human Genetics in Cincinnati, the timing was right. In a background paper for the ELSI group, Benjamin Wilfond and Norman Fost of the University of Wisconsin further substantiated the need for greater attention to policy analysis before large-scale screening programs were put in place.[46] Nancy Wexler presented the ELSI working group's summary statement to the NIH's genome advisory committee in December. The genome advisory committee was leery of getting involved in clinical work, fearing it would create an expectation that the genome office would support field trials of every genetic test developed thereafter. The committee agreed, however, that CF pilot tests were just too important, and the genome center should take the lead if other institutes did not.[56] The full committee endorsed the statement, and urged NCHGR to "take a leadership role in developing support for well designed, cost effective pilot research projects." The genome advisory committee petitioned Watson to pursue support from other parts of NIH and other agencies.[57]

After the meeting, Elke Jordan and Eric Juengst met with acting NIH director William Raub. Raub was supportive, and convened a working group of several institutes. On January 31, 1991, a group of advisers met with staff from the genome office, NIDDK, and the National Institute of Child Health and Human Development to help prepare a request for applications on cystic fibrosis.[58] This led to a request for applications for pilot CF testing, with the genome center taking the lead. NCHGR issued the call and managed review of more than thirty applications.[59; 60] Seven grants were given to six centers to study various approaches to CF testing.[61]

In the CF story, the ELSI working group provided a fulcrum for moving the NIH bureaucracy toward pilot testing. By serving as a forum for discussion linked to but independent of NIH (working group members were not NIH staff), the group became a mechanism to reason toward solutions. Once the genome center made a commitment to pilot projects, other institutes followed. It was an early victory for the program, showing it could have an impact.

The ELSI program exemplified how the foundation of science was broadening. Taxpayers funded the lion's share of basic science, and science intruded ever deeper into daily life, working its way into public consciousness. Science was weaving itself more tightly into the social fabric. Where science was once a cultural embellishment, a luxury for affluent cultures and a hobby for upper-crust patrons, in the post–World War II period it had become the engine for technological change. Technology, for its part, was a major cause of social transformation. The rules had to change. It was no longer sufficient to recount Vannevar Bush's paean to "Science, the Endless Frontier," echoed though the years since he coined the phrase in 1945.[62; 63]

Daniel Koshland, editor of *Science,* wrote an editorial that stood foursquare behind the genome project, invoking the promised benefits of preventing diseases by understanding their genetic causes. He gave special emphasis to mental illness:

The costs of mental illness, the difficult civil liberties problems they cause, the pain to the individual, all cry out for an early solution that involves prevention, not caretaking. To continue the current warehousing or neglect of these people, many of whom are in the ranks of the homeless, is the equivalent of providing iron lungs to polio victims at the expense of working on a vaccine.[64]

The National Foundation for Infantile Paralysis (March of Dimes) had faced precisely this dilemma four decades before. It sensibly opted for iron lungs until the prospects of a successful vaccine looked promising enough to shift resources toward prevention.[65] Faith in science bore fruit in one of the spectacular medical successes of its day—the polio vaccine—but the causal links from poliovirus to poliomyelitis were far simpler and more direct than the connections from genes to homelessness. For disorders clearly involving many genes and complex interactions between person and environment, there were many more opportunities for wrong turns 'twixt gene and effect. Koshland was asking the public to make a leap longer than most were comfortable making. He might well take such leaps of faith; but most Americans seemed inclined to check the landing zone first. Could a Senator Koshland garner votes on a platform espousing "genes for the homeless"? He just might lose, even in Berkeley.

There was nonetheless a kernel of truth in Koshland's rhetoric. Understanding can contribute to the alleviation of human suffering. If genetics helps to clarify the biology of schizophrenia, manic-depressive disorder, and other severe mental illness, it may indeed reduce the number of homeless people by reducing disability. In the near term, however, dissecting genetic factors is more likely to succeed for families like the Wexlers, the Rosses, or the countless others ravaged by genetic diseases traced to one or a few genes. Uncovering genetic factors can also dramatically advance the analysis of risk factors. Families prone to colonic polyps or skin cancers, for example, reveal the weak links in cellular physiology that can lead to cancer. The genetics of familial cancers not only sheds light on the cancers in those families, often providing a welcome technological means to prevent cancer, but also illuminates the general process by which cancer develops in other patients. Knowledge is power. Genetics is the fast track to knowledge, even if it does not run a direct course from a gene to homelessness.

The discovery of new facts, new theories, and new conceptions of the world remains a powerful motivating force for those in science, but does not fully explain the resources devoted to research, both public and private. Science retains its prestige and power to excite cultural pride, but it is also an investment. Biomedical research is regarded as the down payment on future health. Americans are unusually lavish in support of biomedical research, perhaps reflecting a national optimism about the benefits of technology. Today's science is tomorrow's cure or prevention.

The increased scale of the biomedical research, however, carries with it a tendency to self-perpetuation and defensiveness. As the genome program was being formulated, another major theme of biomedical research was the deliberate warping of science for personal benefit—fraud and misconduct. A populist element has expressed skepticism of scientist's motivations, and sees scientists as cold and arrogant fact-seekers oblivious and unaccountable to the world around them, and corrupted by the new-found allures of wealth. A powerful elitist band of scientists indeed has pooh-poohed public controversy over science fraud and misconduct and has seriously underestimated concern for animal rights and protection of human subjects. Between Luddites on one side and arrogant scientists on the other lies a legion of investigators trying to conquer disease, but also concerned about social harms that might result from their technology. Federal sponsorship of bioethics is intended to foster work in this area.

Biotechnology excites awe and distrust. Genetics inspires wonder at its power, but provokes fear of misuse. Inchoate discomfiture is grounded in a sense that genes are inherently important. Genes are tightly linked not only to other genes, but also to personal identity. It is distressing to contemplate losing control over something so intuitively private, something as close to the self as one's genes.

As Patricia King once noted, "we, the public, worry about human control over nature. We are concerned, for example, that advances in genetics will change the nature of humankind, that we will change the genetic structure of human beings." She went on to observe:

> The policy community has been making policy on a range of issues for a very long time and is comfortable with itself. It is the scientific community that has the most at stake, and it is going to be charged with educating the rest of us about its needs, its methodologies, its frameworks, and its values. It seems to me that the burden rests on the scientific and medical communities to educate people like me.[66]

The programs to analyze the social implications of genome research were a means of dealing with public concerns. Scientists did not want those concerns to obstruct science. For some, the ELSI program and its foreign counterparts appeared to be political preemptive strikes intended to thwart criticism of science. Motives matter, and many read the politics as Watson's ploy to protect his research budget. If it was, it was a ploy that could well backfire. It provided an opening for the program's critics by funding social science that could well turn up issues that genome scientists would find uncomfortable. Indeed, that was one of its mandates. Watson's position was consistent with his past actions. In the early 1970s, Watson was almost alone among scientists in supporting a commission on reproductive technology and new biomedical advances. The Senate had a hard time finding a scientist who did not regard such commissions as intrusions onto sacred scientific lands, but Watson spoke out in favor of public deliberation.[67] If there was a protective motive for Watson's support, there was also a long-standing interest.

The ELSI program was not a shield for scientific miscreants. It was an attempt to articulate the values that should govern the research, and to anticipate adverse social consequences of science in time to avert them. The remarkable feature of the ELSI program—and its counterparts in the EC, French, Canadian, German, Russian, and Japanese programs—was not that they came into being, but how quickly policymakers accepted them as the norm despite their absence everywhere else in science. No comparable movement had seized the imagination since the debates about human research subjects and recombinant DNA technology commanded public attention twenty years before.

The ELSI research program was a welcome addition to the NIH, but even as it successfully guided NIH toward a more rational research program related to CF testing, it also faced deeper questions with social implications well beyond its capacity to manage. Many touched on disparate views of political philosophy, justice, and moral values. The genome project attracted the attention of scholars outside of molecular biology, who then began to scrutinize the directions within science and the broader social context within which human genetics was practiced.

The early history of genetics, particularly human genetics, was imbued with an optimism that genetic factors could explain socially important individual traits, such as intelligence, criminal tendencies, and athletic prowess. The eugenics movement was inextricably woven into human genetics, its most public aspect, as eugenics advocates played a role in public policies on immigration, interracial marriage, and mandatory sterilization.[68] Restrictions on U.S. immigration policy during the 1920s, were perhaps driven as much by ethnic politics as by science, but the testimony of eugenicists was avidly sought. They pointed to correlations between low scores on IQ tests, then just coming into use, and the geographic origin of population groups. Data from the 1920s claimed to show Jews were intellectually inferior, for example, yet decades later, it was an article of faith that American Jews did better on standardized tests.[69] The explanation for low scores in the 1920s was genetic; for superior performance in the 1950s and 1960s, cultural and educational. The resort to genetic explanations seemed to depend on more than just the test results or population clusters; it depended as well on an overlay of largely unexamined social theory. The eugenics movement achieved its zenith in the United States in the 1927 Supreme Court decision on *Buck* v. *Bell,* in which Justice Oliver Wendell Holmes justified the mandatory sterilization of Carrie Buck by declaring "three generations of imbeciles are enough."[70]

This decision, like many eugenic initiatives, was based on faulty evidence and ideology masquerading as science. A factual retracing of the case suggests that Carrie Buck was raped by the son of the family for whom she worked, and was remanded to an institution, the same one in which her mother and sister resided, when she became pregnant. Despite expert testimony from some nationally prominent "experts," there is no evidence that she was feeble-minded, and she married twice and lived an unremarkable life after release from the institution. Her mother and sister were also sterilized. Her daughter Vivian, the supposed third generation of "imbeciles," became an honor student before dying in late childhood.[71]

In other nations, eugenics grasped policy with even greater force. In its most infamous embodiment, Nazi eugenics began with the sterilization and then "euthanasia" of those in psychiatric facilities. It then adopted racial overtones culminating in the Holocaust.[72-78] Physicians and geneticists played an active role in promoting the ideology of racial hygiene.[75-78]

Genetic explanation could produce tragic consequences when its reach exceeded its grasp. Evelyn Fox Keller, professor of rhetoric at Berkeley and a historian of genetics, pointed to this tendency in contemporary discourse about molecular biology:

Without doubt, the 1970s was a decade of extraordinary expansion for molecular biology: technically, institutionally, culturally, and economically. My aim is not to question that expansion *per se,* but rather to question the conventional understanding. . . . The concept of genetic disease, enthusiastically appropriated by the medical sciences for complex institutional and economic reasons, represents an expansion of molecular biology far beyond its technical successes. . . . Today we are being told—and judging from media accounts, we are apparently coming to believe—that what makes us human is our genes. Indeed, the very notion of "culture" as distinct from "biology" seems to have vanished.[79]

In his 1991 book *Backdoor to Eugenics,* Berkeley sociologist Troy Duster noted:

It is the halo from the molecular work of the last decades that has helped provide new legitimacy to the old claimants. . . . Those making the claims about the genetic component of an array of behaviors and conditions (crime, mental illness, alcoholism, gender relations, intelligence) come from a wide range of disciplines, tenuously united under a banner of an increased role for the explanatory power of genetics. Relatively few of these claims come from molecular genetics.[69]

The elucidation of genetic mechanisms for specific diseases loaded a layer of race on top of medical genetics. Population groups of different geographic origin, it was argued, are disproportionately prone to some genetic diseases; hemoglobin disorders are more common among those of African, Mediterranean, or Southeast Asian descent; Tay-Sachs disease is more prevalent among Eastern European Jews. These differences can therefore be traced to mutations passed from generation to generation, and only slowly dissipated through intermarriage. The general acceptance of racial differences in disease susceptibility spilled almost imperceptibly into an interest in studies of other traits less clearly "genetic." Success in explaining mechanisms behind a few genetic disorders lent credence to more general claims about mental capacity, gender, and socially desired or unwanted characteristics.

The 1990s began with enthusiasm for genetics, carried on the wings of startling progress in molecular biology. The very real power of new techniques to lay out detailed mechanistic causal chains for specific diseases commingled with studies that projected the medical model into the social realm. Historian of science Daniel Kevles pointed out how the genome project itself grew, in part, out of eugenics and might benefit from its lessons: "In its ongoing fascination with questions of behavior, human genetics will undoubtedly yield information that may be wrong, or socially volatile, or, if the history of eugenic science is any guide, both."[80] As the genome project gathered steam, its natural tendency to rhetorical overreach began to be counterbalanced by sympathetic but critical colleagues in the humanities and social sciences. The ELSI research program would further feed this generally salutary development.

New genetic knowledge seems destined to bring genetic tests that will collide with a growing movement for disability rights. The battleground is likely to be prenatal genetic testing. For a disorder such as Tay-Sachs disease, an unremitting and severe disorder in which children are in essence born dying, prenatal testing is generally accepted. Abortion of a prospective child destined to a short life filled with pain and inability to respond to the world is, to most, a morally acceptable if tragic choice. Abortion for conditions with greater clinical variability, with a mix of genetic and environmental causes, of lesser severity, or with late onset are less obviously beneficial. To those for whom abortion is morally wrong, prenatal diagnosis followed by abortion will not be an option. Prenatal diagnosis may perhaps enable treatment before birth or soon after, or may yield information about what to expect. Theirs is the only moral scenario in which the responses to the new technologies are relatively clear—don't use them or use them only for information and treatment. To women for whom abortion is morally acceptable, the choices are more difficult to sort out. Aborting a fetus with a genetic disease is agonizing and painful, like the death of a wanted child. The new technologies create a more tentative pregnancy, in the words of sociologist Barbara Katz Rothman.[81; 82]

Choosing abortion on the basis of an expected disability raises the specter of choosing what kind of children there should be. The choices implicitly force judgments that echo debates about what lives are worth living, arguments that in an earlier era mushroomed into Nazi atrocities. To some in the newly vibrant disability rights movement, it is an ominous development. Someone born with a disability that is diagnosable before birth can point out that if the diagnostic technology had existed while they were in gestation they would not have been born. Their lives are patently and obviously worthwhile. How can preventing the births of others like them be right? This poses not only a practical, but also a deeply philosophical dilemma. University of Wisconsin philosophers Daniel Wikler and Eileen Palmer have noted:

The charge that medical genetics is a potentially threatening eugenic program begins with the observation that much of medical genetics aims to combat disease not by healing anyone but by preventing the conception or birth of afflicted individuals . . . by picking and choosing among the potential people who might be conceived and born. . . . There are increasing signs, in the United States and Western Europe at least, that some disabled people increasingly identify themselves as a social group. . . . there is a strong disability rights movement, with political influence, there are leaders, and newsletters, and even radical and conservative factions. . . . Advocates of the disabled have urged, for example, that television programs show actors who seem ordinary in most ways but who may be blind or who utilize a wheelchair.

For medical genetics, the disability rights movement is of particular importance. It amounts to an ideological challenge, and it is mounted by the movement's most assertive, radical wing. The point of the radical disability rights critique is that even major disabilities, such as blindness, can under more just social conditions be merely one item in a very large inventory of life circumstances in which an individual might find himself; it is unfair to these people both to fail to create circumstances which minimize the burden of the disability and also to exaggerate the importance of the condition so much that it means that the person is thought of by others primarily in reference to the disability . . . Rather than prevent the birth of these kinds of people, they argue, we should change our attitudes about them, accepting them as equals and as essentially unremarkable.[83]

This is no mild conundrum. Adrienne Asch, a bioethicist long interested in disability questions, has taken a novel tack in considering abortion to avoid a child's future disability. She accepts a woman's *legal* right to choose to terminate any pregnancy, based on wanting or not wanting a child, and thus far remains in the feminist mainstream. But she parts company with many women in questioning the *moral* legitimacy of *selective* abortion for any but the most severely disabling conditions.[84; 85] She accepts abortion to prevent the birth of children with disorders such as anencephaly or Tay-Sachs, but questions abortion of children destined to develop cystic fibrosis or Down's syndrome, for example. She leaves the legal door open to such abortions, but believes that women might be persuaded not to walk through it on moral grounds. It is a subtle argument aimed at women's consciences, not legal rules.

Another tack to address the hard choices about abortion, genetics, and new reproductive technologies is to focus on nurturance. Ruth Schwartz Cowan, a historian of technology at Stony Brook, poses the principle this way:

An embryo cannot become an infant unless it is nurtured; an infant cannot become an adult unless it is nurtured and—at the other end of the developmental spectrum— adults who are ill or disabled cannot continue to live unless they too are nurtured. Nurturance is a continuous, day-to-day, mundane process: feeding, sheltering, protecting, assisting. Its goal is, in the case of embryos, to create an individual who can have a relationship with other individuals, in the case of adults, to maintain and sustain the life of an individual who has relationships. . . .

If this principle were to be taken seriously it would follow that when individuals cannot, for whatever reasons, make decisions for themselves, the person or persons who have the right to make the decisions are those who are nurturing the individual. Whether or not we agree that a fetus is an individual we can still agree that it is not capable of making decisions about itself. This means that decisions about an embryo or a fetus which is *in utero* ought to be made by the person in whose uterus it is developing; this person may or may not be its biological mother or its intended social mother, but certainly won't be its father or a doctor, or the governor of the state in which it happens to be located.[86]

Philosophers Wikler and Palmer take another way out, drawing a distinction between choices about imagined future children as opposed to loving those children actually born, with or without disability. They argue:

. . . A prospective parent is in a quite different context of moral choice than the parent of an actual child. . . . Before the child comes into being, we favor one list of attributes—the healthier ones—over another, if we get to make that choice. This is quite common. But it is unusual to find a parent who wishes that some other parent's child were his, even though each of us knows children in other families who are superior in some respect or other to our own. Thus before the fact we hope for a healthy child, but after the fact we do not regret having the children we do.[83]

They thus apply a concept from moral philosopher Thomas Nagel to the case at hand. Philosophers, social scientists, clinicians, scientists, and those making choices on this ethical frontier cannot help but confront questions for which the answers are but partial and tentative. The thorny and extremely divisive debate about abortion is certain to pervade future debates about biting into the fruits of genome research. Hovering behind the specific controversies about eugenics, disability, abortion, privacy, and other social, legal, and ethical issues is the social history of genetic explanation.

The dangers of genetic deterministic overreach are fed by claims about the power of genetics to explain what we most want to know. For those who toil each day in research laboratories in quest of disease genes, it seems a natural truth that finding genes and their products will illuminate function, and that would be a good thing. Indeed it is, but the public response to the advance of genetics is not received in this context.

Finding a link between Alzheimer's disease and a chromosome region for the Ross family, for example, would be an intriguing scientific clue. It is a long way from finding such a linkage to finding the gene, however, and it is already clear that there are several genes that might cause Alzheimer's disease in families. It is also clear that genes do not wholly determine the disease, as identical twins can differ in age of onset by a decade or more.[87] Environmental factors are at work. Even if all the "genes for" Alzheimer's disease were to be discovered, there is likely to be a long and highly branched causal network. And this for a relatively well-circumscribed biological phenomenon—a disease running in families as a Mendelian trait. How much more complex are other human characters likely to prove?

The explanatory choice between genetic determinism and environmental determinism is a false dichotomy. There are times when a powerful genetic prediction is possible. (Carrying the gene for Huntington's disease, for example, strongly predicts that the disease will ultimately appear. Even here, however there are large variations in severity and age of onset.) Most diseases lie far from this polar extreme, and general characters such as intelligence and athleticism farther still. The point is not that genes don't matter for such characters, or that science will never find "genes for" such characters, but rather that the relative power of the genetic explanation should not be projected from the case of Huntington's, where it is high, to the case of alcoholism or schizophrenia or, worse still, to criminal proclivity or intelligence.

Social analysts differ in their analysis of how dangerous and pervasive genetic determinism will prove to be. The history of eugenics and racial hygiene is enormously disturbing, but it occurred without the countervailing forces of critical scrutiny from inside or outside science and medicine. The genome project seems unlikely to escape such scrutiny, and indeed is nourishing it. Moreover, the Holocaust grew from a political environment fraught with problems much worse than biological determinism. Nazi racial hygiene was fueled by the biology of its day, but the biology did not cause it. As Daniel Kevles once observed, "if a Nazi-like eugenic program becomes a threatening reality, the country will have a good deal more to be worried about politically than just eugenics."[88] He added, even more cogently, that "if we do not use our knowledge wisely, it will be a failure not of science but of democracy."[89] The caution is apt, because we all know of many such failures.

Those who craft public policy, whether from executive offices or legislatures or kibitzing from the academic sidelines, might find little consoling about the fact that totalitarian eugenics would require political apocalypse. A milder but more sustained encroachment on liberties might prove pervasive without authoritarianism. Genetic discrimination and abuse of private genetic data are conceivable, and indeed likely, without policies devised to counteract them. The relative power of science and its critics is far from clear. Howard Kaye, a sociologist at Franklin and Marshall College, has observed:

As our latest attempt at dropping some moral anchor, biology may prove as ambiguous and unsuccessful as previous scientific moralities—and perhaps even more harmful. Our current infatuation with biology, unlike that of a century ago, is occurring at a time when the humanities and social sciences have declared moral bankruptcy, thus depriving us of a vital part of the collective memory we need to regulate and resist our increased capacity for genetic manipulation.[90]

Kaye worries further that "the cumulative effect of the ways such knowledge is likely to be interpreted for and by the broader public will push us, like sleepwalkers, toward the biologizing of our lives in both thought and practice."[90] Genetics might indeed overshoot its actual accomplishments, inserting itself unobtrusively into the unquestioned premises of common culture. Or it might not. Kaye's caricature of 250 million people corralled passively by a thousand or so scientists seems no more accurate a portrait of the future debate, for its pessimism, than the optimistic visions of genome enthusiasts. The critics are also prone to rhetorical excess.

A future in which genetic determinism implodes scientifically as a consequence of its explanatory failures is equally plausible, just as Newtonian mechanics collapsed in the face of the probabilistic physics of quantum mechanics earlier in the century. This seems not merely possible, but likely. Both an expansion of genetic determinism and a weakening of its foundations seems likely to follow, affecting different people in different ways. While those confronting human genetic disease in clinics day by day are unlikely to fall prey to simple genetic determinism, the culture is nonetheless vulnerable to muddle-headed claims about the genetics of intelligence and criminality.

A compact disk containing the DNA sequence of President Abraham Lincoln's genome would tell us very little about the President that we would really want to know. Whether or not he suffered from Marfan's syndrome, a genetic disorder not yet described in his time, would be a minor embellishment in his biography. It is of interest to historians of medicine and human genetics, and might have been of interest to Lincoln himself when choosing whether and how to have children, but the DNA sequence can contribute only a small increment to our understanding of his political ascent and the conduct of his presidency. Blanket generalizations about the worth and danger of genetic information, robbed of their specific social context, render them almost meaningless. And that was the whole point of the genome debate.

Watson's simple dictum to "just do good, and don't care if it doesn't seem good to others" gave way to a recognition that building the scientific foundations required public trust, and a major commitment to systematic exploration of how genetic science would work its way into the world. It was a complex process that would evolve with the science. In hearings for the 1993 NIH genome budget, Watson noted that the ELSI program's budget had risen from 3 percent to 5 percent in 1992, and further indicated, "I would not be surprised that five years from now this area will be 10 percent of our budget."[91] Trust in science involved a renegotiated social contract between scientists and the public that supported their work—those who would bear the brunt of any adverse impacts. The price of intellectual autonomy and support through public monies was continual public scrutiny of the scientific process and its results. The genome project placed genetics under that magnifying glass, where its past would be judged and its future assessed. The ELSI program was an attempt to make the negotiation process open and explicit. Attaching public bioethics to the scientific research program was a new anharmonic in the cacophonous din of democracy.

17
Bioethics in Government

A HEARING ON November 9, 1989, marked Senator Albert Gore's return to the issue of genetics. There had been several hearings in the Senate and House focused on the genome project, and most had touched on ethical issues, but the Senate Subcommittee on Science, Technology, and Space was the first to devote a hearing specifically to social implications of the genome project.[1] Among members of Congress, Gore had long associated himself with ethical issues in human genetics and reproductive technologies. While chairing a subcommittee in the House of Representatives during the early 1980s, he presided over a series of highly publicized hearings on human gene therapy, genetic testing in the workplace, and new reproductive technologies. In 1984, Gore ran successfully for the Senate. His interests in genetics continued unabated, but he was too junior in the Senate to have a platform on which to exhibit them, and a 1988 effort to be the Democrats' presidential nominee took him away from bioethics. His first opportunity to air those concerns came when he assumed the chairmanship of the Science, Space, and Technology Subcommittee of the Committee on Commerce, Science, and Transportation in 1989. Soon after assuming the chair, he scheduled a hearing on the human genome.

The highly public debate about the genome project and well-publicized successes in finding the cystic fibrosis gene and others had rekindled public interest in human genetics. The genome project would clearly result in much greater knowledge about human genes and would produce technologies to make genetic tests faster, cheaper, more accurate, and applicable to many more diseases. The issues of genetic discrimination in employment and insurance and the prospects of backdoor racism through genetic screening and testing became more urgent because of the genome project. Genetic testing and genetic discrimination had been topics of considerable public debate in the 1970s and early 1980s, sparked by genetic screening for sickle-cell disease and the recombinant DNA controversy, but the issues had lain dormant for several years.

In its 1983 report on genetic counseling, the President's Commission for the Study of Ethical Problems in Medicine and Biomedical and Behavioral Research presciently noted that the issues were unavoidable:

Within the next decade screening for cystic fibrosis may be possible. This could be of great benefit. If adequate preparation for its introduction is not made, however, it could also create serious problems. . . . The possible demand for millions—or tens of millions—of tests in a short period of time, and the consequent need for follow-up diagnostic studies and counseling, is daunting in itself. The Commission . . . encourages continued attention to this area by government officials, as well as by people knowledgeable about relevant scientific, ethical, social, and legal concerns.[2]

The President's Commission's reports on gene therapy and genetic screening were aimed at reaching policy guidelines. The President's Commission built on earlier work on genetic testing and screening by the Hastings Center,[3; 4] the National Academy of Sciences,[5] the March of Dimes,[6] and other groups. Soon after releasing its genetic screening report, however, the President's Commission passed out of existence. No federal body existed to monitor implementation of its recommendations.

Its successor, the Biomedical Ethics Advisory Committee and the congressional Biomedical Ethics Board, got stuck in the quagmire of abortion politics. These Siamese twins—a congressional board linked to an outside advisory committee and staff—grew out of a bill introduced by Senator Gore in 1983. Debate about a national commission on ethical and social implications of genome research has had a long history.

The idea for a federal bioethics commission grew out of the remarkable success of two previous bioethics commissions, the President's Commission for the Study of Ethical Problems in Medicine and Biomedical and Behavioral Research and its predecessor, the National Commission for the Protection of Human Subjects of Biomedical and Behavioral Research.[7] When the National Commission was first created in 1973,[8] pundits forecast failure and endless controversy.[9–15] It was created over considerable opposition from scientists and clinical investigators wary of regulatory incursions into research. The Nobel Prize–winning biochemist Arthur Kornberg and the noted cardiac surgeon Christiaan Barnard testified before the Senate that a national commission would hand a license to pen-toting bioethicists who would hold up a healthy research enterprise.[16] When Senator Walter Mondale searched for a famous scientist to support his view that a commission was needed, he found only a few; the one he chose to quote in support was James Watson.

The fear of the day was that methods emerging from embryology would enable the cloning of humans. Watson spoke before a special meeting of the House Committee on Science and Astronautics and discussed recent developments in embryology and genetics. He warned that "any attempts now to stop such work using the argument that cloning represents a greater threat than a disease like cancer is likely to be considered irresponsible by virtually anyone who understands the matter." Having defended the importance of free biomedical research, however, he went on to note the need for public discussion of its applications to humans, especially reproductive technologies:

It is absolutely essential that within the United States, if not in every other country, very important committees be set up basically to know the state of the art . . . and inform the public as a whole. This is a decision not for scientists at all. It is a decision of the general public—do you want it or not? It is not a question for a group of scientists to decide . . . it is a decision which the people as a whole must make. . . . If we do not think about the matter now, the possibility of our having a free choice will one day suddenly be gone.[17]

Senator Mondale began movement toward a commission in 1968 and repeatedly introduced legislation to create one, but it failed in several Congresses for want of House support. In the early 1970s, concerns about heart transplantation and the onslaught of new and powerful genetic technologies intensified concern. The question of fetal tissue research emerged as a national issue. In April 1972, the *Washington Post* reported that NIH scientists were using aborted fetuses for research in Finland, provoking demonstrations and calls for a halt to such research.[18; 19] A series of scandals further indicated to Congress that biomedical researchers could not keep their own house in order. Highly publicized Senate hearings between February and July 1973, before Senator Edward Kennedy, uncovered incontrovertible evidence of research abuse—Tuskegee syphilis trials that left a cohort of four hundred poor black males untreated for decades; hepatitis experiments that inoculated young, mentally infirm residents of the Willowbrook facility with live virus; injection of cancer cells into senile patients at the Jewish Chronic Disease Hospital; use of prisoners to test drugs; whole-body-radiation experiments sponsored by the Department of Defense; testing of hormone analogues among welfare mothers and Mexican-American women.[16; 19; 20] These disclosures undermined those opposed to a commission, and a bill finally passed both houses. President Ford signed it into law on July 12, 1974.[19] The National Commission's first mandated task was to address one of the most contentious issues, fetal research, in a report due three months after it started work.[18] It seemed an impossible task, and well it might have proved to be. The National Commission, however, surprised almost everyone by producing reports with direct policy impact and lasting scholarly value.

The National Commission operated from 1974 to 1978. In its opening gambit on fetal tissue research, the commission was forced to deal immediately with an explosive issue, provoking strong passions, street demonstrations, and opposition from powerful religious groups. The commission cut its teeth on fetal tissue research and went on to produce another seven reports.[21–30] Far from failing, the National Commission became a model of rational policy-making.[19; 31] National Commission reports laid the groundwork for regulations to protect human research subjects.[32; 33] The commission also articulated the principles guiding its approach in the Belmont Report, a landmark in the history of bioethics as a field.[26] The Belmont Report drew out the three principles—beneficence, justice, and respect for persons—that governed the deliberations of the National Commission in its dealings with various problems and that subsequently dominated bioethics scholarship for the next decade. The National Commission went a long way toward establishing that part of bioethics related to public policy. Rather than dying in disgrace, it begat the President's Commission.

In November 1978, Congress created the President's Commission for the Study of Ethical Problems in Medicine and Biomedical and Behavioral Research. Its mandate was broader than the National Commission's, encompassing the protection of human subjects in research but also extending into the delivery of health care. It was to supplant the National Commission, which had expired four years earlier. An Ethics Advisory Board operated in the Department of Health, Education, and Welfare (HEW) during most of this period, but the board's function was mistakenly thought to overlap with the new President's Commission. HEW assented to Congress's diversion of the board's appropriations to support the operation of the new President's Commission, beginning in fiscal year 1980. The new commission had not only a more general mandate, but also elevated presidential status, whereas the National Commission had operated autonomously in HEW.

The President's Commission, created by Public Law 95-622 (1978), operated from 1980 to 1983, issuing eleven reports.[2; 34–43] Two of these dealt with genetics. The 1983 report on genetic screening and genetic counseling was quoted above. At a hearing on human gene therapy in November 1982, Gore presided over the release of the other genetics report, *Splicing Life*. Gore's idea for an independent bioethics commission came directly from recommendations in that report. Alexander Capron, executive director of the President's Commission, was the star at this, the first congressional hearing I ever attended.[43; 44] The proposal for an autonomous bioethics forum was transformed into a congressional body under pressure from Senate conservatives, particularly Jeremiah Denton.

Several Senators—Gordon Humphrey, William Armstrong, Jesse Helms, and James East—were particularly incensed at the President's Commission report *Deciding to Forego Life-Sustaining Treatment*,[39] which asserted that feeding and hydration were like other medical treatments and could thus be stopped in some cases. If there was to be a bioethics commission, these senators wanted it more cognizant of their views. This was the argument that brought the bioethics board under Congress's thumb, modeled on the Office of Technology Assessment. The newly reshaped bioethics forum emerged from a House-Senate conference over the NIH authorization bill in 1984. Anthony Robbins, staff physician for Rep. John Dingell, and David Sundwall, staff physician for Senator Hatch, were the principal architects. This new bioethics body was incorporated into the Health Research Extension Act that was passed over President Reagan's veto in May 1985.[45]

Creation of the congressional Biomedical Ethics Board and Biomedical Ethics Advisory Committee (BEAC) ended the string of successful federal bioethics commissions at two. If the President's Commission and National Commission were home runs, the Biomedical Ethics Board was a strikeout. The board and the advisory committee operated effectively for less than a year, from September 1988 through September 1989, although they existed on paper and consumed considerable energy within Congress from 1985 through 1990. BEAC and the staff it was authorized to hire were responsible for conducting the work and producing reports. BEAC's members were screened and selected by the congressional board in a process that took over two and a half years. BEAC finally met in September 1988, four days before its initial authorization was to expire. Alexander Capron, who had been executive director of the President's Commission, was elected chairman; Edmund Pellegrino, one of the nation's best-known physicians in bioethics, became vice chairman. At the eleventh hour, Congress reauthorized BEAC and its congressional board for another two years. I was hired as acting executive director in December 1988.

The appointed committee members worked well together at both meetings the committee managed to hold before dissolving. A September 1988 meeting focused on election of a chair and vice chair and agreement on operating rules. A February 1989 meeting focused on human genetics. BEAC had a congressional mandate to report on ethical problems related to "human genetic engineering." The committee interpreted this to mean gene therapy as well as uses of genetic testing. Mandates for two other studies had later deadlines and shorter legislative histories, and were politically more complex, so they were moved to the back burner while the committee worked to establish a successful operating style.

LeRoy Walters testified before the committee, as chairman of the NIH subcommittee that oversaw gene therapy. Walters saw little need for yet another report on gene therapy, but much need for thought about uses of genetic testing and screening. He recounted the conclusions of seventeen reports from around the world, all of which agreed that somatic-cell therapy (affecting only the person treated) was morally equivalent to other kinds of therapy. Gene therapy that would cause inherited changes, by affecting sperm and egg cells or early embryos, lacked this consensus, but was technically difficult and unlikely to become practical in the foreseeable future.

Not only was there consensus on policy, but also Walter's committee at NIH as well as the Food and Drug Administration was actively engaged in overseeing the first gene therapy trials. Policy regarding gene therapy was thus thoroughly scrutinized. The same could not be said for uses of genetic tests, where there were many unresolved issues ripe for inquiry.[46] Neil Anthony Holtzman (Johns Hopkins), George Cahill (Howard Hughes Medical Institute), and Daniel Kevles (Caltech), the other invited speakers, concurred.

The committee decided to focus attention on issues of genetic testing and screening and the potential for discrimination in private insurance and in the workplace. The Biomedical Ethics Advisory Committee was poised to start this first project. It was on the verge of commissioning papers when its congressional board blew apart. Distrust among members of the board grew from 1985 through 1988 in the process of appointing BEAC members. The board fell into deadlock when liberal Republican senator Lowell Weicker lost his reelection bid and was replaced by Oklahoma conservative Don Nickles. The advisory committee might have lived on, but one member, the highly respected pro-life lawyer Dennis Horan, died. This left a troublesome vacancy. Gore promised the conservative Senate members that the slot would be filled by a pro-life candidate, but his staff person Gerry Mande tossed out the idea of several strong pro-choice candidates (among them Kate Michelman of the National Abortion Rights Action League and Faye Waddleton of Planned Parenthood) before learning of the deal his boss had made. The miscue proved more than the board could bear, and it broke in two.

A March 7, 1989, meeting of the Senate members devolved into a shouting match between Senators Gore and Humphrey, over the degree to which the pro-life and pro-choice power balance was assured on BEAC. They argued back and forth about who had promised what to whom. Tempers were already hot amid a partisan and acrimonious fight over John Tower's confirmation as Secretary of Defense (a nomination ultimately rejected by the Senate). After this meeting, BEAC was cut from its moorings and set adrift. As I heard the senators yelling at one another, I could vaguely sense my job disappearing and the nation's only national bioethics forum crumbling to dust. My first reaction was fascination that decisions could be made this way, not like my image from eighth-grade civics, with all those impersonal checks and balances of power. Stunned amazement soon collapsed into cynicism, which never entirely dissipated. The good-humored support of chairman Capron and fellow staff person Clair Pouncey were the only redeeming features of a long and frustrating period.

We labored mightily for months to find common ground that might save BEAC. Despite protracted negotiations, the board and the advisory committee died at the end of the fiscal year, and with them the main federal forum to discuss the issues of genetics and public policy. As it turned out, Senate conservatives Gordon Humphrey and Don Nickles killed the agency. If they had not done so, however, Rep. Henry Waxman, a liberal Democrat, was rumored to be waiting in the wings to do likewise. Distrust of the committee was intense from both ends of the political spectrum, with both believing they had conceded too much to their ideological opponents. Congress simply did not trust its own creation.

The power that flows through Washington is the cause of the syndrome known as Potomac fever. The power is real, but evanescent. I will long remember calling an appropriations committee staff member in the waning days of fiscal year 1989 to find out whether I should show up for work the following Monday. The appropriations bill, which was still pending, was no longer a set of abstract words that affected some distant federal agency, but the source of my paycheck four days hence. When the committee staff director did not pick up the thrust of my question, I asked bluntly: "Can I get paid next week?" He replied: "I hadn't thought about that. . . . I guess you're right, you can't." Nickles's staff crafted the language to kill BEAC and did the eleventh-hour maneuvering to insert it into the appropriations bill just days before the end of the fiscal year. They never called to warn me that I might be wise to look for a job. The demise of the BEAC, with a passing reference to my unemployment, was covered in the *New York Times* a few weeks later. Friends from all over the country called to ask what had happened. With a few days' notice, I was a thirty-five-year-old former executive director on the streets. Welcome to Washington.

Gore's November 9, 1989, hearings consolidated NIH's ELSI program and extended it into DOE. If Watson's notion was open to question before, it was thereafter locked in place by clear congressional intent. Watson saw where Gore would likely lead. Watson featured the ELSI program in his opening statement before the subcommittee. Robert Wood, acting director of DOE's Office of Health and Environmental Research (out of which came the genome program), spoke after Watson.

As Wood was reading his prepared statement, Gore pushed aside his microphone and turned to his staff. He asked if DOE had made a commitment of funds to match NIH's ELSI program. Gore then interrupted Wood to ask him directly. Wood began to answer that NIH would address the necessary ethical and legal issues, although DOE was quite concerned about them. Gore came back at him, asking specifically whether DOE had a budget commitment similar to NIH's. Gore suggested strongly that DOE have one. The senator warned of future hearings on the genome project where this would come up. Gore's position was endorsed by Senator John Kerry. It was as clear a signal as Congress could send.[1]

Despite this warning, at the December 1989 meeting of the joint NIH-DOE advisory committee, Ben Barnhart, director of the DOE genome effort, was not certain whether DOE would directly fund "ethics." Watson warned: "If you don't, Congress will chop your head off."[47] Charles Cantor, Robert Moyzis, and Anthony Carrano, all directors of the national laboratory programs supported by DOE, concurred with Watson and expressed genuine interest in joining the NIH ELSI program. Congressional pressure, the interest of Energy Secretary James Watkins, and support within the national laboratories brought DOE back to where DeLisi had left it several years earlier. DeLisi intended to fund bioethical analysis, and he met with LeRoy Walters at the Kennedy Institute of Ethics to discuss the possibility in 1987, long before NIH evinced an interest. DeLisi left DOE, however, before an ethical and social analysis program was in place, and the trail disappeared for two years. Following discussion at the December 1989 meeting, DOE agreed to cosponsor Nancy Wexler's ELSI working group, making it a joint NIH-DOE advisory body.

If the point had not already been made, Congress's concern about social and legal issues was brought home where it really counted, in appropriation hearings. In NCHGR's first appropriations hearings as an autonomous NIH Center, Congressman Obey pointedly raised questions about how insurers and employers might use genetic information to discriminate unfairly against individuals.[48–51] The House appropriations report for the 1991 NIH budget stipulated that NIH come up with a systematic plan to deal with such ethical issues and to develop specific policy options to address those issues.[48] The NIH genome office was thus the first science office with a congressional mandate to mount not only a scientific research effort but also a parallel program to forestall its adverse impacts. With almost no dissent, the appropriations committee ratified Watson's precedent. It also went beyond it, however, and moved in the direction of reestablishing a federal capacity for analyzing bioethics and public policy.

Congressional concern about the implications of genome research reflected ambivalence among the general public. The potential for discrimination on the basis of one's genes emerged as a policy issue just as the genome project was gathering steam. The technical debate about the genome project within science no doubt fueled this movement, but public concern grew even more from successes in mapping specific disease genes.

The utility of the RFLP map was being proved by results. Yet every time a new disease gene was mapped, a potential diagnostic test was also created. New diagnostics informed medical decisions about risks of developing cancer or heart disease or Huntington's disease, but this medical information was also of potential interest to employers and private insurers, among others. Such third-party use of genetic patient data raised a host of difficult questions. The promise of medical benefit was inextricably tied to the prospect of social harm.

While members of Congress and the general public did not partake of the technical debate about the wisdom of the genome project as a scientific program, they could instinctively understand their stakes in its results. The project would unleash a flood of new information about human genetics. In an unjust society, genetic information could be harmful. In a world full of computers, intimate information could fly out of control, with only weak, incomplete, and outdated protections for confidentiality. Once debate about the genome project joined with concern about genetic testing, public reactions to the project were principally channeled through discussions of how increased genetic knowledge would change individual choices. Journalists and teachers used social issues as a hook to draw their readers and students into the science.

Genetic testing and confidentiality of genetic data became grist for the media mill. A 1989 *Time* cover feature on the genome project dedicated two of its five pages to its ethical and social impacts.[52] The Gannett Foundation sponsored a conference in November 1989 at which journalists from around the country listened to experts consider "The New Genetics and the Right to Privacy."[53] *Health* magazine's cover feature wondered about "Tinkering with the Secrets of Life."[54] *Consumer Reports* made concern about genetic screening and health insurance a cover feature in July 1990.[55] Social impact was a prominent theme in a "Mad Scientist" article about Watson in *The New Republic*.[56] Features ran in the Sunday "Outlook" section of the *Washington Post*,[57] *The New York Times Magazine*,[58] the *Wall Street Journal*,[59] and other papers.[60–64] Genes had the *geist* if readers had the *zeit*.

Journalists did not invent public concern, they conveyed it. Dorothy Nelkin and Laurence Tancredi warned of the social power of biological information, especially genetics, in their book *Dangerous Diagnostics*.[65] Radio commentator Paul Harvey expressed the popular distrust of scientific elites:

Genetic engineers are well aware that their science is frightening to a lot of people. Mostly behind closed doors, they have been exploring evidence that human genes can be manipulated to make us taller, healthier, more or less intelligent and more or less likely to commit crimes. . . . Yet, secretive as the genetic researchers try to be, some of their findings are finding their way into the public media. . . . The new technology of genetic probes has led us to the cause of muscular dystrophy and promises to lead us to a cure. Cystic fibrosis and Huntington's disease will be the next targets for molecular geneticists. And Alzheimer's and certain cancers. And—behind those still closed doors— who knows what else?[66]

Within science, there were concerns that the social impact of genetics was larger than could be managed. Liebe Cavalieri of the Sloan-Kettering Cancer Research Center noted how the quantitative change wrought by new genetic technologies would cause qualitative changes in public perceptions, some of them quite worrisome.[67] One ecologist suggested that the public was so woefully ill-informed about genetics that "until a concerted education [effort] is made, even walking along the human chromosomes may be too fast a pace."[68] The group associated with the Council for Responsible Genetics sponsored a symposium on the human genome project at the January 1989 annual meeting of the American Association for the Advancement of Science,[69] where Ray White from the University of Utah was the beleaguered genome supporter in a sea of critics. The council later issued two position papers, one a well-crafted paper on genetic discrimination and the other an incoherent rhetorical blast against genetic determinism.[70; 71]

The interests of private employers and insurers collided with an intuition that people should not be subject to discrimination on the basis of their genes. Like race and gender, genes were well beyond personal control. The dilemma was most acute for private health insurance, although the same principles applied to employment and to other forms of private insurance (such as life insurance, mortgage insurance, disability insurance, automobile insurance, and long-term care insurance).

Genetic discrimination began when physicians started to take family histories. Indeed, stories about insurance denial, among families with Huntington's disease, for example, were well known to genetic disease support groups. Few outside the field of medical genetics knew this, however, and the practice affected only a small fraction of the populace. The practice was also not universal; some insurers did it, but others did not. Moreover, most Americans got their health insurance through their employers, and group policies traditionally had few exclusions. As the 1980s gave way to the 1990s, however, employers became far more attuned to health care costs and insurers ever more concerned about financial risks.

People who knew they would develop a disease, or even those who knew they were far more likely to do so than others, could load up with insurance, throwing off the actuarial tables and saddling an insurer with extra payouts. This undermined the whole premise of insurance, as a mechanism to protect against *unpredictable* events. This was more than a theoretical concern, since private social security firms had gone bankrupt earlier in the century, leaving policyholders bereft of benefits despite years of payments.

The potential conflict between predictive genetic testing and private insurance had been noted many times before. The President's Commission's 1983 report on genetic testing and screening did not address insurance specifically, but it loomed in the background.[2] By the time OTA reviewed issues surrounding genetic testing in an appendix to its 1984 report *Human Gene Therapy*, private insurance was already becoming a more prominent policy issue. Stephen Eckman, a summer fellow at OTA from the Wharton School, called several private insurers and found that they would likely use genetic data to make premium and eligibility decisions if such data were available.[72] Just a few years later, Neil Anthony Holtzman focused his insightful book *Proceed with Caution* on issues surrounding genetic tests and found private insurance one of the most vexing issues.[73]

The rancorous debates about AIDS testing and drug testing during this period sensitized the genetics community to the public policy issues surrounding medical tests. In 1988 and 1989, several new books devoted sections to genetics and private insurance.[65; 74; 75] The rising costs of private health insurance gave employers a financial incentive not to hire those who would incur health care costs. This affected not only the prospective employees themselves, but also any dependents who would be covered under employment-based health plans. Legal scholar Mark Rothstein, from the University of Houston, noted: "The problem with employer-provided health insurance is not that it is employer-funded . . . [but that] employers are increasingly acting as health insurance underwriters. The growth of self-insurance has operated to magnify this problem."[74]

Insurers began to take notice. Robert Pokorski, medical director at Lincoln National Life Insurance, chaired a task force on genetic testing and had edited a "white paper" for the Council.[76] The papers in the collection pointed out the pros and cons of using genetic information, particularly genetic test results, but made no policy recommendations. The white paper was discussed at a CEO-level meeting of the American Council of Life Insurance (ACLI) in July 1989, where I was invited to present my perspective as acting director of BEAC. Ian Rolland of Lincoln National Life wondered aloud whether the industry shouldn't get behind legislation to level the playing field, proscribing insurance use of genetic tests. As it was, if one company used the tests, all companies might have to follow suit or risk losing a competitive edge. If genetic information was not used by any firm, however, then such competitive pressures would not build up. Those with genetic disease were, after all, already accounted for in actuarial tables. Rolland argued that the social contract between insurers and the public might demand that insurers refrain from assessing genetic risk factors. Most of the rest of the other CEOs were more skeptical about whether genetic factors could be factored out.

Within months, the insurance question had "arrived" as an issue. Pokorski and others from private life insurers were constantly on the road giving talks. The Health Insurance Association of America (HIAA) also formed a task force on genetic testing. Jude Payne, who staffed that group, was invited to defend her industry in public almost weekly. Eric Juengst commissioned Larry Gostin of Boston University and Nancy Kass of Johns Hopkins to prepare background papers on the legal issues for the ELSI working group.[77; 78] At a September 1990 meeting where these papers were discussed, the ELSI working group formed an insurance task force cochaired by Tom Murray and Jonathan Beckwith. The purpose was to mediate a productive debate at the national level.

The insurance issue was further complicated by its regulatory framework.[74; 75; 79] Much of the analytical capacity for thinking through public policy resided at the national level, but the relevant statutory law and regulatory power resided in state governments. As the debate intensified over reform of the health care system, genetic illness posed a particularly difficult dilemma. It pitted the interests of those who carried genes they could not control against the fiscal realities of a social policy full of inherent contradictions. A public consensus that health insurance should be an entitlement collided with the reality that health insurance was allocated through a private market that could not be both fair and purely competitive.

HIAA and ACLI issued a joint policy statement in February 1992, companion to an ACLI report on confidentiality.[80; 81] The HIAA-ACLI task force concluded that a more consistent rationale and set of principles would be desirable in dealing with the confidentiality and use of genetic tests.[80; 81] The ACLI document, in particular, noted that "the point here is that in the newly emerging area of genetic test information, adherence to principles rather than complex strictures may well be the preferable approach."[80] While aimed at avoiding an extremely complex and impenetrable regulatory framework, it was even more a call for just the sort of analysis at which the National Commission and President's Commission had excelled.

The critique from policy analysts and academics was overwhelmed for a time by a media blitz orchestrated by activist Jeremy Rifkin. On April 19, 1988, Rifkin held a press conference to propose a "Human Genome Policy Board" and "Human Genome Advisory Committee" modeled on the Biomedical Ethics Board and Advisory Committee.[82–85] The press conference was timed to precede an April 27 hearing before House Energy and Commerce chairman John Dingell, at which OTA's genome report would be released. Rifkin's list of supporters included Judith Areen (dean of the Georgetown Law School), Robert Murray (Howard University), Patricia King (Georgetown Law School), public activist Ralph Nader, Marc Lappe (University of Illinois), and James Bowman (University of Chicago). They were all genuinely concerned about genetic discrimination. Rifkin parlayed their interest into support for his initiative and used their names to lend prestige to his insatiable quest for publicity.

Rifkin next linked the genome project to a suit seeking to block the first human gene transfer experiment. On January 30, 1989, he brought a coterie of disability rights activists to a meeting of the Recombinant DNA Advisory Committee (RAC), which was in the process of reviewing the clinical protocol. He announced his lawsuit and opposition to the genome project until NIH set up an "Advisory Committee on Human Eugenics."[86] The composition of Rifkin's proposed committee bore an uncanny resemblance to his Human Genome Advisory Committee of the year before, despite its different purpose. The RAC "respectfully declined" his suggestion, while agreeing that workplace discrimination and insurance discrimination were issues that needed attention.[87] RAC saw no reason to couple these issues to gene transfer and cancer treatment.

Rifkin linked legitimate policy concerns, assembled an *ad hoc* coalition of distinguished supporters, targeted the salient genetics topic of the day, fanned public fears of eugenics, and kept himself at the center of attention. The press release quoted Rifkin as saying that "if we are not careful, we will find ourselves in a world where the disabled, minorities, and workers will be genetically engineered."[88] Being careful might well mean locking the door and throwing out the key.

According to Rifkin, blocking the threat of genetic discrimination justified stopping an experiment to mark cancer-killing cells in patients with terminal malignant melanoma. Hindering improvements to treat one disability (cancer) was his policy response to speculative dangers that might someday materialize if we slid down a slippery slope toward genetic treatments for other disabilities. A cynic might be forgiven for thinking this was a publicity stunt more than a bona fide attempt to improve public policy.

A copy of the lawsuit to block the experiment was passed out at the RAC meeting, but it was not filed for several weeks. NIH subsequently settled out of court, under terms that were not made public. Rifkin claimed victory, and NIH went on to approve the gene transfer protocol.

In the genome press conference and the suit against gene transfer, Rifkin was repeating earlier forays into the public discussion of genetics. Rifkin was a former class president, cheerleader, and economics major from the University of Pennsylvania who cut his political teeth in the antiwar movement of the late 1960s. He shifted his attention to biotechnology in the 1970s, drawn by the recombinant DNA controversy.[89] In 1977, he published a book with Ted Howard, *Who Should Play God?*[90] Rifkin followed this in 1983 with *Algeny,* a book against gene therapy.[91] Soon after *Algeny* was published, he organized a coalition of prominent clerics to oppose gene therapy of the germ line, which would produce inherited changes, not only in the persons treated but in some fraction of their progeny. (Germ line therapy could ensue from genetic altera-tions of sperm, eggs, their precursors, and cells of an early embryo.) Rifkin sent a proposed resolution to his coalition members. He then forwarded the resolution to Congress, along with a "Theological Letter Concerning the Moral Arguments Against Genetic Engineering of the Human Germline Cells."[92; 93]

Senator Mark Hatfield introduced the resolution in the Senate, where it died.[94] The "Theological Letter" was adapted from sections of Rifkin's book and argued that once any form of gene therapy began, society would be unable to stop it. Gene therapy for a serious genetic disease today, genetic enhance-ment of intelligence and athletic ability tomorrow.

It turned out that the signatories to the resolution had not seen the "Theo-logical Letter" that accompanied it, and many later recanted their support.[95–98] A group of clerics met in August to declare that the June 8 resolution had been "unnecessary and misleading."[99] They continued to be concerned about germ-line interventions, but no scientist was actively proposing them. Some of the clerics surmised they had been duped into promoting sales of Rifkin's book.

In July 1990, Rifkin met Watson briefly at the ABC studios in downtown Washington. They discussed how genetic information needed to remain con-fidential. Rifkin was proposing legislation to safeguard genetic information in the hands of the federal government. On July 24, Rifkin faxed a letter to Watson indicating that "John Fletcher, Tom Murray, Marc Lappe, William French Anderson, and others have all given their input on the bill." He warned that "many congressmen and senators on the Hill have expressed concern about continued funding of the Human Genome Project, worrying that the appropriate genetic privacy legislation needs to be passed 'before' the human genome database and screening information are too far along. . . . Getting this bill passed by next spring may well be key to securing Congressional and public support of the Human Genome Project," hinting that support for the genome project might be bought by support of his privacy legislation.[100] Rifkin's five-page draft bill was translated by congressional staff into H. R. 5612, a twenty-five-page bill introduced by Rep. John Conyers on September 13, 1990.

This bill was a hard issue for Nancy Wexler's ELSI working group to handle. At the January 1991 meeting, Madison Powers of the Kennedy Institute of Ethics spoke about the weak legal protections for confidentiality. Even in medical settings, there were few laws to protect confidentiality, and they had many gaps. The notion of patient autonomy might well be projected from person to information, asserting "informational self-determination" to protect one's interests—including who had access to genetic information—as well as one's body. ELSI group members discussed the Rifkin and Conyers bills and agreed that further legal analysis was necessary. Lori Andrews of the American Bar Foundation agreed to look into mechanisms for developing confidentiality statutes that would more adequately address the major problems than did the Rifkin bill.

Rifkin's contribution was to highlight the weak points of genetics and biotechnology and to focus public attention on them. His flaws were a neglect of homework and poor grasp of policy solutions. He used lawsuits to block government action with some success, but policy changes that required sustained commitment, such as policy analysis and legislation, were generally beyond him. Indeed, he could be something of a liability to his allies. Those who worked with him were often tarred by the association. If Rifkin had taken a position, opponents had a ready-made rhetorical tool. They need merely mention his position to cast doubt on its wisdom.

Policy debate about the confidentiality of genetic information came out from under Rifkin's shadow in 1991. On April 24, 1991, the Conyers bill was reintroduced as H. R. 2045, stripped of its enforcement provisions. The accompanying press release quoted Conyers, who borrowed several phrases from the Nelkin and Tancredi book:

The right to privacy is a personal and basic right protected by the Constitution. That right is now potentially threatened by major and important advances in the biological sciences that are expanding our understanding of the genetic components of human diseases. . . . Because of high-tech developments in gene mapping and screening, genetic privacy could become a major focus of the civil rights movement in the next twenty years. Allowing genetic information outside an individual's personal use threatens to open a "Pandora's box": we may well see genetic information used by the government and the private sector to create a "biological underclass" of those with "inferior" genetic makeups. One's genetic information should never be allowed to be used as a weapon against them.[101]

Sherille Ismail, Conyers's principal staff person for the bill, was initially optimistic about passage, but began to receive feedback from many quarters about its weaknesses. Those who should be friendly to it were finding defects. The intention for the bill changed from seeking passage to using it as a vehicle to provoke discussion about what should be done.[102]

A well-publicized policy problem in France highlighted the complexities faced by privacy laws. A team of researchers culling through family records in Brittany inadvertently discovered a form of glaucoma inherited in some families.[103] (Glaucoma is a treatable eye disease, often detected only after there has already been irreversible damage to vision.) The French team constructed enormous pedigrees and could identify many at risk of going blind. French privacy laws prohibited directly contacting those at risk, however, and the National Commission on Informatics and Liberties (a national privacy commission) decreed that it would also be unwise to list specific individuals when notifying local physicians. Instead, the policy became one of informing physicians in the area to be on the lookout for glaucoma cases, as some families were at increased risk. In the United States, the decision would likely have focused on the individuals' ability to secure information relevant to their own health. The investigators might indeed have been compelled to find those at risk, but in France as in much of Europe, privacy weighed much more heavily.

The investigators had initially approached the privacy commission with a proposal to study several untreatable diseases, including the psychiatric condition manic-depressive disorder as well as glaucoma. The privacy commission was concerned about confidentiality and the possibility of stigma. It was also concerned about intrusions on the privacy of individuals if they were contacted directly by investigators they hardly knew. The privacy commission asked the investigators to drop the work on manic-depressive disorder and to work only through local physicians.[104]

The conflict between individual privacy and public-health case-finding, starkly shown with glaucoma in Brittany, also arose in genetic studies of breast cancer, colon cancer, cholesterol metabolism, and other treatable (or preventable) diseases. It was especially vexing for those studying p53, a protein associated with some cancers. Some families inherited a p53 mutation that made them far more likely to develop cancer. How to study such families, what to tell prospective participants, and how to handle complex familial dynamics were already difficult problems in such families, and became all the more so when children were involved. And yet 20 percent of children with the p53 mutation might die of cancer before reaching the age of majority, when they could make their own choices.[105]

Those studying large pedigrees afflicted with illness confronted the issues every day, but with little policy guidance. Information in such pedigrees was not as simple as other medical information. Genetic data about one member of a family also related to others in the pedigree. Knowing the genetic composition or the clinical status of one family member might well inform others about their risk. When the disease could be treated or prevented, the stakes were especially high. The need for coherent policies based on more than *ad hoc* consideration began to become clear as geneticists discovered more genes predisposing to illness. The question was how to fill the policy gap.

The debate about social impacts was at least as active in Europe and Canada as in the United States, and was emerging in Asia as well. In Germany and German-speaking nations especially, human genetics labored in the shadow of eugenics and "racial hygiene." A spate of books emerged in the late 1980s, detailing how scientists and physicians promoted a racist agenda in the first half of this century.[106–112] The medical model of nondirective genetic counseling became dominant throughout the world in the postwar period,[113] and human genetics as a science explicitly rejected the tenets of eugenics and racial hygiene,[114] but the historical burden could not so easily be removed. Nonscientists were not going to give their trust automatically; scientists would have to earn it.

The Economic Summit nations (the so-called G7 nations) were joined by the European Commission at a meeting in Rome in April 1988 to discuss ethical issues surrounding genome research.[115] Bartha Knoppers from the University of Montreal noted that a new social contract was under negotiation. Individual genetic differences had to be accounted for in legal notions of equality.[116] Knoppers argued for a robust protection of legal equality despite genetic diversity. The connections between a person's genes and notions of human dignity, although murky, were clearly important to conceptions of the individual.[117]

Another contract was also being renegotiated between science and society. Science increasingly carried the mantle of responsibility for how the knowledge it produced was used. The genome project had taken a bold step by folding analysis of the social implications of genome research into the research plan itself. The notion of supporting a social analysis program along with genome research caught hold simultaneously in the United States, Canada, Europe, and Japan.

European efforts built on growing interest in bioethics and public policy. A series of reports stood as landmarks in the evolving debate.[118–129] The most public debate centered on the human genome component of the European Commission (EC) genome program.

The EC successfully launched genome research programs on nonhuman organisms in 1988. Plans were underway for a human genome program as well, but complex European politics came into the picture. Peter Pearson from the University of Leiden chaired a committee charged with formulating plans, until he moved to Johns Hopkins University, whereupon Malcolm Ferguson-Smith from the University of Cambridge took over. The scientific advisers recommended a three-year scientific program funded at 17 million ECU (European Currency Units, around $1.40 at the time). The European Commission plan was routed to the European Parliament. In the Parliament's Energy, Research, and Technology Committee, the bill was assigned to a reader, Benedikt Härlin, a German Grün (Green Party member). The bill had a rough ride through the European Parliament.[130]

When the bill got back to the European Commission, the players had changed. Research commissioner Filip Maria Pandolfi now had authority. He was sensitive about uses of genetic information and held up the proposal for a time.[114] The proposal's name changed from "Predictive Medicine" to "Human Genome Analysis" in its parliamentary transit,[131; 132] signaling a recognition of social concerns.

In 1988, the EC scientific advisory group formed a study group on ethical, social, and legal aspects (ESLA) of the Human Genome Analysis Working Party. Martinus F. Niermeijer of Erasmus University in the Netherlands chaired the ESLA study group, which prepared a series of documents and planned several public conferences and a program of activities to focus on implications of genome research. Inclusion cf a program to consider the ethical, social, and legal aspects of genome research with a 1-million-ECU budget (7 percent of the total program budget) cleared the way for approval of the overall program.[133] It was approved by the council on June 29, 1990, with the proviso that the program would implement confidentiality protections and would explicitly exclude germ-line genetic manipulations.[134] In November, the EC genome program gave out its first batch of eighteen one-year grants from among forty-two submissions.[135; 136]

In Japan, ethical analysis was also incorporated into the Monbusho (Ministry of Education, Science, and Culture) program, headed by Norio Fujiki, a medical geneticist from Fukui Medical School. A 1992 report from the privacy commissioner of Canada, *Genetic Testing and Privacy*, touched on medical testing, how insurers and employers might use genetic tests, and DNA forensics.[137] The Canadian genome program announced in 1992 also included a minimum 7.5 percent of its budget earmarked for analysis of social, ethical, and legal issues.[138] The development of bioethics studies in parallel to genome research was becoming an international phenomenon, but it was given the most resources in the United States.

Many months after its appropriations authority died, the Biomedical Ethics Advisory Committee reared its head again briefly, but only because of the ELSI working group. During 1989, the Americans with Disabilities Act was under debate in the House and Senate. Senator Tom Harkin and Representative Steny Hoyer introduced this sweeping revision of federal disability law, the first in well over a decade, into their respective chambers. Robert Silverstein, Harkin's staff director handling the bill, met with me briefly outside his office just before the Labor Day recess in 1989. We discussed whether the bill covered genetic disabilities and genetic testing. I also spoke with Chai Feldblum, an American Civil Liberties Union lawyer involved in drafting the bill. She noted that genetic disease had not been a major issue in deliberations about the bill, but the scope of its definition should encompass those with genetic disease.

The Americans with Disabilities Act was mentioned in passing at the Gore hearings on November 9, 1989,[1] and again at a Williamsburg, Virginia, ELSI meeting in February 1990, when Adrienne Asch pointed out its relevance to protection from genetic discrimination by employers. At that point, I dusted off some background memos from BEAC, written a year before.

As the bill neared final passage, I belatedly struck upon the idea of having the Biomedical Ethics Advisory Committee act, in its assigned role as adviser to Congress. The committee could not expend federal dollars, but it still existed on the books and retained its original mandate. The members had never resigned, nor had I done so as acting executive director. (They just stopped paying me.) BEAC's chair, vice chair, and acting director—Alexander Capron, Edmund Pellegrino, and I—sent a letter to Senator Harkin and Rep. Hoyer seeking clarification about whether the ADA covered genetic testing.

By the time we acted, the bill had passed both houses, but was held in a conference committee to resolve differences between the House and Senate bills.[139] The Biomedical Ethics Advisory Committee sent copies of its letter to congressional staff and outside groups working on the bill. The Council for Responsible Genetics sent a well-argued position paper along with collected cases of alleged genetic discrimination to Rep. Hoyer and Senator Harkin just a few days later.[71; 140] Congressmen Owens, Edwards, and Waxman noted how the ADA should prevent genetic discrimination in endorsement statements for the House, and Orrin Hatch did so in the Senate.[141; 142] Rep. Hoyer sent a letter in reply to the BEAC letter, indicating that genetic testing was never considered explicitly during debate, but the language of the statute was broad enough that courts would likely cover the situations of potential concern. Hoyer noted that implementation and interpretation of the statute would need to be monitored closely.[143] How right he was.

Mark Rothstein was an attorney at the University of Houston Law Center. His book on medical testing and the cost of employee health benefits included a section on genetic testing and insurance.[74] Rothstein followed the Americans with Disabilities Act closely. He urged the office implementing the statute, the Equal Employment Opportunity Commission in the Department of Labor, to interpret the statute so as to protect against genetic discrimination,[144] and he discussed his recommendations at a meeting of the ELSI working group in January 1991. Rothstein argued the Act should cover those expected to become disabled and also parents who were carriers of disease-associated genes. Employers might have incentives to discriminate against those who had a child with genetic disease, or were at risk of having one, because of expenses incurred under employee family health benefits.

If a prospective employee was a CF carrier married to another carrier, for example, employers might choose not to hire him or her, fearing the costs of medical care for a child born with CF. Rothstein argued that such discrimination should be proscribed, and was certainly within the intent of the ADA.

In draft regulations issued by the EEOC on February 28, 1991, genetic testing was not mentioned, but a category of "conditional offeree" was created.[145] This was a person who had been tentatively offered employment, but was then subject to a broad range of tests and medical inquiries by the employer. By this stroke, the EEOC undermined the ADA. Rothstein was appalled by several interpretations of the statute embodied in the regulations.[146] He was invited to address the ELSI working group again at its April 1991 meeting in Los Alamos.

The first day of the meeting was the last day of the comment period on the draft EEOC regulations. Rothstein presented his opinion, and after some confusion about the group's authority to comment on draft regulations from another part of the federal bureaucracy, the ELSI working group quickly prepared a statement and faxed it to Washington just minutes before the comment period closed. The group urged EEOC to explicitly protect those carrying deleterious genes, but not themselves affected by them, from genetic discrimination. The statement also asked that the regulations reject the subterfuge of conditional employment offers, or at least narrow the range of what information could be gathered about job applicants.[147] Those tested should be told the results of their tests, and EEOC should erect safeguards to protect the confidentiality of medical information gathered about job applicants. The working group also encouraged good data-management practices to restrict access to only those who processed employee health claims, so that only those with a need to see personal data had access to them.[148; 149]

Nancy Wexler presented the ELSI working group statement to the NIH / DOE joint subcommittee on the human genome on June 15, and it unanimously endorsed the statement. Paul Berg and Sheldon Wolff, subcommittee cochairs, sent a letter to EEOC chairman Evan Kemp that summarized the central points.[150] These efforts had no impact on the final regulations, which were promulgated on July 26, 1991.[151] There was a significant irony, in that Evan Kemp had previously made strong statements about the dangers of genetic discrimination in another context.

Kemp supported Rifkin's proposed bill in 1989, soon after he took the reins of the EEOC from Clarence Thomas as the equal employment opportunity commissioner. (This office achieved national notoriety when Thomas was nominated to the Supreme Court in 1991 and allegations of Thomas's sexual harrassment of Anita Hill at EEOC became a national news sensation.) In 1989, Kemp noted that "the terror and risk that genetic engineering holds for those of us with disabilities are well grounded in recent events. Baby Doe was not an isolated case. Our society seems to have an aversion to those who are physically and mentally different. Genetic engineering could lead to the elimination of the rich diversity in our peoples. It is a real and frightening threat."[86] Two years later, Kemp's EEOC chose to interpret the Americans with Disabilities Act differently from the lawyers who wrote it, thus missing an opportunity to prevent genetic discrimination of a different kind.

The ADA experience was important because it indicated limitations in the

ELSI working group's ability to analyze and formulate policy. The working group was operating in a mode of short statements separated by three- or four-month intervals of inaction. A policy analysis group would have met far more often, and would have much more actively monitored political events and relevant developments in academia. Most important, the short working-group statements said what to do, but did not lay out the reasons why. The statements were prepared in haste and lacked the coherence borne of sustained deliberation and systematic data-gathering. Much of the value of previous federal bioethics commissions came from the documentation of reasons, which in turn came out of a complex feedback loop including discussions, solicitation of new information and commissioned papers, and further analysis. Repeated meetings of the commissions had characterized previous bioethics commissions, in the United States and abroad. A capacity to do policy analysis depended on several full-time staff and an active committee.

Beyond these process limitations, and arguably more important, the working group was buried several layers down in the NIH and DOE bureaucracies. It was advisory to a joint NIH-DOE subcommittee, in turn advisory to the main outside advisory bodies to NIH and DOE. The ELSI working group was advisory to an advisory group to an advisory group to two different government departments. If a recommendation got to the parent organizations, it might still have to transit several more layers en route to the outside world. NIH and DOE officers had not hindered the working group's freedom to make statements, nor was there any indication they would do so in the future. But if a working-group statement touched on policies of either parent agency, recommendations might well require review through the Department of Health and Human Services and through DOE's hierarchy. It seemed likely that at some time, the working group would encounter an issue that touched directly on sensitive policy matters, such as abortion. The group's hard-earned freedom might then disappear. The working group lacked a clear mandate from its executive sponsors or Congress. It also lacked official standing, such as a congressional warrant, to comment on the policies of other agencies.

The working group had directly promoted a new NIH policy on CF pilot testing, a research program within the purview of one of its two parent agencies. In research policy, it might thus exercise influence, but when it came to the first area of social policy, the result was less impressive. The EEOC had safely ignored the working group's comments on ADA. EEOC did not have to respond to the working group more than to any other member of the public. The lack of clout was not due to the competence or intentions of NIH or DOE officials. It was an intrinsic structural problem. While ample staffing and institutional support might increase the working group's stature, nothing could substitute for a congressional charge.

The experience also brought to mind another recent failure in public bioethics—the 1988 Fetal Tissue Transplantation Research Panel.[152–155] This panel was convened by the NIH director's office in September, October, and

December 1988 to deliberate on whether NIH should fund research to trans-plant fetal tissue. Preliminary experiments using fetal brain cells to treat Par-kinson's disease and pancreatic cells to treat diabetes had been done abroad, and to a limited extent in the United States with private funds. A 1987 Uni-versity of Wisconsin grant proposed to use federal funds for pancreatic cell transplantation. A survey of federal grants uncovered more than one hundred that employed fetal tissue over the decades, but NIH director Wyngaarden judged that the contemplated experiments that would employ fetal cells in transplantation would attract more notice and were likely to be controversial. He sought approval from his boss, the assistant secretary for health and direc-tor of the Public Health Service, then Robert Windom.

Windom reacted by imposing a moratorium on such research and re-quested that NIH convene a panel to scrutinize how use of fetal tissue in transplantation might relate to abortion—specifically whether it might en-courage women to seek abortions. His staff drafted ten questions for the NIH panel to address, the first few of which focused on whether and to what degree tissue donation from a fetus might constitute an inducement to abortion.

As it turned out, the policy rationale turned on question 2: "Does the use of the fetal tissue in research encourage women to have an abortion they might otherwise not undertake?" This was a poor question to ask a group of "ex-perts," as it was obviously an empirical one, subject to assessment by sophisti-cated survey methods, but the method chosen was deliberative. The question could only be answered by data, but no data were gathered, and committee discussion was sought as a substitute. In the end, the panel voted on a series of answers to Windom's questions. A report was filed, with three statements concurring with the majority positions, two dissenting statements, and a letter that was passably close to a dissent. The panel's report was considered by the NIH Director's Advisory Committee, which approved the majority posi-tion.[156]

Windom had imposed the moratorium in hopes of forestalling a more permanent restriction from the White House. He intended to lift the morato-rium when "lo and behold, the hand from above denied me that privilege."[157] He later explained: "I thought at the time that by getting the interested parties to analyze the complexities of the issue before approval for the first implant would solve the problem once and for all; then we could go full speed ahead without hindrance. I also felt that if I did approve the initial request the White House would have overruled, and there might not have been the proper scrutiny of the issue, such that approval might never come."[158]

The process that started under Secretary Otis Bowen and Robert Windom in 1988 was then passed to Secretary Louis Sullivan and Public Health Service director James Mason in the new Bush administration. The new guard in the department was faced with a tough decision. In his Senate confirmation hear-ings, Secretary Sullivan had been given a rough ride over the abortion question by Senator Bill Armstrong, a strong pro-life advocate. Faced with the majority recommendation from the NIH panel, Mason and Sullivan demurred, and extended the research moratorium indefinitely.

The deliberations of the panel had in essence been considered, but subordinated to political judgments that could have been made without benefit of such a panel. Part of the flaw was the topic at hand. Part of the problem was time, but the NIH panel took almost as long as the National Commission in its first report, but with remarkably different policy impact. The National Commission's report was almost directly implemented, and its report had a long shelf life, while the Fetal Tissue Transplantation Research Panel produced a series of reports with a confusing set of conflicting views. Another element was the rising role of interest group politics, and the significance of this cannot be fully judged, but is certainly significant. But of most relevance to the dilemma facing the ELSI working group—how to formulate policy options from a position within NIH—the process was also flawed.

In contrast to the National Commission, the Fetal Tissue Transplantation Research Panel was starved of staff and other resources. The first National Commission report lists sixteen staff, most of whom had direct training in law, ethics, or some field of substantive relevance to the commission's mandate. Many were well-recognized national experts. The 1988 report from NIH lists the panel members, but no staff. Several staff in the NIH director's office and a couple in the Office of the Assistant Secretary for Health were indeed focused on the fetal tissue effort, but they also had other duties, and bioethics was but a part of their job descriptions. A few outside papers were commissioned for the 1988 panel, most notably a legal background paper prepared by the Poynter Center of the University of Illinois, and many additional documents were contributed to it for consideration. This was a far cry, however, from the spate of reports prepared at the request of the National Commission. One important difference is that the National Commission sought papers after its meetings to decide what it needed to know. The NIH background papers were requested as preparation for the first meeting, and no further papers were ever sought after the panel actually met.

The 1988 panel had been expected initially to achieve consensus in a single meeting, perhaps by analogy to the format of NIH consensus development conferences. Such conferences were, however, most successful when convened around technical questions rich with data that were ripe for expert analysis. No bioethics commission operated in this manner, and for good reason.

It quickly became apparent that a second meeting of the fetal tissue transplantation panel was necessary. The expectation of a single meeting precluded systematic planning about what data to gather and which topics might warrant examination by hired consultants. Since the panel did not expect to meet again, there was no preparation of an agenda to find out what facts should be gathered, what views solicited. If at least two or three meetings had been anticipated at the outset, the first meeting might have been spent deciding what questions to address and how best to take advantage of outside experts, taking

pressure off the initial meeting and also deferring discussion of recommendations until the group had begun to work together. When the second meeting
in October adjourned in near-chaos, a third meeting was scheduled for December.

Bureaucratic autonomy was just as important as the expectation of an
extended series of meetings. Previous bioethics commissions had been given a
mandate to report independently to Congress and the executive branch. They
were handed topics, but not told how to address them. Much of the creative
hurly-burly of policy analysis came from finding new approaches to old problems. In contrast, the 1988 panel was given a fixed list of ten questions to
address, fixing the deliberations in a tight frame and precluding the best hope
of creative consensus formation. Not only was the agenda set from above, but
also the unrealistic schedule. Previous bioethics commissions depended critically on time to gather facts, to prepare background papers, and to discuss
options among the commissioners. All these elements were rendered impossible by the framing of the questions and the expectation of immediate resolution.

Beyond these process flaws, and in part because of them, the product of
the deliberations was ill suited to achieve its given ends. As one commentary
in an Institute of Medicine report noted:

Did [the Department of] Health and Human Services, and the public generally, get
what it most needed from the panel's report? I would argue that it did not. What was
most needed was not only a cogent, clarifying discussion of the issues by nonmedical
experts but also a rhetorically and aesthetically attractive report. When one enters the
field of public policy debate on issues that are as strongly controversial as abortion, one
must find a language and a set of images that will help a polarized community begin to
build a consensus. . . . What was lacking was a document of the style that is needed
today: an eloquent, appealing, quotable report that can assist the decision maker both
in making and later in defense of difficult policy decisions.[159]

Patricia King, who served on the NIH panel as well as the Institute of
Medicine committee that reviewed its process, wrote another commentary:

The drive to achieve consensus was central to the panel's work, and, indeed, consensus
was achieved. Yet I believe that ultimately the product is not particularly persuasive.
The fact that the panel's recommendations were not adopted by the Department is not
the test of their persuasiveness . . . it failed to make clear how persons holding radically
different views about abortion could nonetheless agree that the use of fetal tissue from
induced abortion is "acceptable public policy" under specified conditions. It was probably necessary to *describe the process that resulted in acceptance* of this point rather than
merely stating it [italics added].[160]

Some argued that the process had produced just what department officials
wanted, a delay into the next administration and a confusing policy document.
It produced a document that could be used initially to delay removing the
research moratorium, and then because of its internal inconsistencies could be
safely ignored, leaving the moratorium in place indefinitely. In conversations
with staff and panel members involved, however, there was little evidence for

this cynical view. Rather the failure seemed more one of topic, process, staff, and timing than deliberate subterfuge. In bureaucratic terms, the lesson seemed to be that grappling with policy questions would require more time, more meetings, more staff, more commissioned papers, and more institutional autonomy.

The House Subcommittee on Government Information, Justice, and Agriculture explored the ELSI working group's policy analysis capacity in a hearing on October 17, 1991.[161] Rep. Bob Wise chaired the subcommittee, and the hearing was largely organized by subcommittee counsel Robert Gellman. The subcommittee had an abiding interest in issues related to privacy and data-management practices, tracing back to before the Privacy Act of 1974. The subcommittee had convened an earlier hearing on general privacy matters, including access to computer records, on April 10. The October hearing was focused on the genome program, and specifically the ELSI working group and the ELSI grants programs in NIH and DOE.

The subcommittee heard testimony from NIH director Healy, NIH genome center director Watson, Nancy Wexler, and David Galas, director of the life sciences research section of DOE. Jeremy Rifkin was there to promote his privacy bill, which had initially been a focus of the hearing, but had since been recognized as wanting by committee members and staff. Rep. Wise asked several questions about the ELSI grants program and ELSI working group procedures, the main theme of which focused on whether the welter of grants and contracts would congeal into useful policy options for legislators and executive branch officials. He was clearly skeptical that the ELSI working group could operate autonomously where it was currently situated. Healy detailed how a policy proposal would go from the genome office to NIH to the Health Department through OMB to the White House. Wise pressed Healy and Watson about how policy recommendations would rise out of a program of research grants, however competently carried out and administered by NIH and DOE. The gist of the replies was that it was too soon to tell. Healy complimented OTA on its capacity for policy analysis, but questions about policy analysis by the ELSI working group were left largely unanswered. Wise later came back to questions of structure. Healy then noted her experience on the fetal tissues transplantation research panel:

NIH handled it in, I think, a very responsible way by bringing together groups that represented all parties and came up with a series of very well-thought-out policy recommendations. . . . the debate was elevated to the highest level. . . . it certainly is something that was not hidden in some back closet.[161]

Rep. Wise then asked if any witnesses cared to comment on whether genome research might not provoke passions surrounding abortion, because of the link to prenatal genetic testing. Healy replied:

We do not believe for a minute that the human genome project relates to the issue of abortion. And abortion is also not an NIH issue. In fact, I will tell you, Mr. Chairman, that one of the difficulties that NIH has faced historically is that people believe abortion is an NIH issue. It is an issue that is much broader within our society. It is an issue for the Congress; it is an issue for the White House. But it is not an NIH issue.[161]

Robert Gellman worked on a policy report in the wake of the hearing. His report was ready by April 1992.[162] Rep. Wise sent it to the Secretaries of Health and Human Services (for NIH) and Energy, accompanied by a cover letter concluding that "the existing ethical, legal, and social implications (ELSI) programs at the National Institutes of Health and Department of Energy are principally designed to support academic research. . . . these programs are not capable of developing or presenting policy recommendations and there is no existing policy process. . . ." The letter went on to recommend that NIH and DOE "jointly establish an Advisory Commission on the Ethical, Legal, and Social Implications of the Human Genome Project" and hammered home the point urging that it be established *immediately* and that it be in operation within four months (emphasis in original).[163]

The report was a distillation of successful federal commissions, including not only the National Commission and President's Commission in bioethics but also the Advisory Committee on Automated Personal Data Systems (1972–1973), the Privacy Protection Study Committee (1975–1977), and presidential commissions on AIDS.[162] (Notably, it did not describe the less successful BEAC or other bioethics efforts of the late 1980s.) The report passed lightly over the grants program mounted by NIH and DOE. These were immensely useful and likely to contribute more than the committee might appreciate. Had the President's Commission and National Commission had a fertile field of grant-supported investigators to consult with, their harvest of new ideas would likely have been richer and their job easier. Moreover, the agencies could farm out policy analysis on specific topics, even if they could not conduct it themselves. The Institute of Medicine, for example, was pursuing just the kind of systematic policy analysis of genetic testing that the committee called for, through a contract with NIH and DOE.

The committee's point was nonetheless valid. The thrust of the NIH and DOE programs was indeed on academic research, and that did not equate to policy analysis. If Congress wanted policy analysis, it might have to craft a new instrument. The ELSI program was established to thwart misuse of genetic information, and that would require policy analysis. A structure to sustain that analysis was needed.

NIH's initial response to the letter from Wise was not promising. The proposal sent to the department was for a structure seemingly modeled on the fetal tissue transplantation research panel. This corroborated initial press reports of the policy analysis unit being established in director Healy's office.[164] Daryl Chamblee was hired from the Washington law firm of Steptoe & Johnson to head the bioethics component.[165] Chamblee had been interested in bioethics for several years, centered on work with the Columbia Hospital for Women. The ethical analysis component would be part of a general science policy shop answering to deputy director Jay Moskowitz, who had overseen the fetal tissue transplantation research panel. The proposal was to have the ELSI advisory commission on the genome report to the NIH director's advisory committee, with staffing from the staff pool of the director's science policy analysis center. House subcommittee staff were unimpressed with the proposed structure, mainly because it lacked the reporting autonomy they believed necessary to function effectively.

The NIH draft strategic plan stated that "the aim of confronting these social, legal and ethical problems in research is not to promulgate new regulations or create another layer of research review. Rather, the aim is to provide the research community with relevant guidance for the conduct of research and to assure the public's understanding of the social benefits and consequences of science."[166] There were indeed science policy questions that might be handled well within the NIH, centering on peer review, standards for scientific research, and administrative process. For such questions, the direct link into the director's office, and thence into the administrative hierarchy, would be a great advantage. Other questions that touched on broad social policy well beyond the confines of NIH, but directly linked to science policy decisions, would require an analytical engine on a separate track. A commission's credibility in addressing such questions would hinge on independence from NIH (or DOE), and its influence would derive from reporting directly to Congress, the President, and the nation, free of bureaucratic clearance procedures.

While the House subcommittee was crafting its report, the Senate was also interested in bioethics, but from an independent starting point. Senator Mark Hatfield had long been concerned about advances in genetics. In 1983, he introduced Jeremy Rifkin's resolution and "theological letter" on germ line gene therapy.[94] In that statement, he noted how he had visited Hiroshima soon after the atomic bomb was dropped. He worried about whether technology was not racing ahead of the capacity to control it, and worried specifically about genetics. In 1981, the biotechnology company Cetus raised $120 million in the fifth-largest public stock offering in history. Genetic advances might produce a Brave New World, and "return to slavery is possible unless public and private institutions enter into a dialogue with vigor."[94]

In 1989, Senator Hatfield was the only member of Congress, other than Biomedical Ethics Board members, to support the Biomedical Ethics Advisory Committee. He viewed its work on human genetics as important, and said so in support of funding for BEAC before the Senate Appropriations Committee, on which he sat as ranking minority member.

In April 1992, Hatfield's concerns were rekindled by an NIH effort to patent 2,700 DNA sequences (see Chapters 19 and 20). He announced an intention to call for a moratorium on patenting DNA sequences and genetically engineered animals until a study of the ethical issues could be conducted. He rose to address the Senate:

The research conducted by the National Institutes of Health is, in my opinion, one of the Federal Government's greatest gifts to the Nation; it is the gift of improving the human condition by alleviating the pain and suffering associated with disease. As deep as my respect is for the medical research community, Mr. President, I stand here today deeply concerned about the future use of research findings."[167]

Senator Hatfield intended to propose that BEAC be brought back to life, with a new congressional board and new BEAC members, to conduct a study of whether patents should be issued. He contemplated introducing an amendment to impose a moratorium on such patents until the study was complete. Senators DeConcini and Kennedy, working with Senator Domenici, cut a deal with him. Hatfield expressed his concerns on the floor, but did not offer his amendment. In return, Senators Kennedy and DeConcini agreed to host hearings before the Judiciary Committee, on patenting issues, and in the Labor and Human Resources Committee, on other ethical and social issues related to genetics.[167] Out of this agreement, the Office of Technology Assessment was requested to begin a project on DNA patenting and to hold a "bioethics summit" to review the history of bioethics commissions, with an eye to laying out options for future efforts. OTA released its report in fall 1993.[169]

Hatfield's proposed amendment provoked a frenetic lobbying effort against it.[168] A letter from Louis Sullivan, Jr., Secretary of Health and Human Services, argued that reconstituting the BEAC was a redundancy, once again betraying a confusion between the research program sponsored by NIH and a capacity for conducting policy analysis of that research, similar to the President's Commission or National Commission:

Much of the information that this board would collect is already being collected in other venues. The National Center for Human Genome Research has devoted approximately 5 percent of its budget to the study of ethical, legal, economic, and social problems associated with the mapping and sequencing of the human genome.[170]

The point about the value of the research effort was true, but the incapacity to synthesize it into policy was clearly not appreciated. As the genome project neared the end of its fifth year, the social issues raised by the coming deluge of genetic information continued to fester. NIH and DOE mounted research programs to support policy analysis and to provoke a broad debate. These programs were unprecedented in science, and were auspicious institutional innovations within both agencies. At the same time, however, the agencies failed to take account of the process, staffing, and funding requirements to nurture policy analysis, in particular the general and principled guidance that might assist Congress to craft rules of law. A renewed interest in bioethics was clear in both the House and Senate. Because the executive branch failed to appreciate the chasm between the ELSI research program and policy analysis, it invited Congress to fill the gap.

18
Wizards of the Information Age

ONE DAY IN 1950, the Polish mathematician Stanislaw Ulam had an idea about how to set off a thermonuclear fusion bomb, or hydrogen bomb. His idea became reality with the first explosion of a fusion device in October 1952.[1] Ulam came to the United States in December 1935, at age twenty-six, at the invitation of John von Neumann, who later became his link to the Manhattan Project. Ulam went to Los Alamos in February 1944, several months after the boys' school there was commandeered to become the scientific hub of the project. When the war ended, Ulam moved to the University of Southern California, where he felt intellectually isolated. According to Gian-Carlo Rota, a longtime colleague of Ulam's, "suddenly he found himself in the middle of an asphalt jungle, teaching calculus to morons."[2] He weathered a bout with encephalitis, which left him with a distaste for Los Angeles and a persistent insecurity about whether his mind worked as well as it had before. A fascination with the workings of the brain was perhaps the most significant legacy of Ulam's brush with death.

Ulam returned to Los Alamos in May 1946 and began to work on the next secret Los Alamos project, a fusion bomb, or "Super." In 1949, the Soviet Union detonated a fission bomb, similar to those exploded over Hiroshima and Nagasaki. This revived Edward Teller's "Super" project. Whether the United States should proceed to produce a fusion bomb more powerful than the fission bombs used at the end of the war was a central policy question of the day. Scientists debated its feasibility, and they joined others in questioning the wisdom of such a project. The Soviet successes fueled paranoia in Washington and tipped the political balance toward Edward Teller's position of enthusiastic support for a fusion bomb: President Truman plunged into developing a more powerful weapon, and Ulam was where most of the action was.

Ulam's first contribution to the hydrogen bomb was to show that Teller's scheme would not work.[3] Ulam met Teller during the Manhattan Project, and the two became linked in a collaboration of immense historical significance, but leavened with mutual distaste at the personal level. Rota ascribed Ulam's enthusiasm to personal motives: "Stan Ulam was out to get him [Teller] by proving that his plans for the new bomb would not work."[2] Two years of

elaborate and tedious hand calculations by Ulam and C. J. Everett were corroborated by one of the first major uses of a digital computer, under von
Neumann at Princeton, using a different modeling method.

Having shot Teller's idea down, Ulam floated one of his own. "Adding
insult to injury, Stan, in a sudden flash of inspiration, came upon a trick to
make the first hydrogen bomb work."[2] Ulam's notion was "an iterative idea"
that he developed with Teller into a report that "became the fundamental basis
for the design of the first successful thermonuclear reactions and the test in the
Pacific called Mike."[1] These were two separate events, with the first test of a
sixty-five-ton device named Greenhouse on Eniwetok Atoll in May 1951, and
the Mike test in November 1952.[4] Ulam's critical role in the fusion bomb is
highlighted in documents recently brought to light.[5]

The mind that devised the scheme for the hydrogen bomb also spun out
ideas central to the mathematical analysis of DNA sequences. Ulam was drawn
into biology by his younger colleagues at Los Alamos, particularly George Bell
and Walter Goad, and by Leonard Lerman and Theodore Puck from Colorado, who had collaborative ties with Los Alamos. Lerman later became a
wizard of technological development in molecular biology, and Puck one of
the major figures in gene mapping and tissue culture of mammalian cells
(growing cells from the body in petri dishes, much like bacteria or fungi).

Many roots of computational biology that began to blossom in the 1970s
and 1980s trace to the group surrounding Ulam at Los Alamos. To analyze
the information contained in the DNA code, one needed a metric, a formal
measure of the similarity between two DNA sequences. Ulam suggested that
the metric might be formulated as the least number of changes (substitution
of one base for another or addition or deletion of a base) necessary to transform
one sequence into another. This idea emerged from a conversation between
Ulam and Temple Smith in 1968, soon after Ulam moved to Colorado. Ulam
sketched these ideas in a 1972 paper.[6] Smith and Ulam worked with William
Beyer and Myron Stein of Los Alamos to extend these ideas in 1974.[7]

A rigorous measure for sequence similarity fed into a growing movement
to analyze protein and DNA sequence data using mathematical tools. The
mathematical work built on several comparative studies in biochemistry and
molecular biology. Emile Zuckerkandl and Linus Pauling unexpectedly found
in 1962 that the amino acid sequence information in proteins could be used as
a "molecular clock" to trace evolutionary history. They published the seminal
paper in the field three years later.[8] Walter Fitch and E. Margoliash similarly
constructed an evolutionary tree based on similarities among amino acid sequences of cytochrome c proteins in different species.[9] During the mid-1960s,
Margaret Dayhoff of Georgetown University began to assemble an *Atlas of
Protein Sequence and Structure*.[10; 11] Dayhoff devised rules for calculating sequence similarities among proteins, based on the chemical similarity of amino
acids.[12] Her catalog of proteins was organized according to functional and

evolutionary types. Meanwhile, work on DNA sequences paralleled protein sequence analysis.

In 1970, Saul Needleman and Christian Wunch developed an algorithm, easily adapted to computers, to compare the similarity of two sequences, basically by counting how many steps it would take to transform one into the other.[13] This method was quite useful, but was not mathematically rigorous until Peter Sellers of Rockefeller University formalized Ulam's idea of a metric in 1974.[14; 15] Temple Smith and Michael Waterman generalized the measure in a mathematically rigorous way that could account for small insertions and deletions.[16]

Mathematicians and statisticians thus greatly expanded the sophistication of techniques to derive meaningful information from DNA and protein sequences.[17] Indeed, the originator of the DOE genome project, Charles DeLisi, came from this field of mathematical biology. His mind was thus prepared for the notion that DNA sequence information could become the raw data for an entirely new field of theoretical genetics, in which the genome project would be a major step.[18]

The linear sequence of information in the order of amino acids constituting a protein, or of nucleotides making up a strand of DNA, was a natural target for computer analysis. Molecular techniques came of age at just the time that digital computers were becoming faster, cheaper, and more portable. "Tools needed to store, search, and analyze the new data grew up alongside the tools necessary to generate the data."[11] Through the 1970s, a small group of individuals began to realize that computers and sequence information were a natural marriage. Bride and groom struggled to overcome vast cultural differences. Computer scientists and molecular biologists traced their lineage through different tribes, with vastly different norms, and only a few hardy souls could converse in both languages and command respect in both communities. The databases that stored sequence data became their meeting ground.

A meeting held at Rockefeller University in March 1979 was a watershed. According to one participant, "no one questioned but that computers and sequence data were made for each other. Transmitting a long, seemingly random sequence of four letters from one person to another without errors is hardly possible except by putting the information on a computer-readable medium. The need for a data bank was in the air."[19] The meeting brought together an eclectic collection of people interested in applying computers to molecular biology. There were groups that had worked on the NIH-funded PROPHET project in Boston, which grew up around those associated with MIT and the private firm Bolt, Beranek & Newman (BBN). Two projects that brought together computers and chemistry at Stanford—SUMEX (for medical applications) and Molgen (for organic chemistry)—were also represented. Temple Smith and Michael Waterman attended from Los Alamos. The two had not worked together before, but both returned from the meeting enthusiastic about the idea of a database and DNA sequence analysis center at Los Alamos.[20; 21] Five other groups were possible candidate sites for a database: the group surrounding Dayhoff at the National Biomedical Research Foundation, the MIT-BBN group, the Stanford group, another group led by Olga Kennard and Frederick Sanger in Cambridge, England, and a group at the European Molecular Biology Laboratory (EMBL).[11]

By the following April, several groups had assembled pilot DNA sequence repositories that were discussed at a meeting in Schönau, Germany. There was further discussion about the need for a national or international database, and consternation among Americans about inaction at the National Institutes of Health. A workshop was held at NIH on July 14, organized by Elke Jordan of the National Institute of General Medical Sciences (NIGMS). Within a month, a flurry of proposals came in to NIH for databases, DNA sequence analysis methods, or both. On October 26, Elke Jordan convened an *ad hoc* advisory group, which on December 7 elaborated a plan. A public national database for DNA sequence information was to be established as Phase I, followed by a center for development of analytical methods in Phase II. Phase I would establish the capacity to collect data, Phase II would develop mathematical and computational methods to analyze the information. On April 2, 1981, NIH released a "sources sought" document, inviting expressions of interest in applying for funding. Meanwhile in Heidelberg, Gregg Hamm began operating the EMBL Data Library, thus getting a six-month jump on American efforts.[22]

Late in 1981, NIH requested contract proposals for the database part, or Phase I. Phase II was dropped because of cost, politics, and lack of clarity about what a DNA analysis center would be.[19] Three proposals were sent to NIH. One came from Dayhoff's group. Los Alamos was part of the other two. In both proposals, Los Alamos was the collection and storage center, but the distribution plans were different. One was a collaboration with a company formed out of the Stanford group, IntelliGenetics; the other involved the MIT-BBN group. NIH announced its award of the contract to the BBN–Los Alamos group on June 30, 1982, and the database began to operate in October.[11] An August meeting on computational biology in Aspen brought together many members of this small community for two weeks,[23] in the wake of NIH's announcement but before GenBank began to operate. David Lipman there introduced his idea of a "hash code," a way of aggregating data to expedite searches of massive databases.[11; 24; 25] (Six years later, Lipman became director of the National Center for Biotechnology Information at the National Library of Medicine.) Margaret Dayhoff died soon after the workshop.

A different part of NIH, the Division of Research Resources, later funded BIONET at IntelliGenetics, which grew out of the Stanford group's work. This was initially to serve both as a software resource and as a center for development of analytical methods, overlapping with the Phase II plan abandoned by NIGMS. The BIONET grant was not renewed in 1989. It linked small laboratories to software they otherwise would lack, but never developed into the major center for devising new methods that some had hoped.[11; 26] Indeed no systematic program to cultivate new analytical methods for molecular analysis ever developed, in the judgment of many, until the Human Genome Project came along to shine a spotlight on the problem again.

GenBank and the EMBL Data Library fell quickly into arrears. During the early years, most entries were typed in by hand at GenBank or EMBL. The two database centers divided journals into two roughly equal sets and split responsibility for entering sequence data. The DNA sequence databases were quickly inundated by the flood of data. In their first four years, they grew twenty-five-fold, far faster than projected. Yet GenBank was funded by a five-year budget that was already insufficient two years into operation. In mid-1986, only 19 percent of the sequence data published in 1985 were entered.[22] Some sequences were more than two years old before entry, and the databases contained barely half the sequence information published to date.[27] At an acrimonious meeting of the GenBank advisory committee, scientists were divided over how to cope with the emergency. The result was a new policy of entering only raw sequence data, leaving out most features known about the sequence. The number of "unannotated entries" rose quickly, and began to drop only in 1988, after a new contract substantially increased funding for GenBank.[28]

In this forced marriage of computer and DNA, there was little time to enjoy a honeymoon. GenBank was a patched-together raft riding out a tsunami. As originally proposed in December 1989, GenBank was to build on a data structure known as a relational database. In essence, data would be organized into tables. The rows of the table would be individual entries—the sequence data from a gene or chromosomal region. The columns would be biological features, information about the DNA's region of origin, information about the gene, the species of origin, and other details relevant to interpreting the sequence.[29] The advantage of the relational framework was its logical structure, which also simplified search strategies. As actually implemented, GenBank instead stored the data as a "flat file," basically units of text with punctuation marks not organized into tabular form. The computer and storage hardware were also far less powerful than originally proposed.

Both GenBank and EMBL Data Library were politically complicated. GenBank was operated at a DOE-owned national laboratory but funded by a group of NIH institutes, the Department of Defense, the National Science Foundation, and the Department of Energy. EMBL's database was funded by various European governments and the Commission of the European Communities. Both databases were struggling to secure resources, very much stepchildren in the biomedical research family. Molecular biologists had little notion of the difficult issues facing any large public database. They were impatient to use the data, but unenthusiastic about paying enough for its storage. The genome debate was a glass slipper for the databases, making apparent how important they were to the future of biology, and making them into Cinder-

ellas within their respective bureaucracies. GenBank's second five-year contract, negotiated in 1987, had a budget of $17.5 million, compared to the
$3.5 million contract that took effect in October 1982. In fiscal year 1993,
GenBank prepared for another transition, as its management was slated for
transfer to the National Library of Medicine.

The new reliance on computers and communications technologies melted
yet another group into the genome pot. Databases, computers, and mathematical algorithms proved as important as DNA sequencing, cloning, and
other more obviously biological techniques. As geneticists produced a deluge
of data during the 1990s and beyond, those who understood hardware and
software would play an increasingly important role.

The multitude of genome research programs in the United States, Japan,
the USSR, and Europe all promised to generate massive amounts of data. The
flow of information became the focus for negotiations among scientists and
science administrators because this flow was of benefit to every nation. The
inherent need to ensure smooth flow of data and the benefits of ensuring some
consistency of interpretation necessitated extensive international collaborations that nucleated around genome databases, as databases became brokers
for international information exchange.

At a 1987 meeting in Heidelberg, the DNA Database of Japan (DDBJ),
funded by the Science and Technology Agency, was brought into the fold,
joining the EMBL Data Library and GenBank. DDBJ had served as a distribution node since 1984, but agreed to begin gathering data for Japan and
other parts of Asia. An international advisory committee, chaired by Dieter
Soll of Yale University, met in February 1988. The advisory committee ber-
ated the database managers in stern language, urging them to make their
databases compatible, so that editing, annotation, and corrections entered in
one database did not have to be repeated at the other centers.[30]

The features that made databases complex to maintain politically, with
their disparate and multicentric funding structures, also made them natural
foci for organization at the international level. Having already solved the
problem of gathering funds and scientists together, they were natural centers
to coordinate other research activities. An odd *Nature* editorial appeared soon
after the Alberts report of the National Research Council was released in
February 1988. It urged an expeditious genome project, but cautioned against
organizing a biological Apollo project. It tipped its hat to databases:

If the databanks were a necessity when they were first created five years ago, improved
versions of them are surely even more necessary now. . . . This is where the organiza-
tion-building should begin. . . . The job should be tackled internationally so as to win
the greatest benefits from the data gathered in widely scattered laboratories.[31]

But *Nature* was unwilling to use its clout to promote the flow of data into
databases. John Maddox, editor of *Nature,* balked at the notion that journals,
least of all his, should play the role of traffic cop. Lennart Philipson noted

Maddox's inconsistency in advising against formal coordination of the genome project while promoting an "internationally coordinated mega-databank."[32] Philipson noted that EMBL, GenBank, and DDBJ were jointly "close to an international data source in biology" and urged that *Nature* join the journals mandating sequence submission to databases. *Nucleic Acids Research,* the *Journal of Biological Chemistry,* and other journals adopted policies requiring such submission, but Maddox held out noisily. He argued against coercive editorial policies again in 1989, asserting that some laboratories did not have ample computer facilities, that sequence databases were inaccessible in the Soviet Union, and that it was not the role of editors to serve as enforcers.[33]

Richard Roberts of Cold Spring Harbor Laboratory countered in a letter to *Nature* that journals were uniquely positioned to ensure timely submission of data to databases.[34] Sequence data were essential to verifying statements in articles, and the only meaningful way to interpret DNA sequence data was to use a computer: "A computer is every bit as necessary as a centrifuge to today's molecular biologist." Roberts called Maddox's arguments "a potpourri of excuses for inaction."[34] Thomas Koetzle noted that the crystallographic research community faced similar dilemmas and favored mandated submission, with the option of one year's delay from time of publication to public release.[35] As EMBL weighed in,[36] Maddox capitulated, but his was not an unconditional surrender.

Maddox argued, "Difficulties arise when contributors are asked to satisfy conditions that have *nothing to do* with the content of what they have to say" (italics added).[37] Maddox's point was strained, to say the least. Conclusions in the papers were based on the sequence data in question or there was little point in publishing them. Sequence data were of little use as letters on a page without the ability to compare to other sequences or to analyze them with computer software. The problem was even more acute if DNA sequence data were merely contributory to a publication. If the sequence data were not published and never contributed to a database, how would others know the reliability of the work? How would others compare their sequences against the new sequence? Maddox could not imagine why his journal "founded 120 years ago to bring the record of research to a wide audience, could fall in with the idea that its readers' appreciation of what they read will hang crucially on the accessibility of a databank in Heidelberg, Los Alamos, or Mishima."[37] *Nature* depended instead on mail circulation, which one might doubt was as widespread in the 1870s as computers were in the 1990s. Thomas Kuhn observed that scientific revolutions sometimes required dinosaurs, the waning generation of scientists, to die off before being supplanted by mammals.[38] Add journal editors kneeling before the altar of the printed page, and the printed page alone, to the list of species headed for extinction. Where there was DNA sequence information, there would be computers.

Sequence databases were not the only ones established to deal with genetics. Another group of databases centered on gene maps and genetic diseases. The most famous among these grew out of human genetics. Victor McKusick at Johns Hopkins began to keep track of genetic disorders and variants in 1960. In 1962, he published the first catalog of human genes, a hundred-page compilation of data on the X chromosome.[39] He also monitored other chromosomes, and produced the first edition of his book *Mendelian Inheritance in Man* in 1966.[40] McKusick published seven editions of his book, and then put his list of disorders and genes on-line in 1987, in a collaboration with the National Library of Medicine and with funding from the Howard Hughes Medical Institute. McKusick continued to publish the book version every two years, making the data more portable and also surveying the field in a foreword of increasing breadth and length with each edition.[41; 42] The book version became a periodic snapshot of the computer database, "arguably the most valuable compendium of human genetic disease information currently available."[43] The database was an "encyclopedia of genes" in the human genome.[41] By the ninth edition in 1990, the catalog contained 37,987 references written by 54,623 authors, and yet it covered "perhaps only 5 to 10 percent or less of all the structural genes of man."[42]

Beginning with the first human gene mapping meeting in New Haven, Connecticut, in 1973, those engaged in mapping human genes began to catalog their progress at periodic meetings. The data pooled at these human gene mapping workshops cried out for consolidation into a database. Frank Ruddle and others at Yale University stepped forward and began to catalog the information about gene location, chromosome structure, literature citation, and other relevant information. The Human Gene Mapping Library was cross-tabulated to McKusick's catalog of genes and disorders.

Yale's Human Gene Mapping Library was supported by an NIH grant initially. When the grant was up for renewal, an NIH review panel recommended against continued funding, citing problems with the underlying technical approach and other factors.[20] HHMI stepped in to preserve the database with several years' funding, starting in 1985. This arrangement persisted until 1989, when HHMI decided to move toward a system that was easier to use and whose underlying database structure was more up to date than the SPIRES system (an IBM software package) used at Yale. HHMI shifted support to Peter Pearson, who was recruited to Johns Hopkins from Leiden to run a new Genome Database at the Welch Library, already home to McKusick's catalog of human genes.[44]

The Genome Database was for gene mappers and contained information on genetic linkage maps and physical maps. The focus was on gene location on the chromosomes, with information about genes, disease-associated regions, and markers. The Genome Database released a mission statement in September 1989[45] and debuted at Oxford in September 1990, at a meeting to prepare for the international Human Gene Mapping Workshop the following year in London.[46] By January 1991, three thousand users were registered from around the globe and two thousand people logged onto the database each month.[47] HHMI funding established the Genome Database, but government funding would be necessary to make it permanent. Through 1991 and into 1992, NIH and DOE negotiated funding for it.[48] Self-supporting centers were established as nodes in the UK, Australia, and Germany, and establishing a center in Japan was a part of the 1992 genome program in Japan.[49; 50]

The information in the Genome Database largely complemented the contents of GenBank, which stored information about DNA sequence; and also the McKusick catalog, which focused on pedigree information, clinical description, and mode of inheritance. For some applications, melding information from the three different sources could be difficult. It was not possible to retrieve DNA sequences known to be taken from a particular chromosome region, for example, and yet this was a common need for those seeking mapped genes.[51] Similarly, the data on physical maps under construction were not initially integrated into either the Genome Database or GenBank. Devising ways to link the various databases and to consolidate information derived from disparate mapping techniques loomed as a major objective for software development.[51] Jacqueline Courteau, who wrote the appendix on databases for the 1988 OTA report, revisited the topic in 1991 in a pull-out section of *Science*'s annual theme issue on genome research.[52] The databases had grown enormously in three years, and with this growth came political and intellectual complexity in linking them.

Databases became an essential element in the pursuit of genetic knowledge. There were many relevant databases, however, and they were managed differently, located throughout the world, used different software and hardware, and contained different types of information. Making the databases more readily accessible and linking them to analytical software packages to assist in analysis became another set of problems to solve. The National Library of Medicine (NLM), through the National Center for Biotechnology Information, became one integrative force. GenBank was moved administratively from the National Institute of General Medical Sciences to NLM in 1992. There was also a minor industry devoted to improving database access and analytical software. GenBank had always been housed at Los Alamos National Laboratory, but the information had been distributed through a commercial firm (Bolt, Beranek & Newman from 1982 to 1987, and then IntelliGenetics from 1987 to 1992). Several small firms had grown up making software to analyze genetic data, and many sold their software with somewhat modified packages of GenBank data. The transition back into the federal fold provoked a major controversy that endangered these companies. They retaliated by threatening NLM's budget.[53]

Some of the software vendors believed that NLM was developing competitive software with a heavy subsidy, undercutting the markets for their products. Several of the software companies' directors complained to David Lipman, director of the NLM effort. Their discomfiture began in 1989 and 1990, when discussions about moving GenBank were being negotiated within NIH.[54] They hired a lobbyist, who taught them how to go for the jugular. On June 18, just weeks before the appropriation figures would be finalized, they sent a letter to William Natcher, chair of the subcommittee that appropriated NIH funds, including the NLM. Staff for the subcommittee were concerned that federal dollars would go for research that could be done by private industry; the message was clear to NLM that its program was in serious jeopardy. Matters had gone further than the instigators intended. Moreover, they had failed to inform other companies that also had a stake in the outcome, and these firms then wrote letters to Natcher disavowing the original letter.

The conflict stemmed from disturbing the sociology of software development. This was territory where both federally sponsored investigators and private vendors grazed, and there was bound to be continuing tension. There was an enormous amount of work to be done, and there was money to be made. Much like the looming conflict over patenting of DNA sequences, the question of who would control the territory was unsettled. Analytical software and database access were among the many disputed borders within the genome project. The companies and NLM reached an agreement for the time being, but the issue would surely surface again.

The purpose of the databases was to pool data from disparate sources, so that information would be readily available to others who could use it. Indeed, if the goal of the genome project was to construct maps, those maps were inevitably built from information contributed to databases. Maps were irreducibly collective endeavors. Once made, they enabled analysis of information in an entirely new way. The analysis of all the data spawned a new field. Computers and mathematical techniques were turned loose on the data to construct theories of biological structure and function. The first glimmerings of Walter Gilbert's dream, a theoretically driven science of molecular biology, could be seen on the horizon.

Databases were tools to find and to understand genes. Mathematical methods were an important part of making the stored information biologically meaningful, and this circled back to the interests of Stanislaw Ulam's heirs.

One cluster of activities centered on the interpretation of genetic linkage— statistical methods for relating the inheritance of characters to physical locations on the chromosomes. Analyzing pedigrees for statistical evidence of gene transmission was a complex art, and a small core of scientists traversed the treacherous trail. A few biologists comfortable with both the molecular biology laboratory and mathematical reasoning became the luminaries of a small and highly mathematical field, devising new ways to find genes floating on oceans of data.[17; 55–60] Entirely different fields of mathematical genetics flourished, for example the analysis of amino acid sequence patterns in proteins, used to predict their structure and to sort protein regions into classes with similar functions.[61–65] Mathematical tools were also applied to physical mapping. Eric Lander and Michael Waterman, for example, made a straightforward mathematical model of how cloned DNA fragments could be assembled into physical maps, thus providing a theoretical benchmark against which those constructing maps could compare their experimental results.[66]

Mathematical methods were clearly destined to play an increasing role in understanding the complex phenomena of biology. It became clear that sequence comparisons would become a powerful tool in assessing biological function. Russell Doolittle and his collaborators shocked the world of molecular biology by finding a sequence similarity between the *sis* oncogene and a cellular growth factor.[67] The oncogene came from a cancer-causing virus of primates. The growth factor was one of many that transmitted signals controlling cellular functions. The growth factors bound to the outer surfaces of cells and were translated by cellular machinery into a cascade of events, most often leading to proliferation or differentiation of cells. Doolittle's finding of a structural similarity was entirely unexpected and came from simply scanning sequences in a database. He had not determined the sequences, nor had he done the underlying biochemistry that illuminated the two proteins' functions. His contribution was to notice that two proteins hitherto thought entirely unrelated were structurally similar, and likely to be functionally related. By this stroke, he discovered a biological relationship that had eluded those working on the biochemistry of the oncogene protein and other groups focused on the growth factor. A computer comparison thus unified two separate subfields. The receptor for another growth factor and a different oncogene were found related a year later, lending credence to theories that linked cancer to aberrant control of cell growth.[68] The importance of building databases to enable the detection of such similarities began to dawn on molecular biologists, and the analytical tools to detect the similarities developed apace.[69]

In a 1985 report, a committee of the National Research Council (NRC) emphasized the importance of computational methods.[70] One spinoff of the report was a month-long meeting, the Matrix of Biological Knowledge Workshop, at the Santa Fe Institute from July 13 to August 14, 1987.[71] Another NRC committee released a report in 1987, emphasizing the importance of computer models to understand the structure and function of proteins, DNA, and RNA.[72] The theoretical methods that had so long brought coherence to physics and chemistry began to seep into the crevices of molecular biology, which for a generation had been an almost purely experimental science.

Many molecular biologists were uncomfortable with the shift of power to the mathematicians and analysts. Some even resented forays such as Doolittle's into their territory. Why should he be able to publish a major discovery that came from just sitting at a computer terminal? That wasn't biology, was it? From 1950 to 1980 it had not been molecular biology; beginning in 1983, it was. The intrusion of computers into molecular biology shifted power into the hands of those with mathematical aptitudes and computer savvy. A new breed of scientist began to rise through the ranks, with expertise and molecular biology, computers, and mathematical analysis.

In 1860, Charles Darwin's champion, Thomas Henry Huxley, debated the Bishop of Oxford, Samuel ("Soapy Sam") Wilberforce, before the British Association for the Advancement of Science.[73] Before a packed room of seven hundred eager to watch a rhetorical bloodbath, the bishop made a major tactical error while belittling the theory of evolution. He asked whether Huxley traced his descent from the apes through his mother or his father. Huxley responded, in a riposte taught to several generations of biology students, "If the question is put to me 'Would I rather have a miserable ape for a grandfather, or a man highly endowed by nature and possessed of great means and influence, and yet who employs these faculties and that influence for the mere purpose of introducing ridicule into a grave scientific discussion,' I unhesitatingly affirm my preference for the ape."[73; 74]

Debate about human evolution in the ensuing century and a third was a series of similar imbroglios. Liberal theology incorporated science by interpreting religious texts as analogical rather than literal. Science was accepted as the factual base, and theology the spiritual guide. Fundamentalists could not abide such concessions, and thus brooked direct confrontation with science. Bad blood coursed between biology and theology when evolution was the topic. Molecular genetics made matters worse.

In 1975, 115 years after the Huxley-Wilberforce debate, Mary-Claire King and Allan Wilson from the University of California at Berkeley published a paper synthesizing various analyses of DNA and protein similarities between humans and chimpanzees. Soapy Sam got horrid news. Genetic similarities between man and chimp were as close as sibling species of mammals (various members of the cat family, or canines, for example). King and Wilson noted that "the average human polypeptide [protein] is more than 99 percent identical to its chimpanzee counterpart."[75] Differences in DNA sequence were somewhat greater, as expected, but genetically speaking, the two species were amazingly similar.

For centuries, humans led by the clergy had struggled to distance themselves from the rest of creation, to build a qualitative wall between themselves and the other creatures of the animal kingdom. Humans placed themselves in a separate genus and family from other hominid apes. A part of this was chauvinism, well documented in separatist theology and philosophy. The classification scheme also acknowledged, however, enormous biological differences. Humans alone spoke, had written language, engaged in moral discourse, and built civilizations. The genetic similarities in the face of such dramatic anatomic and behavioral differences led King and Wilson to postulate that the changes must affect regulatory elements, perhaps those controlling the expression of genes during skeletal and brain development. They invoked subtle changes in when and how genes were expressed to bridge the chasm between genetic similarity, on the one hand, and behavioral and anatomic difference, on the other. The structural alterations would affect not which genes (and proteins) there were, but when they turned on and off, how long they acted, and how much they produced.

The genetic similarity of man and chimp was but one of many results that came from applying molecular methods to evolutionary biology. In the related field of paleontology, Wilson and colleagues used DNA sequence data from mitochondria (subcellular structures with a residual genome of their own) to trace the origin of modern humans to Africa several hundred thousand years ago.[76] The initial analysis proved more tenuous than imagined, subject to foibles of computational strategy, and suggested alternatives to human populations radiating out of Africa.[77] The use of DNA sequencing as a tool for paleontology was no longer in doubt, however, and augmented the traditional methods of bone analysis and archaeology. Luigi Luca Cavalli-Sforza and his colleagues at Stanford analyzed data from several chromosomal locations and found that DNA-based data corroborated archaeological and linguistic reconstructions of human migratory patterns.[78] Molecular biology invaded the fields of evolutionary biology, paleontology, and population genetics, yielding a wealth of data that awaited the new technologies for analyzing DNA structure.[79]

Comparing protein sequences among different species revealed several distinct classes of proteins.[80] Some were quite ancient, and were shared with many bacteria and other primitive organisms. These proteins were involved in metabolic processes essential to all life forms on earth. At the other end of the spectrum were proteins of more recent vintage, which showed evidence of having been assembled in bits and pieces by exchanges of blocks of DNA. Other proteins fell in between. Astronomers had long accustomed themselves to looking at the distant past. Their telescopes and detection instruments often captured light that left its source billions of years ago, so direct observation of distant objects was the same as looking into the remote past. By studying sequence data, the detritus accumulated over the course of evolution, biologists could now similarly glimpse into the distant biological past.

Analysis of DNA also permitted the study of contemporary human populations. Luca Cavalli-Sforza and Walter Bodmer surveyed variations among human populations in a 1976 textbook, *Genetics, Evolution, and Man*,[81] published at the dawn of the recombinant DNA era. Using DNA analysis and protein sequence analysis, the historical migrations of human populations in prehistoric periods could be inferred. Results from DNA analysis corroborated in general outline the conclusions from linguistic analysis, through the study of how languages were related to one another. Most of the conclusions remained the same, but the base of data was greatly expanded, and there was a promise of greater precision in the future as the genetic gold was mined. As methods improved, however, one factor necessary to assess human genetic variation diminished. The optimal populations to produce useful data, those that had been relatively isolated and geographically stable for a long time, were either dying off, being assimilated into surrounding populations, or migrating in the face of war and internal conflict. Anne Bowcock and Cavalli-Sforza noted in a 1991 article:

A number of populations of considerable interest are rapidly disappearing. Large geographic areas are being exploited and developed, changing rapidly and irreversibly the tribal worlds that still survive in every continent. The loss of traditional lifestyles destroys established communities, and their Diaspora makes it practically impossible to sample them. It is only from knowledge of the gene pools of these populations that we can hope to reconstruct the history of the human past. But humans are an endangered species from the point of view of genetic history.[82]

The need to secure the resources necessary to interpret human evolutionary history and to interpret contemporary data on population genetics brought together groups long known for word-to-word combat in the scientific trenches. A series of documents called for international mobilization to sample aboriginal populations around the globe, so that genetic variation could be assessed now and in the future as better methods developed.[82–87] UNESCO agreed to assist in the effort at a June 1991 meeting in Paris, and HUGO appointed a committee to set forth the appropriate scientific strategy. On June 5, 1992, the National Science Foundation, DOE, and NIH funded a $150,000 two-year grant to Marcus Feldman, working with Luca Cavalli-Sforza at Stanford, to support three workshops intended to plan at a much larger subsequent effort.[88] The overall project might cost more than one hundred times as much and would involve sampling populations from around the globe in search of our collective genetic history. Allan Wilson died in 1991, but his ideas did not.

Charles Darwin founded the science of biology on a theoretical footing. He closed the most important book in biology, *The Origin of Species,* on a philosophical note:

From the war of nature, from famine and death, the most exalted object which we are capable of conceiving, namely, the production of the higher animals, directly follows. There is grandeur in this view of life, with its several powers, having been originally breathed by the Creator into a few forms or into one; and that, whilst this planet has gone cycling on according to the fixed law of gravity, from so simple a beginning endless forms most beautiful and most wonderful have been, and are being evolved.[89]

The genome project has its sights aimed at the biological stuff that mediated the process of evolution. DNA was the structure that conferred inheritance and permitted small incremental changes to pass into new generations, while ensuring sufficient inherited stability to carry on life. Over the millennia, the instructions in the genetic code were modified not only in humans but in every living thing on the planet. The genome revealed relics of this evolutionary history. The biological revolution had, in many senses, been a continuous one over more than a century. Another major revolution in mathematics was of more recent origin.

At the turn of the century, Alfred North Whitehead and Bertrand Russell attempted to place mathematics on solid footing, to build its foundation on intellectual bedrock. Together, Russell and Whitehead wrote *Principia Mathematica,* one of the intellectual monuments of this century,[90–92] a culmination of thought that developed in the previous century. In this, they followed the traditions of Gottlob Frege, David Hilbert, and others in the field of mathematical logic.[93]

The foundations cracked in 1930 and 1931, when Kurt Gödel wrote a series of papers that demonstrated some statements in mathematics could not be proved.[94–96] He constructed a sentence based on the rules of arithmetic that could be proved only if it was wrong. Using the tenets of number theory, the part of mathematics concerned with the addition and multiplication of numbers, Gödel showed that this essential core of mathematics, basic arithmetic, was either self-contradictory or there were things within it that simply could not be proved. He then extended these findings by showing that the problem could be solved only by borrowing from another theory based on stronger, and thus less reliable, assumptions.[97] Gödel shattered the dreams of generations of mathematicians.

Gödel's methods drew on a new field of mathematics concerned with iterative processes. Mathematics after Gödel attempted to synthesize what he had cast asunder. Ulam, von Neumann, and countless others worked to bring coherence to information theory and related fields. The computer was a natural partner, and became an integral part of such research. This field of mathematics resonated in harmony with another, seemingly unrelated, field—molecular biology.

DNA passed through countless generations of organisms since the beginning of life on earth. Perhaps DNA became the keeper of inherited information only after RNA or some other molecule. DNA emerged as the dominant, if not exclusive, mediator of inheritance. Gregory Chaitin from IBM's Thomas J. Watson Research Center closed his book *Algorithmic Information Theory* with observations that clothed Darwin's conclusion in the garb of modern mathematics:

I would like to end with a few speculations on the deep problem of the origin of biological complexity, the question of why living organisms are so complicated, and in what sense we can understand them, i.e., how do biological "laws" compare with the laws of physics? . . . Biological evolution is the nearest thing to an infinite computation in the limit that we will ever see: it is a computation with molecular components that has proceeded for [a billion] years in parallel over the entire surface of the earth. . . .[98]

Exploring the structure of DNA was more than a practical matter; it was science aimed at the informational core of life. As genome researchers revealed that information, they were not only discovering genes that caused disease, they were also generating data to contend with the core hard questions facing all of science.

The revolution that took place in mathematics after Gödel and the parallel revolution in physics that replaced the mechanics of Newton with the probabilistic physics of quantum mechanics presaged the future of biology. Simplistic reductionism had to give way to a richer, if less predictive, science. The simple model of a broken gene causing Huntington's disease in a one-to-one correspondence would have to be embellished and adapted to the complexities of the human organism. Most genetic contributions to disease were not so simple. The causal theory had a kernel of truth, but disorders such as Huntington's were the unusual simple case, and even here the biology confounded simplicity. Even that prototype genetic disease showed diverse symptoms, and the age of onset could be determined by genetic "imprinting" (subtle changes in inheritance when chromosomes were inherited from the father rather than the mother). Most other diseases were far more complex.

The information from the genome project would accumulate most rapidly for human disease genes, and for organisms useful in studying human disease, because that is how humans would deploy their resources. Even at this first level, genetic diversity was impressive. As samples were compared from variant human populations, the richness of biology would inevitably come to the fore. With comparison to other organisms, the variety would become overwhelming. Approaching biology from the genome was destined to become a central strategy in penetrating the maze. The unifying force was evolutionary history; the core strategy was comparison of sequences. Shared genetic structure implied similar function.

In writing the biography of great men and women, printing the DNA sequence of their genome would not be a good place to start. Studying the structure of DNA could not explain how Beethoven created his music or how Einstein thought about physics. Genetics offered only a new tool with which to approach the problems confronting medicine and biology—for example, it might shed light on the gene responsible for Alzheimer's disease in the Ross family, perhaps eventually helping to relieve that part of their suffering. At the root of these discoveries, one would find computers running programs based on the work of mathematicians, comparing DNA sequences.

19
DNA Goes to Court

C OMPUTER HACKERS were not the only infiltrators in the gene wars. Another, more virulent pest, the lawyer, also began to infest molecular biology. DNA and the technologies to manipulate and analyze it became new frontiers for patent and copyright law. DNA methods were also used to link suspects to crime, mainly rape and murder, thus drawing the pristine science of genetics into the courtroom battles between prosecutors and defense attorneys. Adapting genetics to social functions through law required accommodation on both sides.

In November 1983, residents of Leicester County, England, found the dead body of fifteen-year-old Lynda Mann by a path. She had been raped and killed by an unknown assailant. Traditional forensic methods were used, but the case was still not solved when fifteen-year-old Dawn Ashworth, from a nearby town, was also found raped and murdered in late July 1986. Richard Buckland, a worker at a local psychiatric facility, was arrested. The police attempted to link Buckland to the victims, and contacted geneticist Alec Jeffreys of Leicester University.

Jeffreys was a world figure in the development and analysis of human genetic markers. He was interested by the request and agreed to help out. He used DNA typing on material from vaginal swabs of the victims and compared them to suspect Buckland's DNA. Jeffreys concluded that the two young women had indeed been raped by the same man, but it was not Buckland. Buckland was released, despite having made a dramatic confession. Buckland became the first person exonerated by DNA testing.[1]

The police then began a "genetic sweep" of the population in January 1987, intending to determine the DNA type of all young males in the vicinity. Colin Pitchfork was scared. He resorted to subterfuge, enticing a coworker to substitute for him when blood samples were drawn, so Pitchfork's DNA was not typed. By May, more than 3,600 DNA typings had been performed, but there was still no match to samples taken from the victims. In August 1987, a coworker admitted having substituted for Pitchfork, and six weeks later the police were notified. Pitchfork confessed to both murders and was convicted.

DNA typing had earlier been used to establish relatedness among individ-

uals (to resolve disputed paternity, to enforce child support, or to allow immigration into the United Kingdom), but the Leicester County case was more
widely publicized and promised far broader application to forensic testing.
The Leicester case inaugurated a new technology for identifying individuals in
criminal proceedings and touched off a battle that raged for several years. By
the end of 1989, DNA typing had been used in at least eighty-five cases in
thirty-eight states in the United States alone,[2] and in spring 1991, every state
had used forensic DNA testing.[3]

For prosecutors, DNA typing was especially effective in rape and murder
cases. DNA typing might enable them to link criminals to the scene of the
crime as reliably as standard fingerprinting, but without requiring that the
criminal leave a good fingerprint. In most crimes, there was a struggle, with
bloodstains to be analyzed, semen from a rapist, or hair or skin tissue inadvertently left behind. If the perpetrator did leave behind a bit of hair, blood,
semen, saliva, urine, or other tissues that could be typed by DNA analysis,
then they could be identified. For defense lawyers, DNA typing could be an
overwhelming exculpatory technology if there was no match. As in the Leicester case, DNA typing was more convincing than a false confession.

The power of the technology came from linking a person to the scene, not
proving that the defendant committed a crime. (In rape, for example, a match
between suspect and semen type indicated that intercourse took place, but not
that it took place against the victim's will.) DNA typing was clearly a powerful
new tool for law enforcement, but important questions remained about how
to use it properly.

The techniques were quite similar to those used in pedigree studies for
genetic linkage, and indeed used many of the same reagents. There were major
differences, however, that emerged as more cases appeared in court. In genetic
linkage, a genetic marker is followed through a family. The value of a marker
is that it enables one to trace inheritance from one generation to another, to
discern which marker was inherited from the mother and which from the
father for each member of a pedigree. Typically, in pedigree research, fresh
blood samples are taken from those in the pedigree.

In forensic investigations, however, there is less information to start with,
since there are no family ties to help interpret the DNA findings. The critical
question to answer is: How likely is it that these tissues—such as blood, hair,
or semen—came from this particular suspect? The amount of material may be
quite small, the sample is unlikely to be fresh, and it may be mixed with tissue
from other individuals, as in cases of multiple rape, or mixed samples of both
perpetrator and victim. Blood samples are often dried, and the DNA may be
partially degraded. Sample DNA may be completely used up in the analysis,
precluding reanalysis if the test fails and eliminating the possibility of further
tests if initial tests are not definitive. Tests must be performed adequately by
the laboratory, so that samples are not switched and criteria for calling a match
are reliable.

Most important, however, one is not merely comparing DNA type between parent and child within a pedigree, but rather trying to assess the likelihood that it comes from a particular person, the suspect. That assessment, in turn, depends on how often that DNA typing pattern occurs in the entire population, not just in a family pedigree. The probability of a match thus depends on statistical analysis of DNA typing patterns across the population and knowledge of how prevalent a given DNA type is. The statistical power of DNA typing thus ultimately rests on data that are expensive to collect, requiring systematic survey of the population with DNA typing of very large numbers of individuals.

As DNA typing entered the courtroom, questions about how adequately it had been performed and interpreted began to arise. The process of introducing the new evidence hinged on satisfying the *Frye* standard, a set of legal criteria that grew out of a 1923 murder case. A court was faced with deciding whether to admit evidence taken from one James Alfonso Frye, a young African-American accused of having murdered a white man in Washington, D.C. The prosecution proposed to introduce into evidence data about how his blood pressure responded to questions about the crime, as a measure of his veracity in a primitive precursor of the polygraph test. The court agonized, but ultimately rejected the proffered evidence, noting:

Just when a scientific principle or discovery crosses the line between experimental and demonstrable stages is difficult to define. . . . while courts will go a long way in admitting expert testimony deduced from well-recognized scientific principle or discovery, the thing from which the deduction is made must be sufficiently established to have gained general acceptance in the particular field in which it belongs.[4]

The court thus established a two-tiered sociological standard for the acceptance of scientific evidence. A court must decide the field whence it arose, i.e., identify a scientific community, and must determine that it was accepted within that community. These criteria were bulwarks against admitting evidence from new scientific techniques until the Supreme Court cast down the Frye standard in 1993. The reason for special caution was a belief that scientific data might unduly sway judges and juries. The *Frye* criteria, however, were rather vague. Just how to define a field and how to assess consensus was far from clear. The alternative to the *Frye* standard, under federal rules of evidence, was premised on relevance to the matter at hand—admitting into evidence anything that helped the court to assess the facts. The federal rules were developed under precepts outlined in a 1975 statute and subsequent amendments.[5] Here also, while not so rigid as the *Frye* criteria, the judgment of admissibility turned on whether expert testimony would be useful in ascertaining or understanding the facts and required a judgment of the qualifications of experts. Rule 702 specified that expert status could be inferred from "knowledge, skill, experience, training, or education."

In the United States, most early DNA typing was performed by two private firms. One, GenMark, was affiliated with the British chemical giant ICI, and licensed the methods developed by Alec Jeffreys. Lifecodes was a small, independent firm based in Valhalla, New York. These private firms initiated forensic typing on a fee-for-service basis for prosecutors or defense attorneys.

DNA tests were first brought into American courts in 1986, but really caught hold in 1988. The Federal Bureau of Investigation began to focus on the promise of DNA testing as the technology was introduced into courtrooms throughout the states. The FBI set up a laboratory in Quantico, Virginia, to perform tests on request and to train those who wished to learn about the technology from state crime laboratories. The FBI also proposed to standardize the methods used so that results could be compared from one state to another and DNA typing profiles could be matched at the federal level, comparing samples analyzed by laboratories in different states. A California investigation might turn up a match to a Colorado serial killer, for example, using only the limited data from a computer code for DNA typing. The FBI could then notify police in both states to contact one another to pool their evidence.

As work on the OTA report on the genome project was winding down early in 1988, the number of criminal cases using DNA forensics rose quickly. It became clear that an assessment of forensic typing could also be useful. OTA thus began an assessment that produced a separate report in July 1990.[2] The National Research Council of the National Academy of Sciences also formed a committee to assess forensic uses of DNA. The committee, chaired by Victor McKusick, began work in January 1990 and released its report in April 1992.[6]

DNA evidence was first accepted as evidence in *Florida* v. *Andrews*;[7] Tommy Lee Andrews was accused of having raped and slashed several women. DNA typing was used by the prosecution, and he was convicted in November 1987. In October 1988, the Florida State Court of Appeals for the Fifth District upheld the admission of DNA typing evidence.[2] The prospects for DNA forensics looked rosy, but then some sloppy work showed how it could be troublesome.

The watershed case that cast doubt on how well DNA forensic testing was being performed was the highly publicized *New York* v. *Castro*. This case was tried in the same court caricatured in Tom Wolfe's novel *Bonfire of the Vanities*.[8] Lifecodes had performed the DNA typing, concluding that a bloodstain on Jose Castro's watchband matched the blood types of a woman and daughter murdered in the building where he was a janitor. Lifecodes claimed that the likelihood of the match they found was one in 738 trillion.[9] When expert witnesses scrutinized the evidence, it turned out that Lifecodes had ignored two bands in the DNA typing pattern, had failed to run appropriate controls, and had not applied its own quantitative criteria for matches.[7;9] Lifecodes had thus interpreted its evidence in what could be charitably be called a creative fashion. These lapses called into question the entire enterprise of DNA forensics.

The expert witnesses called by both prosecution and defense took the unusual step of going outside the courtroom to confer among themselves, and they prepared a report for the judge. The judge ruled that the evidence could be introduced to exculpate the suspect, but not to corroborate his guilt. The court clearly indicated that DNA testing was, in theory, admissible as evidence to identify the suspect positively, but doubts about laboratory procedure and interpretation in the current case made it inadmissible for that purpose. Castro pled guilty, and while the status of DNA testing in the case was thus not crucial to its outcome, the exposure of some pitfalls in DNA forensics became a lasting residue.

In other cases, Lifecodes presented statistics suggesting that the chances of a match were one in several hundred million or in the billions. The claims were outrageous given the paucity of the population-genetic database, which was held as a proprietary secret. Beyond the insufficiency of the population genetic data, there was always a possibility that the laboratory inadvertently switched samples, or that the person had a twin and did not know it (DNA typing, unlike standard fingerprinting, could *not* distinguish identical twins). The likelihood of such errors was obviously much higher than the figures being quoted in court. The *Castro* case ushered in a debate about laboratory practices, consistent band-matching criteria, and standards for interpreting the statistics.

The courtroom conflict between prosecution and defense began to spill over into the scientific community. A relatively small number of human geneticists, particularly those who were knowledgeable about both DNA typing and population genetics, were called as expert witnesses in many cases, but they did not agree among themselves about how to interpret the tests. The center of the controversy was the degree and significance of population substructure.

If the accused person came from a population that often had a DNA-type profile unusual in other groups, and if this group had few or no members in the population database used for interpretation, then the result could be highly misleading. Suppose, for example, that the suspect was a Basque and that Basques had type Z very frequently but other groups did not. There might be hundreds of thousands of Basques with that type. The database would not reveal this fact because it would include few Basques and would lump them with Caucasians. In the total database, the Basque pattern would appear quite rare. Moreover, Basques might be expected to cluster in the same neighborhoods, say the one where a murder or rape took place. If DNA types did indeed vary among subpopulations, errant statistics could make it seem that a match was far more significant than it actually was. Another possible source of errant interpretation was if two traits were assumed to be independent, but were actually associated with each other. Nordics, for example, might often have blond hair and blue eyes, but if the probabilities of blue eyes and blond hair, both uncommon traits, were multiplied together, it would seem extremely unusual for individuals to have this combination. There were few data to assess how often this kind of bias was present for the new DNA markers, and so there was ample room for divisive scientific combat.

The scientific controversy spilled onto the pages of *Science*. Two groups of highly competent population geneticists took opposite sides on this question. One article cast doubt on how forensic tests were being employed and asserted a need for considerably more data about the frequency of DNA types among disparate populations before the technique should be used to decide the fates of the accused.[10] A companion piece, commissioned by editor Daniel Koshland, doubted that population substructure was large enough to mislead juries and judges.[11]

Those interpreting DNA forensics typically used different base statistics, depending on the race of the suspect. This practice was troubling from both social and technical points of view. It was inherently disturbing to use race overtly in criminal proceedings. Moreover, the "racial" categories corresponded only poorly to population-genetic knowledge. The Federal Bureau of Investigation used a category of "Hispanic," for example, but this could refer to a person from a Caribbean island, someone from Aztec, Inca, or Mayan extraction, or someone whose ancestors came from Spain and Portugal—a hopeless mishmash.

The task of sorting out the technical arcana fell to the NRC committee. The NRC report was caught in a crossfire between the FBI and prosecutors on one side and defense attorneys on the other. The battle was joined by geneticists. The intensely adversarial ethos of the courtroom seared the professional egos of many, as their motives were impuegned, inconsistencies amplified, and characters flayed not only before the jury but also in the public media.[12] A network of prosecutors and a countervailing network of defense lawyers resorted to tactics of intimidation and persistent annoyance vastly more aggressive and personal than the usual intellectual fencing within science. Some scientists, although not the most prominent scientific authorities, and not a few lawyers made a living on the introduction of DNA forensics. Most of the fights centered on how to interpret laboratory results.

The 1990 OTA report noted this controversy and called for population geneticists to formulate standards.[2] The NRC committee, as an expert body, attempted to do just that, to find consensus about how to interpret the results of DNA forensic tests. As the report approached release, the *New York Times* broke a story that concluded the report would recommend a moratorium on DNA forensics until there were better standards.[13] This forced the committee to schedule a press conference in great haste, to dispel the call for a moratorium recounted in the story.[14]

The NRC report made a series of significant recommendations. It called for an independent expert body, outside the FBI, to monitor laboratory practices and to make recommendations on how DNA forensic testing should be performed. The committee's most significant contributions but also its greatest vulnerability, came from recommendations about how to handle the population-genetic analysis. The committee reviewed evidence that population substructure was probably not a major source of errant matches, but it allowed that "population substructure may exist." It acknowledged that existing statistical databases were insufficient to determine the extent of population substructure, and argued that "the solution, however, is not to bar DNA evidence, but to ensure that estimates of the probability that a match between a person's DNA and evidence DNA could occur by chance are appropriately conservative."[6] The databases should be made better, and this could be done by directly measuring the extent of population substructure among ethnic groups.

The controversy over DNA forensics thus reinforced the need for better data about human population genetics. This had already been raised as an urgent priority among anthropologists and paleontologists who hope to use genetics to understand human origins and historical migratory patterns. The need for robust forensic databases gave the same data a decidedly practical twist, with lives hanging in the balance. Controversies about how best to interpret population-genetic data that had long been obscure and of only academic interest were suddenly directly relevant to the fates of suspected criminals, and to the pursuit of justice. Whether data would dispel the fractiousness of the population-genetics community was open to doubt, but if not, there was a long future for careers in expert testimony. The first step, and the best hope, was to collect empirical data on human populations by going out and sampling them.

A recommendation to stop using race-specific analyses was also a major advance. The committee suggested that the frequency of any given DNA marker pattern should be interpreted to the benefit of the suspect, under a "ceiling principle." The number used to assess the likelihood of a match should be taken from the population group with the highest frequency. This default assumption was a clever way to get around the troublesome process of determining racial origin. Instead of trying to decide which "race" the suspect came from and applying different statistics for each group, the suspect's racial background would be irrelevant, and the suspect would be given the benefit of the doubt for each marker tested. If prosecutors needed more statistical power, they could order more markers to be tested, possible in many but not all cases. This proposal would bring down the probabilities from the ludicrous range of one in millions or billions, but the technique would still generally be far more reliable than eyewitness identification or blood-protein tests. The committee suggested setting arbitrary conservative probabilities until data began to flow in from the empirical population surveys.

The committee thus cut the Gordian knot, hoping to preserve the admissibility of the evidence, to shore up the regulatory framework, and yet to interpret the evidence in a conservative and scientifically defensible manner. It was not clear, however, how judges would react. In what was purportedly a review of books about the genome project but proved more a platform to air his views, Harvard geneticist Richard Lewontin noted that the NRC report might not resolve the population-genetic controversy. Judges might focus on the report's ambiguity, calling for empirical data, rather than accept the committee's interim solution of ceilings and arbitrary marker frequencies.[15] A cautious response would favor waiting for more data about the markers being used in specific populations. The courts predictably differed in how tightly they embraced DNA forensics. Some accepted DNA forensic evidence, while others awaited resolution of the population genetic issues.[16] While the eventual acceptance of DNA forensics was not in doubt, the speed with which it would become routine was highly uncertain and appeared likely to differ markedly among jurisdictions.

The controversy refused to die, and even as some courts began to admit DNA forensic evidence more readily, another controversy broke out in *Science* with the publication of an article critical of the NRC report and a news feature on the same topic.[17; 18] This time, the NRC committee was lambasted for erring too far on the side of conservatism. A group of population geneticists cast doubt on the significance of population substructure for those markers being used in forensic work. They disagreed with the logic behind the ceiling principle and asserted that most data suggested that marker frequencies could generally be multiplied together—the procedure that produced such astoundingly large odds ratios. They pointed out the need for much larger samples of population groups than those suggested in the NRC report to get sufficient data; a survey of the proposed size would generate relatively unreliable and unduly high estimates of marker frequency because the number of individuals sampled would be low and the margin of error correspondingly high. The upshot was that courts were being misled into a too conservative stance on DNA forensics by the NRC committee, and that the empirical surveys intended to solidify the basis for interpretation would not be sufficiently robust to restore balance. The NRC report was thus being attacked from both sides. Some critics claimed it was unduly conservative, while others contended the NRC committee had too readily accepted DNA forensics. The NRC commenced a second DNA forensic study in the summer of 1993, hoping to finally quell the controversy.

The courtroom entry of a new genetic technique, derived from gene mapping efforts, was noisy and slow. The problems emerged only as specific cases provoked scrutiny of existing practices. Abstruse questions of population genetics, a mathematically complex and relatively small academic subspecialty, were suddenly exposed to the harsh realities of the criminal justice system. Science collided with an adversarial court tradition, and the result was five years of turmoil, entailing hundreds of hearings, thousands of hours, and millions of dollars. In the wisdom of hindsight, the sources of controversy could have been resolved by empirical research, standardization of methods, and conservative race-neutral interpretation. But like the field from which it arose, DNA forensic testing took a bumpier road to acceptance. Yet it was not the only area where science and law collided. There was also the matter of who owned the genetic terrain now being explored.

New relationships between industry and academia proliferated in molecular biology soon after the invention of gene splicing in the mid-1970s. Molecular cloning and fusing of cells bred commercial biotechnology.[19; 20] The commercialization of molecular biology coincided with a shift in government policies to promote U.S. economic competitiveness. Values within biomedical research groups in universities shifted from suspicion of commercial attachments to active promotion of technological spinoff.

Industry began to fund more biomedical research, particularly work related to development of new drugs. Government funding grew, but at a pace well behind that of pharmaceutical firms, which significantly increased their research and development efforts as a competitive strategy. In the mid-1980s, private funding for biomedical research and development surpassed NIH's; by the end of the decade, it was greater than federal funding from NIH and all other agencies combined.[21] Catching hold of the best in new science became an important element in drug discovery, driving pharmaceutical firms to heavy research investments as a matter of financial survival.[22]

Changing intellectual property law was a prominent feature of this policy turnabout—the new age of molecular genetics came in with a patent. When Herbert Boyer and Stanley Cohen patented their method for splicing DNA in 1976,[23] it caused quite a stir. In 1980, the U.S. Supreme Court ruled that a microorganism, a living thing, could be patented in *Diamond* v. *Chakrabarty*.[24] The trend to promote commercial applications of biomedical research continued under a series of new public laws and executive orders throughout the 1980s. These consistently encouraged patents by U.S. institutions receiving federal grants and contracts, by conferring substantially greater authority to those receiving federal funds.[25] Congress was presented with evidence that when the government owned patent rights, it did not foster commercial applications. Several studies suggested that the government was not aggressive in pursuing patents and did not license them or otherwise ensure their translation into useful products. It seemed reasonable to assume that if those doing the research had a stake in patents, they would pursue commercial gain more assiduously.

The statutory changes began with the Patent and Trademarks Amendments of 1980[26] and continued in the Trademark Clarification Act of 1984[27] and the Technology Transfer Act of 1986.[28] President Reagan issued Executive Order 12591 in April 1987 to promote technology transfer out of federal laboratories. Congress and the President sent strong signals that they wanted the investment in science to pay off in the form of patents held by institutions of all types that received taxpayers' research dollars. Universities were expected to license their patents to commercial firms, thus harnessing biomedical research to commercial enterprise. American firms were given patent rights in most cases, and foreign firms and research institutions were given incentives to involve US manufacturers. Just how to transform biomedical research mavericks into team horses, however, was a more complex matter.

Intellectual property law would be at least part of the rigging, but in biotechnology, intellectual property was in a tangle. As universities and small dedicated biotechnology companies developed new techniques and new products, the scientific races for priority spilled over into battles to secure patent rights, with different rules in the United States and other countries.[29; 30] The number of biotechnology patent applications reached 6,900 in January 1988.[31; 32] The Patent and Trademark Office (PTO) in the Department of Commerce disposed of 2,200 biotechnology patent applications in 1987, but received over 3,100 new ones that same year.[31] The PTO was deluged with applications, and it staggered under the burden.[33] In 1988, the PTO created a new unit to handle biotechnology patents in response to the growing demand.[31] Universities and companies continued to complain about the patent process nonetheless: the flow was too slow, the level of expertise of patent examiners insufficient to mete out the rewards of innovation, staff turnover was too high, and the decisions of the office unreliable and thus inclined to exacerbate rather than abate the surge of costly patent litigation.[34]

Approval of a patent application was just the start. If there was enough money at stake, residual uncertainties would later surface in suits about who was infringing whose patent, to be decided case by case, in years of costly litigation. Patent rights regarding therapeutic pharmaceuticals, many of them based on patented human genes, were disputed in many hard-fought legal battles. These were of great consequence to biotechnology in general but only tangentially related to the genome project. Some patent uncertainties, however, arose in matters directly connected to genome research. The technique of mapping with RFLPs was itself the subject of a patent application filed in the early 1980s, first by a Utah group and then pursued by Stanford University. The polymerase chain reaction was the battleground for a patent dispute between the chemical company Du Pont and the biotechnology firm Cetus, recounted in Chapter 4. By far the greatest area of uncertainty, however, was how the coming flood of DNA sequence information would work its way into the legal framework for deciding how far property rights extended and how well they would protect an inventor.

Uncertainty about patenting gene maps and DNA sequence information was apparent from the early debates about a genome project. The criteria for granting patents and registering copyrights were clear, in principle.[20; 35] How to apply the general criteria to new technologies and new scientific strategies for defining biological functions was not, however, immediately obvious. Walter Gilbert was among the shock troops invading this hotly contested legal territory.

When Walter Gilbert embarked on raising funds to start the Genome Corporation in 1987, he claimed he would copyright the information and sell it to companies or researchers who wanted it. They would purchase access to the information because he could generate it more quickly and cheaply than they could themselves. This gesture toward a commercial venture provoked an outcry among molecular biologists. As Lennart Philipson and John Tooze put it, commenting from a European perspective, "the prospect of private capital financing this work and then keeping secret the sequence information and restricting access to the libraries of clones from which it was obtained, in order to generate corporate profits, is too obscene to find many supporters."[36] Gilbert replied that biologists had also complained about buying restriction enzymes and laboratory glassware when they were first produced commercially.[37]

The 1988 OTA genome report waded into the morass of intellectual property law and surveyed the noisome debate about how it would apply to genome studies. Susan Rosenfeld, a lawyer from New York, prepared a background paper,[25] and OTA convened a workshop in June 1987.[38] *Science* summarized the meeting, asking "Who Owns the Human Genome?"[39] Those around the table agreed that one could not patent sequences *per se,* and while Gilbert's idea of a copyright seemed plausible, those assembled cast doubt on whether copyright protection was sufficient to protect a massive private investment.

Irving Kayton argued in 1982 that clones and sequences could be copyrighted, like computer programs or photographs.[40] Susan Rosenfeld argued this was unlikely to prove true in fact, and even if a database could be copyrighted, the subsequent uses to which the data were put could not be controlled. A later OTA report cited unofficial statements from the Register of Copyrights and the Copyright Royalty Tribunal that DNA molecules or sequences would not be protected under copyright in the same way as art, literature, computer programs, or electronic media.[35] The 1987 workshop surveyed disagreements about other intellectual considerations related to genome research without reaching any conclusions. Various legal experts held different opinions about whether scientific data were property[41] and about the degree to which trade secrets would be effective in biotechnology.[42; 43]

The matter of intellectual property law was raised time and again at genome conferences. At DNA sequencing conferences, there was debate about just what information needed to be kept secret before filing a patent application. At ethics conferences, there was concern about balancing the advantages of intellectual property protection against the need to pool information expeditiously, to reduce duplication, and to broaden access to important data. As George Cahill from the Howard Hughes Medical Institute put it, "What is the bucks-to-ethics ratio here?"[38; 39] While the 1987 workshop raised questions about whether DNA sequences could be patented in general, as part of a general mapping of the genome, it was equally clear that genes that were isolated and manipulated could indeed be patented. Rebecca Eisenberg, from the University of Michigan Law School, analyzed the question of just what could be patented in a 1990 law review article.[44] She argued that the critical criterion was "whether the claimed invention is the result of human intervention," adding that "if the DNA sequence is identical to a sequence that exists in nature, it may still [be patented] if the patent applicant has made the sequence available in an isolated or purified form that does not exist in nature." Many genes had been found and sequenced, and case law upheld the patent claims for new drugs based on those genes.

Eisenberg noted a potential conflict:

. . . The patent system rests on the premise that scientific progress will best be promoted by conferring exclusive rights in new discoveries, while the research scientific community has traditionally proceeded on the opposite assumption that science will advance most rapidly if the community enjoys free access to prior discoveries."[44]

Eisenberg then sketched out the possibilities of defining research uses exempt from patent protection, permitting research to proceed without worry of being sued for infringement. This would preserve the right to use the information for noncommercial purposes, such as constructing genome maps in the public domain, but would preserve the power of patent protection in the commercial realm. Another possible solution was suggested by Dennis Karjala and others who helped prepare an outline of legal issues confronting the genome project.[45] A new form of intellectual property protection might be tailored to biological technologies. This option would attempt to retain the incentive for private research investments while acknowledging the intuitive disharmony between the notion of an "invention," with its image of a machine to be protected by a traditional patent, and the newfound commercial power of biology.

Despite continued debate, the issue failed to provoke a policy response until it came up at a Senate meeting in July 1991. Matthew Murray, who helped arrange a previous 1990 hearing on the genome project for Senator Pete Domenici, returned to Capitol Hill for a short stint to set up a progress review meeting on the genome project, a senatorial "annual checkup." Domenici's main interest remained commercial spinoff of genome research.

The review of scientific results was upbeat. The chromosomal defect underlying the fragile X syndrome, a common cause of mental retardation mainly affecting males, had recently been uncovered, and Baylor University geneticist Tom Caskey was present to tell the story, as one of the stalwarts in a massive international collaboration. The discovery had used the methods promoted by the genome program and had involved several laboratories supported by genome project monies.[46]

Domenici's meeting sparked an exchange about how and when to file patent applications on findings from large-scale sequencing projects. John Barton from Stanford Law School pointed out that while the general patent criteria were generally accepted, their interpretation in the specific case of finding genes through sequencing was a gray area. At the root of the uncertainties was the new scientific process of discovering DNA sequences before knowing their function. It seemed intuitively clear that sequences, like mountains, could only be discovered, not patented. Yet it was equally clear that patents already protected the isolation and purification of genes and development of protein products encoded by them. The patents rested on a legal distinction between discovering a gene and producing it in a new and useful form, which constituted the invention. DNA sequence information could be the initial step in a long journey leading to diagnostic tests and might even might lead to new treatments in some cases. The open questions were how and to what extent DNA itself would be a patentable subject matter.

NIH scientist J. Craig Venter announced at the Domenici meeting that he had been isolating DNA fragments from brain tissue, corresponding to stretches that coded for proteins, and that NIH filed a June 1991 patent application to cover the sequences. Venter's group was collecting such fragments, isolating them, and determining short stretches of DNA sequence as gene indices. In consultation with the NIH Office of Technology Transfer, NIH filed patent applications on several hundred of them, listing Craig Venter and his coworker Mark Adams as the inventors.

Most of us in the room were startled by the disclosure. Watson was aware of the patent application, but did not support it. Adler and Venter had conferred with Robert Strausberg in the genome office as they frenetically helped prepare the patent application, but within the genome office, there was considerable doubt that the sequences were patentable. Adler and several lawyers with whom he worked, however, thought that the sequences were very likely patentable. From their perspective, NIH ran a serious risk if it failed to patent the gene sequences. If another group later succeeded in securing similar patents, NIH would certainly be criticized. Congress would surely excoriate NIH if, for example, a Japanese firm grabbed the patent rights for genes, when NIH might have been able to confer a preference for American manufacture through licensing its patent rights. Moreover, failure to patent might also make it difficult to patent full genes when they were found. If NIH's data were published without patents and scientific groups later isolated the genes in a more useful form, the prior publication of NIH data might make the subsequent gene purification seem obvious, and thus unpatentable. If no patent application was filed, NIH would irreversibly foreclose its future options.

Venter worked as an intramural scientist in the National Institute for Neurological Disorders and Stroke (NINDS). NINDS was bureaucratically separate from, but scientifically linked to the NIH genome center. When Venter mentioned his patent application at the Domenici meeting, Watson was lying in wait and took aim with heavy artillery. Watson asserted that it was sheer lunacy to patent such incomplete information. He objected strenuously that the automated sequencing machines "could be run by monkeys," and if sequences could be "locked up" by the first person to sequence a part of a gene, without knowing its function or even the sequence of a significant fraction of a gene, biomedical research could be tied up in knots by patent litigation. The exchange initiated a running battle that culminated in Watson's resignation nine months later. After the meeting, Watson mentioned privately that he had "been too hard on Craig," and that Venter was only following the advice of others at NIH, but he did not back down from his stance that the patenting idea was fundamentally wrongheaded.

J. Craig Venter and Leroy Hood were instrumental in devising methods to automate the sequencing of DNA. Venter's work on a system for rapid DNA sequencing, originally done at the National Institutes of Health, led to a controversial patent application that pitted NIH against a formidable array of opponents. Hood, who initially headed a major research group at the California Institute of Technology, is now at the University of Washington. *Courtesy Institute for Genomic Research*

The conflict grew from small beginnings, intimately intertwined in the early history of large-scale DNA sequencing efforts. Venter ran one of the largest sequencing laboratories in the world. His work centered on understanding genes for molecules involved in transmitting signals between nerve cells—enzymes that made neurotransmitters and receptors for intercellular communication. This work drew him into sequencing the genes encoding the proteins. He made a commitment to DNA sequencing as a major scientific strategy to understand the process of neural communication, several years before this approach became generally accepted. Venter believed in the technology. He organized the first international DNA sequencing conferences, which became annual events, and also helped plan an early international meeting on ethical aspects of genome research, which took place in Valencia, Spain (the second Valencia meeting, in November 1990).

On the technology front, Venter's laboratory had a cooperative research and development agreement with the California instrumentation company Applied Biosystems. His laboratory secured early access to instruments in development, including state-of-the-art DNA sequencers, and his laboratory group was a proving ground to help the company work out technological kinks. While Venter had stable research funding through the National Institute on Neurological Disorders and Stroke, he applied for funding from the National Center for Human Genome Research. After discussing the possibilities for almost a year, Venter agreed to send in a proposal to be considered for funding. Watson initially asked him to submit a brief description, but the rules changed when the genome center decided that Venter's intramural funding request should be considered by the same peer review group considering university applications for large-scale sequencing. Several projects proposals had been submitted in late 1989 and early 1990. Venter initially proposed to sequence the terminal stretch of the X chromosome. It was an audacious proposal to mount an assault on a region known to contain a relatively dense cluster of at least thirty disease-associated genes.

The review of sequencing grants took place as acceptance of automated sequencing instruments was rapidly shifting. Some scientists were simply convinced that the current generation of machines would never prove useful for large-scale sequencing. Venter's laboratory had begun to turn out solid data with greater reliability and at a faster rate than perhaps any other group in the world using the Applied Biosystems instruments, but the early record of difficulty in getting automated sequencers to work reliably had tarnished their reputations. Some had taken to calling the machines "$100,000 paperweights."[47] Unlike the other investigators rejected in the first review, Venter's group never secured genome project funding. The reasons were complex, and partially due to Venter's changing scientific direction. A growing frustration with delays in getting funding from the genome center also contributed to the decision.

Venter's X chromosome sequencing proposal had evolved from earlier discussions, and would continue to evolve. Venter already managed a large and growing research team without genome project monies, as he had direct funding from a separate NIH institute. He did, however, want genome funds to expand the scale of his work. Watson was initially quite enthusiastic, and said he would devote $5 million to Venter's work. The politics of sequencing, however, were intense, as there was considerable opposition to large-scale sequencing among molecular biologists, even for model organisms such as yeast, nematodes, and fruit flies. The opposition was even stiffer when it came to sequencing human DNA, which had a lower density of genes, a far less elaborate genetic map, a greater depth of ignorance about gene location and function, and less powerful genetic tools for analysis. There was also considerable disagreement about the proper sequencing method (whether it should be done by hand or by automated sequencers) and the proper strategy (whether it should start from unedited chromosomal DNA—genomic sequencing—or only from those edited sections known to code for protein). The fierce opposition forced Watson to retrench on his initial commitment.

Venter worked directly at NIH, and so Watson could have directly funded the work with an internal review of its merit. Sensitive to opposition within the biomedical research community, the NIH genome center decided on a cautious approach, and Venter's project was put on hold along with other major sequencing proposals, while the policy was sorted out. NIH and DOE appointed a joint working group on DNA sequencing that met in July and September and recommended a special "request for applications," soliciting grants to do large-scale sequencing, and specifying minimum criteria to meet in those applications. Watson had initially asked Venter to submit a short summary of his ideas, but Venter was now asked to submit a more formal application to be judged with others from outside NIH.

The process entailed a special request for applications and a review process, causing another six to eight months' delay in a discussion that had been going on for over a year. In the review, Venter's group was the only one from an NIH-based intramural group. Having an intramural proposal reviewed by much the same group considering grant applications from outside NIH was an unusual move. Venter's group alone held a face-to-face meeting with the study section—the scientific review panel charged with assessing the scientific soundness of the proposals. It was a difficult meeting, and the study section assigned the proposal, along with most others, a priority below that needed to obtain funding. Walter Gilbert's proposal to sequence a bacterial genome and Leroy Hood's proposal to sequence large stretches of DNA containing genes for immune function also had a rough ride, although they were eventually funded in later funding cycles. While Venter's review was relatively unfavorable, the genome office nonetheless told Venter he, like others, was likely to be funded if he revised his approach in light of the reviewer's comments. Venter began to formulate a proposal to continue ongoing work on several different chromosomal regions—those known to contain the genes for Huntington's disease (chromosome 4), a part of chromosome 19, and regions linked to disorders on chromosomes 17 and 15.

The new strategy was settled between Watson and Venter at a meeting on Hilton Head Island, South Carolina, at Venter's 1990 DNA sequencing meeting. The first day of the conference coincided with the date that Watson had chosen as the official start date for the genome project, October 1, 1990.[48]

Before this proposal was fully evaluated, however, Venter withdrew his request for funding, in an April 23, 1991, letter to Watson. In that letter, Venter noted:

Had we started over two years ago, when we first discussed automated sequencing, we probably would have completed 1-2 megabases of contiguous sequence by now. This has represented another major frustration for me. I am concerned that the bureaucracy that is a necessary part of the grant review process cannot keep pace with the rapid developments in the genome area.[49]

The letter also laid out his scientific reasons for wanting to take a different scientific tack, focusing on coding regions rather than genomic sequencing of unedited DNA from human chromosomes. Venter decided to pursue a line that promised to produce results quickly and that would link sequencing to functional clues and gene mapping directly. Venter's change of heart came partly from his frustration with bureaucratic unresponsiveness and partly from a belief that the tedium of direct genomic sequencing yielded data that were too hard to interpret.[50] He argued it would be more efficient to first sequence protein coding regions, which would in any case be essential to interpreting any genomic sequence derived directly from chromosomal DNA. The new strategy dropped the idea of sequencing a stretch of the X chromosome and instead focused on protein-coding throughout the genome.

The idea of focusing on protein-coding regions had been proposed time and again in the genome debate, indeed even before and during the June 1986 contretemps at Cold Spring Harbor Laboratory,[51] but had never become policy at the NIH genome center. Sydney Brenner championed this strategy for the initial genome efforts in the United Kingdom, and it was also pursued by groups in Japan, France, and elsewhere. In the United States, the Department of Energy decided to include a similar component in its genome research program beginning in 1990. Venter was an adviser on this effort, and his laboratory was among those funded by DOE.

Venter's main innovation was the nature of the indexing system for the gene catalog, based on automated sequencing to identify short stretches of DNA. Venter's work on brain molecules made this a natural strategy. Brain cells produced a far wider variety of proteins than any other organ, and so were logical sources from which to start a gene catalog. The global strategy to determine DNA sequences from protein-coding regions was a natural extension of his brain research and required no funds from Watson's genome center. Venter's group published initial results in June 1991[52], the same month they filed the patent.

Starting a sequencing effort from protein-coding regions of DNA was a quick way to identify new genes and begin to index them. It was also an efficient way to find functional clues about newly discovered human genes, by comparing sequence similarity to genes of known function from other organisms.[49; 52–55] The index could be created readily, but would be incomplete.

Short sequences from expressed regions could indicate where genes were located on a chromosome map and could be used to fish them out of the genome.

The short sequence "tags" would not, however, be a complete inventory
of protein-coding regions. Some genes were difficult to isolate and did not
appear in the standard collections that were the starting point for Venter's
effort. Moreover, since only a short stretch of each gene's DNA was known,
an investigator with only a part of a gene sequence looking for his or her gene
could miss a match until the entire gene sequence was logged into the catalog.
The catalog would thus be relatively easy to construct, and would be quite
useful for identifying candidate genes in a given region, but the information
was not archival or comprehensive.

The great advantage of the approach was its speed, its ability to suggest
functional clues based on sequence similarity to known genes, and its identification of previously undiscovered genes of unknown function. These unknown genes were truly "Terra Incognita," and the gene catalog could, with a
bit more work, include their location on the human chromosomes. As such a
map neared completion, a group hunting for a disease gene might simply go
to the catalog and look for tagged genes from that region, whether or not the
genes' functions were known. A group that managed to link Alzheimer's disease to a chromosomal region, for example, would turn to the gene catalog to
find the list of genes in that region. This list would, in turn, enable them to
study the DNA sequences from those known genes, hoping to find a mutation
correlating with the presence of disease.

Watson did not oppose the concept of work on protein-coding regions,
but he wanted to ensure that the overall genome maps, the first goals of the
genome project, were completed. The genome office did not fund Venter's
work, or other similar proposals from others. Their view was that partial gene
sequence catalogs were not substitutes for the global and complete maps that
were the goal of the genome project, but rather were useful resources that
could later be integrated into more complete databases.The past tensions between Venter and the NIH genome center blossomed into a public conflict.
Venter did not regard his effort as a substitute for the genome project, and
himself noted that his approach did not "eliminate the need for the Human
Genome Project."[56] Venter told *Science,* however, that his approach was "a
bargain by comparison to the genome project."[56] His choice of phrases was
unfortunate, seeming to offer a contrast rather than a complement to "the
genome project." For its part, the genome center did not go out of its way to
welcome Venter's effort. Watson never responded to Venter's letter, for example. The damage was done; the knights were off on separate paths in the
quest for the Holy Grail.

At the Domenici meeting in July 1991, Venter was personally stung by
Watson's attack in front of the senator and the science press. Watson and
others at the genome center were irked, in part, because they were not directly

involved in the patent applications. Reid Adler spoke with Robert Strausberg of the genome office as he was preparing the patent application, and there was a formal meeting to discuss its implications a week after it was filed. The decision to file the patent application was made by Venter's group, senior administrators at his institute (the National Institute on Neurological Disorders and Stroke), the Patent Policy Board at NIH, and the NIH Office of Technology Transfer, headed by Adler. The genome office was notified, and did not object, but neither did it endorse the action.

Venter did not initially think about filing a patent application. Adler learned about Venter's work through a letter from Max Hensley, senior patent counsel at the biotechnology firm Genentech, who urged him to talk to Venter about whether he should apply for a patent.[57; 58] Adler had never met Venter, but happened to bump into him in the hall when Venter introduced himself to ask for directions. Venter had an article accepted for publication in *Science* just a month or so hence, which meant that NIH would have to move very quickly. To preserve foreign patent rights, NIH would have to file the patent application before the publication date. Having decided the sequences might be patentable, it might be incumbent on NIH to file a patent application to comply with federal law.

Adler argued that "it was worth filing the application if for no other reason than not to miss the boat."[59] He was breaking new ground in intellectual property law; Venter was pursuing a scientific strategy that promised quick payoff. Others were less enthused. The American Society of Human Genetics drafted a policy statement against Venter's patent practices and asked for a preemptive declaration by the Patent and Trademark Office.[60; 61] The Human Genome Organization followed suit.[62] The heads of the NIH and DOE advisory committees wrote to NIH director Bernadine Healy asking that the patent applications be made public so they could be openly debated. The committees were "unanimous in deploring the decision to seek such patents."[63]

The nature of what was patentable had been continuously expanding during the 1970s and 1980s, especially in the United States. The general patent criteria were novelty (having made something new), nonobviousness (having done something that would not be readily apparent to those "with ordinary skill in the art"), and utility (having found something that had commercial potential). If the isolated sequences were indeed newly discovered, they would be judged novel. Whether the process was obvious or not depended on the details in the application and a difficult judgment about the state of the field at the time the patent application was filed. Utility claims were even more difficult to predict. The standard in most of the world required a specified commercial prospect. Some countries held a patent valid only for those uses listed in the patent claims. In the United States, however, the scope of utility claims had become progressively more permissive. One generally needed only a reasonable prospect of some use.

Patent criteria might be clearly specified by statute and case law, but applying them to DNA sequences was fraught with uncertainty.[64] The initial NIH
patent application in 1991 was very broad, claiming not only the known
sequence, but also its gene, the protein produced by the gene, and any anti-
bodies raised against that protein. In a standard patenting tactic, a February
12, 1992, modification (a "continuation in part") added 2,375 sequences to
the list claimed and narrowed the claims to the sequence and its corresponding
gene.[53; 54]

The process for finding the fragments was taken out of the patent application and separately filed as a statutory invention registration, which was by
definition dedicated to the public. Such registration did not secure a monopoly
right, but if granted would preclude others from patenting this part of the
gene-hunting process. It was a defensive move to preclude others from asserting a monopoly later. The 1992 continuation in part restarted the patent clock
and rekindled the public debate.

The issue was not about access to the sequence information or to the DNA
being sequenced, although many of the objections to the patent application
mistakenly focused on these issues. Venter immediately released his sequence
information to GenBank and sent the DNA fragments to a repository once the
patents were filed. Making the information public after filing did not endanger
the patent.

NIH did make attempts to solicit views on its action. The Office of Technology Transfer held a meeting within weeks of when the patent application
was filed, and another public meeting in November 1991. Between the June
and November meetings, the patent application had become a major biomedical research policy issue, debated widely in the science press. NIH was vigorously attacked for making a preemptive move that could confound domestic
collaboration on the genome project and would complicate international cooperation. A decision that had been made in the emerging routine of technology transfer policy within NIH was thus elevated to the level of the NIH
director, Bernadine Healy.

Healy first learned of the patent application in the fall of 1991, when
reporters began to call the director's office for its views on what Watson had
said at the Domenici meeting two months earlier. Healy then conferred with
Adler and Venter. She also had a conversation with Watson, in which he
expressed his reservations about the patent application, but said he understood
that it was prudent to have filed it as an interim policy. In a series of interviews
and at a May 1992 meeting at the National Academy of Sciences, sponsored
by the White House Office of Science and Technology Policy, Healy defended
the patent application on several grounds. First, it would protect NIH's options, in case the patents were issued. Failing to file would irretrievably sacrifice
any patent protection. She argued that the current decision might not be the
best policy, but it protected future options. She conceded that "this is not a

statement that we believe that patenting this material is the proper thing to do now or for the future."[65]

Adler stated that the goal of the initial patent filing was a desire to advance the debate, not an "act of NIH economic imperialism."[57] He viewed the patent application as an experiment to see what discussion would follow.[57; 66] What he had in mind was quite different from what actually ensued, however. The standard policy-making process was to solicit views from NIH's Patent Policy Board and to take actions, anticipating comments from those companies, trade groups, university technology transfer offices, and others who closely followed the relatively obscure field of biotechnology patent law. His office would typically take actions and then respond to comments coming in from this relatively technical and narrow constituency. The patent application on Venter's work, however, exploded far beyond this audience. NIH's intent was, in part to get comments on its action; it was overwhelmingly successful in this respect.

When Watson resigned as director of NIH's genome center in April 1991, debate about the NIH patent application became more separable from the intense drama unfolding between Watson and Healy. In August 1992, *Science* ran three companion pieces analyzing the NIH patent application. All agreed that the NIH decision to file patent applications had been reasonable, but the authors differed markedly in their tone and thrust.

Reid Adler reviewed the history of the decision and examined the precedents indicating that a patent might issue. He noted that early discussions about the genome project had failed to take account of how technology transfer policies and DNA sequence data would interact.[67] Responding to Watson's reference to monkey labor, Adler pointed out that the amount of effort involved in making a discovery was not necessarily a criterion for issuing a patent, and that in the United States, at least, patent protection was for the thing patented, not for any particular use. The central worry about failing to patent Venter's gene fragment sequences was that inaction might make subsequent patents on the complete genes impossible to obtain. By having a portion of the gene in the public domain, NIH might inadvertently thwart future patents on genes that might lead to important drugs and to gene therapy.

The Venter effort was scientifically formulated with one main purpose, to assemble a catalog useful in finding genes and discerning functional clues, not clearly a commercially viable use, but obviously a step in the direction of finding new therapeutics and other discoveries with commercial potential. The sequence from protein-coding regions also had other potential uses.[68] To obtain patent protection, NIH did not need to specify the ultimate use, but only a plausible one. The patent application claimed " 'enriched' or 'purified' full-length polynucleotide sequences, which are related to genes that do not exist naturally in this form." In a *Science* article that discussed the patent application, Adler argued that "when full-length coding sequences can be obtained through even a dozen or more conventional sequencing steps without undue experimentation, a patent application disclosing partial gene sequences should entitle their discoverers to patent the full cDNA [complementary DNA] coding sequence."[69] Aside from the meaning of "full-length," clear enough.

The crux of the rationale, however, was that failure to patent could foreclose future commercial options: "If partial or full cDNA sequences without apparent biological function enter the public domain through publication, the sequences themselves would remain unpatentable even if applications were discovered later to genetic therapy or other emerging DNA-based therapies."[69] That is, if NIH did not patent its partial gene sequences, others might later be precluded from patenting any protein pharmaceuticals or other products related to those genes.

The trade organizations representing industrial biotechnology, the groups for whose benefit NIH's policy was crafted, all agreed the NIH patent application had a salutary effect in generating a policy debate. They differed, however, in whether patents should be permitted to issue and what policies NIH should pursue. In correspondence with Health and Human Services Secretary Sullivan and with Healy, the Pharmaceutical Manufacturers' Association opposed patenting of sequences with unknown utility, but urged NIH to pursue the existing filing until an international agreement on data sharing could be forged.[70-73]

The Industrial Biotechnology Association (IBA), representing mainly pharmaceutical firms and large biotechnology firms, commended NIH for filing the patent applications, but urged NIH to place into the public domain any patents for only partially sequenced genes and to license patents only when the complete coding region and biological function were known. IBA noted: "It is perceived as unfair to permit the Government to exercise complete control over a product to whose development the Government contributed little." IBA also pointed out that if the NIH patents issued, as well as similar ones from other research efforts, multiple parties could hold patents to different parts of the same gene, resulting in a thicket of infringement actions.

The scale of the patent applications was also troublesome. How would a small company know whether it needed a license or not? Each company would have to determine DNA sequences for products under development and compare them continually to the NIH set. If, as seemed quite plausible, patents issued in the United States, but not abroad, the patents could actually prove a disadvantage. The Cohen-Boyer patent, for example, obligated U.S. firms to pay a fee, but foreign firms did not because the patent was not valid there. Having to pay when foreign firms did not was hardly a competitive advantage. A countervailing argument, however, relied on control of gene discoveries for U.S. manufacture. Adler argued that patents, combined with a licensing policy giving preference to U.S. manufacturers, could at least give U.S. firms some advantage in the domestic market.[58]

IBA also expressed concern that the NIH patent application would displace action on other pending patents, while Patent and Trademark Office staff were diverted to examine the thousands of claims in the NIH application. Finally, IBA offered several options to assuage the policy dilemmas posed by NIH's patent application.[74]

The Association of Biotechnology Companies (ABC), another trade association with greater representation of small firms and also counting patent law firms among its members, supported the patent filing as "the only responsible course under existing federal law" and encouraged NIH's pursuit of similar claims in the future.[75] ABC focused on licensing, which should be given to one firm exclusively only when a full sequence and function were known, and should be nonexclusive otherwise.[75] ABC also sent letters to President George Bush and the Patent and Trademark Office.[76; 77]

In a companion article, Rebecca Eisenberg noted that one distinctive feature of the controversy was that it was NIH, a federal agency, seeking the patent rather than a corporation whose private investment was at stake. She judged:

The specific argument that patenting these inventions will promote investment in product development rests on two premises, both questionable but neither clearly wrong . . . [first] that NIH is entitled to claim patent rights that are broad enough to provide effective monopolies for firms . . . [second] that unless NIH obtains patent rights now, firms interested in marketing related products will not be able to secure effective monopolies in the future.[78]

She raised the chilling prospect that future patent rights might be undermined by the patent application if "NIH's disclosure is inadequate to satisfy the enablement standard for the broad claims in the application, yet revealing enough to render subsequent related inventions obvious and therefore unpatentable."[78] In other words, it could backfire. Disclosure in the patent application, however, would presumably be no more damaging to future patent rights than open publication would have been.

Patent lawyer Thomas Kiley was less circumspect in his analysis. His article did not condemn the NIH patent application, but instead urged that it become the vehicle for exposing deficiencies in the law, stating bluntly that "the trend of patent law in biotechnology is toward the debilitation of science."[79] He added:

The NIH proposal for patents is only an extreme example of a widespread practice in biotechnology that seeks to control not discoveries themselves but the means of making discoveries. Patents are being sought daily on insubstantial advances far removed from the marketplace. These patents cluster around the earliest imaginable observations on the long road toward practical benefit, while seeking to control what lies at the end of it.[79]

NIH acted reasonably in filing the applications before a broad public debate, because there was simply no time to carry on such a debate. Moreover, if

NIH had not filed the patent application, under the 1986 Technology Transfer Act, Venter himself could do so if NIH waived its rights. This was more a theoretical concern than the driving force behind the decision in the case at hand, but it might prove more important in future decisions.

More generally, Kiley questioned the wisdom of such patents, especially the utility claims justifying them:

To speak plainly, these are utilities concocted to carry the patents until someone finds out what the DNA is really good for. Since the real purpose of the applications is to control individual DNAs and thereby commerce in the proteins they encode, this approach, in my opinion, amounts to a cynical resort to deficiencies in the law concerning what utility is sufficient for patents.[79]

He argued that NIH had but one option to improve public policy, to "use them [patent claims] as a vehicle to ask the Supreme Court . . . if minimal contributions will continue to merit the grant of substantial monopolies."[79] Kiley also urged Congress to clarify the research exemption, so that NIH or another patent-holder could not shut down an area of research. While no university laboratory had ever been sued to block research, the increasing commercial attachments of such laboratories made such action more likely in the future, and the law did not explicitly provide any protection. NIH itself had a policy of permitting research uses, but that policy could change. In a more likely scenario, one university might sue another over patent infringement; universities and small firms might not be as restrained as NIH in pursuing their interests. The research exemption was created in case law and had narrow, but fuzzy boundaries. Kiley urged Congress to broaden and sharpen them.

Congress could also follow European precedent, allowing patents for new uses of known substances, "eliminating altogether NIH's excuse for its patent claims. . . . Here the work would be done by the group that did the hard work of inventing something more beneficial to the public than a mere catalog of mystery DNAs."[79]

Dr. Healy herself defended the NIH policies in the *New England Journal of Medicine*:

NIH has taken the interim steps of publishing, and simultaneously applying for a patent to protect, the series of more than 2,000 partial gene sequences discovered in its laboratory. The rationale is not to make money, but rather to promote and encourage the development and commercialization of products to benefit the public and to do so in a socially responsible way.[80]

Healy noted that as a matter of policy, NIH would not charge for licenses for those engaged in research, as opposed to commercial development. Healy expressed NIH's willingness to seek patents only on well-characterized sequences if an international agreement could be forged to ensure that subsequent patents "for the full gene, its expression products, and their method of use" would not be endangered.[80]

A week before Healy's article was published, the Patent and Trademark Office (PTO) rebuffed NIH's patent application in a thirty-page document.[81] The patent claims were rejected on all three grounds—nonobviousness, novelty, and utility—and whether the description of the process was sufficient to enable others to produce the claimed sequences (enablement). The rejection had been subject to unconfirmed rumors,[82] but NIH did not make the response available or publicly acknowledge its receipt until Dr. Healy confirmed it at a hearing before the Senate Judiciary Committee on September 22.[82–85] At that same hearing, Craig Venter recounted how protecting the future patentability of genes was the main reason NIH had sought patents in the first place, and he proposed language that would make the NIH patent unnecessary by declaring in statute that disclosing part of a gene's sequence would not preclude later patents of the whole gene.[86] Healy supported this suggestion.[87]

One reason for the secrecy surrounding the PTO patent rejection was a battle raging within the Department of Health and Human Services. The department's chief counsel, Michael Astrue, had taken the unusual step of sending a letter to the PTO, asking that it suspend examination of the NIH patent application, because "I have concluded that a large portion of the applications do not satisfy the threshold legal requirements of an invention because they do not describe the function or use of the sequence."[88] The PTO sensibly pointed to the importance of pursuing the patent application to its logical conclusion, so that legal uncertainty could be reduced, and rejected Astrue's call for a halt to patent examination.[89; 90] The PTO also appeared to chide Astrue for the way in which he raised the matter.[90] Once the PTO rejection was received by NIH, Astrue had also ruled that NIH could not respond, effectively killing the patent application if his orders were obeyed. (NIH was given three months to respond on the cover sheet of the August 20 PTO rejection.)[81] Healy was strongly opposed to Astrue's position, and Secretary Sullivan had not yet decided which faction to support.[91; 92] The internal dispute between Astrue and Healy became public in early October, when *Science* and *Nature* ran news stories on the controversy.[91; 92]

The rejection of claims from an initial patent application was not unexpected, and indeed Adler had predicted in his *Science* article that the initial decision would likely be a rejection.[69] NIH was taking considerable flak, as rumors of the PTO action spread through Washington, and yet NIH silence on the matter persisted. Theories of a devastating rejection of the patent application proliferated wildly, and suspicions of NIH's motives were rampant. Healy was in an awkward position, being ordered from above not to take action, but taking political heat for being surreptitious about her policymaking.

Healy's disclosure of the rejection was not in her written testimony at the September 22 Senate hearing, but she did volunteer the information in her oral statement. This action was taken despite removal of language about the PTO action during departmental and OMB review of her statement. Healy thus courageously defied attempts to muzzle her and skated out on bureaucratic thin ice.

Prospects for the patent application itself were uncertain, and the subject of considerable speculation. The PTO rejected all twenty-four patent claims, saying they were "vague, indefinite, misdescriptive, incomplete, inaccurate, and incomprehensible."[81] But aside from that, they were fine.

Opinions about whether the PTO language was merely routine, whether it was actually an invitation to respond so as to move the application into an appeals process, or whether it was a devastating blow to prospects for the patent could all be heard. An initial rejection was quite common even when patents were subsequently issued. Ned Israelsen, the patent lawyer handling the application for NIH, believed that while the PTO rejection was longer and more detailed than usual, it appeared actually to be an invitation to respond, particularly in the sections covering utility claims. He concluded that the DNA sequences, and probably also their corresponding genes, "are patentable over the prior art."[93] Adler was convinced that ultimately the patent office would be obligated to follow the trend of U.S. law and issue a patent.[58] Several individuals who read the patent office's document interpreted it as far more than the routine initial rejection and read it as dooming any ultimate patent. Leslie Roberts of *Science* faxed the PTO document to several patent attorneys; none believed the PTO objections were insurmountable.[91] NIH did seek reversal of the PTO rejection, once Astrue left government; it also filed a new patent application for another 4,448 sequences on September 25, 1992.

Beneath the debate was a paucity of empirical data about the value of patents. Quite simply, no one knew and no one could really know. Only the results of years of patent decisions, litigation, and the complex workings of the global economy could answer the fundamental questions. The legal analysts, by and large, focused on the scope and economic return of the patent monopoly, but neglected the very high transaction costs of patent proliferation, with its toll of costly patent application, licensing and cross-licensing, and defense of patent rights. Every dollar that a research university spent for these purposes was likely to detract ultimately from the flow of dollars going into the research itself. With the prospect of 100,000 or more human genes and all the technologies to find them, characterize them, and manipulate them, the costs of obtaining the patents alone might ultimately be high, and the costs of defending those patents against infringement daunting indeed. The costs would hinge critically on the stringency of review in the patent examination process, the amount of subsequent litigation, and the ultimate scope of research subject to patent protection, all highly unpredictable factors. The costs might be low or high, but no one could predict this.

Research dollars fueled biomedical research. Most of the initial legal analysis seemed like naval strategy based on ship counts, with little attention to the importance of oil to move the ships around. If fuel didn't matter, the Japanese navy would have been far more effective in World War II. The neglect of transaction costs promised to loom large in future discussions of patent policy. In the meantime, the captains steered in the dark.

Amid the furor over whether his scientific results could be patented, Venter took his science elsewhere. He resigned his NIH post on July 13, 1992, to head up a new Institute for Genomic Research.[94] He left NIH on good terms, responding to an opportunity to greatly expand his enterprise.[95; 96] The new institute was funded under a $70 million, ten-year agreement with a corporate partner, Human Genome Sciences, Inc. The corporation, owned by the venture capital firm HealthCare Investment Corp., would retain commercial rights to discoveries emanating from Venter's research, although Venter would have rights to publish. Patent rights would go to Human Genome Sciences, and the laboratory would seek patents, in Venter's words, "on genes where we have substantial information . . . and where we feel there is a reasonable chance they will play an important role in diagnosis and therapy."[95]

The institute would go upscale even from the French Généthon, using state-of-the-art automation on a massive scale to zip through the genome. The institute ordered fifty DNA sequencing machines immediately, along with a host of Sun computer workstations and Macintosh computers, a supercomputer, and an enormous array of automated equipment.[50] It was the most audacious attack on the genome yet, replete with the highest of high-tech wizardry. Wallace Steinberg, who footed the bill as chairman of the board of HealthCare Investment Corp., referred to the dangers of international competition in guiding his decision. He hoped that the NIH patents, at least that fraction with unknown function, would be rejected, as it would remove any threat to subsequent patents for therapeutics and other products. But he justified the private-sector investment in Venter's work in nationalistic terms. He judged that NIH could not invest sufficient capital quickly enough to move as fast as the nation should demand:

My God, if this thing doesn't get done in a substantive way in the United States, that is the end of biotechnology in the U.S. . . . . There is a tremendous effort in France, England, and Japan. . . . If this becomes a race and if gene fragments become proprietary, then it is in the best interests of the U.S. and entities of the U.S. to file for patents."[94]

Venter planned to take several key staff members with him, so it seemed likely that the pace of research producing future continuations on the patent application would abate at NIH itself. But the issue would clearly surface elsewhere. In the meantime, one lasting legacy of the controversy was Watson's acrimonious departure from the NIH genome center.

Exodus: The End of the Beginning

THE CONFLICT BETWEEN James Watson and Bernadine Healy pitted two of the most powerful figures in biomedical research against each other. Watson was the most famous molecular biologist of his day and Healy the most powerful biomedical research administrator. Each was propelled by strongly held views, style, and personality into a battle from which each would emerge wounded, and with little to show in the way of improved policy. Watson left federal service in a conspicuous furor—no hollow man, he left with a bang, not a whimper.

Like Gettysburg, it was a battle that neither general had planned; they did not anticipate it would exact such a toll. Watson and Healy were drawn into battle by the force of events and timing. Controversy over patenting partial sequences of human genes unexpectedly became the battleground. On policy grounds, the fight was avoidable, but for the main characters there seemed no way out. The conflict was argued as a policy disagreement, but in the end it was not policy but information flow and personal style that drove Watson and Healy apart. Common ground could have been found—there were many positions that could accommodate both of their views—but mutual distrust obstructed communication and amplified disagreements. Healy's gaze fixed on the commercial promise of genome research and the increasingly strong mandate to NIH from Congress and the administration to make science into technology and economic power. Watson was determined to prevent a genetic gold rush that could undermine collaboration among research groups, not only within the United States but also, and more noisily, among genome projects internationally.

In the end, the battle was more important for its drama and symbolism than for any lasting impact on the success or failure of the genome project, whether measured scientifically or commercially. It was a transient focus for both Watson and Healy, although likely in the long run to prove only a footnote in either's career. The most important steps to establish the genome research agenda had already been taken. As Stanford geneticist David Botstein observed, "It was really crucial for the first four years to have Watson lead. . . . It was during this period that the agenda, plan, style, and funding level for the project were established. . . . Now Dr. Watson's leadership is not as crucial."[1]

Watson had indicated from the start that he intended to direct the NIH genome office for only five years or so. He indicated publicly, at a genome advisory committee meeting four months before his resignation, that he was thinking seriously about stepping down. At that point, the DNA patenting controversy was smoldering, but it had not yet burst into flames. Watson disliked the pressures of his NIH job from the beginning, and he found them

Bernadine Healy was appointed NIH director in 1991, after the Human Genome Project was launched but while it was still a major source of controversy. Her dispute with James Watson over a complex web of issues led to his resignation as head of the NIH genome effort in 1992. With the advent of the Clinton Administration in the following year, she was relieved of her duties at NIH. *Courtesy National Institutes of Health*

particularly trying in 1991 and early 1992 as international tensions mounted, recurrent budget battles raged, and NIH's internal politics intensified in anticipation of and then with the reality of a new NIH director. Moreover, the NIH genome center's program was getting sufficiently large to demand the attention of a full-time director on site. The Watson-Healy contretemps prematurely ended Watson's federal career, but only by a year or so.

One positive aspect of the controversy was that it focused attention on the director's position at the NIH genome center and made finding Watson's replacement an important objective. By resigning, Watson became powerless to direct the selection. Yet Healy's reputation would be judged, in part, by whom she could attract to replace him. Watson's acrimonious exodus upped the ante, drawing attention to selecting a Moses for the genome project. This positive aspect, however, was overwhelmed by the far greater damage done to both Healy and Watson in the exchange.

The first public spat between Healy and Watson took place in 1985, pre-dating the genome project. Watson complained about Reagan administration policy on regulations governing genetic technologies, making his point by noting that within the White House, "the person in charge of biology is either a woman or unimportant. They had to put a woman someplace."[2] He was referring to Bernadine Healy, deputy director for biomedical affairs in the Office of Science and Technology Policy.

Healy learned of Watson's remarks when staff from the Delegation for Basic Biomedical Research, of which Watson was an active member, called to apologize. As one of ten female students in her class at Harvard Medical School, she had experienced sexism directly. She had also been the victim of a cruel sexual joke while working years later at Johns Hopkins, at an especially vulnerable time amid a divorce, and she had pursued redress relentlessly.[3] Healy termed Watson's remarks "an offense to both men and women," while Watson replied that "anyone who heard me would know I meant it as a slap at the Reagan administration, not at Bernadine."[2] This became the first instance of a pattern, with Watson and Healy communicating their dissonance through the pages of public media rather than face to face.

If the full fury of the reaction had been known in advance, NIH might have chosen a different tack on the DNA patent application. While federal law would indeed require a patent application for an obviously patentable inven-tion, there was arguably a weaker obligation for NIH to push the frontier of what might be patentable. Disagreement among competent patent lawyers about the NIH patent application meant either choice could be justified. It was one matter to abide by the law, quite another to push its limits. Yet if failing to patent might preclude patents on genes subsequently found, then the prudent course was to err on the side of filing the patent applications. Pursuing the patent application was arguably more a policy decision and less an exercise of mere obedience to federal law, but as an interim policy, it was likely to command wide support in the end.

Reid Adler, director of NIH's Office of Technology Transfer, asked an audience of genome researchers, "What better strategic message could you send to Congress to embellish your own funding requests than evidence of your commitment" to commercial application?[4] Healy was pursuing technol-ogy transfer at NIH as a major policy thrust. She had a long-standing interest in commercial applications of biomedical research and hoped to raise NIH's awareness of its importance. She served as chair of a panel that advised OTA in preparing five biotechnology policy reports from 1987 to 1990, and she remained interested in technology transfer issues and the role of biotechnology in the emerging global economy. The Bush administration was focusing atten-tion on economic competitiveness, and biotechnology was one of its darlings. Healy could not afford to lose on the patent application issue. Watson did not believe he could back down either.

The biomedical research community divided into camps over the patent application. Venter was not part of the genome group close to Watson. Adler's office conferred with the genome office, but the policy decision already had momentum when the genome office was initially notified. Concern about future patent consequences drove the decision, not concerns about the patent application's impact on genome politics. Healy was not privy to the initial decision to file the patent application, as Adler considered it a matter of routine technology transfer and did not expect such public controversy. Healy was brought in only when the press began to raise the matter as a policy issue.

Healy later wondered why Watson did not inform her about the patent issue if it so upset him, and why he would go public first to complain about it. Watson acknowledged that he did not bring it to her attention, assuming that the Office of Technology Transfer must surely have done so for such a novel policy initiative, and judging that the patents were so unlikely to issue that the matter would dissipate of its own accord. The miscues were symptomatic of a general breakdown of communication. When Healy appointed Craig Venter, on the other side of the patent issue from Watson's genome center, to head up the NIH team to plan an intramural genome research effort at NIH in October 1991, Watson and others near him read it as an indication of whom Healy consulted for most genome advice. While Venter, Adler, and Healy all pointed to a broader group advising Healy on genome matters, including the genome center, those working with Watson did not feel they were consulted to the same degree as Venter or Adler. While planning and consultation were cordial and relatively smooth at the staff level, it was clear that communication between Watson and Healy was strained far more than usual for an NIH director and a center director.

Once established, this cleavage deepened into a chasm. Watson chafed and complained publicly about the patent application decision until he met with Healy in the fall of 1991, a few months after the Domenici meeting. He then agreed to desist from public dispute over DNA sequence patents, as the policy was still unsettled.[5] Watson deliberately chose to avoid Washington, not wanting to exacerbate tensions. Both Healy and Watson acknowledged a long period from fall 1991 into spring 1992 when they conferred little. Watson began to make private attacks on Healy. At many meetings, he railed about the lunacy of NIH's policy to his friends. Healy inevitably learned of Watson's attacks. Finding a policy accommodation became increasingly unlikely as the conflict escalated and became personal.

Beyond the issue of domestic technology transfer—the main concern of Adler and the NIH Office of Technology Transfer—a series of questions about adverse impacts on international collaboration flowed directly from deciding to file a patent application. One rationale for patent protection was to preclude a private firm or university, particularly a foreign one, from grabbing all the patent rights on human genes. A domestic company (or the domestic arm of a foreign one) could indeed file a patent claim, but the main use of the sequence tags was to find genes or to give clues about gene function, achievements several steps removed from commercial application. There would be many more steps to commercial application, generally requiring substantial invest-ments, before diagnostic and therapeutic products became practical. In other nations, patenting the sequence tags was unlikely to be possible. The main battleground would be over rights in the U.S. market.

The NIH patent application touched off an international firestorm. Alan Howarth, Britain's science minister, announced that the UK would file patent applications to hold its territory in the face of the American decision.[6] The UK Medical Research Council had previously considered patent applications, but had rejected this course under advice that the patents would not issue.[7,8] When the Americans filed an application, however, the UK felt compelled to do likewise. How the American and UK applications would be judged was an open question. It was clear that there was overlap between the fragments in the UK collection and Venter's collection.[4] How would such conflicting pat-ent claims be sorted out? The rules governing such conflicts were generally established, but the number of such conflicts arising from a few patent appli-cations was novel.

The UK and French governments joined forces to urge an international agreement that would stave off a patent rush.[6] Japanese scientists made clear they would not pursue patents for two thousand gene sequences deposited in their sequence database,[9] but the decision applied only to university scientists funded by the Ministry of Education and might not apply to industrial re-searchers or others funded from different agencies. Analysts believed patent applications for sequence tags would not pass muster in Japan and many other countries.[10] While the sentiment seemed to be dim for foreign patents along the lines of the U.S. application, there was nonetheless concern, often incoher-ently voiced, that U.S. rights alone would be sufficient to provoke a gold rush. Consternation over the NIH patent application initiated a regular traffic of diplomatic cables from U.S. embassies in Paris and London and from the U.S. representative at the Organization for Economic Cooperation and Develop-ment in Paris,[11] and it was suggested as a topic to be considered for a high-level multilateral treaty. The First South-North Human Genome Conference in Caxambu, Brazil, in May 1992, urged that "consideration be given to avoiding the patenting of naturally occurring DNA sequences. The protection of intellectual property should, in our opinion, be based on uses of sequences rather than on the sequences themselves."[12] This recommendation from sci-entists, however, flew in the face of American patent law.

Hubert Curien, French minister for research and technology, vigorously opposed the NIH patent application and obtained assurances from the Euro-pean Patent Office that Venter's work could not be patented there. NIH nonetheless took steps to protect its options to subsequently seek European patents,[13] by filing a patent application under the Patent Coordination Treaty on June 19, 1992. That application expressed the intention to seek patents in Europe, Japan, and other nations. In a letter to *Science,* Curien cautioned that "attempts to commercialize basic data from the study of the human nucleotide sequence could be the death warrant of one of the most prodigious projects the scientific world has known."[14] Or it might not. The use of sequence information in securing patents of genes and gene products was well established. Human sequences were part of many patents already granted.

Just how to deal with the international tensions elicited by the NIH patent application was not clear. Watson proposed to hold an international meeting to discuss options, but Healy directly ordered him not to do so. Europeans urged an international treaty, and the Japanese were utterly mystified by the mixed signals emanating from America.

NIH lawyer Reid Adler explained that "it would be unfortunate if misconceptions about the patent system lead to a self-fulfilling prophecy that international research cooperation will be impaired,"[15] but NIH's patent application was an act that necessitated a foreign response. Adler pointed out that *research* need not be impeded, as NIH's policy was to permit unrestricted use for research purposes. To the degree that other nations were investing in genome research even partly in hopes of commercial promise, however, the basis for international conflict was real. Cordoning off research use as a free zone did not avoid the policy dilemma facing foreign governments if U.S. researchers sought patents and foreign scientists did not.

Bernadine Healy addressed the international implications directly, and acknowledged the need for international agreement.[16] The international objections to NIH's policy were couched in sanctimonious rhetoric, and may even have betrayed a misunderstanding of what was at stake, but the danger to international scientific cooperation was nonetheless genuine. This was not because patents would shroud the genome in secrecy, but because each country wished to translate genome research into commercial payoff. The genome project was held out as a scientific and *technical* enterprise with commercial spinoff.

If one country controlled patents, even if only in the United States (the largest single market), others could not. Every nation viewed biomedical research as linked to commercial development in biotechnology. To the extent that the genome project was supported as pure science, and if an international agreement could be forged to enable free sharing of data, then collaboration could indeed be preserved. But absent such an agreement, each nation had strong incentives to file patent applications independently, and not to share the DNA sequence tag data until such applications were filed. The incentive for a gold rush hinged on the act of filing a patent application, not whether the patents eventually issued. To preserve their interests, foreign competitors had to assume that patents might issue to the NIH or others, and the only defense was to file applications of their own—and to cover the territory more swiftly than U.S. investigators. Filing patents in their home country would establish the date for any future disputes over U.S. patent rights.

On the home front, Senator Al Gore questioned NIH's wisdom:

These patent applications by NIH are not a defensive maneuver, they smack of a first strike, a preemptive strike that has predictably caused counterattacks by other governments and possibly by private researchers as well. . . . Unfortunately for the future of the human genome project and international cooperation in science, NIH's actions speak much louder than its words. The very act of filing these applications . . . is universally viewed as an attempt to corner the market on human genetic information.[17]

The cost of doing the science itself could go up. Unfettered competition, with a delayed flow of data while intellectual property rights were staked out, had real dollar costs. The expense of duplicating gene mapping efforts, when different groups did much the same work in competition—with all the spoils going only to the winner—was being demonstrated in the cases of corn and rice genomes. The overall cost of deriving the same amount of information was many times higher, because everyone had to do the entire genome independently. The fragile cooperative framework for the genome project had prevented much wasted effort, by putting groups in touch with one another for the yeast, *C. elegans,* bacterial, mouse, human, and fruit fly genomes.

The importance of patenting to commercial biotechnology, however, was undeniable even in the absence of solid empirical data. While there were few studies of how patents influenced private research and development investment or subsequent product development, it was clear from the history of the pharmaceutical industry that patents were critical. This still left open the question of how best to preserve intellectual property rights while enabling research collaboration.

An international agreement was desirable, but a daunting prospect, as it would require resolving in treaty language just those points of uncertainty that provoked the controversy in the first place, not just within the confines of U.S. law, but among nations. A narrow provision that enabled patents on a full-length sequence even if part of the sequence had previously been published or patented—along the lines proposed to fix the patent dilemma domestically—might solve the immediate problem at the international level as well, but there were much broader issues at stake. International patent standards pertaining to molecular genetics would also have to address the scope of the research exemption, criteria for establishing priority of patent rights, publication practices after filing, whether there was a grace period after public disclosure before applying for a patent, and other vagaries of different nations' patent policies. The DNA patent issue seemed likely to become but a small part of a much larger effort to harmonize international patent policies. An international agreement was desirable, but it was unlikely to be forged quickly, and might not come in time to forestall a gene patent rush.

Scientists helped define the problem, but they could contribute little more than background technical information relevant to the legal rules. The law and not science would decide how technical information fit into the intricate structure of the national and international economy. Congress would write the rules, patent examiners would grant or reject patent claims, patent lawyers would litigate, and judges would decide individual cases. National governments might meet to craft agreements. Over time, the outcome might become clear, but it would be quite some time, and scientists would for the most part be observers and advisers, not policymakers.

The genome project was sold, in part, as a huge international collaboration, and coordination with researchers abroad was a major preoccupation of NIH's genome office. The force driving decisions about the NIH patent applications, however, was technology transfer. The genome office and the NIH director's office came into conflict in part because they were attending to different problems.

Disagreement over DNA patenting was the most conspicuous irritant, but there were many other sources of conflict between Watson and Healy. Frederick Bourke, a former squash champion from Connecticut whose business interests turned to the commercial promise of DNA sequencing, helped to dislodge Watson. In fall 1991 and into 1992, Bourke made overtures to several genome researchers. He hoped to set up a company in Seattle, Washington, to do genome research on a massive scale with high technology. Bourke aspired to do with private American funds, and with an eye to future commercial benefit, what the French had pioneered with Généthon and Craig Venter would months later begin to establish at the Institute for Genomic Research in Maryland.

Bourke's basic idea was to use pilot projects to develop sequencing capacity, and to use that sequencing capacity to pursue commercial leads. Bourke began to negotiate with John Sulston in the UK and Robert Waterston in St. Louis about doing the *C. elegans* sequencing project under his patronage. This project had been among the first large-scale mapping and sequencing efforts under the genome banner and was perhaps the most successful transatlantic collaboration, producing results at an impressive clip.

Watson met Bourke for the first time on January 24, 1992, and the two quickly developed a strong mutual distaste.[18] Bourke characterized Watson as "reactionary"[19] and Watson privately professed his strong distrust of Bourke. Watson interpreted Bourke's overtures to Sulston, Waterston, and their colleagues as a direct threat to a highly successful transatlantic collaboration, endangering a genome research project likely to bear early fruit. Watson viewed Bourke's efforts as a torpedo aimed at his flagship.

At the same time, a genome research center affiliated with the University of Washington in Seattle was forming, based on a $12 million donation from William Gates, cofounder and CEO of Microsoft Corporation.[20] Microsoft's assets skyrocketed over two decades, and it quickly became one of the wealthiest corporations in the nation. Genome research became one Microsoft's beneficiaries. Gates, who had quickly become an immensely wealthy business leader, was intrigued by the natural alliance between computers and DNA analysis and put up the donation to bolster research at the major university nearest to Microsoft's headquarters in Redmond, Washington.

Gates's Seattle venture was entirely separate from Bourke's, established through the university. It proved more successful, netting two giants of genome research—Maynard Olson from Washington University in St. Louis and Leroy Hood from Caltech. Applied Biosystems, Inc., was to supply the instrumentation. (Applied Biosystems merged with Perkin Elmer months later, bringing together two of the most important biotechnology instrumentation companies.)

Hood initially helped found the Bourke venture and was already serving as a Bourke adviser.[20] Bourke's separate institute was to be started with $50 million, collaborating with the University of Washington genome center. Young scientists would be attracted to the facility, working shoulder to shoulder with two of the field's luminaries in a high-tech genomics heaven on the shores of Lake Washington.

Watson was working on both sides of the Atlantic to preserve the existing *C. elegans* collaboration. He spoke with representatives of the Wellcome Trust in London, the UK Medical Research Council, and his extensive set of British contacts. Government officials and private science philanthropies were also brought into the fray; British science administrators, in particular, did not wish to see further UK-to-U.S. brain drain, with another highly touted research team leaving England. Aaron Klug, Nobelist and director of the MRC molecular biology laboratory in Cambridge, regarded Bourke's offer as a hostile takeover bid, and other MRC officials complained loudly.[19] As Bourke continued discussions with the *C. elegans* researchers, the government grant that supported work in Cambridge expired. Bourke agreed to Sulston and Waterston's demand that all their work would be in the public domain, but distrust of the venture nonetheless ran high among outside observers.

With his London contacts, Watson discussed the need to keep a vital and open *C. elegans* collaboration between the United Kingdom and United States, along with other issues unrelated to the *C. elegans* collaboration—the NIH patent application, how to share support of databases, and revitalization of international coordination more generally. The Wellcome Trust, a large private philanthropy already involved with genome research through various scientific contacts and the Human Genome Organization, stepped in with a £50 million, five-year grant to support the *C. elegans* project and to expand British genome research on other organisms and on informatics.[21; 22] In the end, Sulston and Waterston continued their collaboration, with more resources and strengthened international reputations. The Wellcome Trust and MRC took steps to establish a major new facility in Hinxton Park, south of

Cambridge, fittingly named the Sanger Centre. (Fredrick Sanger, whose philosophy was to approach molecular biological function through the study of ever larger structures, was a pioneer of protein and DNA sequencing—see Chapter 4. His approach not only helped guide research in the UK, but was also carried through Maynard Olson's voice on the National Research Council committee—see Chapter 10.) Sulston was to direct a substantial genome research group at the Sanger Centre, which also hoped to become the major informatics center for genetics in Europe. Watson had won the battle, but Bourke struck back.

Bourke spoke to Healy, and followed up with a letter detailing his complaints about Watson. In the February 25, 1992, letter, Bourke recapitulated: "In our recent conversation, we discussed the resistance I have encountered. . . . I believe that much of this resistance originated with Dr. James Watson of your staff."[18] Bourke cited a conversation with C. Thomas Caskey, who had expressed reservations about joining Bourke after speaking to Watson. Bourke also invoked the names of Leroy Hood, Charles Cantor, and John Sulston as corroborating instances of Watson's interference. Leroy Hood also called Healy to complain about Watson's lobbying to scuttle the Bourke venture. Others also began to call Healy's office to complain about Watson, many alleging conflicts of interest.[5]

Bourke asserted that his commercial interests were thwarted by Watson and questioned whether Watson had the best interests of the nation at heart.[18] Bourke cited Caskey's account of Watson's rendition of a meeting with Glaxo officials to torpedo the venture. This was not straight from the horse's mouth, and its implication was false. Watson had indeed spoken with Glaxo officials, to whom he was a regular adviser, but he had addressed the issue of DNA patenting, before he was even aware of Bourke's interest in the *C. elegans* project.[23; 24] While Bourke's torpedo may have passed wide of the mark in its first pass, it ultimately circled back and found Watson's hull.

Bourke's allegations were short on proof and long on hearsay, but his letter was enough to precipitate a Healy inquiry. Healy was quite sensitive on matters of financial conflict of interest, having herself been publicly accused of a conflict for holding stock in the biotechnology company Genentech. Healy was one among many clinicians and scientists whose ownership of stock in Genentech raised eyebrows regarding clinical trials of its blood-clot-dissolving drug, tissue plasminogen activator (TPA). Healy did not purchase the Genentech stock until several years after the TPA trial ended, and she had left the sponsoring institution, so it could not affect the results; but she did purchase the stock before her former colleagues published the trial's results.[25] The episode made an impression, and Healy was subsequently instrumental in putting together strict guidelines to prevent conflicts of interest in clinical trials.

After Bourke's letter and other calls about Watson's possible financial conflicts, Healy asked for the files on Watson's financial holdings, inspected them, and forwarded them to James Mason, the assistant secretary for health.[5; 26–30]

A few months later, Watson was called into the office of Jack Kress, who handled conflict-of-interest issues for Mason. Watson first saw the Bourke letter in Kress's office. Watson had disclosed his holdings in several previous reviews,[27-29] but the Bourke letter provoked another look. In Watson's files, Healy found an electronic mail message from NIH counsel Rob Lanman to her predecessor, acting Director William Raub, that raised questions in her mind.[30;31] This memo was referred to Kress, who did not act on it.[32]

Watson was outraged that he first heard about a letter from Kress rather than Healy. Healy responded that she was merely complying with departmental policy on matters of ethics, on Mason's advice. After his meeting with Kress, Watson began a series of calls to a circle of confidants announcing his intention to resign. He believed that the way the Bourke letter had been handled and the inspection of his financial background were bureaucratic moves to get rid of him.

Kress assured Watson that there was no conflict of interest as far as he was concerned, although there were a few matters of concern. Kress said he intended to recommend that Healy sign a waiver enabling Watson to retain his holdings. Healy's office, however, expressed great concern about a potential conflict of interest to *Nature*[33] and told the *New York Times* that Healy "would rather not resolve the matter by giving Dr. Watson a waiver."[34] Healy did not want to sign waivers, but was willing to have her superior, James Mason, do so; Mason was not willing to sign waivers that Watson's direct supervisor, the NIH director, would not.[5] Michael Astrue, chief counsel for the Department, met with Watson and advised him "it would be wise for him to resign."[23; 35]

Kress openly defended Watson to reporters, saying that "this is very common, nothing out of the ordinary. . . . I made it very clear to him that in no way, shape, or form did I find anything improper about anything he was doing,"[36] and "after talking it over with Dr. Watson, I was satisfied that there was no conflict . . . there is no ethical reason for him to leave."[33] Yet on the same day, Johanna Schneider from Healy's office told the *Washington Post* that "Dr. Healy does not have the luxury of ignoring ethical questions, even for a Nobel Prize winner."[37] Accounts emanating from two points in the same department of government were in clear opposition.

Watson's interpretation that he was being sacked turned on several factors. One was the fact that Kress, who usually handled ethics matters and to whom Healy had referred the matter, believed there was no conflict of interest but Healy apparently did. Watson also called a fellow Nobel laureate, Daniel Nathans of Johns Hopkins. Healy had discussed the situation with Nathans, whom she knew from her John Hopkins days. Soon after Nathans's conversation with Healy, Watson contacted him. Watson concluded from this conversation that he was being subtly told to leave. He told the *Washington Post* that Healy "does not want me."[38] Another part of the crescendo was a March 25 hearing before the House appropriations subcommittee.

Watson and Healy appeared together to justify the genome budget request. The process of formulating a budget for the genome center had been frustrating for all concerned. It was Healy's first full budget cycle as NIH director. She had also launched a strategic planning exercise for the entire NIH. Healy's initiative to bolster trans-NIH planning was in line with a pair of reports on NIH structure and management, prepared by Institute of Medicine committees in 1984 and 1988.[39; 40] The strategic planning process was controversial as it unfolded, for reasons unrelated to genome politics. Rancor centered on the process rather than the need or intent. The research community was highly suspicious of a process they perceived as guided by NIH bureaucrats rather than scientific experts. Regardless of the outcome or which faction was closer to the truth, one feature of Healy's directorship was clear— she would play a much stronger role in budgeting of the individual institutes than had her predecessors.

The genome budget was prepared amid the patent policy disagreement between Watson and Healy and further hampered by lack of communication. A symptom of the difficulty was a briefing set up by Healy's office, which disrupted a meeting planned between Watson and Rep. John Dingell, chairman of the committee that authorized NIH (including the genome center). In preparation for House hearings on NIH's budget, Healy scheduled briefing sessions with senior managers at each NIH institute, center, and division. The meetings fell off schedule the morning the genome office was on the roster. Watson and the genome center staff were put on hold for several hours, which forced Watson to cancel the Dingell meeting, an appointment he had labored to secure for many months. The press of time before the appropriations process limited flexibility in Healy's schedule. Watson was only in town for a day, and the Dingell meeting could not be readily rescheduled. While it was understandable and clearly not deliberate, the schedule conflict was just another reminder to Watson of his uncharacteristically subordinate position in the federal hierarchy. From Healy's perspective, it was another instance of the difficulty of having a major program directed by a person who was only occasionally present.

Watson avoided reporters for many months, having agreed in the fall not to publicly criticize NIH policy on the patent controversy[41] and to clearly distinguish his personal views from official NIH policy when commenting on it privately. At the March 25 hearing, Representative William Natcher, chairman of the subcommittee, asked Watson point-blank: "What do you think of NIH's decision to seek patents on several thousand gene sequences?"[42] Watson began to temporize: "The patent law doesn't really cover DNA. It was invented before DNA was discovered. There is the possibility that, in fact, DNA can be patented. That will be decided by lawyers and judges." But he then answered directly: "The second question is, if it is patentable, is that a good thing for the human genome [project and] the biotechnology industry? This is a debatable issue. I think it would be better if we did not patent sequences that you don't understand. Once you understand what it does, then I am in favor of patenting."

Watson then opened the floor for his boss. "I am not a lawyer. I am not responsible for the decision. I think you should ask Dr. Healy her views on this point."

Natcher did. Healy responded with a statement of interim policy:

I think the debate on DNA patenting is inevitable. Every time we have moved into the issue of patenting some aspect of genes—transgenic animals, genetically engineered microbes, and so forth—there has been an enormous debate and question. I think a debate on DNA, for which we have limited knowledge, was inevitable. NIH policy has been, after many months of considerable discussion and review of the issue, to take what we view as a protective posture. It is not to make a statement as to whether or not it is good or bad to hold a patent on this material under these circumstances, but rather a position that until we have a position of legal harmony and legal certainty; until we have international agreement on what is patentable and what is not; and until we know what the consequences of the patenting or licensing would mean, and what it means if this information is put in the open literature without any kind of intellectual protection—until those issues are resolved, NIH is taking a protective posture.[42]

This was a dangerous matter to bring unresolved before the House appropriations subcommittee. Either policy choice—to seek or not to seek patents—could be defended; the process that produced the interim NIH policy, however, had not yielded consensus among the major players, and indeed Watson and Healy were on sufficiently bad terms that they had not discussed the patent application for many months. It festered as an open sore, a major controversy in the scientific press, brought untreated before the congressional group with enormous power over NIH's budget. Given the indirect routes of communication between them, it is not surprising that Watson read Healy's motives as hostile, and vice versa.

NIH could have found policies to accommodate both future commercial opportunities and the need for information flow and scientific collaboration. Options for such accommodation abounded—international agreements, statements about how the patents would be licensed if issued, a clearer definition of what research uses would be exempt from the patent monopoly right, or agreement to pursue a new form of intellectual property protection. Each of these avenues could satisfy both Watson's and Healy's policy goals. Indeed, Adler's article explaining NIH's patent decision suggested that "perhaps patenting and licensing optimally should be pursued only for complete coding portions of a gene for which a generalized biological function seems apparent,"[15] a position indistinguishable from Watson's except for the "perhaps." Adler's point was that this could be decided only after careful review, and that failure to file applications on the partial gene sequences would have been irreversible. Adler raised the issues with senior genome advisers within NIH and in the university community after the controversy hit, but found them too hostile to have a meaningful discussion. Adler believed NIH could still control when and whether the patents finally issued; it could exercise no control if patent applications were never filed.[15] There was common ground, but it was being fought over, not cultivated. They beat their plowshares into swords.

On March 26, 1992, a day after the House appropriations hearing, Watson called a circle of confidants, telling them he had to resign. He privately vowed never to appear before a congressional committee again under conditions where he would feel personally compromised. The conflict had crossed a threshold, violating his personal sense of integrity. He felt that juggling his financial holdings to eliminate Healy's concerns about conflict of interest would merely delay his departure. He told reporters from the *Washington Post* he was willing to sell his stocks: "I could divest most of them, but it would be point-less."[38] He believed another bureaucratic burr would be placed under his saddle, and then another, until he left. Watson felt that while Kress had been careful to say there was no irresolvable conflict of interest, Healy's office had set him up for a public flaying.[24; 35; 43] Healy felt that the conflicts were real and required either Watson's resignation or divestiture.

From Watson's perspective, the alternative to his resignation was a pro-tracted and public legal battle. He believed he would continue to be attacked in newspapers. Many of his friends urged him to stay on through the budget cycle, and he vacillated for several weeks, while press reports speculated on his imminent resignation.[33; 36; 3 8] Following the meeting with the head counsel for the Department of Health and Human Services, Michael Astrue, he decided to resign very quickly. On a Friday at 1:00 P.M., April 10, 1992, Watson resigned.[41; 44-47] No face-to-face Watson-Healy meeting ever took place to discuss the resignation. In a move pregnant with symbolism, Watson resigned by fax from his office at Cold Spring Harbor Laboratory, his safe haven.

In the wake of Watson's resignation, press reports dealt with the conflict-of-interest issue, but centered on Watson's contribution to the project and the rough treatment he had received at Healy's hands. *Science, Genetic Engineering News,* and a *Nature* editorial were largely laudatory of Watson.[1; 41; 46-48] A *Nature* editorial, taking a more pro-Watson stance than its accompanying news articles, ventured that "Dr. Bernadette [sic] Healy . . . has wanted Watson out of the Human Genome Project. . . . But Healy will find she has damaged herself more than she has hurt Watson."[20; 44; 48]

Within hours of Watson's resignation, Healy released a diplomatically phrased statement expressing regret that Watson had resigned and naming geneticist Michael Gottesman, from the intramural research program at the National Cancer Institute, acting director. She had invited Gottesman into her office to offer him the position and had made it clear he had to make his mind up. In her public statement, Healy reiterated regret at Watson's decision to resign. This was repeated several weeks later at a May 5 press conference,

staged to introduce the press to Dr. Gottesman and to clarify NIH's continuing commitment to the genome project just before the annual genome research
meeting at Cold Spring Harbor Laboratory.

In the end, both Healy and Watson emerged diminished. Watson was
stripped of his official capacity as head of the NIH genome program. Healy
alienated a powerful figure in science. The genome project itself carried on,
now robust enough to withstand such buffeting. The Watson-Healy rift seemed
likely to heal at least partially. Watson, in his annual report for Cold Spring
Harbor, acknowledged the disagreement with Healy over DNA patenting,
but attributed his inclination to resign more to his "inability to be the active
manager the Project now needs."[49] He proffered an olive branch, noting that
"there is every indication that Dr. Healy will desire to quickly appoint a
scientist of major accomplishments to replace me. Naturally, I will continue to
remain a strong proponent of genome programs and if asked, will enthusiastically give the new director my assistance."[49] For her part, Healy expressed
personal affection for Watson, although giving no ground on the conflict-of-
interest issue. She was convinced that the genome project was the better for
her actions. In the intricate web of personalities, ideas, and issues that created
the genome project, Watson's resignation was yet another *Rashomon*.

The Watson era of the genome project ended as it began, subject to the
complex interplay of scientific objectives, positions of political power over
biomedical research, and contending visions. The purpose of the science was
to create precise information about human genes and technologies to explain
genetic mysteries. Pursuing that purpose, however, was an inherently political
process. It involved individuals vying for power to make decisions—players in
the drama by dint of their positions in the federal government and in the
scientific community.

The science of the genome project built on facts; its history, on stories.

Epilogue

THE JANUARY 1, 1993, issue of *Science* announced that Francis S. Collins of the University of Michigan had agreed to direct the NIH genome program,[1] confirming rumors that had persisted since midsummer. Collins agreed to join NIH on condition that a significant intramural genome research capacity be created on the NIH campus in Bethesda, Maryland, so he would not have to give up active laboratory work. Collins agreed to make the move despite a cut in pay and the disruption of one of the most secure scientific empires in human genetics. Several members of his group, including himself, were funded by the Howard Hughes Medical Institute. This highly prestigious and financially stable base was buttressed by an NIH-funded genome research center linked to the University of Michigan.

The University of Michigan combined one of the best state-supported universities with a medical school that had chosen soon after World War II to emphasize human genetics. Stanford geneticist David Botstein, for one, believed that his unusually broad and deep training at Michigan—with exposure to excellent molecular biology and world-class human genetics, including population genetics—gave him the requisite background to prepare him for the 1978 insight about the importance of a human genetic linkage map.[2] Construction of just such a map helped spawn the revolution in human genetics that began in the 1980s.

Why would Collins leave such an enviable position to direct a federal program? His answer: "Because there is only one human genome program. It will only happen once, and this is that moment in history. The chance to stand at the helm of that project and put my own personal stamp on it is more than I could imagine."[1]

Recruiting Collins was a major coup for NIH director Bernadine Healy.[3] She had to go to considerable lengths to secure precious laboratory space on NIH's campus, displacing other research groups and thus engendering strong antipathy among those who had waited for years to get it. Healy's own future became quite cloudy with the election of President Bill Clinton, and she announced on February 26, 1993, that she would leave the NIH directorship by June 30.[4] In her statement, Healy singled out among the major initiatives she hoped would continue after the end of her two-year tenure the NIH strategic plan, women's and minorities' health initiatives, recruitment of scientific talent, and an expanded Human Genome Project. It seemed likely that attracting Francis Collins, as part of the expansion of the genome project, would be an important part of her legacy.

Collins was on everyone's short list for the job, probably the only person about whom that could be said. His scientific qualifications were unques-

Francis S. Collins was successfully recruited by Healy to head the NIH genome program, following Watson's resignation. Collins, a leading researcher in the field, had been at the University of Michigan, where he directed the team that found genes for several hereditary diseases, including cystic fibrosis. *Courtesy National Center for Human Genome Research*

tioned. Together with Lap-Chee Tsui of the University of Toronto, Collins had directed the team that first found the cystic fibrosis gene,[5-7] and his was one of two teams that found the gene for neurofibromatosis, type I.[8-11] (The other group was directed by Ray White at the University of Utah; while the two groups had initially collaborated, they parted company, only to cross the finish line almost simultaneously.) Collins was an integral part of the collaborative team organized by Nancy Wexler to search for the Huntington's disease gene,[12] and his group was in the hunt for an early-onset familial breast cancer gene mapped to chromosome 17, near the neurofibromatosis gene. Collins and his group were thus in the thick of some of the most conspicuous gene quests. That work, in turn, was coming to define a deep current in the mainstream of biomedical research.

Collins continued to work in medical genetics and genetic counseling, one of but a few first-rank molecular biologists to maintain clinical skills. The clinical work gave him an instinctive feel for the impact of genetic information on families. He had a keen appreciation of and support for the ethical, legal, and social issues program. He had followed the ELSI working group's efforts on cystic fibrosis pilot-testing programs. People who worked closely with him at the University of Michigan were tracking several issues in close collaboration with the ELSI research program, especially those related to Huntington's disease and breast cancer.

Collins was also rare among genome scientists in accommodating religious interests. He was comfortable speaking publicly about his religious beliefs, as when he spoke to a group of theologians at a March 1990 conference, noting how his Christian values reinforced his commitment to biomedical research.[13] This breadth of clinical and scientific background and appreciation for the broader context in which the science was being performed made him an ideal candidate to direct the NIH genome research effort.

Even as Collins began to grab the reins, the NIH part of the genome project was on its way to becoming a fully entrenched part of the bureaucracy. The NIH authorization bills, S. 1 in the Senate and H. R. 4 in the House, both formally authorized the National Center for Human Genome Research. The genome center had been created by administrative action within the Department of Health and Human Services, with agreement of the appropriation committees. The new authorization statute gave the genome center more permanent status, so that a future Secretary of Health and Human Services could not simply dissolve it. The bills also mandated that at least 5 percent of the budget go to the ELSI program. The House and Senate NIH bills were put on the fast track in 1993, largely because they had been vetoed by President Bush the previous year over provisions concerning fetal-tissue research. Several members of Congress promised during the 1992 presidential campaign to pass the law early in a Clinton administration if Bush lost the election. Bush did lose, and the NIH bill did indeed receive early attention from Congress. The Senate went so far as to introduce NIH authorization as its first bill for the new Congress. President Clinton signed it into law as Public Law 103-43 on June 10, 1993.

This special attention to NIH authorization ironically posed a problem for Collins. He aspired to transform the National Center for Human Genome Research into the National Institute for Genomics and Medical Genetics, making clear its broad mandate and conferring full institute status upon it. Collins was not yet the director of the genome center as the bills were transiting Congress, however, and AIDS research and fetal-tissue research provisions commanded almost all the political energies of the bills' congressional champions. He was thus poorly positioned to succeed in effecting last-minute changes in the NIH bill. The NIH genome center would have to wait a few more years, until the next authorization cycle, before it could become an NIH institute by statute. (An agreement to pursue institute status through action within the administration, however, was part of the recruitment package that brought Collins to NIH.)[14] When Healy resigned as NIH Director on June 30, Ruth Kirschstein replaced her as Acting Director on July 1, 1993. As the transition began at NIH, David Galas at DOE also announced he was departing, to

U.S. budget for genome research grew sharply from 1987 onward, before leveling off at a total of about $170 million per year, as shown in this bar chart, representing the two main U.S. programs. Figures for the Department of Energy (DOE) are for its Human Genome Initiative. The 1987 budget reflects funds reprogrammed from other areas; earmarked budgets within DOE began in 1988. The National Institutes of Health (NIH) figures for 1988 and 1989 correspond to earmarked funds spent at the National Institute for General Medical Services and coordinated by the Office for Genome Research, part of the NIH Director's office. The National Center for Human Genome Research was established at the beginning of fiscal year 1990 and acquired its own budget. Figures for 1993 are estimates based on congressional appropriations. Figures for 1988–1992 are based on appropriations, adjusted for actual expenditures.

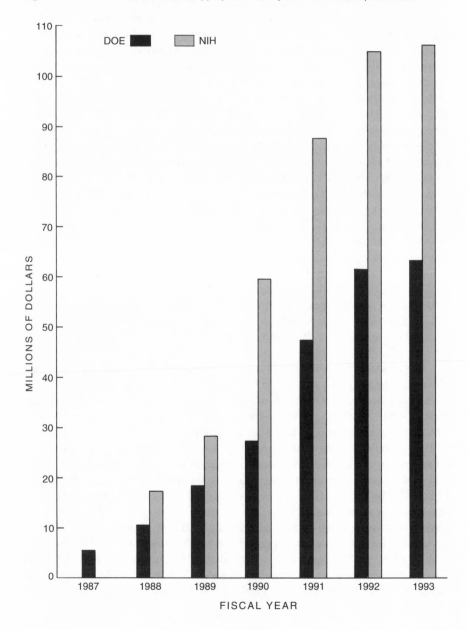

become part of the scientific team at Darwin Molecular, Inc., a new biotech-nology company based in Seattle. In August, President Clinton announced his nomination for NIH Director: Harold Varmus, a Nobel laureate cancer re-searcher from the University of California, San Francisco.

While the NIH genome center was attaining statutory sanction, the ge-nome project as a whole became the exemplar of yet another major policy debate, this time over commercial investments. Private corporate investment in genome research became fashionable in 1992 and 1993. The prospects for attracting private capital had changed dramatically in five years. Whereas Wal-ter Gilbert had great difficulty in finding venture capital to launch the Genome Corporation in the spring of 1987, a symposium devoted to solicit interest among pharmaceutical firms, organized by Craig Venter and Gilbert, drew a respectable audience in September 1990.[15] Translating intellectual interest into direct financial investment, however, took several more years. By early 1993, in contrast, many of the most prominent genome researchers were being approached by venture capital firms, major pharmaceutical houses, and other sources of private funding. The Institute for Genomic Research in Gaithers-burg, Maryland, was but the largest among many new privately funded ven-tures with working capital of over $70 million. It was a nonprofit entity attached to several for-profit corporations under a corporate umbrella. The for-profit arm was Human Genome Sciences, Inc., which in May 1993 ap-pointed William Haseltine CEO, and announced an alliance with Smith-Kline Beecham valued at over $130 million. Other ventures were organized as for-profit companies to do genome research. Frederic Bourke, whose interest in genome research indirectly contributed to Watson's resignation, resurfaced as a major investor, and there were dozens more.

Science devoted a feature article to the emergent private investments in genome research, raising questions about whether the genome project could be both a public good and a recipient of substantial private investments.[16] Could the most prominent genome researchers disclose their data quickly and also honor their commercial commitments? Questions about conflict of inter-est had become much more prominent throughout biomedical research, as the scope of commercial research investment grew and scientists shed their vows of poverty. While the genome feature article was only one among a half-dozen pieces that *Science* ran on conflict of interest and commercial aspects of biomed-ical research, it nonetheless placed the genome project once again in the spot-light of an emerging controversy. As had happened in so many other policy areas, the genome project became the focal point in a more general debate. The ELSI grant money distinguished the genome research effort from some of the other foci of attention over conflict of interest, as it enabled NIH to give a grant to David Blumenthal of Massachusetts General Hospital to gather empirical data about the extent and impact of private financing on genome research.

The influence of private funding was by no means confined to the United

States. The United Kingdom, the only other country to make commensurate genome research commitments, also had a heavy infusion of private funds. Indeed, the private support of the Wellcome Trust far outstripped the government funding through the Medical Research Council. The group charged with preparing an analysis of UK genome research included representatives from the three largest pharmaceutical firms—Glaxo, ICI, and Smith-Kline Beecham—and the Wellcome Trust funding came from the foundation arm of pharmaceutical giant Burroughs-Wellcome.[17] In France, the most conspicuous genome effort was the collaboration between Généthon and AFM, the muscular dystrophy association, both largely privately funded, although not directly tied to corporate interests. The nature of most private investment abroad, therefore, was different in intention, although there were signs that other governments hoped to entice investment by pharmaceutical firms. Canada and the UK, in particular, were clearly making overtures to private drug companies.

The increasing role of private funding of genome research, both corporate and nonprofit, was an indicator of developments throughout biomedical research. Because genome research was conspicuous, and entailed state-of-the-art instrumentation and first-rate talent, and because finding genes might be a short path to possible new pharmaceuticals, genome research attracted venture capital.

At the same time, biomedical research in general, with genome research as a specific instance, had attained sufficient national importance to become a political issue. The genesis of the genome project was itself a demonstration of this fact. It was created not by citizens concerned about cancer or heart disease or even genetic disorders, but rather by scientists who argued that a concerted research program was an expeditious way to improve research on all diseases. This was a subtle departure from traditional biomedical politics, in which those affected by a disorder generally lobbied for funds to stop their suffering. The rationale was still ultimately linked, and legitimately so, to preventing the suffering caused by disease. The genome project was initially presented by scientists, however, not disease-group advocates, and its impetus came from technology rather than a specific disorder. The successful launching of the genome project demonstrated that biomedical research as an enterprise could flex its political muscles.

By 1993, biomedical research consumed more than $10 billion per year in federal funding and somewhat more in private funding, mainly from pharmaceutical firms and biotechnology companies. The pharmaceutical industry alone invested $6.6 billion in 1990 and employed almost thirty thousand research and development workers.[18] Between public and private funding, there were more than 100,000 people who made their living in biomedical research, more than half deriving most of their funding from NIH—a large enough group to function as a political interest group, or minor government-dependent industry, with its eye on jobs and fiscal stability.

The explosive growth of biomedical research in the period since World War II presented a deep irony. The wealth of scientific knowledge flowing out of this public investment made it abundantly clear just how immensely difficult it would be to conquer chronic disease. President Nixon's War on Cancer, begun in 1971, was likened to the Manhattan Project. This frontal assault on a dread disease promised to produce a cure by 1976. By the 1980s and 1990s, such audacity seemed reckless. Naiveté of this magnitude is almost impossible to understand in retrospect, given the complexity of human biology. Many have come to suspect that the promises were deliberately overstated in order to extract a federal boost in research dollars, a cynical ploy undertaken by cancer research enthusiasts with full knowledge that there would never be a full accounting.

The War on Cancer had indeed succeeded in expanding biomedical research, but had not led to a cure for cancer. Instead, it helped fuel the work that led to recombinant DNA technology, DNA sequencing, and the other remarkable advances in molecular biology during the 1970s. These, in turn, spawned the new biotechnology and its industrial applications. The gnawing question was not whether good had come from the federal infusion of funds—clearly it had—but whether the scientific community had obtained it under false pretenses. Watson had called the War on Cancer "lunacy" in 1973,[19] and was careful not to promise more than he thought the genome project could actually deliver. He was quite enthusiastic about it, but did not deliberately mislead Congress. Others danced closer to the edge, but it is clear from the record and from interviews with those controlling NIH funding that Congress and budget officers in the executive branch understood the genome project to be a genetics infrastructure project. This did not alter the underlying policy problem—the difficulty of sustaining long-term government commitments.

Understanding human biological function will cost a lot of money, and will consume the careers of thousands of investigators for many generations. To approach the goals of today, biomedical research will require billions of dollars for many decades at least. Indeed, the quest will prove endless, as humans will always die of some cause. As today's scourges are eradicated or their effects softened, new diseases will rise to prominence. The twentieth century has seen an extraordinary shift in patterns of mortality. The leading causes of death in developed nations have changed from infectious disease to chronic disease. Tuberculosis, typhoid, pneumonia, polio, meningitis, small-pox, yellow fever, and other infections have given way to cancer, heart disease, stroke, and Alzheimer's disease. Medical technology has not been the only cause of this shift, which began before the antibiotic era, but technology has accelerated the trend and amplified its magnitude.

Those diseases that were most amenable to a "technical fix," through anti-biotics or surgery, have been greatly reduced in incidence, prevalence, and severity, dropping several rungs on the ladder of public health problems. The future is likely to see a similar phenomenon, with those disorders that yield to

the day's scientific capacity falling first. This is not an excuse for inaction—far
from it. There is plenty of suffering to relieve in families like the Rosses, or in
people confronting schizophrenia, cancer, stroke, heart disease, or AIDS. The
fact that humans must die of something merely means that the road is long.
One of the main lessons of modern biomedical research is that science is not
very far along it. One does not take a break one mile into a marathon. The
inevitability of death is thus hardly an argument to reduce attention to those
conditions that wreak havoc among the living. If someday it might become
difficult to justify replacing one disease with another of later onset, that day is
still a long way off.

At the same time that biomedical researchers have grown sufficiently in
numbers to become a political interest group, their mission to eradicate disease
has been complemented by a new mission to create wealth and jobs and to
promote the national economy through innovation. Health care, the market
for most innovations devised through biomedical research, has become an ever
larger fraction of the national economy. Health expenditures rose from 5
percent of the Gross National Product just after World War II to 12.1 percent
in 1991[20; 21] and are projected to reach 18.1 percent of Gross Domestic Product by the year 2000 (and 32 percent by 2030).[22] Health care was grabbing a
larger share of the total economy and doing so significantly faster in the United
States than in any other major industrial country.[21] Were this the computer
industry or automobiles, such growth would have been regarded as auspicious,
the mark of an economic sector with remarkable potential for continued expansion.

Health services are highly labor-intensive, and thus a major source of new
jobs, but the service components cannot be exported and the government is a
major payer. Health costs are thus not only a source of jobs but also a drag on
the economy; given the choice, people would rather spend their money on
something else. Health expenditures do not result in possession of enduring
material goods or have great entertainment value. Moreover, health goods and
services are largely immune from normal market forces, particularly regarding
price discipline. People are not in a position to comparison-shop or to sift
through price-sensitive business calculations about when and how much health
care to purchase; only a fraction of medical expenses are paid out-of-pocket,
dramatically reducing price sensitivity. This reflects a deep and laudable desire
not to have economic forces determine life-and-death decisions, but it plays
havoc with economic rules. Expenditures in the health sector have consistently
outstripped those in the rest of the U.S. economy since 1947, particularly in
the 1970s and 1980s.[20; 21]

Some parts of the health care industry manufacture high-value-added goods,
such as pharmaceuticals and medical devices, that can be exported and sold as
commodities. They are more similar to other goods in this respect than are
hospital services, physician visits, or nursing homes. But they too are increas-

ingly regarded as taking undue advantage of their government-sanctioned monopoly (patent rights) to extract unseemly profits from people who depend on drugs and medical devices for their well-being, or even survival. Many controversies have erupted over drug pricing for AIDS treatments, a treatment for the rare genetic condition Gaucher's disease, and new growth-factor drugs to combat chronic kidney failure and other conditions.[18]

The cost of pharmaceuticals is a small fraction of total health expenditures, accounting for roughly 8 percent,[23] but it is a highly profitable sector. The return to pharmaceutical research and development is highly dependent on government policies—regulatory approval processes, research and development tax credits, tax subsidies for manufacture in U.S. offshore possessions such as Puerto Rico, orphan drug provisions, and payment for drugs through government health entitlement programs. Taxpayer support for the scientific and technical engine that drives much biomedical innovation—NIH research—is another major area of federal policy that directly affects the future of pharmaceuticals.

New technology growing out of biomedical research is by no means solely responsible for the cost escalation of health care expenditure, and the pharmaceutical sector accounts for only a small fraction of increased costs. Indeed, there have been clear examples of cost-saving pharmaceutical innovations, such as the reduction of iron lung use after the polio vaccine or the shortened length of hospital stays that followed the introduction of antibiotics. The advent of recombinant DNA pharmaceuticals shared some attributes of earlier innovations. Following treatment with the growth factor erythropoietin, for example, some patients with kidney failure can live who would have died in the recent past. Developing that single drug put the biotechnology firm Amgen on the pharmaceutical landscape. While pharmaceuticals made using recombinant DNA techniques may be life-saving and highly innovative, their price tends to be quite high. This is partly explained by the considerable new front-end investment in biological science, adding to the already costly process of discovering and testing a new drug.[18]

Prices for prescription drugs rose faster than general inflation and higher even than inflation in the health sector through the 1980s. Some of this increase was due to the higher cost of developing a new drug,[18] and some was due to improved quality and introductions of entirely new kinds of pharmaceuticals.[24] (This is analogous to tracking only the price of new cars, without taking into account better gas mileage, higher reliability, and safety improvements. Many new drugs, by analogy, are like cars that can traverse rivers or fly.) Despite the escalating costs of doing business, however, it appears that investments in pharmaceuticals still enjoy a higher rate of return than other industries, even when adjusted for the high risks.[18] Molecular biology has transformed the pharmaceutical industry; but the industrial applications of molecular biology are also transforming the process of biomedical research.

Biomedical science has become ever more tightly tethered to an industry, and a highly profitable one at that. Questions about how industrial funding of university research might corrupt the pure motives and independent inquiry of science arose even as the biotechnology revolution began in the late 1970s, when the magnitude of such investment was relatively low. By 1993, the Office of Technology Assessment, in discussing this policy issue, concluded that the evidence appeared to indicate that those few scientists with substantial industrial funding agreements published more and taught more, and so "commitment to the academic institution appears not to be a big problem"; the potential for conflicts of interest arises in only "a very small minority of cases."[18] While conflict of interest for individual investigators might not often conflict with science, the ethos of publicly funded academic research nonetheless confronted deep systemic problems.

The 1980s clearly brought a dramatic shift in the role of biomedical research, and yet public perceptions and public policy have not yet adjusted. Federal policies on technology transfer are contradictory. One faction of Congress promotes industrial applications of research, while others are deeply suspicious of industrial ties. Both pristine science and vigorous technology transfer are laudable goals, but they come into conflict from time to time.

A string of laws, beginning in 1980, gave patent rights to universities doing research, but universities are now beginning to make agreements with corporations that could result in taxpayers giving hefty research subsidies to private firms. For a few tens of millions of dollars, a company might wrap up the patent rights to university work paid for by federal grants, thus leveraging a public research investment many times larger than the private one. The main beneficiaries may be the universities or research centers and the investigators. This promises to emerge as a major issue in the 1990s.

The decade that culminated with the human genome project also saw a powerful alignment between biomedical research, commercial biotechnology, and pharmaceutical innovation. Much good can come from this synergy, but promotion of biotechnology must inevitably collide with the other public policy goal of constraining how much health care encroaches on the rest of the economy. Moreover, the public is likely, slowly but ineluctably, to become more suspicious of biomedical research that is financially rewarding as well as a purely academic pursuit aimed at mitigating suffering.

The public is sure to sense the emerging power of a research-industrial complex that is growing in size, much as the military-industrial complex that President Eisenhower warned of in the late 1950s. The question is whether biomedical scientists, and genome researchers in particular, can keep their hands clean in an environment that consumes more and more resources, relentlessly increases in scope and scale, and depends on the federal government for succor.[25; 26] Research funded by private foundations, by other philanthropies, or from venture capital is far less at issue than research that taxpayers fund. Private funding sources must make difficult decisions about the purposes of their research, how to allocate patent rights, how much their funds can be used to support industrially relevant research, and when to share data. Unlike federally funded research, however, this is a private matter rather than a public policy problem.

The values that have merged science as the dispassionate pursuit of truth with the profit motive are a volatile mix. Early critics worried that the inflated rhetoric supporting the genome project was due to the personal aspirations of its promoters. How much more weight might their concerns have carried had there also been financial motives? The monetary rewards of genome research are becoming more apparent. The genome project is destined to be a proving ground for the new rules governing science.

The future of the genome project will clearly be caught up in an abstruse technology-transfer debate about the industrial uses of its information, but the broad social impact of genome research will depend on the degree to which it can recast the debate about genetic determinism. The twentieth century began with the emergence of genetics as a science, and quickly got caught in a simplistic interpretation of inheritance that bred the eugenics and racial hygiene movements. These were virulent ideologies that provoked a backlash, casting a long shadow over the science itself. Both genetic determinism and its equally flawed antagonist, environmental determinism, are utterly incapable of explaining biology. As genetics turns up more and more knowledge about the role of genes in diseases and other traits, will the same simple-minded interpretations of genetics continue to dominate public discourse? Will "it's genetic" continue to mean "we can't do anything about it" in cocktail party prattle? Will genetics continue as the stalking horse for racist ideology and ethnic prejudice? Surely we can move beyond these vacuous ideologies to a richer understanding that embraces both genetic and environmental factors in the complex dance of life. The interesting question is not whether it is nature or nurture, but how they interact.

In the 1970s, molecular biologists imposed a moratorium on themselves while they debated the safety of recombinant DNA research. Historical interpretations differ on whether it was science or self-interest that played the leading role in ending the moratorium,[27-29] but the fact that the moratorium took place and was self-imposed is not in question. This gives molecular biology a social cast unlike the secrecy of the Manhattan Project. The birth of the research program on ethical, legal, and social issues (ELSI) follows this precedent in a new form, attending explicitly to the social impact of science. One of the most daunting tasks of the ELSI program is to change the social framework in which genetics is cast. The public debate need not repeat historical mistakes premised on genetic determinism or cling to a reactive environmental dogmatism. How genes influence biology and behavior is an extremely complex phenomenon that science has only begun to understand. Immense beauty resides in understanding it; immense joy in finding it out. At its best, genome research can replace caricature with nuance, and provide a richer vocabulary for understanding genetics.

In the end, nations must decide how to spend their public dollars. Biomedical research is a public good, and the genome project is intended to make that research go faster and probe deeper. Is this more or less important than a new highway? New weapons? Health care? Social Security? In the grand scheme of things, the answers to such questions return to the lives of citizens. The Ross family lives with Alzheimer's disease every day, but it also enjoys movies and protection by the armed forces. How much is it worth to them to get rid of Alzheimer's disease? The answer is clearly a lot, but not everything. Most families can find a similar dread disease somewhere in their pedigree. Producing new knowledge through the discipline of science, building on the work of others towards a universally laudable goal, is a noble pursuit—or at least it can be. Few jobs can be more gratifying than discovery. A biomedical researcher lays small bricks in a growing edifice, but one whose foundation is far more stable than those of most other professions. If it can preserve its noble aims and promote social policies to thwart the demonstrably destructive power of genetic information, the genome project can build a permanent monument of new knowledge—a solid structure of great beauty but also immense practical significance. Understanding ourselves better can benefit everyone.

Chronology

Genesis of the Human Genome Project

(This chronology, which runs from May 1985 through the project's "official" beginning in October 1990, lists political events, not scientific or technical accomplishments.)

1985 *May*   Robert Sinsheimer convenes the first meeting on sequencing the human genome at the University of California, Santa Cruz.

      October   Renato Dulbecco introduces his idea of sequencing the human genome at a talk organized by the Italian embassy in Washington, DC

              Charles DeLisi of the U.S. Department of Energy (DOE) has the idea of mapping and sequencing the human genome while reading a draft report on heritable mutations from the congressional Office of Technology Assessment (OTA). DeLisi and David Smith, working at DOE headquarters in Germantown, Maryland, hatch plans for a Human Genome Initiative in the days just before Christmas.

1986 *February*   Sydney Brenner sends a letter to the European Commission in Brussels, urging a concerted program to map and sequence the genomes of various organisms.

      March   A group from Los Alamos National Laboratory, led by Mark Bitensky, convenes DOE's first meeting on sequencing the human genome, held in Santa Fe, New Mexico.

              Renato Dulbecco publishes a commentary on sequencing the human genome in *Science*.

      June   James Watson organizes a rump session on the genome project at a Cold Spring Harbor Laboratory symposium on the molecular biology of *Homo sapiens*. Watson invites Paul Berg and Walter Gilbert to cochair the session, which reveals considerable opposition to the DOE program among molecular biologists.

      July   The Howard Hughes Medical Institute convenes an Informational

Forum on the Human Genome on the campus of the National Institutes of Health (NIH), the first international meeting devoted to the genome project. While there is strong opposition to a mindless sequencing project, the process of redefining the genome project to encompass mapping begins.

September   The governing board of the National Research Council, National Academy of Sciences, approves a study of mapping and sequencing the human genome, which leads a report of the same name. The study is quickly funded by the James S. McDonnell Foundation. Bruce Alberts of the University of California, San Francisco, is appointed chairman. John Burris is study director.

At the same hour of the same day, the congressional Technology Assessment Board approves an OTA project, which leads to the report "Mapping Our Genes: Genome Projects—How Big? How Fast?" LeRoy Walters of Georgetown University chairs the advisory panel. Staff includes Robert Cook-Deegan as project director.

October   NIH director James Wyngaarden convenes the fifty-fourth meeting of the Director's Advisory Committee. This is the first public NIH-sponsored meeting devoted to the genome project.

1987 *February –March*   House and Senate appropriations committees hold hearings on the NIH and DOE budgets for fiscal year 1988, the first year that congressionally earmarked genome research funds are set aside in the NIH and DOE budgets. DOE proposes its Human Genome Initiative. Funds for NIH come from questions put to NIH director James Wyngaarden by House subcommittee chairman Rep. William Natcher.

March   Charles DeLisi and Leroy Hood testify at a hearing before a subcommittee of the House Committee on Science and Technology. Eileen Lee and other staff organize the hearing for Rep. James Scheuer, whose subcommittee authorizes funding for the DOE life sciences program. The hearing constitutes the most vulnerable moment for the DOE program, as congressional staff are uncertain whether to condone DeLisi's actions.

April   DOE's Health and Environmental Research Advisory Committee releases a report recommending a DOE-led fifteen-year Human Genome Initiative, with a budget reaching $200 million per year.

May   Renato Dulbecco, Paolo Vezzoni, and others launch the Italian genome program.

James Watson, David Baltimore, and Bradie Metheny, representing the Delegation for Basic Biomedical Research, meet with House and Senate Appropriations Committee members and staff (including Rep. William Natcher, chairman of the House subcommittee that funds NIH, and Senator Lowell Weicker, former chairman of the corresponding Senate subcommittee). The main purpose of the meeting is to promote more AIDS research funding, but Watson also requests $30 million for genome research at NIH.

Senator Pete Domenici hosts a meeting on the future of DOE's national laboratories at the U.S. Capitol. Donald Fredrickson, Jack McConnell, congressional staff, and senior officials at DOE and the national laboratories attend. Fredrickson suggests genome research as a productive new direction for DOE research. McConnell and Domenici's staff draft legislation to promote technology transfer that includes a human genome component.

December   Alexander Bayev and Andrei Mirzabekov first present the idea for a Soviet genome program to government officials in Moscow.

1988 *February*   The National Research Council (NRC) releases its report "Mapping and Sequencing the Human Genome," which recommends a concerted genome research program with a budget to reach $200 per year.

February –March   James Wyngaarden convenes the *Ad Hoc* advisory Committee on Complex Genomes in Reston, Virginia, to draft plans for a genome research program along the lines recommended by the NRC report. Wyngaarden announces his intention to form the Office of Human Genome Research at NIH and to appoint an associate director of NIH to head it. Several scientists urge Wyngaarden to appoint Watson.

April   Rep. John Dingell releases the OTA report at a hearing before the House Committee on Energy and Commerce. Lesley Russell, staff biologist for Dingell, organizes the hearing, which includes testimony from OTA, Bruce Alberts, James Watson, Maynard Olson, and others.

The Human Genome Organization (HUGO) is founded at Cold Spring Harbor Laboratory. Victor McKusick, Sydney Brenner, James Watson, Leroy Hood, and others urge that scientists, rather than administrators or politicians, direct international collaboration, and suggest HUGO as the mechanism.

May   First scientific meeting on human genome mapping and sequencing at Cold Spring Harbor Laboratory, organized by Maynard Olson, Charles Cantor, and Richard Roberts.

June   Reps. James Scheuer and Douglas Walgren, chairmen of two subcommittees of the House Committee on Science and Technology, convene a joint hearing on NIH-DOE collaboration. The committee subsequently approves a bill similar to one already passed by the Senate that is sponsored by Senators Lawton Chiles, Pete Domenici, and Edward Kennedy. The House Energy and Commerce Committee, which shares House jurisdiction over the bill and constitutes the principal remaining congressional obstacle to passage, threatens to pass the bill also, unless NIH and DOE agree to cooperate. NIH and DOE sign a Memorandum of Understanding that fall.

September	NIH director James Wyngaarden appoints James Watson NIH associate director for human genome research. Watson hires Elke Jordan and Mark Guyer as the first employees.
October	Santiago Grisolia convenes the First International Workshop on Collaboration for the Human Genome Project in Valencia, Spain.

1989	*February*	Sir James Cowan, secretary of the Medical Research Council, and Sir Walter Bodmer, head of the Imperial Cancer Research Fund, officially launch a genome research program in the United Kingdom. The program supports research along the lines suggested by Sydney Brenner.
	April	Japan's Ministry of Education, Science, and Culture (Monbusho) commences a genome research program, based on a framework drafted by Kenichi Matsubara and others.
	June	The European Commission's Human Genome Analysis Programme is approved in Brussels, Belgium.
	August	An NIH-DOE planning retreat at the Banbury Center, Cold Spring Harbor Laboratory, lays the groundwork for the first five-year plan.
	September	Nancy Wexler chairs the first meeting of the NIH Working Group on Ethical, Legal, and Social Issues (ELSI) at the Cloisters building, National Institutes of Health. The working group drafts a mission statement. In December, the working group becomes joint with DOE, which begins its own ELSI program under pressure from Senator Albert Gore.
	October	Louis Sullivan, Secretary of the Department of Health and Human Services, elevates NIH's Office of Human Genome Research to become the National Center for Human Genome Research, with spending authority. James Watson is appointed director and Elke Jordan deputy director.

1990	*February –August*	Martin Rechsteiner (University of Utah) begins a letter-writing campaign opposing the genome project. Michael Syvanen (University of California, Davis) mounts a similar campaign via computer networks. Bernard Davis and colleagues from the department of microbiology at Harvard Medical School publish a letter opposing the genome project in *Science*.
	April	NIH and DOE release their first joint five-year plan.
	June	Hubert Curien, minister of science and technology, announces the French government's intention to commence a genome research program. The program is officially launched in October.

Acknowledgments

THIS BOOK began in a 1987 conversation with Sandra Panem of the Alfred P. Sloan Foundation. The makings of a good story—the immensely interesting personalities already deeply engaged in supporting and opposing the genome project—gave several people the idea for a book. Sandra was interested in a book with a science policy slant, one that focused on how a highly conspicuous decision was made in the highest reaches of government. This book results from her interest. Sandra left the Sloan Foundation in 1988, but the foundation, in the person of Michael Teitlebaum, continued to support the project, despite changes in my plans too numerous to recount.

The book was delivered in draft form to W.W. Norton, where it was published in its finally revised form under the able and patient direction of editor Joseph Wisnovsky. In retrospect, as well as in prospect, this was clearly the right choice. Alan Iselin did the illustrations, and Ted Johnson did a meticulous job of copyediting. I also thank Sydney Cohen for preparing the index, which should make the book much more useful to future scholars. In addition, I must thank Joan Bossert of Oxford University Press, Howard Boyer of Harvard University Press, and Elizabeth Knoll of University of California Press for their interest and assistance. While they did not end up publishing the book, their comments and attention to the review process were greatly appreciated.

While I strive to remain objective, I was close to the action as it took place. From 1986 until 1988, I followed the genome project as the director of a team at the Office of Technology Assessment, U.S. Congress. The other members of that team were Patricia Hoben, Jacqueline Courteau, David Guston, Teresa Schwab Myers, and Gladys White. The result of the OTA project was a 1988 report, *Mapping Our Genes*.[1] Every member of the team has since proved through other accomplishments what they demonstrated on the OTA project—that they are an extraordinarily talented lot.

Our task at OTA was to report to Congress on what the genome project was, whether or not it was worth funding, and how its underlying bureaucracy should be constructed. This vantage point placed me in the midst of three groups: scientists debating the merits of genome projects, administrators working in federal agencies responsible for funding the scientists, and members of Congress who had the constitutional authority to specify how much and in what way federal funds could be used.

Our function at OTA was to gather information, filter it, and direct it to the people who wrote laws. Congress was involved because Article 1 of the Constitution gives it sole authority to tax and spend. OTA's proximity to political power ensured ready access to the nation's best expertise; its location in the legislative branch of government insulated it from the crossfire of inter-agency scuffles, a particularly important advantage in this case. Gretchen Kols-rud directed OTA's Biological Applications Program while I worked there, and it was her vision of the future of biotechnology that made OTA a national center of policy-making in that field. Gary Ellis, Val Giddings, Kathi Hanna, Robyn Nishimi, Rand Snell, Kevin O'Connor, Gladys White, Nanette New-ell, Susan Clymer, Geoffrey Karny, and many others created a charged atmosphere and set high standards for work on genetics and biotechnology; work of distinction was the inevitable by-product. Jack Gibbons and Roger Herd-man established the institutional values that made OTA the finest place I have ever worked.

I left OTA in December 1988 to become acting executive director of the Biomedical Ethics Advisory Committee, whose mandate included an assessment of policy implications of human genetics. The venue was again Congress, but the congressional leaders associated with it were entangled in a highly charged abortion debate. The Biomedical Ethics Advisory Committee died as a result of internecine wars among its congressional patrons, closing its doors in late September 1989. It officially expired on paper a year later. I thus had the glorious task of helping to create a new federal agency, serving as acting executive director, but this happy duty later turned sour.

I then worked from December 1989 until August 1990 at the National Center for Human Genome Research, the administrative hub of the genome project within the National Institutes of Health. The center was two months old when I arrived, and I was there to watch it grow from infancy into adolescence. I have since moved on to the Institute of Medicine, National Academy of Sciences, into a field only tangentially related to genome research, although I remain involved in the debate about social, ethical, and legal issues.

The human genome debate was a passionate affair. Champions and detractors saw it as a critical battle in the history of biology; high stakes amplified differences among factions. A self-conscious historicism pervaded arguments about the project from the beginning. Proponents and antagonists argued to influence members of Congress, opinion leaders of molecular biology, and administrators of the largest life science programs in the world. While proximity to events enhanced access to information, it also distorted what I acquired. I had direct contact with the principal actors, and talked with them many times even as the projects began to unfold. I attended many of the meetings and had unparalleled access to the records of various bureaucracies. I followed up with interviews of major players.

James D. Watson figures prominently in this book. In a book on the Human Genome Project, he naturally looms as a towering figure. I first met him after a lecture in Biochemistry 10, when I was a Harvard undergraduate in 1972. As a devotee of physical chemistry at the time, I did not deign to partake of biochemistry, but I did want to hear the man who wrote *The Double Helix*.[2] I had decided he was a jerk after reading the book; his lecture confirmed that impression.

It was not the last time I was wrong. I learned a great deal from Jim Watson; it was an honor to work for him. I have followed federal science policy for roughly a decade now, and have never seen a scientist so quickly adapt to the policy process. I was immensely impressed by how he took the reins of the genome project. From the time of our first interview in 1987, when he was obviously trying to learn about Congress and federal policy, I was astonished at his energy, the degree to which he was willing to learn new skills, and the facility with which he began to amble through the halls of power.

Francis Crick noted in another book supported by Sandra Panem and the Sloan Foundation: "Rather than believe that Watson and Crick made the DNA structure, I would rather stress that the structure made Watson and Crick."[3] Discovering the double helix did indeed manufacture Watson's public persona, but it was his choice how to use it. His stature was a policy tool to promote genome research that he used with zest. He did not need the genome project nearly so much as the genome project needed him.

Many other groups were generous with additional support for this book. The National Science Foundation (NSF) funded a grant to archive historically relevant documents and to create an oral history of the genome project at the National Reference Center for Bioethics Literature, Georgetown University. Most of the papers, videotapes, and other materials that I gathered are available there, along with prepublication drafts of this book deposited so that others might use them before it came out. Anita Nolen and Doris Goldstein, co-investigators on the NSF grant at Georgetown, did most of the work collecting my chaotic files into a useful archive. I dumped; they organized. Staff at NSF were remarkably free of territorial reflexes and gave me faith that committed resistance can tame a federal bureaucracy. NSF and Georgetown created a resource that should make it easier for future historians of science to correct my mistakes.

Through the efforts of Victor McKusick and Joanna Strayer Amberger, Academic Press, Inc., donated a computer to the Johns Hopkins University for my use as special features editor for the genetics journal *Genomics*. This book was largely written on that computer. I particularly thank Alan Kay,

Steve Jobs, and the software wizards at Microsoft (Word) and Niles and Associates (EndNote), who made the Macintosh computer a source of pleasure as well as power.

I owe another debt to former coworkers at the National Center for Human Genome Research, National Institutes of Health (NIH). NIH provided an unrivaled opportunity to learn how the "working" side of government operates, as opposed to the legislative side. NIH is the preeminent biomedical institution in the world, and it was a privilege to work there. I was welcomed by Elke Jordan, Mark Guyer, Jane Peterson, Pam Lokken, Anita Brooks, Linda Engel, Eric Juengst, Bettie Graham, Leslie Fink, and dozens of others. Those in the NIH genome office shared a commitment to making the genome project a success. I was offered unequaled access to the information that flowed naturally into the world's largest administrative center for genome research.

Late in 1989, Kathi Hanna, then at the Institute of Medicine (IOM), asked me to write a policy history of the genome project for an IOM committee to study decision-making. The committee used the genome project as one of several case studies of how health and science policy decisions are made.[4] I and other authors attended an extremely stimulating March 1990 meeting at the Beckman Center in Irvine, California. Paul Berg, a member of the committee, warmly recounted his own personal experience of the genesis of the genome project. Kathi Hanna's invitation to write the IOM case study forced me to commit my thoughts to paper months earlier than I would have otherwise. It also presented an unforeseen opportunity for external review, comment, and corrections. When Ted Friedmann of the University of California, San Diego, generously invited me to write a slightly longer account to open the premier issue of the book series *Molecular Genetic Medicine,* I had another such opportunity.[5] This was another connection to Academic Press, which publishes those volumes.

As I worked to complete the book, I began a new position at IOM. I thank Queta Bond, Kenneth Shine, Ruth Bulger, Gary Ellis, Jane Fullarton, Elaine Lawson, Richard Rettig, Mark Randolph, Carolyn Peters, Gail Spears, and others who saw bits and pieces of the manuscript trekked in from home. Jack Barchas and William Bunney from my board were particularly supportive, plowing through a complete draft, and Betsy Turvene from IOM's editorial office also gave invaluable advice.

Akihiro Yoshikawa identified the importance of the genome project in U.S.–Japan relations long before others. He provided initial insights about the Japanese policy process and obtained information that I could never have obtained without his help. His Russian Hill home in San Francisco was the site of more than a few late-night discussions about genome politics. Aki and Nancy also saved me many nights' hotel charges, and permitted me to sleep at the foot of Lombard Street, with a breathtaking view of one of the most beautiful cities on earth. Ken-ichi Matsubara was of inestimable help in keeping me up to date about Japan, and Akiyoshi Wada was amazingly open and generous about the early workings of the Japanese genome project. Japan was not at all a closed society when it came to exchanging information about genome doings.

Academicians Andrei Mirzabekov and Alexander A. Bayev were my pipeline into the USSR, at a time of endless complexities. I must also give special thanks to Vladimir Larionov and Natalya Kouprina for permanent memories of white nights on the Neva and friendship in Maryland. Diane Hinton, first with the Howard Hughes Medical Institute and then the Human Genome Organization, knew everyone and everything. She was generous with her knowledge and, more important, her friendship. James Wyngaarden was a central figure, first at NIH, then through the Human Genome Organization, and finally at the National Academy of Sciences. He granted open access at all three sites, and provided clarifications along the way.

Several people at the Department of Energy were generous with access to their files and willing to recount stories and "soft" Washington information. David Galas, Charles DeLisi, David Smith, and Daniel Drell merit special mention among many.

Those at genome research centers, particularly the NIH-supported genome centers and Department of Energy-supported national laboratories, were also quite helpful. I attended dozens, perhaps hundreds, of site visits, meetings, and other events as the genome story unfolded, and there are far too many people to list individually, so I have made a longer list below.

And what can I say about Nancy Wexler that others have not? She is as wonderful as I make her out to be in Chapter 16.

In addition to these sources of support, I was indirectly aided by many institutions. Many paid for my travel to give lectures or to attend meetings—opportunities to keep abreast of science and politics among the various genome projects—my greatest source of information. Among the groups that helped out were:

American Association for the Advancement of Science
American Council of Life Insurance
Association of Academic Health Centers
Beckman Center for the History of Chemistry
Berlex Laboratories
Biomedical Ethics Advisory Committee, U.S. Congress
California Institute of Technology
California State Polytechnic University, Pomona
Ciba-Geigy Pharmaceuticals
Commission of the European Communities Council of International Organizations of Medical Sciences
Dibner Foundation
George Washington University
Georgetown University
Harvard University
Hobart and William Smith Colleges
Institute for Advanced Studies, Valencia
Institute of Medicine, National Academy of Sciences
Lippoldt Trust and University of Central Florida
National Institutes of Health
National Science Foundation

Office of Technology Assessment, U.S. Congress

Ortho Diagnostic Systems, Inc.

Park Ridge Medical Center, Rochester, N.Y.

Public Responsibility in Medicine and Research, Inc.

Ruhr-Universität, Bochum

Salk Institute

San Francisco State University

Serono Symposia on Reproductive Medicine

Southern Medical Association

Student Pugwash

Tenth Federal Judicial Circuit

United Nations Educational, Scientific, and Cultural Organization (UNESCO)

University of California, Berkeley

University of California, Irvine

University of California, Los Angeles

University of California, San Francisco

University of California, Santa Cruz

University of Maryland

University of Oslo

University of Rochester

University of Southern California

University of Wisconsin

USSR Academy of Sciences

Vahl Lectures, University of Rochester

Virginia Polytechnic Institute

Virginia Polytechnic Institute and State University

Woodrow Wilson National Fellowship Foundation, Princeton, N.J.

I depended on interviews with the principal actors to reconstruct some events, particularly those for which there was little or no documentation. Most of these interviews took place informally at meetings or over the phone in the context of other work. Some, however, have been taped and transcribed for future use. Transcripts of formal interviews are available through the National Center for Bioethics Literature at Georgetown University, and through a grant from the National Science Foundation will be available on-line, subject to any restrictions imposed by those interviewed. A listing of the documents that I cited at some stage of writing this book (not all citations survived the editing process) has been deposited there as well. It is available in electronic form for scholars, in EndNote format (Niles and Associates, Berkeley, Calif.) for both PC and Macintosh computers.

The following people deserve special thanks for their contributions, either as sources of information, as means of confirming information, or for reviewing parts of the book. I have tried to be fair and comprehensive in this list, but I have no doubt inadvertently left out some individuals who helped me a great deal. I apologize for such oversights. No thanks are adequate for an undertaking so complex, when the people involved are so bright, talented, committed, and wonderfully diverse.

Mark Adams

Reid Adler

Bruce Alberts

Duane Alexander

John Alwen

Christopher Anderson

Leigh Anderson

Norman Anderson

W. French Anderson

Lori Andrews

Wilhelm Ansorge

Judith Areen

Adrienne Asch

Francisco Ayala

David Baltimore

Zbigniew Bankowski

Benjamin J. Barnhart

John Barry

Jack Bartley

Alexander A. Bayev

John Beatty

Jonathan Beckwith

George Bell

Paul Berg

Beverly Berger
Mark Bitensky
Walter Bodmer
Queta Bond
Judith Bostock
David Botstein
Anne Bowcock
Elbert Branscomb
Sydney Brenner
Ruth Bulger
David Burke
Christian Burks
John Burris
Andrew Bush
George Cahill
Mark Cantley
Charles Cantor
Alexander Morgan
    Capron
Anthony Carrano
C. Thomas Caskey
James Cassatt
Luigi Luca Cavalli-Sforza
Andrew Chen
Ellson Chen
Masato Chijiya
Lawton Chiles
Ralph Christoffersen
George Church
Mary Clutter
Susan Clymer
Sharon Cohen
Chris Coles
Francis Collins
P. Michael Conneally
Cheryl Corsaro
Alan Coulson
Charles Coulter
Jacqueline Courteau
Charles Coutelle
David Cox
Robert Crawford
Barbara Culliton
Teri Curtin
Jean Dausset
Ronald Davis
Larry Deaven
Charles DeLisi

Anthony Dickens
David Dickson
Pete Domenici
Helen Donis-Keller
Russell Doolittle
Janet Dorigan
Daniel Drell
Renato Dulbecco
Troy Duster
Irene Eckstrand
Clark Elliott
Gary Ellis
Linda Engel
Mark Evans
Glen Evans
Sherman Finesilver
Leslie Fink
John C. Fletcher
Clarissa Formosa-Gauci
Norman Fost
Maurice Fox
Donald Fredrickson
Lawrence Friedman
Tracy Friedman
Theodore Friedmann
Asao Fujiyama
Jane Fullarton
Patrick Gage
David Galas
Bernard Gert
Raymond Gesteland
Jack Gibbons
Walter Gilbert
Jonathan Glover
Doris Goldstein
Albert Gore, Jr.
Beatrice Gorman
Patricia Gossel
Michael Gough
Bettie Graham
Loren R. Graham
Santiago Grisolía
Francois Gros
David Guston
Mark Guyer
Benedikt Härlin
J. Michael Hall
Kathi Hanna

Bernadine Healy
John Heilbron
Pam Henson
Roger Herdman
C. Edgar Hildebrand
Diane Hinton
Patricia Hoben
Neil Anthony Holtzman
Leroy Hood
Jerry Hsueh
Henry Huang
Ruth Hubbard
Mike Hunkapiller
Tim Hunkapiller
Susanne Huttner
Francois Jacob
Jose Jaz
Albert Jonsen
Bertrand Jordan
Elke Jordan
Christopher Joyce
Horace Freeland Judson
Eric Juengst
Mariko Kakunaga
Steven Keith
Evelyn Fox Keller
Daniel Kevles
Mary-Claire King
Patricia King
David Kingsbury
Ruth Kirschstein
Paula Kokkonen
Arthur Kornberg
Natalya Kouprina
Sheldon Krimsky
Laura Kumin
Nancy Lamontagne
Eric Lander
Elinor Langfelder
Vladimir Larionov
Elaine Lawson
Joshua Lederberg
Eileen Lee
Rachel Levinson
Roger Lewin
Reidar K. Lie
Donald A. B. Lindberg
M. Susan Lindee

Bronwen Loder
Lee Loevinger
Pamela Lokken
William Lowrance
Jerry Mande
Jean-Louis Mandel
Betty K. Mansfield
Daniel Masys
Ken-ichi Matsubara
Svetlana Matsui
Federico Mayor
George Mazuzan
Jack McConnell
Keith McKenney
Victor A. McKusick
Diane McLaren
Everett Mendelsohn
Mortimer Mendelsohn
Bruce Merrifield
Bradie Metheny
Jerome Miksche
Henry Miller
Akihiko Mine
Andrei Mirzabekov
Matsura Miyaka
Happei Miyakawa
Bernadette Modell
Wataru Mori
Jay Moscowitz
Robert Moyzis
Robert Murray
Thomas Murray
Teresa Schwab Myers
Rick Myers
DeLill Nasser
David L. Nelson
Michael R. Nelson
Elizabeth A. Newell
Peter Newmark
M. F. Niermeijer
Robyn Nishimi
Anita Nolen
Harry Noller
Michio Oishi
Maynard Olson
Ronald Overmann
Joseph Palca

Jon Palfreman
Robert Palmer
John Parascandola
Diane Paul
Mark Pearson
Peter Pearson
Dan Perry
Carolyn Peters
Jane Peterson
Betty Pickett
Maya Pines
Robert Pokorski
Blair Potter
Claire Pouncey
Theodore Puck
Robert Rabin
Paul Rabinow
Lisa Raines
Mark Randolph
William Raub
Judith Reagan
Philip Reilly
Esther Reiswig
Richard Rettig
Jeremy Rifkin
Bo Andreassen Rix
Jerry Roberts
Leslie Roberts
Richard Roberts
Ian Rolland
Leon Rosenberg
Susan Rosenfeld
Karen Rothenberg
Mark Rothstein
Frank Ruddle
Lesley Russell
Richard Saltus
Helen Willa Samuels
Susan Sanford
Mona Sarfaty
Hans-Martin Sass
David Schlessinger
Charles Scriver
Kenneth Shine
Greg Simon
Klaus Singer
Robert Sinsheimer

Mark Skolnick
David Smith
Lloyd Smith
Robert Smith
Temple Smith
Rand Snell
Eichi Soeda
Jan Helge Solbakk
Dieter Soll
Gail Spears
Nigel Spurr
Michael A. Stephens
Robert Stephenson
Irene Stith-Coleman
John Sulston
David Swinbanks
Kathleen Sykes
Michael Teitlebaum
Arnold Thackray
Larry Thompson
Shirley Tilghman
Ignacio Tinoco
David Trickett
Kevin Ulmer
Charles Van Horn
Craig Venter
Tony Vickers
Akiyoshi Wada
LeRoy Walters
Michael Waterman
Robert Waterston
James Dewey Watson
Carol A. Westbrook
Nancy Wexler
Gladys White
Raymond L. White
Thomas J. White
Daniel Wikler
Benjamin Wilfond
Robert Williamson
Lois Wingerson
Robert Wood
John Wooley
Ronald G. Worton
James Wyngaarden
Kunio Yagi
Michael Yesley

| Akihiro Yoshikawa | Frank Young | Norton Zinder. |
| Philip Youderian | Vladimir Zharov | |

Every author, by reflex, notes the missed weekends, evenings, and other disruptions of family life that a book entails. As this book neared completion, our friend Marji Balzer asked how it was coming, a question posed every few months between her treks to eastern Siberia for anthropological fieldwork. It was an innocent question. I began to answer, but my wife Kathryn intervened, noting that, in our family, "book" had become a four-letter word. Like certain other four-letter words, including life and wife, this one had to be experienced to be appreciated. Through such experiences, life acquires meaning.

References and Notes

A COMPLETE SET of 1,431 references cited in this book, or in preliminary drafts of it, is available from the National Center for Bioethics Literature (NRCBL), Georgetown University, Washington, DC 20057 (1-800-MED-ETHX, or 202-687-3885). Sharon J. Durfy and Amy E. Grotevant of NRCBL also prepared a brief annotated bibliography (Scope Note 17, December 1991). Michael S. Yesley of Los Alamos National Laboratory is maintaining a bibliography of Ethical, Legal, and Social Implications of the Human Genome Project, to be periodically published by the Department of Energy. The first such bibliography was published as DOE/ER-0543T, available to DOE contractors from the Office of Scientific and Technical Information, PO Box 62, Oak Ridge, TN 37831; available to the public through the National Technical Information Service, 5285 Port Royal, Road, Springfield, VA 22161.

Accounts of early events by some of the major players can be found in:
Watson, JD. (1990). The Human Genome Project: Past, Present, and Future. *Science* **248**:44–49.
Sinsheimer, R. (1989). The Santa Cruz Workshop, May 1985. *Genomics* **5**:954–956.
DeLisi, C. (1988). The Human Genome Project. *American Scientist* **76**:488–493.
Mullis, KB. (1990). The Unusual Origin of the Polymerase Chain Reaction. *Scientific American* **262** (April):56–65.

I have previously published abbreviated descriptions of the early history in articles and book chapters, although this book corrects some inaccuracies and oversights in those accounts. The earlier accounts include:
Watson, JD, and Cook-Deegan, RM. (1991). Origins of the Human Genome Project. *FASEB Journal* **5**(January):8–11.
Cook-Deegan, RM. (1991). The Human Genome Project: Formation of Federal Policies in the United States, 1986–1990. In *Biomedical Politics*, K Hanna, Ed., pp. 99–168. National Academy Press, Washington, DC.
Cook-Deegan, RM. (1991). The Genesis of the Human Genome Project. In *Molecular Genetic Medicine*, Vol. 1, T Friedmann, Ed., pp. 1–75. Academic Press, San Diego.

French scientist Bertrand Jordan spent a year traveling around the world, visiting genome research laboratories in different countries. He wrote a series of "genome chronicles" during this experience, which were later published in book form in both French and English. This is the most useful book for capturing how scientists thought about their work and how it would be done at the time.
Jordan Bertrand. (1993). *Voyage Autour du Genome: Le Tour du Monde en 80 Labos*. John Libbey Eurotext, Montrouge, France. Also available in English as *Traveling Around the Human Genome: An In Situ Investigation*, from the same publisher.

One of the most significant early meetings predated the concept of the genome project, but brought together those involved in the technological developments that gave rise to it, summarized in:
Cook-Deegan, RM. (1989). The Alta Summit, December 1984. *Genomics* **5**:661–663.

Several other books on the genome project also review some of its historical origins:
Bishop, JE, and Waldholz, M. (1990). *Genome: The Story of the Most Astonishing Scientific Adventure of Our Time—the Attempt to Map All the Genes in the Human Body*. Simon & Schuster, New York.

Davis, J. (1990). *Mapping the Code: The Human Genome Project and the Choices of Modern Science.* Wiley & Sons, New York.

Lee, TF. (1991). *The Human Genome Project: Cracking the Code of Life.* Plenum, New York.

Shapiro, Robert. (1991). *The Human Blueprint: The Race to Unlock the Secrets of Our Genetic Script.* St. Martin's Press, New York.

Wingerson, L. (1990). *Mapping Our Genes.* Dutton, New York.

Wills, C. (1991). *Exons, Introns, and Talking Genes.* Basic Books, New York.

Many collections of essays have also appeared, particularly those focused on ethical, legal, and social issues. One of the earliest is also among the best:

Kevles, DJ and Hood, L, Eds. (1992). *The Code of Codes: Scientific and Social Issues in the Human Genome Project.* Harvard University Press, Cambridge, MA.

The journal *Los Alamos Science* has produced two special issues of particular interest. A November 1992 edition, No. 20, is devoted to the genome project, emphasizing the Department of Energy program, but including informal musings of many of the principal actors in the NIH program also. No. 15, from 1987, centered on Stanislaw Ulam, and was republished as *From Cardinals to Chaos: Reflections on the Life and Legacy of Stanislaw Ulam* (1989), Cambridge University Press, New York.

1. WHY GENETICS?

1. Katzman, R. (1976). The Prevalence and Malignancy of Alzheimer's Disease: A Major Killer. *Archives of Neurology* 33:217–218.

2. Walton, J, Ed. (1977). *Brain's Diseases of the Nervous System,* 8th ed. Oxford University Press, p. 1183, New York.

3. Flugel, FE. (1929). Zur Diagnostik der Alzheimerschen Krankheit. *Gesamte Psychiatrie* 120:783–787.

4. Schottky, J. (1932). Uber Präsenile Verblödungen. *Gesamte Neurologie und Psychiatrie* 140:333–397.

5. von Braunmuhl, A. (1932). Kolloidchemische Betrachtungsweise Seniler und Präseniler Gewebsveranderungen. *Gesamte Neurologie und Psychiatrie* 142:1–54.

6. Cook-Deegan, RM, and Winters-Miner, LA. Appendix: Research Bibliography on Familial Alzheimer's Disease. In *Familial Alzheimer's Disease: Molecular Genetics and Clinical Perspectives,* Miner, GD, Richter, RW, Blass, JP, Valentine, JL, and Winters-Miner, LA, Eds. Marcel Dekker, New York.

7. The search is much more thoroughly described in Pollen, D. (1993). *Hannah's Heirs: The Quest for the Genetic Origins of Alzheimer's Disease.* Oxford University Press, New York.

8. McKusick, VA. (1988). *The Morbid Anatomy of the Human Genome: A Review of Gene Mapping in Clinical Medicine.* Howard Hughes Medical Institute, Bethesda, MD.

9. Saint George-Hyslop, PH, Tanzi, RE, Polinsky, RJ, Haines, JL, Nee, L, Watkins, PC, Myers, RH, Feldman, RG, Pollen, D, Drachman, D, Growdon, J, Bruni, A, Foncin, J-F, Salmon, D, Frommelt, P, Amaducci, L, Sorbi, S, Piacentini, S, Stewart, GD, Hobbs, WJ, Conneally, PM, and Gusella, JF. (1987). The Genetic Defect Causing Familial Alzheimer's Disease Maps on Chromosome 21. *Science* 235 (20 February):885–890.

10. Goate, AM, Haynes, AR, Owen, MJ, Farrall, M, James, LA, Lai, LYC, Mullan, MJ, Roques, P, Rossor, MN, Williamson, R, and Hardy, JA. (1989). Predisposing Locus for Alzheimer's Disease on Chromosome 21. *Lancet* I for 1989 (18 February):352–355.

11. Schellenberg, GD, Bird, TD, Wijsman, EM, Moore, DK, Boehnke, M, Bryant, EM, Lampe, TH, Nochlin, D, Sumi, SM, Deeb, SS, Beyreuther, K, and Martin, GM. (1988). Absence of Linkage of Chromosome 21q21 Markers to Familial Alzheimer's Disease. *Science* 241 (16 September):1507–1510.

12. Saint George-Hyslop, PH, Haines, JL, Farrer, LA, Polinsky, R, and others. (1990). Genetic Linkage Studies Suggest That Alzheimer's Disease Is Not a Single Homogeneous Disorder. *Nature* 347 (13 September):194–197.

13. Martin, GM, Schellenberg, GD, Wijsman, EM, and Bird, TD. (1990). Dominant Susceptibility Genes. *Nature* 347 (13 September):124.

14. Goate, AM, Chartier-Harlin, M-C, Mullan, M, Brown, J, Crawford, F, Fidani, L, Giuffra, L, Haynes, A, Irving, N, James, L, Mant, R, Newton, P, Rooke, K, Roques, P, Talbot, C, Pericak-Vance, M, Roses, A, Williamson, R, Rossor, M, Owen, M, and Hardy, J. (1991). Segregation of a Missense Mutation in the Amyloid Precursor Protein Gene with Familial Alzheimer's Disease. *Nature* 349 (21 February):704–706.

15. Murrell, J, Farlow, M, Ghetti, B, and Benson, MD. (1991). A Mutation in the Amyloid Precursor Protein Associated with Hereditary Alzheimer's Disease. *Science* 254 (4 October):97–99.

16. Schellenberg, GD, Anderson, L, O'dahl, S, Wisjman, EM, Sadovnick, AD, Ball, MJ, Larson, EB, Kukull, WA, Martin, GM, Roses, AD, and Bird, TD. (1991). APP717, APP693, and PRIP Gene Mutations Are Rare in Alzheimer Disease. *American Journal of Human Genetics* 49:511–517.

17. Pericak-Vance, MA, Bebout, JL, Gaskell, PCJ, Yamaoka, LH, Hung, W-Y, Alberts, MJ, and Walker, AP. (1991). Linkage Studies in Familial Alzheimer Disease: Evidence for Chromosome 19 Linkage. *American Journal of Human Genetics* 48:1034–1050.

18. Schellenberg, GD, Bird, TD, Wijsman, EM, Orr, HT, Anderson, L, Nemens, E, White, JA, Bonnycastle, L, Weber, JL, Alonso, ME, Potter, H, Heston, LL, and Martin, GM. (1992). Genetic Linkage Evidence for a Familial Alzheimer's Disease Locus on Chromosome 14. *Science* 258 (23 October):668–671.

19. Breitner, JCS, Murphy, EA, and Folstein, MF. (1986). Familial Aggregation in Alzheimer Dementia—II. Clinical Genetic Implications of Age-Dependent Onset. *Journal of Psychiatric Research* 20:45–55.

20. Breitner, JCS, Folstein, MF, and Murphy, EA. (1986). Familial Aggregation in Alzheimer Dementia—I. A Model for the Age-Dependent Expression of an Autosomal Dominant Gene. *Journal of Psychiatric Research* 20:31–43.

2. MAPPING OUR GENES

1. Botstein, D, White, RL, Skolnick, M, and Davis, RW. (1980). Construction of a Genetic Linkage Map in Man Using Restriction Fragment Length Polymorphisms. *American Journal of Human Genetics* **32**:314–331.

2. Sutton, WS. (1902). On the Morphology of the Chromosome Group in *Brachystola magna*. *Biological Bulletin* **4** (December):24.

3. Sutton, WS. (1903). The Chromosomes in Heredity. *Biological Bulletin* **4** (April):231.

4. Grisolia, S. (1982). *Recollections* (Research Week 1982 Highlights: Research as the Basis for Modern Health Care). University of Kansas Medical Center, College of Health Sciences and Hospital.

5. Kevles, DJ. (1985). *In the Name of Eugenics*. University of California Press, Berkeley, CA.

6. Mittwoch, U. (1973). *Genetics of Sex Differentiation*. Academic Press, New York.

7. Carlson, EA. (1991). Defining the Gene: An Evolving Concept. *American Journal of Human Genetics* **49**:475–487.

8. McKusick, VA. (1988). *The Morbid Anatomy of the Human Genome: A Review of Gene Mapping in Clinical Medicine*. Howard Hughes Medical Institute, Bethesda, MD.

9. Provine, WB. (1971). *The Origins of Theoretical Population Genetics*. University of Chicago Press, Chicago.

10. Johannsen, W. (1909). *Elemente der Exakten Erblichkeitslehre*. G. Fischer, Jena.

11. Stubbe, H. (1972). *History of Genetics. From Prehistoric Times to the Rediscovery of Mendel's Laws*. MIT Press, Cambridge, MA.

12. Sturtevant, AH. (1965). *A History of Genetics*. Harper and Row, New York.

13. Bearn, AG, and Miller, ED. (1979). Archibald Garrod and the Development of the Concept of Inborn Errors of Metabolism. *Bulletin of the History of Medicine* **53** (Fall):315–328.

14. Childs, B. (1970). Sir Archibald Garrod's Conceptions of Chemical Individuality: A Modern Appreciation. *New England Journal of Medicine* **282** (#2, 8 January):71–77.

15. Garrod, AE. (1909). *Inborn Errors of Metabolism*. Oxford University Press, Oxford.

16. Wilson, EB. (1911). The Sex Chromosomes. *Archiv für Mikroskopie und Anatomie Entwicklungsmech* **77**:249–271.

17. Donahue, RP, Bias, WB, Renwick, JH, and McKusick, VA. (1968). Probable Assignment of the Duffy Blood Group Locus to Chromosome 1 in Man. *Proceedings of the National Academy of Sciences* **61**:949–955.

18. Mohr, J. (1951). Search for Linkage Between Lutheran Blood Group and Other Hereditary Characters. *Acta Pathologica e Microbiologica Scandinavica* **28**:207–210.

19. Weiss, M, and Green, H. (1967). Human-Mouse Hybrid Cell Lines Containing Partial Complements of Human Chromosomes and Functioning Human Genes. *Proceedings of the National Academy of Sciences (USA)* **58**:1104–1111.

20. Ruddle, F, Bentley, KL, and Ferguson-Smith, MA. (1987). *Physical Mapping Review*. Contract report prepared for the Office of Technology Assessment, U.S. Congress.

21. Ruddle, FH. (1989). Tribute to Torbjörn Caspersson. *American Journal of Human Genetics* **44** (#4, April):439–440.

22. Caspersson, T, Lomakka, C, and Zech, L. (1970). Fluorescent Banding. *Hereditas* **67**:89–102.

23. Caspersson, T, Zech, L, and Johansson, C. (1970). Differential Banding of Alkylating Fluorochromes in Human Chromosomes. *Experimental Cell Research* **60**:315–319.

24. Caspersson, T, Zech, L, Johansson, C, and Modest, EJ. (1971). Quinocrine Mustard Fluorescent Banding. *Chromosoma* **30**:215–227.

25. Caspersson, TO. (1989). The William Allan Memorial Award Address: The Background for the Development of the Chromosome Banding Technique. *American Journal of Human Genetics* **44** (#4, April):441–451.

26. Avery, OT, MacLeod, CM, and McCarty, M. (1944). Induction of Transformation by a Desoxyribonucleic Acid Fraction Isolated from Pneumococcus Type III. *Journal of Experimental Medicine* **79** (1 February):137–158.

27. Watson, JD, and Crick, FHC. (1953). Genetical Implications of the Structure of Deoxyribonucleic Acid. *Nature* **171** (30 May):737–738.

28. Schrödinger, E. (1967). *What Is Life?* Cambridge University Press, Cambridge, UK.

29. Neel, JW. (1949). The Inheritance of Sickle Cell Anemia. *Science* **110**:64–66.

30. Pauling, L, Itano, HA, Singer, SJ, and Wells, IC. (1949). Sickle Cell Anemia: A Molecular Disease. *Science* **110**:543–548.

31. Ingram, VM. (1957). Gene Mutation in Human Hæmoglobin: The Chemical Difference Between Normal and Sickle Cell Hæmoglobin. *Nature* **180**:326–328.

32. Nathans, D, and Smith, HO. (1975). Restriction Endonucleases in the Analysis and Restructuring of DNA Molecules. In *Annual Reviews of Biochemistry*, EE Snell, Ed., pp. 273–293. Palo Alto, CA: Annual Reviews, Inc.

33. Smith, HO. (1979). Nucleotide Specificity of Restriction Endonucleases. *Science* **205**:455–462.

34. Southern, EM. (1975). Detection of Specific Sequences Among DNA Fragments Separated by Gel Electrophoresis. *Journal of Molecular Biology* **98**:503–517.

35. Rigby, PW, Dieckmann, M, Rhodes, C, and Berg, P. (1977). Labeling Deoxyribonucleic Acid to High Specific Activity *in Vitro* by Nick Translation with DNA Polymerase I. *Journal of Molecular Biology* **113** (15 June):237–251.

36. Maniatis, T, Jeffrey, A, and Kleid, DG. (1975). Nucleotide Sequence of the Rightward Operator of Phage Lambda. *Proceedings of the National Academy of Sciences, USA* **72**:1184–1188.

37. Jeffreys, AJ, and Flavell, RA. (1977). A Physical Map of the DNA Regions Flanking the Rabbit Beta-Globin Gene. *Cell* **12** (October):429–439.

38. Grodzicker, T, Williams, J, Sharp, P, and Sambrook, J. (1974). Physical Mapping of Temperature-Sensitive Mutants of Adenoviruses. *Cold Spring Harbor Symposia on Quantitative Biology* **39** (Tumor Viruses):439–446.

39. Petes, TD, and Botstein, D. (1977). Simple Mendelian Inheritance of the Reiterated Ribosomal DNA of Yeast. *Proceedings of the National Academy of Sciences (USA)* **74** (November):5091–5095.

40. Goodman, HM, Olson, MV, and Hall, BD. (1977). Nucleotide Sequence of a Mutant Eukaryotic Gene: The Yeast Tyrosine-Inserting Ochre Suppressor SUP4-o. *Proceedings of the National Academy of Sciences (USA)* **74** (December):5453–5457.

41. McKusick, VA. (1989). Historical Perspectives: The Understanding and Management of Genetic Disorders. *Maryland Medical Journal* **38** (November):901–908.

42. Kan, YW, and Dozy, AM. (1978). Polymorphism of DNA sequence adjacent to Human Beta-Globin Structural Gene: Relationship to Sickle Mutation.

Proceedings of the National Academy of Sciences (USA) **75**:5631–5635.

43. Jeffreys, AJ. (1979). DNA Sequence Variants in the G-γ, A-γ, ∂, and ß Globin Genes of Man. *Cell* **18**:1–10.

44. Solomon, E, and Bodmer, WF. (1979). Evolution of Sickle Cell Variant Gene (letter). *Lancet* **I for 1979** (April 28):923.

45. Simon, M, Pawlotsky, Y, Bourel, M, Fauchet, R, and Genetet, B. (1975). Hemochromatose idiopathique. Maladie Associée à l'Antigène Tissulaire HL-A3? *Nouvel Presse Médicale* **4**:1432.

46. Simon, M, Pawlotsky, Y, Bourel, M, Fauchet, R, and Genetet, B. (1976). Association of HLA-A3 and HLA-B14 Antigens with Idiopathic Hemochromatosis. *Gut* **17**:332–334.

47. Simon, M, Bourel, M, Genetit, B, and Fauchet, R. (1977). Idiopathic Hemochromatosis. Demonstration of Recessive Transmission and Early Detection by Family HLA Typing. *New England Journal of Medicine* **297**:1269–1275.

48. Kravitz, K, Skolnick, MH, Cannings, C, Carmelli, D, Baty, B, Amos, B, Johnson, A, Mendell, N, Edwards, CQ, and Cartwright, G. (1979). Genetic Linkage between Hereditary Hemochromatosis and HLA. *American Journal of Human Genetics* **31**:601–619.

49. Edwards, CQ, Griffen, BA, Goldgar, D, Drummond, C, Skolnick, MH, and Kushner, JP. (1988). Prevalence of Hemochromatosis Among 11,065 Presumably Healthy Blood Donors. *New England Journal of Medicine* **318** (21):1355–1362.

50. Bishop, JE, and Waldholz, M. (1990). *Genome: The Story of the Most Astonishing Scientific Adventure of Our Time—the Attempt to Map All the Genes in the Human Body.* Simon & Schuster, New York.

51. Wingerson, L. (1990). *Mapping Our Genes.* Dutton, New York.

52. Fogle, L. (1992). Rest, Relaxation, and RFLPs. *Journal of NIH Research* **4** (January):66–71.

53. Botstein, D, 1987, Interview, Genentech, January.

54. Botstein, D, 1988, Interview, Genentech, 19 August.

55. Davis, R, 1987, Interview, Stanford University, January.

56. White, R. (1988). Chromosome Mapping with DNA Markers. *Scientific American* **258** (February):40–48.

57. Botstein, D, 1991, Electronic mail note, Stanford University, 26 February.

58. Fox, M, 1989, Dinner with Evelyn Fox Keller and Diane Paul, home of Maurice Fox.

59. White, R, 1987, Interview, University of Utah, January.

60. White, RL. (1979). *Human Linkage Mapping with DNA Polymorphism* (R01-GM27313). University of Massachusetts, Worcester. Proposal signed and dated 27 February; logged in at the National Institutes of Health 1 March.

61. Wyman, AR, and White, RL. (1980). A Highly Polymorphic Locus in Human DNA. *Proceedings of the National Academy of Science (USA)* **77**:6754–58.

62. Donis-Keller, H, 1988, Interview, Cambridge, MA, 14 July.

63. Donis-Keller, H, 1991, Telephone interview, Washington University, St. Louis, MO, 8 March.

64. Barinaga, M. (1987). Critics Denounce First Genome Map as Premature. *Nature* **329** (15 October):571.

65. Roberts, L. (1987). Flap Arises over Genetic Map. *Science* **238** (6 November):750–752.

66. Ackerman, S. (1988). Taking on the Human Genome. *American Scientist* **76** (January–February):17–18.

67. Donis-Keller, H, Green, P, Helms, C, et al. (1987). A Genetic Linkage Map of the Human Genome. *Cell* **51** (October):319–337.

68. Crow, JF, and Dove, WF. (1988). Anecdotal, Historical, and Critical Commentaries on Genetics: A Diamond Anniversary: The First Chromosome Map. *Genetics* **118** (January):1–3.

69. Roberts, L. (1991). The Rush to Publish. *Science* **251** (18 January):260–263.

70. Dausset, J, Cann, H, Cohen, D, Lathrop, M, Lalouel, J-M, and White, RL. (1990). Centre d'Etude du Polymorphisme Humain (CEPH): Collaborative Genetic Mapping of the Human Genome. *Genomics* **6** (March):575–577.

71. Marx, JL. (1985). Putting the Human Genome on a Map. *Science* **229** (12 July):150–151.

72. Gusella, JF, Wexler, NS, Conneally, PM, Naylor, SL, Anderson, MA, Tanzi, RE, Watkind, PC, Ottina, K, Wallace, MR, Sakaguchi, AY, Young, AM, Shoulson, I, Bonilla, E, and Martin, JB. (1983). A Polymorphic DNA Marker Genetically Linked to Huntington's Disease. *Nature* **306**:234–238.

73. Davies, KE, Pearson, PL, Harper, PS, and others. (1983). Linkage Analysis of Two Cloned Sequences Flanking the Duchenne Muscular Dystrophy Locus on the Short Arm of the Human X Chromosome. *Nucleic Acids Research* **11**:2302–2312.

74. Reeders, ST, Breunig, MH, Davies, KE, Nicholls, RD, Jarman, AP, Higgs, DR, Pearson, PC, and Weatherall, DJ. (1985). A Highly Polymorphic DNA Marker Linked to Adult Polycystic Kidney Disease on Chromosome 16. *Nature* **317**:542–544.

75. Cavenee, WK, Hansen, MF, Nordernskjold, M, and others. (1985). Genetic Origin of Mutations Predisposing to Retinoblastoma. *Science* **228**:501–503.

76. Tsui, L-C, Buchwald, M, Barker, D, Braman, JC, Knowlton, R, Schumm, JW, Eiberg, H, Mohr, J, Kennedy, D, Plavsic, N, Zsiga, M, Markiewicz, D, Akots, G, Brown, V, Helms, C, Gravius, T, Parker, C, Rediker, K, and Donis-Keller, H. (1985). Cystic Fibrosis Locus Defined by a Genetically Linked Polymorphic DNA Marker. *Science* **230** (29 November):1054–1057.

77. Wainwright, BJ, Scambler, PJ, Schmidtke, J, Watson, EA, Law, H-Y, Farrall, M, Cooke, HJ, Eiberg, H, and Williamson, R. (1985). Localization of Cystic Fibrosis Locus to Human Chromosome 7cen-q22. *Nature* **318** (28 November):384–385.

78. White, RL, Woodward, S, Leppert, M, O'Connell, P, Hoff, M, Herbst, J, Lalouel, J-M, Dean, M, and Vande Woude, G. (1985). A Closely Linked Genetic Marker for Cystic Fibrosis. *Nature* **318**:382–384.

79. Knowlton, RG, Cohen-Haguenauer, O, van Cong, N, and others. (1985). A Polymorphic DNA Marker Linked to Cystic Fibrosis in Located on Chromosome 7. *Nature* **318** (28 November):381–382.

80. Begley, S, with Katz, SE, and Drew, L. (1987). The Genome Initiative. *Newsweek* (August 31):58–60.

81. Royer, B, Kunkel, L, Monaco, A, Goff, S, Newburger, P, Baehner, R, Cole, F, Curnutte, J, and Orkin, S. (1987). Cloning the Gene for an Inherited Human Disorder—Chronic Granulomatous Disease—on the Basis of Its Chromosomal Location. *Nature* **322**:32–38.

82. Koenig, M, Hoffman, EP, Bertelson, CJ, Monaco, AP, Feener, C, and Kunkel, LM. (1987). Complete Cloning of the Duchenne Muscular Dystrophy (DMD) cDNA and Preliminary Genomic Organization of the DMD Gene in Normal and Affected Individuals. *Cell* **50** (July):509–517.

83. Lee, W, Bookstein, R, Hong, F, Young, L-J, Shew, J-Y, and Lee, EY-HP. (1987). Human Retinoblastoma Susceptibility Gene: Cloning, Identification,

and Sequence. *Science* 235:1394–1399.

84. Friend, SH, Bernards, R, Rogelj, S, Weinberg, RA, Rapaport, JM, Albert, DM, and Dryja, TP. (1986). A Human DNA Segment with Properties of the Gene That Predisposes to Retinoblastoma and Osteosarcoma. *Nature* 323 (16 October):643–646.

85. Rommens, JM, Iannuzzi, MC, Kerem, B-S, Drumm, ML, Melmer, G, Dean, M, Rozmahel, R, Cole, JL, Kennedy, D, Hidaka, H, Zsiga, M, Buchwald, M, Riordan, JR, Tsui, L-C, and Collins, FS. (1989). Identification of the Cystic Fibrosis Gene: Chromosome Walking and Jumping. *Science* 245 (8 September):1059–1065.

86. Riordan, JR, Rommens, JM, Kerem, B-S, Alon, N, Rozmahel, R, Grzelczak, Z, Zielenski, J, Lok, S, Plavsic, N, Chou, J-L, Drumm, ML, Iannuzzi, MC, Collins, FS, and Tsui, L-C. (1989). Identification of the Cystic Fibrosis Gene: Cloning and Characterization of Complementary DNA. *Science* 245 (8 September):1066–1072.

87. Kerem, B-S, Rommens, JM, Buchanan, JA, Markiewicz, D, Cox, TK, Chakravarti, A, Buchwald, M, and Tsui, L-C. (1989). Identification of the Cystic Fibrosis Gene: Genetic Analysis. *Science* 245 (8 September):1073–1080.

88. Roberts, L. (1988). The Race for the Cystic Fibrosis Gene, and *The Race for the CF Gene Nears End*. *Science* 240 (8 April and 15 April):141–144 and 282–285.

89. Dean, M. (1988). Molecular and Genetic Analysis of Cystic Fibrosis. *Genomics* 3:93–99.

90. Davies, K. (1989). The Search for the Cystic Fibrosis Gene. *New Scientist* (21 October):54–58.

91. Stephens, JC, Cavanaugh, ML, Gradie, MI, Mador, ML, and Kidd, KK. (1990). Mapping the Human Genome: Current Status. *Science* 250 (12 October):237–244, also The Human Genome Map 1990 center section.

92. Jasny, BR. (1990). The Human Genome Map 1990. *Science* 250 (12 October):Pull-out insert.

93. Culliton, BJ. (1990). Mapping Terra Incognita (Humani Corporis). *Science* 250 (12 October):210–212.

94. Roberts, L. (1992). Two Chromosomes Down, 22 More to Go. *Science* 258 (2 October):28–30.

95. NIH / CEPH Collaborative Mapping Group. (1992). A Comprehensive Genetic Linkage Map of the Human Genome. *Science* 258 (2 October):67–86.

96. Goodfellow, P. (1992). Variation Is Now the Theme. *Nature* 359 (29 October):777–778.

97. Weissenbach, J, Gyapay, G, Dib, C, Vignal, A, Morissette, J, Millasseau, P, Vaysseix, G, and Lathrop, M. (1992). A Second-Generation Linkage Map of the Human Genome. *Nature* 359 (29 October):794–801.

98. Pines, M, Editor. (1991). *Blazing a Genetic Trail*. Howard Hughes Medical Institute. May.

3. OF YEASTS AND WORMS

1. Schrödinger, E. (1967). *What Is Life?* Cambridge University Press, Cambridge, UK.

2. Maniatis, T, Hardison, RC, Lacy, E, Lauer, J, O'Connell, C, Quon, D, Sim, GK, and Efstratiadis, A. (1978). The Isolation of Structural Genes from Libraries of Eucaryotic DNA. *Cell* 15 (October):687–701.

3. Olson, MV, Dutchik, JE, Graham, MY, Brodeur, GM, Helms, C, Frank, M, MacCollin, M, Scheinman, R, Frank, T. (1986). Random-Clone Strategy for Genomic Restriction Mapping in Yeast. *Proceedings of the National Academy of Sciences* (USA) 83 (October):7826–7830.

4. Coulson, A, Sulston, J, Brenner, S, and Karn, J. (1986). Toward a Physical Map of the Genome of the Nematode *Caenorhabditis elegans*. *Proceedings of the National Academy of Sciences (USA)* 83 (October):7821–7825.

5. Botstein, D, and Fink, GR. (1988). Yeast: An Experimental Organism for Modern Biology. *Science* 240 (10 June):1439–1443.

6. Kenyon, CJ. (1988). The Nematode *Caenorhabditis elegans*. *Science* 240 (10 June):1448–1453.

7. Brenner, S. (1973). Genetics of Behavior. *British Medical Bulletin* 29:269–271.

8. Brenner, S. (1974). Genetics of *Caenorhabditis elegans*. *Genetics* 77 (1):71–94.

9. Brenner, S. (1988). Foreword. In *The Nematode Caenorhabditis elegans*, W. B. Wood, Ed., pp. ix–xiii. Cold Spring Harbor, NY: Cold Spring Harbor Laboratory.

10. Roberts, L. (1990). The Worm Project. *Science* 248 (15 June):1310–1313.

11. Sulston, JE, and Horvitz, HR. (1977). Post-Embryonic Cell Lineages of the Nematode *Caenorhabditis elegans*. *Developmental Biology* 56:110–156.

12. Sulston, JE. (1983). Neuronal Cell Lineages in the Nematode *Caenorhabditis elegans*. *Cold Spring Harbor Symposia on Quantitative Biology* 48:443–452.

13. Sulston, JE, Schierenberg, E, White, JG, and Thompson, JN. (1983). The Embryonic Cell Lineage of the Nematode *Caenorhabditis elegans*. *Developmental Biology* 100:64–119.

14. White, JG, Southgate, E, Thompson, JN, and Brenner, S. (1986). The Structure of the Nervous System of the Nematode *Caenorhabditis elegans*. *Philosophical Transactions of the Royal Society of London* Series B, Vol. 314:1–340.

15. Sulston, JE, and Brenner, S. (1974). The DNA of *Caenorhabditis elegans*. *Genetics* 77 (1):95–104.

16. Sulston, J, 1990, Conversation at DNA Sequencing Meeting, II, Hyatt Regency, Hilton Head, SC, 3 October.

17. Burke, DT, Carle, GF, and Olson, MV. (1987). Cloning of Large Segments of Exogenous DNA into Yeast Artificial-Chromosome Vectors. *Science* 236:806–808.

18. Nasmyth, K, and Sulston, J. (1987). High-Altitude Walking with YACs. *Nature* 328 (30 July):380–381.

19. Lander, ES, and Waterman, MS. (1988). Genomic Mapping by Fingerprinting Random Clones: A Mathematical Analysis. *Genomics* 2:231–239.

20. Coulson, A, Waterston, R, Kiff, J, Sulston, J, and Kohara, Y. (1988). Genome Linking with Yeast Artificial Chromosomes. *Nature* 335 (8 September):184–186.

21. Cook-Deegan, RM, Guyer, M, Rossiter, BJF, Nelson, DL, and Caskey, CT. (1990). The Large DNA Insert Cloning Workshop. *Genomics* 7 (August):654–660; based on Olson's summary on p. 655.

22. Sulston, J, Du, Z, Thomas, K, Wilson, R, Hillier, L, Staden, R, Halloran, N, Green, P, Thierry-Mieg, J, Qiu, L, Dear, S, Coulson, A, Craxton, M, Durbin, R, Berks, M, Metzstein, M, Hawkins, T, Ainscough, R, and Waterston, R. (1992). The *C. elegans* Genome Sequencing Project: A Beginning. *Nature* 356 (5 March):37–41.

23. Schwartz, DC, and Cantor, CR. (1984). Separation of Yeast Chromosome-Sized DNAs by Pulsed Field Gel Electrophoresis. *Cell* **37**:67–75.

24. Mlot, C. (1991). A Well-Rounded Worm. *Science* **252** (21 June):1619–1920.

25. Vollrath, D, Foote, S, Hilton, A, Brown, LG, Beer-Romero, P, Bogan, JS, and Page, DC. (1992). The Human Y Chromosome: A 43-Interval Map Based on Naturally Occurring Deletions. *Science* **258** (2 October):52–59.

26. Foote, S, Vollrath, D, Hilton, A, and Page, DC. (1992). The Human Y Chromosome: Overlapping DNA Clones Spanning the Euchromatic Region. *Science* **258** (2 October):60–66.

27. Little, P. (1992). Mapping the Way Ahead. *Nature* **359** (1 October):367–368.

28. Chumakov, I, Rigault, P, Guillou, S, Ougen, P, Billaut, A, Guasconi, G, Gervy, P, LeGall, I, Soularue, P, Grinas, L, Gougueleret, L, Bellanné-Chantelot, C, Lacroix, B, Barillot, E, Gesnouin, P, Pook, S, Vaysseix, G, Frelat, G, Schmitz, A, Sambucy, J-L, Bosch, A, Estivill, X, Weissenbach, J, Vignal, A, Riethman, H, Cox, D, Patterson, D, Gardiner, K, Hattori, M, Sakaki, Y, Ichikawa, H, Ohki, M, Le Paslier, D, Heilig, R, Antonarakis, S, and Cohen, D. (1992). Continuum of Overlapping Clones Spanning the Entire Human Chromosome 21q. *Nature* **359** (1 October):380–387.

4. SEQUENCE UPON SEQUENCE

1. Gamow, G. (1954). Possible Relation Between Deoxyribonucleic Acid and Protein Structures. *Nature* **173** (13 February):318. I thank Maynard Olson for pointing out this reference.

2. Sulston, J, Du, Z, Thomas, K, Wilson, R, Hillier, L, Staden, R, Halloran, N, Green, P, Thierry-Mieg, J, Qiu, L, Dear, S, Coulson, A, Craxton, M, Durbin, R, Berks, M, Metzstein, M, Hawkins, T, Ainscough, R, and Waterston, R. (1992). The *C. elegans* Genome Sequencing Project: A Beginning. *Nature* **356** (5 March):37–41.

3. Oliver, SG, van der Aart, QJM, Agostoni-Carbone, ML, Aigle, M, Alberghina, L, Alexandraki, D, Antoine, G, Anwar, R, Ballesta, JPG, Benit, P, Berben, G, Bergantino, E, Biteau, N, Bolle, PA, Bolotin-Fukuhara, M, Brown, A, Brown, AJP, Buhler, JM, Carcano, C, Carignani, G, Cederberg, H, Chanet, R, Contreras, R, Crouzet, M, Daignan-Fornier, B, Defoor, E, Delgado, M, Demolder, J, Doira, C, Dubois, E, Dujon, B, Dusterhoft, A, Erdmann, D, Esteban, M, Fabre, F, Fairhead, C, Faye, G, Feldmann, H, Fiers, W, Francingues-Gaillard, MC, Franco, L, Frontali, L, Fukuhara, H, Fuller, LJ, Galland, P, Gent, ME, Gigot, D, Gilliquet, V, Glansdorff, N, Goffeau, A, Grenson, M, Grisanti, P, Grivell, LA, de Haan, M, Haasemann, M, Hatat, D, Hoenicka, J, Hegemann, J, Herbert, CJ, Hilger, F, Hohmann, S, Hollenberg, CP, Huse, K, Iborra, F, Indge, KJ, Isono, K, Jacq, C, Jacquet, M, James, CM, Jauniaux, JC, Jia, Y, Jimenez, A, Kelly, A, Kleinhans, U, Kreisl, P, Lanfranchi, G, Lewis, C, van der Linden, CG, Lucchini, G, Lutzenkirchen, K, Maat, MJ, Mallet, L, Mannhaupt, G, Martegani, E, Mathieu, A, Maurer, CTC, McConnell, D, McKee, RA, Messenguy, F, Mewes, HW, Molemans, F, Montague, MA, Muzi Falconi, M, Navas, L, Newlon, CS, Noone, D, Pallier, C, Panzeri, L, Pearson, BM, Perea, J, Philippsen, P, Pierard, A, Planta, RJ, Plevani, P, Poetsch, B, Pohl, F, Purnelle, B, Ramezani Rad, M, Rasmussen, SW, Raynal, A, Remacha, M, Richterich, P, Roberts, AB, Rodriguez, F, Sanz, E, Schaaff-Gerstenschlager, I, Scherens, B, Schweitzer, B, Shu, Y, Skala, J, Slonimski, PP, Sor, F, Soustelle, C, Spiegelberg, R, Stateva, LI, Sttensma, HY, Steiner, S, Thierry, A, Thireos, G, Tzermia, M, Urrestarazu, LA, Valle, G, Vetter, I, van Vliet-Reedjk, JC, Voet, M, Volckaert, G, Vreken, P, Wang, H, Warmington, JR, von Wettstein, D, Wicksteed, BL, Wilson, C, Wurst, H, Xu, G, Yoshikawa, A, Zimmermann, FK, and Sgouros, JG. (1992). The Complete DNA Sequence of Yeast Chromosome III. *Nature* **357** (7 May):38–46.

4. Eijgenraam, F. (1992). Yeast Chromosome III Reveals a Wealth of Unknown Genes. *Science* **256** (8 May):730.

5. Anderson, A. (1992). Yeast Genome Project: 300,000 and Counting. *Science* **256** (24 April):462.

6. Judson, HF. (1979). *The Eighth Day of Creation: The Makers of the Revolution in Biology.* Simon & Schuster, New York.

7. Molecular Genetics Today. (1990). *Genesis 2000 Uncovers Cystic Fibrosis Gene* (Vol. 2, #1). Du Pont. January.

8. Sanger, F. (1988). Sequences, Sequences, and Sequences. *Annual Reviews of Biochemistry* **57**:1–28.

9. Edman, P. (1950). A Method for the Determination of the Amino Acid Sequence in Peptides. *Acta Chemica Scandinavica* **4**:283–293.

10. Hewick, RM, Hunkapiller, MW, Hood, LE, and Dreyer, WJ. (1981). A Gas-Liquid Solid Phase Peptide and Protein Sequenator. *Journal of Biological Chemistry* **256**:7990–7997.

11. Hunkapiller, MW, and Hood, LE. (1983). Protein Sequence Analysis: Automated Microsequencing. *Science* **219**:650–659.

12. Holley, RW, Agpar, J, Everett, GA, Madison, JT, Marquissee, M, Merrill, SH, Penswich, JR, and Zamer, A. (1965). Structure of Ribonucleic Acid. *Science* **147**:1462–1465.

13. Holley, RW, Everett, GA, Madison, JT, and Zamir, A. (1965). Nucleotide Sequences in the Yeast Alanine Transfer Ribonucleic Acid. *Journal of Biological Chemistry* **240**:2122–2128.

14. Wu, R, and Taylor, E. (1971). Nucleotide Sequence Analysis of DNA. II. Complete Nucleotide Sequence of the Cohesive Ends of Bacteriophage Lambda DNA. *Molecular Biology* **57**:491–511.

15. Judson, HF. (1987). *Mapping the Human Genome: Historical Background* (Mapping Our Genes contractor reports, Vol. 1, NTIS Order No. PB 88-160-783 / AS). Office of Technology Assessment, US Congress. September 1987.

16. Sanger, F, and Coulson, AR. (1975). Rapid Method for Determining Sequences in DNA by Primed Synthesis with DNA-Polymerase. *Journal of Molecular Biology* **94**:441–448.

17. Sanger, F. (1975). The Croonian Lecture, 1975: Nucleotide Sequences in DNA. *Proceedings of the Royal Society of London* **B 191**:317–333.

18. Sanger, F, Nilken, S, and Coulson, AR. (1977). DNA Sequencing with Chain-Terminating Inhibitors. *Proceedings of the National Academy of Sciences (USA)* **74**:5463–5468.

19. Gilbert, W, and Maxam, A. (1973). The Nucleotide Sequence of the *lac* Operator. *Proceedings of the National Academy of Science (USA)* **70** (12):3581–3584.

20. Gilbert, W, Maxam, A, and Mirzabekov, A. (1976). Contacts Between the *lac* Repressor and DNA Revealed by Methylation. In *Ninth Alfred Benzon Symposium, Copenhagen: Control of Ribosome Synthesis.*

NO Kjeldgaard and O Maaloe, Ed., pp. 139–143 and discussion 144ff. New York: Academic Press.

21. Kolata, GB. (1980). The 1980 Nobel Prize in Chemistry. *Science* **210**:887–889.

22. Gilbert, W, 1988, Interview, Harvard University Biological Laboratories, 11 November.

23. Maxam, AM, and Gilbert, W. (1977). A New Method for Sequencing DNA. *Proceedings of the National Academy of Sciences (USA)* **74** (February):560–564.

24. Barrell, B. (1990). DNA Sequencing: Present Limitations and Prospects for the Future. *FASEB Journal* **5** (January):40–45.

25. Phillips, K. (1988). Biologist Hood Uses Loose Reins to Guide "Gang of 70." *The Scientist* (25 July):26.

26. Ciotti, P. (1985). Fighting Disease on the Molecular Front. *Los Angeles Times Magazine* (20 October):18–19, 38, 50, 56.

27. Hunkapiller, T, 1988, Interview, California Institute of Technology, July.

28. Interviews, 1988–1991, Leroy Hood (17 July 1988; 1 November 1989; 13 January 1991; and various scientific meetings); Tim Hunkapiller (14 October 1991 and 13 January 1991); Michael Hunkapiller (January 1987 and 24 August 1988); Lloyd Smith (January 1987; 2 March and 25 March 1991) and Henry Huang (26 March 1991).

29. Huang, H, 1991, Telephone interview, Washington University, St. Louis, MO, 26 March.

30. Smith, LM, Rubenstein, JLR, Parce, JW, and McConnell, HM. (1980). Lateral Diffusion of M-13 Coat Protein in Mixtures of Phosphatidylcholine and Cholesterol. *Biochemistry* **19** (25):5907–5911.

31. Smith, LM, McConnell, HM, Smith, BA, and Parce, JW. (1981). Pattern Photobleaching of Fluorescent Lipid Vesicles Using Polarized Laser Light. *Biophysical Journal* **33** (January):139–146.

32. Smith, LM, Sanders, JZ, Kaiser, RJ, Hughes, P, Dodd, C, Connell, CR, Heiner, C, Kent, SBH, and Hood, LE. (1986). Fluorescence Detection in Automated DNA Sequence Analysis. *Nature* **321** (12 June):674–679.

33. Smith, LM, Fung, S, Hunkapiller, MW, Hunkapiller, TJ, and Hood, LE. (1985). The Synthesis of Oligonucleotides Containing an Aliphatic Amino Group at the 5' Terminus: Synthesis of Fluorescent DNA Primers for Use in DNA Sequence Analysis. *Nucleic Acids Research* **13**:2399–2412.

34. Prober, JM, Trainor, GL, Dam, RJ, Hobbs, FW, Robertson, CW, Qagursky, RJ, Cocuzza, AJ, Jensen, MA, and Baumeister, K. (1987). A System for Rapid DNA Sequencing with Fluorescent Chain-Terminating Dideoxynucleotides. *Science* **238** (16 October):336–341.

35. Roberts, L. (1987). New Sequencers Take on the Genome. *Science* **238** (16 October):271–273.

36. EG&G Biomolecular, 1987, Interviews, Watertown, MA, May.

37. Hunkapiller, M, 1991, Interview, Applied Biosystems, Inc., Foster City, CA, 18 March.

38. Anderson, N, 1986, Informational Forum on the Human Genome, National Institutes of Health Conference Room, sponsored by Howard Hughes Medical Institute, August.

39. Wada, A. (1984). Automatic DNA Sequencing. *Nature* **307** (12 January):193.

40. Wada, A, and Soeda, E. (1986). Strategy for Building an Automatic and High Speed DNA-Sequencing System. In *Proceedings of the 4th Congress of the Federation of Asian and Oceanic Biochemists*, pp. 1–16. London: Cambridge University Press.

41. Wada, A. (1987). Automated High-Speed DNA Sequencing. *Nature* **325** (26 February):771–772.

42. Wada, A. (1987). Japanese Super DNA Sequencer Project. *Science and Technology in Japan* **6** (22):20–21.

43. Wada, A, and Soeda, E. (1987). Strategy for Building an Automated and High Speed DNA-Sequencing System. In *Integration and Control of Metabolic Processes*, OL Kon, Ed., pp. 517–532. New York: Cambridge University Press.

44. Wada, A. (1987–1990). Interviews at the American Type Culture Collection, Rockville, MD (July 1987); Office of Technology Assessment, Washington, DC (1988); and University Club, Tokyo (August 1990).

45. Wada, A. (1988). Future Prospects of Automated and High Speed DNA Sequencing. *Proceedings of the International Conference on Bioethics, Rome, Italy* (10–15 April):41–52.

46. Wada, A. (1988). The Practicability of and Necessity for Developing a Large-Scale DNA-Base Sequencing System: Toward the Establishment of International Super Sequencer Centers. In *Biotechnology and the Human Genome: Innovations and Impact*, AD Woodhead and BJ Barnhart, Ed., pp. 119–130. New York: Plenum.

47. Ansorge, W, 1989, Interview, Genome Mapping and Sequencing Symposium, Cold Spring Harbor Laboratory, May.

48. Smith, LM, and Hood, LE. (1987). Mapping and Sequencing the Human Genome: How to Proceed. *Bio/Technology* **5** (September):933–939.

49. Hood, LE, Hunkapiller, MW, and Smith, LM. (1987). Automated DNA Sequencing and Analysis of the Human Genome. *Genomics* **1**:201–212.

50. Hood, L, and Smith, L. (1987). Genome Sequencing: How to Proceed. *Issues in Science and Technology* **3**:36–46.

51. Hood, LE. (1988). Biotechnology and Medicine of the Future. *Journal of the American Medical Association* **259** (25 March):1837–1844.

52. Selvin, P. (1992). Science Innovation '92: The San Francisco Sequel. *Science* **257** (14 August):885–886.

53. Hunkapiller, T, Kaiser, RJ, Koop, BF, and Hood, L. (1991). Large-Scale and Automated DNA Sequence Determination. *Science* **254** (4 October):59–67.

54. Olson, MV, Hood, L, Cantor, C, and Botstein, D. (1989). A Common Language for Physical Mapping of the Human Genome. *Science* **245**:1434–1435.

55. Caskey, CT, 1990, Introduction of Kary B. Mullis, DNA Sequencing, II, symposium, Hilton Head Island, SC, 3 October.

56. Mullis, KB. (1990). The Unusual Origin of the Polymerase Chain Reaction. *Scientific American* **262** (April):56–65.

57. Appenzeller, T. (1990). Democratizing the DNA Sequence. *Science* **247** (2 March):1030–1032.

58. White, T. (1992). Interview with Paul Rabinow, Department of Anthropology (deposited at the Human Genome Archives, National Reference Center for Bioethics Literature, Georgetown University). University of California, Berkeley, presentation at Rice University. 23 January.

59. Saiki, RK, Scharf, SJ, Faloona, FA, Mullis, KB, Horn, GT, Erlich, HA, and Arnheim, N. (1985). Enzymatic Amplification of Beta-Globin Genomic Sequences and Restriction Site Analysis for Diagnosis of Sickle Cell Anemia. *Science* **230** (20 December):1350–1354.

60. Mullis, KB, Faloona, F, Scharf, S, Saiki, R, Horn, G, and Erlich, H. (1986). Specific Enzymatic Amplification of DNA *in Vitro*: The Polymerase Chain Reaction. *Cold Spring Harbor Symposia on Quantitative Biology*, Vol. LI:263–273.

61. Mullis, KB, and Faloona, FA. (1987). Specific Synthesis of DNA *in vitro* via a Polymerase Catalyzed Chain Reaction. In *Methods in Enzymology*, Vol. 155, R Wu, Ed., pp. 335–350. San Diego: Academic Press.

62. Koshland, DE. (1989). The Molecule of the Year. *Science* **246** (22 December):1541.

63. Guyer, RL, and Koshland, DE. (1989). The Molecule of the Year. *Science* **246** (22 December):1543–1544.
64. Kornberg, A. (1960). Biologic Synthesis of Deoxyribonucleic Acid. *Science* **131**:1503–1508.
65. Kornberg, A. (1980). *DNA Replication.* W.H. Freeman, New York.
66. Saiki, RK, Gelfand, DH, Stoffel, S, Scharf, SJ, Higuchi, R, Horn, GT, Mullis, KB, and Erlich, HA. (1988). Primer-Directed Enzymatic Amplification of DNA with a Thermostable DNA Polymerase. *Science* **239**:487–491.
67. Erlich, HA, Gelfand, DH, and Saiki, RK. (1988). Specific DNA Amplification. *Nature* **331** (4 February):461–462.
68. Erlich, HA, Gelfand, D, and Sninsky, JJ. (1991). Recent Advances in the Polymerase Chain Reaction. *Science* **252** (21 June):1643–1651.
69. White, TJ, Madej, R, and Persing, DH. (1992). The Polymerase Chain Reaction: Clinical Applications. *Advances in Clinical Chemistry* **29**:161–196.
70. Special Technology Section. (1991). Methods and Materials. Amplification of Nucleic Acid Sequences: The Choices Multiply. *Journal of NIH Research* **3** (February):81–94.

71. Barinaga, M. (1991). Biotech Nightmare: Does Cetus Own PCR? *Science* **251** (15 February):739–740.
72. Barinaga, M. (1991). And the Winner: Cetus Does Own PCR. *Science* **251** (8 March):1174.
73. Gershon, D. (1990). Cetus Battles with DuPont. *Nature* **343** (25 January):301.
74. McGourty, C. (1989). DuPont Battles with Cetus. *Nature* **342** (2 November):9.
75. Price, JS. (1989). PCR Origins. *Nature* **342** (7 December):623.
76. Evangelista, P, Grace, J, and Russell, K. (1991). *Hoffmann-La Roche Acquisition of PCR Technology Completed. Roche and Perkin-Elmer Form Strategic Alliance.* Hoffmann-La Roche, Inc. 11 December.
77. Gershon, D. (1991). Is Cetus Selling the Family Silver? *Nature* **352** (1 August):364.
78. Evangelista, P. (1992). *Hoffmann-La Roche Expands Access to PCR Technology for Diagnostic Testing Services—Move Expected to Continue to Accelerate Pace of PCR Development.* Roche Diagnostics Group, Hoffmann-La Roche, Inc., Nutley, NJ. 11 February.
79. Perkin Elmer Cetus. (1990). *PCR Bibliography* (Volume 1, No. 5). Perkin Elmer Cetus. June.

5. PUTTING SANTA CRUZ ON THE MAP

1. Sinsheimer, R, 1988, Interview, California Institute of Technology, 20 July.
2. Smith, TF. (1990). The History of the Genetic Sequence Databases. *Genomics* **6** (April):701–707.
3. Wade, N. (1981). The Complete Index to Man. *Science* **211** (2 January):33–35.
4. Anderson, N, 1990, Phone conversation, Large Scale Biology, Inc., September.
5. Anderson, NG, and Anderson, NL. (1985). A Policy and Program for Biotechnology. *American Biotechnology Laboratory* (Sept/Oct):1–3.
6. Hall, SS. (1988). Genesis: The Sequel. *California* (July):62–69.
7. Sinsheimer, R. (1989). The Santa Cruz Workshop, May 1985. *Genomics* **5**:954–956.
8. Sinsheimer, R, 1990, Interview, Hope Ranch, Santa Barbara, California, 25 July.
9. Sanger, F. (1988). Sequences, Sequences, and Sequences. *Annual Reviews of Biochemistry* **57**:1–28.
10. Sinsheimer, R. (1984). Note to Harry Noller, Robert Edgar, and Robert Ludwig. University of California, Santa Cruz. 23 October.
11. Sinsheimer, R. (1984). Letter, Robert Sinsheimer to David Gardner. University of California at Santa Cruz. 19 November.
12. Edgar, R, and Noller, H. (1984). *Human Genome Institute: A Position Paper.* Biology Board of Stud-

ies, University of California at Santa Cruz. 31 October.
13. Sinsheimer, RL. (1985). Note to Professors Robert Edgar and Harry Noller. Office of the Chancellor, University of California, Santa Cruz. 8 March.
14. Sanger, F. (1984). Letter to Harry Noller. Laboratory of Molecular Biology, Medical Research Council, Hills Road, Cambridge, UK. 22 November.
15. Human Genome Workshop. (1985). *Notes and Conclusions from the Human Genome Workshop, May 24–26, 1985.* Reprinted in Sinsheimer, *Genomics* 1989, ref. 7. University of California, Santa Cruz. 5 June.
16. Sinsheimer, R. (1985). Letters to Donald Sharp Fredrickson, President, Howard Hughes Medical Institute. University of California, Santa Cruz. 3 April and 5 June.
17. Fredrickson, DS. (1985). Letter to Robert L. Sinsheimer, Chancellor, University of California, Santa Cruz. Howard Hughes Medical Institute. 3 July.
18. Beckman, AO. (1985). Letters to Robert L. Sinsheimer, Chancellor, University of California, Santa Cruz. Arnold and Mabel Beckman Foundation. 12 July and 21 August.
19. Sinsheimer, R, 1988, Letter to author, University of California, Santa Barbara, 7 September.
20. UC Santa Cruz Review. (1990). Sinsheimer Labs Dedicated at Last. *UC Santa Cruz Review* (Spring):3.

6. GILBERT AND THE HOLY GRAIL

1. Gilbert, W (1988). 60th Birthday Videotape. Taken at "James D. Watson: Celebrating 60 Years" symposium. Cold Spring Harbor Laboratory. 16 April.
2. Judson, HF. (1979). *The Eighth Day of Creation: The Makers of the Revolution in Biology.* Simon & Schuster, New York.
3. Gilbert, W, and Muller-Hill, B. (1966). Isolation of the *Lac* Repressor. *Proceedings of the National Academy of Sciences (USA)* **56**:1891–1898.
4. Pendrell, E. (1968). *The Scientist.* Man and His Universe (television documentary series), ABC News. 29 November.
5. Ptashne, M. (1986). *A Genetic Switch: Gene Control and Phage λ.* Blackwell, Palo Alto, CA.

6. Gilbert, W, 1988, Interview, Biological Laboratories, Harvard University, 11 November.
7. Gilbert, W, and Maxam, A. (1973). The Nucleotide Sequence of the *lac* operator. *Proceedings of the National Academy of Science (USA)* **70** (12):3581–3584.
8. Hall, SS. (1987). *Invisible Frontiers: The Race to Synthesize a Human Gene.* Atlantic Monthly Press; paperback Tempus Press, New York.
9. Edgar, RS. (1985). Letter to Walter Gilbert. University of California, Santa Cruz. 27 March.
10. Gilbert, W. (1985). Letter to Robert Edgar. River Associates, Inc. 27 May.
11. Gruskin, KD, and Smith, TF. (1987). Molecular

Genetics and Computer Analyses. *CABIOS* **3** (No. 3):167–170.

12. McAuliffe, K. (1987). Reading the Human Blueprint. *U.S. News & World Report* (December 28, 1987, & January 4, 1988):92–93.

13. Begley, S, with Katz, SE, and Drew, L. (1987). The Genome Initiative. *Newsweek* (August 31):58–60.

14. del Guercio, G. (1987). Designer Genes. *Boston Magazine* (August):79–87.

15. Beam, A, and Hamilton, JO. (1987). A Grand Plan to Map the Gene Code. *Business Week* (April 27):116–117.

16. Holzman, D. (1987). Mapping the Genes, Inside and Out. *Insight* (May 11):52–54.

17. Kanigel, R. (1987). The Genome Project. *New York Times Magazine* (December 13):44, 98–101, 106.

18. Palfreman, J. (1989). *The Book of Man* (Video documentary). Horizon, BBC; and NOVA, WGBH.

19. Gilbert, W. (1987). Genome Sequencing: Creating a New Biology for the Twenty-first Century. *Issues in Science and Technology* **3**:26–35.

20. Hood, L, and Smith, L. (1987). Genome Sequencing: How to Proceed. *Issues in Science and Technology* **3**:36–46.

21. Gilbert, W. (1986). Two Cheers for Human Genome Sequencing. *The Scientist* (October 20):11.

22. Bodmer, WF. (1986). Two Cheers for Genome Sequencing. *The Scientist* (October 20):11–12.

23. Roberts, L. (1990). A Meeting of the Minds on the Genome Project? *Science* **250** (9 November):756–757.

24. Roberts, L. (1990). Large-Scale Sequencing Trials Begin. *Science* **250** (7 December):1336–1338.

25. Gilbert, W. (1978). Why Genes in Pieces? *Nature* **271**:501.

26. Dorit, RL, Schoenbach, L, and Gilbert, W. (1990). How Big Is the Universe of Exons? *Science* **250** (7 December):1377–1382.

27. Gilbert, W. (1991). Towards a Paradigm Shift in Biology. *Nature* **349** (10 January):99.

7. GENES AND THE BOMB

1. Office of Energy Research. (1984). *Health and Environmental Research: Summary of Accomplishments* (DOE / ER-0194). Department of Energy. April.

2. Compton, AH. (1956). *Atomic Quest: A Personal Narrative.* Oxford University Press, New York.

3. Sekimori, G. (1986). *Hibakusha.* Kosei, Tokyo.

4. Hersey, J. (1985). *Hiroshima.* Vintage, New York.

5. Lindee, MS. (1990). *Mutation, Radiation and Species Survival: The Genetics Studies of the Atomic Bomb Casualty Commission in Hiroshima and Nagasaki, Japan* (UMI Order Number 9018152, Ann Arbor, Michigan). Ph.D. dissertation, Cornell University.

6. Roesch, WC. (1987). *US-Japan Joint Reassessment of Atomic Bomb Radiation Dosimetry in Hiroshima and Nagasaki.* Hiroshima: Radiation Effects Research Foundation.

7. Muller, HJ. (1946). Muller, Biologist Wins Nobel Prize. *New York Times* cited by John Beatty, "Genetics in the Atomic Age: The Atomic Bomb Casualty Commission, 1947–1956," in *The Expansion of American Biology,* K. Benson et al. (Eds.), Rutgers University Press (1991) (1 November):284–324.

8. Sturtevant, A. (1947). Social Implications of the Genetics of Man. *Science* **120**:407.

9. Weart, SR. (1988). *Nuclear Fear. A History of Images.* Harvard University Press, Cambridge, MA.

10. Sankaranarayanan, K. (1988). Invited Review: Prevalence of Genetic and Partially Genetic Diseases in Man and the Estimation of Genetic Risks of Exposure to Ionizing Radiation. *American Journal of Human Genetics* **42**:651–662.

11. Neel, JV, and Schull, WJ. (1991). *The Children of Atomic Bomb Survivors: A Genetic Study.* National Academy Press, Washington, DC.

12. Beatty, J. (1991). Genetics in the Atomic Age: The Atomic Bomb Casualty Commission, 1947–1956. In *The Expansion of American Biology,* K Benson, Ed., Rutgers University Press.

13. Genetics Study Conference. (1984). *Report.* Hiroshima: Radiation Effects Research Foundation. 20 April.

14. Olson, M, 1988, Interview with Leslie Roberts and author at Genome Mapping and Sequencing Symposium, Cold Spring Harbor Laboratory, 28 April.

15. Cook-Deegan. (1989). The Alta Summit, December 1984. *Genomics* **5**:661–663.

16. Mendelsohn. (1985). *Informal Report of Meeting on DNA Methods for Measuring the Human Heritable Mutation Rate* (UCID-20315). Lawrence Livermore Laboratory. January.

17. Mendelsohn, ML. (1986). Prospects for DNA Methods to Measure Human Heritable Mutation Rates. In *Genetic Toxicology of Environmental Chemicals, Part B: Genetic Effects and Applied Mutagenesis,* Alan R. Liss, Ed., pp. 337–344.

18. Delehanty, J, White, RL, and Mendelsohn, ML. (1986). Approaches to Determining Mutation Rates in Human DNA. *Mutation Research* **167**: 215–232.

19. US Congress. (1986). *Technologies for Detecting Heritable Mutations in Human Beings* (OTA-H-298, Washington, DC, Government Printing Office). Office of Technology Assessment. September.

20. DeLisi, C, 1986, Interview, November.

21. DeLisi, C. 1987, Interview, January.

22. DeLisi, C. (1987–1988), various meetings and phone conversations.

23. DeLisi, C. (1988). The Human Genome Project. *American Scientist* **76**:488–493.

24. Bitensky, M, Burks, C, Goad, W, Ficket, J, Hildebrand, E, Moyzis, B, Deaven, L, Cramm, S, and Bell, G. (1986). Memo to Charles DeLisi, DOE (LS-DO-85-1.14–132). Los Alamos National Laboratory. 23 December.

25. Laurence, W. (1959). *Men and Atoms: The Discovery, the Uses, and the Future of Atomic Energy.* Simon & Schuster, New York.

26. Rhodes, R. (1986). *The Making of the Atomic Bomb.* Basic Books, New York.

27. Blow, M. (1968). *The History of the Atomic Bomb.* American Heritage, New York.

28. DeLisi, C. (1985). Note to David Smith. Office of Health and Environmental Research, US Department of Energy. 24 December.

29. Smith, DA. (1985). Note to Charles DeLisi. Office of Health and Environmental Research, US Department of Energy. 30 December.

30. DeLisi, C. (1985). Note to David Smith. Office of Health and Environmental Research, U.S. Department of Energy. Undated, but likely 30 or 31 December.

31. Bitensky, MW. (1986). Memo to Los Alamos Planning Group (LS-DO-86-1.8-3). Los Alamos National Laboratory. 7 January.

32. DeLisi, C. (1986). Letter to Mark W. Bitensky. Department of Energy. 24 February.

33. Bitensky, MW. (1986). Letter to Charles DeLisi

(LS-DO-86-1.2). Los Alamos National Laboratory. 2 April.
34. Bitensky, M. (1986). *Sequencing the Human Genome.* Santa Fe, NM: Office of Health and Environmental Research, US Department of Energy (published by the University of California under contract W 7405-ENG-36, Los Alamos National Laboratory, NM), Santa Fe Workshop, March 3–4.
35. Collected papers and letters, 1986, from March 1986 Santa Fe Workshop, DOE Headquarters, Germantown, MD.
36. Smith, HO. (1986). Letter to Mark W. Bitensky. Johns Hopkins University. 6 March.
37. Carrano, AV, and Branscomb, EW. (1986). Letter to Mark Bitensky. Lawrence Livermore National Laboratory. 13 March.
38. Comings, DE. (1986). Letter to Mark Bitensky. City of Hope National Medical Center. 11 March.
39. DeLisi, C, 1991, Letter to author, 20 February.
40. Botstein, D. (1987). *Implications for Basic Research* (Advances in Genomic Analysis lecture series). National Institutes of Health. 3 November.
41. Palfreman, J. (1989). *The Book of Man* (Video documentary). Horizon, BBC; and NOVA, WGBH.
42. DeLisi, C. (1986). Memo to Alvin Trivelpiece. Office of Health and Environmental Research, Department of Energy. 6 May.
43. DeLisi, C. (1986). Second memo to Alvin Trivelpiece. Office of Health and Environmental Research, Department of Energy. 6 May.
44. Bostock, J, 1988, Interview, Office of Management and Budget, New Executive Office Building, 8th floor, 23 September.
45. Bostock, J, 1989, Interview, Office of Management and Budget, New Executive Office Building, 8th floor, 4 April.
46. OMB Briefing, 1986, The budget briefing documents for the OHER-OMB meetings included budget projections for fiscal years 1987–90 of $5.64, $11.55, $18, and $22 million. The cover sheet for the DOE document to OMB specified a four-year project starting October 1, 1987, extending to September 30, 1991, and costing $95 million. By simple arithmetic, the fiscal year 1991 budget must have been projected at $40 to $45 million. Decisions about a Phase II budget were to be made in 1990 and 1991.
47. Subcommittee on the Human Genome. (1987). *Report on the Human Genome Initiative.* Health and Environmental Advisory Committee, for the Office of Health and Environmental Research, Department of Energy. April.

48. DOE initial cost estimates. (1986). The DOE-OMB agreement is dated 18 December 1986. DeLisi had briefed OMB in July and got a preliminary go-ahead signal on 5 September. DeLisi thus cut his deal some time before HERAC reported. Indeed, he reported to the HERAC subcommittee at its meeting on 6 and 7 November (in Denver) that he had reprogrammed several million dollars in fiscal year 1987 and expected a $20 million annual budget two years hence. While they did not influence the initial DOE plans, the HERAC subcommittee's budget figures were to figure in the larger budget debate that developed later.
49. Hall, SS. (1988). Genesis: The Sequel. *California* (July 1988):62–69.
50. Office of Health and Environmental Research. (1986). Background documents for briefing with the Office of Management and Budget. US Department of Energy. 18 December.
51. Cahill, GF. (1986). Memo to Donald Fredrickson. Howard Hughes Medical Institute. 7 November.
52. Dupree, AH. (1985). *Science and the Federal Government.* Johns Hopkins University Press, Baltimore, MD.
53. Guston, DH. (1990). *Congress and the History of Science: The Allison Commission, 1884–1886.* American Political Science Association Annual Meeting, San Francisco. 31 August.
54. Erdheim, E, 1987, conversation in staff offices, Rayburn House Office Building, Subcommittee on Natural Resources, Agriculture Research, and Environment of the Committee on Science, Space and Technology, US House of Representatives, March.
55. US House of Representatives. (1987). *Fiscal Year 1988 DOE Budget Authorization: Environmental Research and Development* (No. 58). March 19, 1987.
56. Watson, JD, 1987, Interview, Cold Spring Harbor Laboratory, director's office, July.
57. McConnell, J, 1988, Interviews, March and October.
58. McConnell, J, 1989, Interview, July.
59. McConnell, J, 1990, Interview, his home, Hilton Head Island, SC, September–October.
60. McConnell, J, 1990, Conversation at Workshop on International Cooperation on the Human Genome Project, II, Ethics, Valencia, Spain, 12–14 November.
61. Charles, D. (1988). Labs Struggle to Promote Spinoffs. *Science* **240** (13 May):874–876.
62. DeLisi, C, 1990, Letter to author, 9 March.
63. DeLisi, C, 1990, Letter to author, 11 October.

8. EARLY SKIRMISHES

1. Dulbecco, R. (1986). A Turning Point in Cancer Research: Sequencing the Human Genome. *Science* **231**:1055–1056.
2. Mannarino, E, 1991, Letter to author, Italian Embassy, Science Attaché, Washington, DC, 7 March.
3. Dulbecco, R, 1990, Interview, Salk Institute, 16 December 1990.
4. Dulbecco, R. (1987). A Turning Point in Cancer Research: Sequencing the Human Genome. In *Viruses and Human Cancer,* Alan R. Liss, Ed., pp. 1–14.
5. Cooper, SG, 1991, Note to author, CSHL Communications Office director, 22 January.
6. Dulbecco, R, 1987, Interview, Salk Institute, January.
7. Watson, JD. (1986). Foreword. *Cold Spring Harbor Symposia on Quantitative Biology, the Molecular Biology of Homo sapiens* **51**:xv–xvi.

8. Bodmer, WF. (1986). Human Genetics: The Molecular Challenge. *Cold Spring Harbor Symposia on Quantitative Biology, the Molecular Biology of Homo sapiens* **51**:1–13.
9. McKusick, VA. (1986). The Gene Map of *Homo sapiens:* Status and Prospectus. *Cold Spring Harbor Symposia on Quantitative Biology: the Molecular Biology of Homo sapiens* **51**:15–27.
10. Lewin, R. (1986). DNA Sequencing Goes Automatic. *Science* **233** (4 July):24.
11. Berg, P, 1990, Note to author, commenting on a meeting of the Committee to Study Decisionmaking, Institute of Medicine, Beckman Center, Irvine California, March.
12. Gilbert, W, and Watson, JD, 1989, Conversations at Engelhardt Institute of Molecular Biology, Moscow, June.
13. Gilbert, W, 1986, Comments taped by C. Thomas Caskey at session on the Human Genome project,

symposium on the Molecular Biology of *Homo sapiens*, Cold Spring Harbor Laboratory, 3 June.

14. Botstein, D, and Berg, P, 1986, Comments taped by C. Thomas Caskey at session on Human Genome Project, at the symposium on the Molecular Biology of *Homo sapiens*, Cold Spring Harbor Laboratory, 3 June.

15. Walsh, JB, and Marks, J. (1986). Sequencing the Human Genome. *Nature* 322 (14 August):590.

16. Joyce, C. (1987). The Race to Map the Human Genome. *New Scientist* (5 March):35–39.

17. Weinberg, RA. (1987). The Case Against Gene Sequencing. *The Scientist* (16 November):11.

18. McAuliffe, K. (1987). Reading the Human Blueprint. *US News and World Report* (December 28, 1987 & January 4, 1988):92–93.

19. Gall, JG. (1986). Human Genome Sequencing. *Science* 233 (26 September):1367–1368.

20. Schmidtke, J, Krawczak, M, and Cooper, DN.

(1986). Human Gene Cloning: The Storm Before the Lull? *Nature* 322 (10 July):119.

21. Maddox, J. (1992). Ever-Longer Sequences in Prospect. *Nature* 357 (7 May):13.

22. Roberts, L. (1990). Large-Scale Sequencing Trials Begin. *Science* 250 (7 December):1336–1338.

23. Roberts, L. (1990). A Meeting of the Minds on the Genome Project? *Science* 250 (9 November):756–757.

24. Sinsheimer, R, 1990, Interview, Santa Barbara, Hope Ranch California, 25 July.

25. DeLisi, C, 1990, Conversation before a Senate Hearing, Dirksen Senate Office Building, November.

26. Lewin, R. (1986). Molecular Biology of *Homo sapiens*. *Science* 233 (11 July):157–160.

27. Lewin, R. (1986). Proposal to Sequence the Human Genome Stirs Debate. *Science* 232 (27 June):1598–1600.

9. THE ODD LEGACY OF HOWARD HUGHES

1. Thorn, GW. (1978). Howard Hughes Medical Institute. *New England Journal of Medicine* 299 (23):1278–1280.

2. Howard Hughes Medical Institute. (1987). *Annual Report*. Howard Hughes Medical Institute.

3. Howard Hughes Medical Institute. (1983). *Thirtieth Anniversary Report*. Howard Hughes Medical Institute.

4. Thompson, L. (1987). Buying the Best in Science. Washington, DC: Health Section 12–15, *Washington Post*, 31 December.

5. Schrage, M, and Henderson, N. (1986). Hughes Institute Woos Science's Best and Brightest. Washington, DC: Washington Business, pp. 1 and 27, *Washington Post*, 4 August.

6. Cahill, G, 1991, Letter to the author, Dartmouth College, 1 March.

7. Desruisseaux, P. (1987). Hughes Will Spend Extra $500 Million on Grants to Science over Next 10 Years. *Chronicle of Higher Education*. March.

8. Culliton, BJ. (1989). GM, Hughes Settle Stock Fight. *Science* 245 (10 March):1283.

9. Howard Hughes Medical Institute. (1989). *Annual Report*. Howard Hughes Medical Institute.

10. Powledge, TM. (1987). Donald Fredrickson: Spending Hughes' Legacy. *The Scientist* (26 January):16–17.

11. Culliton, BJ. (1987). Hughes Settles with IRS. *Science* 235:1318.

12. Culliton, BJ. (1987). Fredrickson's Bitter End at Hughes. *Science* 236 (12 June):1417–1418.

13. Boffey, PM. (1987). Medical Institute Shaken in Furor over Its President. New York City: Science Times, C1 and C3, *New York Times*, June 23.

14. Botstein, D, 1990, Interview, Genentech Headquarters, South San Francisco, October.

15. Cahill, G, 1988, Interview, Howard Hughes Medical Institute, Bethesda, MD, 26 July.

16. Scriver, C, 1988, Interview, Hyatt Regency Hotel, Bethesda, MD, 19 September.

17. Lewin, R. (1986). Shifting Sentiments over Sequencing the Human Genome. *Science* 233 (8 August):620–621.

18. Guston, DH. (1986). Notes from Informational Forum on the Human Genome, National Institutes of Health, Bethesda, MD, Howard Hughes Medical Institute, 23 July.

19. Pines, M, 1988, Letter to the author, Howard Hughes Medical Institute, Bethesda, MD, 22 September.

20. Pines, M. (1986). *Shall We Grasp the Opportunity to Map and Sequence All Human Genes and Create a "Human Gene Dictionary."* Bethesda, MD: Howard Hughes Medical Institute, Prepared for a meeting of the Trustees of the Howard Hughes Medical Institute.

21. Cook-Deegan, R, 1986, Notes from Informational Forum on the Human Genome, National Institutes of Health, Bethesda, MD, sponsored by Howard Hughes Medical Institute, 23 July.

22. Pines, M, 1988, Interview, Howard Hughes Medical Institute, Bethesda, MD, August.

23. Metheny, B, 1988, Interview, Monocle Restaurant, Capitol Hill, Washington, DC, Delegation for Basic Biomedical Research, 17 May.

24. Metheny, B, 1988, Note to the author, Tricom International and the Delegation for Basic Biomedical Research, 31 May.

25. Pines, M. (1987). *Mapping the Human Genome* (Occasional Paper Number One). Howard Hughes Medical Institute, December.

10. THE NAS REDEFINES THE PROJECT

1. Dupree, AH. (1985). *Science and the Federal Government*. Johns Hopkins University Press, Baltimore, MD.

2. National Academy of Sciences. (1990). *Questions and Answers About the National Academy of Sciences, National Academy of Engineering, Institute of Medicine and the National Research Council*. National Academy of Sciences. April.

3. National Academy of Sciences. (1990). *Staff Orientation Booklet*. National Academy of Sciences.

4. Burris, J, and Walton, F. (1986). *Mapping / Sequencing the Human Genome*. National Research

Council, Commission on Life Sciences, Board on Basic Biology. 3 July.

5. Burris, J. (1986). Draft Minutes of 3 March meeting, Board Room, NAS Building, Washington, DC. Board on Basic Biology. 24 March.

6. Burris, JE. (1986). Memo regarding 5 August Woods Hole meeting. Board on Basic Biology, Commission on Life Sciences, National Research Council. 10 July.

7. Schmeck, HM. (1986). Rapid Advances Point to the Mapping of All Human Genes. The Science: New Research in Decoding DNA Is Leading to Drugs and

Treatments. New York City: Science Times, C1ff. *New York Times,* 15 July.
8. Boffey, PM. (1986). Rapid Advances Point to the Mapping of All Human Genes. The Debate: Would Accelerated Project Distort Usual Methods of Biology? New York City: Science Times, C1ff. *New York Times,* 15 July.
9. Bitensky, MW. (1986). Letter to Charles DeLisi (LS-DO-86-1.2). Los Alamos National Laboratory. 2 April.
10. Watson, JD, Cantor C, Gilbert, W, Hood, L, White, R, Ruddle, F, Kingsbury, D, Press, F, and members and staff of the Board on Basic Biology, 1986, Participants in Woods Hole meeting, National Academy of Sciences, 5 August.
11. Burris, J. (1986). Minutes of Board on Basic Biology Meeting, 5 and 6 August, 1986. National Research Council, Commission on Life Sciences, Board on Basic Biology.
12. Burris, JE. (1986). Proposal: Mapping and Sequencing the Human Genome. National Research Council, Commission on Life Sciences, Board on Basic Biology. 24 September.
13. Burris, J, 1991, Electronic mail note to the author, National Academy of Sciences, Washington, DC. The check arrived from the McDonnell Foundation two days after the project was approved by the NAS Governing Board, 28 February.
14. Alberts, BM. (1985). Limits to Growth: In Biology, Small Science Is Good Science. *Cell* **41:**337–338.
15. Olson, M, Davis, R, Davies, K, Patterson, D, and Weissenbach, J, 1987, invited speakers to the Committee on Mapping and Sequencing the Human Genome, Board on Basic Biology, National Academy of Sciences, 19 January.
16. Church, G, Efstradiatis, A, and Chen, E, 1987, invited speakers to the Committee on Mapping and Sequencing the Human Genome, Board on Basic Biology, National Academy of Sciences, 19 January.
17. Alberts, B, and Burris, J, 1991, Joint interview, National Academy of Sciences, 28 February.
18. Alberts, B, 1990, Interview, University of California at San Francisco, 12 October.

19. National Research Council. (1988). *Mapping and Sequencing the Human Genome.* National Academy Press, Washington, DC.
20. American Society for Biochemistry and Molecular Biology. (1987). Council Policy Statement on Mapping and Sequencing the Human Genome. 7 June.
21. Alberts, B, 1988, Interview, University of California at San Francisco, 18 August.
22. Report reviewer, 1988, Interview with individual who reviewed the NRC committee's report, whose name is withheld to preserve anonymity, Office of Technology Assessment, January.
23. NRC Committee members, 1988, Phone and personal interviews, Office of Technology Assessment, January–March.
24. Roberts, L. (1988). Academy Backs Genome Project. *Science* **239** (12 February):725–726.
25. The Economist. (1988). The Mapping of Man. *Economist* (20 February):91–92.
26. Cherfas, J. (1988). A Guide to Being Human. *New Scientist* (25 February):30–31.
27. The Scientist. (1988). NAS Report: Full Speed Ahead on Human Gene Sequencing (excerpts from NRC report). *The Scientist* (7 March):12.
28. Zurer, PS. (1988). Molecular Biologists Backing Effort to Map Entire Human Genome. *Chemical and Engineering News* (14 March):22–26.
29. Merz, B. (1988). National Research Council Endorses Human Gene Mapping Project. *Journal of the American Medical Association* **259** (11 March):1433–1434.
30. Hammond, A. (1988). $3 Billion Gene Map. *Science Impact Letter* **1** (March):1 and 8.
31. Fox, JL. (1988). $200 Million a Year for Human Genome. *Bio / Technology* **6** (April):370.
32. Weiss, R. (1988). Report Adds to Gene Map Momentum. *Science News* **133** (20 February):117–118.
33. US House of Representatives (1988). Departments of Labor, Health and Human Services, Education, and Related Agencies Appropriations for 1989. Subcommittee on Labor, Health and Human Services, Education and Related Agencies of the Committee on Appropriations. Part 4A. 3 March.

11. THE NIH STEPS FORWARD

1. US Senate. (1982). Nominations. Committee on Labor and Human Resources. 21 April.
2. Wyngaarden, J, 1988, Interview, NIH Director's Office, Building 1, National Institutes of Health, 19 September.
3. Dupree, AH. (1985). *Science and the Federal Government.* Johns Hopkins University Press, Baltimore, MD.
4. Strickland, SP. (1989). *The Story of the NIH Grants Program.* University Press of America, Lanham, MD.
5. Shannon, JA. (1987). The National Institutes of Health: Some Critical Years, 1955–1957. *Science* **237:**865–868.
6. Culliton, BJ. (1989). NIH: The Good Old Days. *Science* **244** (23 June):1437.
7. Strickland, SP. (1972). *Politics, Science, and Dread Disease: A Short History of United States Medical Research Policy.* Harvard University Press, Cambridge, MA. See esp. Chapter 3: The Rise of the Research Lobby, pp. 32–54.
8. Patterson, JT. (1987). *The Dread Disease: Cancer and Modern American Culture.* Harvard University, Cambridge, MA.
9. Rettig, RA. (1977). *Cancer Crusade: The Story of the National Cancer Act of 1971.* Princeton University Press, Princeton, NJ.

10. Fredrickson, DS. (1987). Challenge of Change in Biology. *Nature* **329** (22 October):686–687.
11. Healy, BP. (1988). Tour of Duty: A Medical Academic in the White House. *Harvard Medical Alumni Bulletin* **61** (Spring):49–53.
12. National Institutes of Health. (1989). *NIH Data Book 1989* (NIH Publication Number 90-1261). National Institutes of Health. December.
13. Marston, RQ. (1987). Dilemmas of Decision-Making. *Nature* **329** (22 October):683–685.
14. Wyngaarden, J, 1991, Interview, Office of International Affairs, National Academy of Sciences, July.
15. Kirschstein, RL. (1986). Memo to NIH Director re sequencing the human genome. National Institutes of Health. 2 July.
16. Wyngaarden, JB. (1986). Letter to Alvin Trivelpiece. National Institutes of Health. 10 July 1990.
17. Kirschstein, RL. (1986). Letter to Charles DeLisi. National Institute of General Medical Sciences. 9 July.
18. Baltimore, D. (1987). Genome Sequencing: A Small-Scale Approach. *Issues in Science and Technology* **3:**48–50.
19. Cook-Deegan, R, 1986, Personal notes from meeting, National Institutes of Health, 16 and 17 October.
20. Office of Program Planning and Evaluation. (1987).

The Human Genome. 54th Meeting of the Advisory Committee to the Director, National Institutes of Health.

21. Palca, J. (1986). More Actors Apply for Parts. *Nature* **323** (23 October):660.

22. NIH Working Group on the Human Genome. (1986). *6 November Meeting Summary.* National Institutes of Health.

23. NIH Working Group on the Human Genome. (1986). *16 December Meeting Summary.* National Institutes of Health.

24. National Institutes of Health. (1987). New Approaches to the Analysis of Complex Genomes. *NIH Guide to Grants and Contracts* **16** (#18, 29 May):11–14.

25. National Institutes of Health. (1987). Computer-Based Representation and Analysis of Molecular Biology Data. *NIH Guide for Grants and Contracts* **16** (#18, 29 May):9–11.

26. Wyngaarden, JB. (1987). The National Institutes of Health in Its Centennial Year. *Science* **237** (21 August):869–874.

27. US House of Representatives. (1987). *Departments of Labor, Health and Human Services, Education, and Related Agencies Appropriations for 1988* (Part 4A). National Institutes of Health.

28. Stephens, M, 1989, Meeting in committee reception area, US Capitol, Committee on Appropriations, US House of Representatives, May.

29. Medical Research Council. (1991). *Annual Report, April 1990–March 1991.* Medical Research Council, UK.

30. Metheny, B, 1988, Interview, Monocle Restaurant, Capitol Hill, Washington, DC, Delegation for Basic Biomedical Research, 17 May.

31. Metheny, B, 1988, Note to the author, Tricom International and the Delegation for Basic Biomedical Research, 31 May.

32. Watson, JD. (1990). The Human Genome Project: Past, Present, and Future. *Science* **248**:44–49.

33. Byrnes, M, 1990, Phone conversation, National Commission on AIDS, referring to period when Dr. Byrnes was staff to Senator Lowell Weicker, Committee on Appropriations, US Senate, June.

34. Peterson, J, 1989, Notes from presentation at Genes and Machines, II (New Hampshire conference on DNA and computers), National Center for Human Genome Research, August.

35. Eckstrand, I, and Greenberg, J, 1989, Meeting at Westwood Building, Bethesda, MD, National Institute of General Medical Sciences, September.

36. Gray, RM. (undated). Note on conversation with Lesley Russell. National Institutes of Health.

37. Russell, L, 1987, Phone conversations and meetings in B-33, Rayburn House Office Building, Committee on Energy and Commerce, US House of Representatives, 6 March and other dates.

38. National Research Council. (1988). *Mapping and Sequencing the Human Genome.* National Academy Press, Washington, DC.

39. US House of Representatives. (1988). *Departments of Labor, Health and Human Services, Education, and Related Agencies Appropriations for 1989* (Part 4A). Subcommittee on Labor, Health and Human Services, Education and Related Agencies of the Committee on Appropriations. March 3, 1988.

40. National Institutes of Health. (1988). Gene Mapping / Sequencing: Questions and answers prepared by NIH staff for House Appropriations hearings. Undated, but likely March or April.

41. Barry, J, 1990, Phone and in-person conversations, free-lance writer working on profile of James D. Watson, May.

42. Wyngaarden, JB. (1988). Trials and Tribulations of Being Director of the NIH. *The Scientist* (11 July):13–14.

43. Institute of Medicine. (1984). *Responding to Health Needs and Scientific Opportunity: The Organizational Structure of the National Institutes of Health* (Washington, DC: National Academy Press). Institute of Medicine, National Academy of Sciences. October.

12. TRIBES ON THE HILL

1. Squires, S. (1986). Putting Genes on a Map: World's Scientists Begin Planning a Detailed Description of Human DNA. Washington, DC: *Washington Post,* 30 July.

2. Brown, GE. (1986). Working Together, Benefiting All. *Congressional Record* (31 July):E 2668–2669.

3. Weatherford, MM. (1981). *Tribes on the Hill.* Rawson, Wade, New York.

4. Thompson, L. (1987). In Gene Mapping, an Opening Gambit. Washington, DC: Health Section, p. 9. *Washington Post,* 21 July.

5. Palca, J. (1987). Human Genome Sequencing Plan Wins Unanimous Approval in US. *Nature* **326** (2 April):429.

6. Roberts, L. (1987). Agencies Vie over Human Genome Project. *Science* **237** (31 July):486–488.

7. US House of Representatives. (1988). *Departments of Labor, Health and Human Services, Education, and Related Agencies Appropriations for 1989* (Part 4A). Subcommittee on Labor, Health and Human Services, Education and Related Agencies of the Committee on Appropriations. March 3, 1988.

8. Health and Environmental Research Advisory Committee. (1987). *Report on the Human Genome Initiative.* Office of Energy Research, US Department of Energy.

9. Cook-Deegan, RM. (1986). Memo on possible OTA project. Office of Technology Assessment. 14 July.

10. Office of Technology Assessment. (1986). Proposal for an OTA Assessment of Mapping the Human Genome. Office of Technology Assessment, US Congress. 24 September.

11. US Congress. (1988). *Mapping Our Genes—Genome Projects: How Big? How Fast?* Office of Technology Assessment, OTA-BA-373, Washington, DC: Government Printing Office; reprinted by Johns Hopkins University Press. April.

12. Domenici, P. (1987). S. 1480. (10 July):S 9706–9720.

13. Chiles, L, and Snell, R, 1988, Joint interview, Room 250, Russell Senate Office Building, 6 August.

14. Chiles, L, Kennedy, E, and Domenici, P. (1988). S. 1966, Biotechnology Competitiveness Act of 1988. (17 June):S 8061–8074, S 8097–8101, D781.

15. Keith, S, and Sarfaty, M, 1988–1990, Personal and phone interviews, Subcommittee on Health, Committee on Labor and Human Resources, US Senate.

16. Bush, A, Greenlaw, D, and Gilman, P, 1987–1990, Phone and personal interviews, Office of Pete Domenici, US Senate.

17. McConnell, J, 1988, Interviews, March and October.

18. McConnell, J, 1989, Interview, July.

19. McConnell, J, 1990, Interview, 12–14 November.

20. Snell, R, 1988–1990, Personal interviews and phone conversations, Office of Senator Lawton Chiles (1988); Office of Technology Assessment, US Congress (1989–1990).

21. Hall, M, 1988 and 1989, Phone interviews, Subcommittee on Labor, Health and Human Services, Education and Related Agencies, Committee on Appropriations, US Senate, October 1988 and July 1989.

22. Byrnes, M, 1990, Phone conversation, National Commission on AIDS, referring to period when Dr. Byrnes was staff to Senator Lowell Weicker, Committee on Appropriations, US Senate, June.

23. DeLisi, C. (1990). DeLisi left DOE employ 1 September 1987, two years after joining it, to chair a department of mathematical biology at Mt. Sinai Medical Center in New York City. He left that post in January 1990 to become dean at the College of Engineering, Boston University.

24. US Senate. (1987). *Workshop on Human Gene Mapping* (Committee Report No. 100-71). Committee on Energy and Natural Resources. August 31.

25. US Senate. (1987). *Department of Energy National Laboratory Cooperative Research Initiatives Act* (S. Hrg. 100-602, Pt. 1). Subcommittee on Energy Research and Development of the Committee on Energy and Natural Resources. September 17.

26. FDC Reports. (1988). Human Genome Project: Interagency Task Force. *The Pink Sheet* (16 May):7.

27. Palca, J. (1988). Another Report Smiles on Human Genome Sequencing Project. *Nature* 332 (28 April):769.

28. Roberts, L. (1988). Genome Projects Ready to Go. *Science* 240 (29 April):602–604.

29. Thompson, L. (1988). Mapping the Human Genes: Is the Megaproject Politically a "Go"? Washington, DC: Health Section, p. 8. *Washington Post,* 16 February.

30. Industrial Biotechnology Association. (1987). Memo, Results of the IBA Membership Survey on Mapping the Human Genome. Government Relations Committee, IBA. 7 September.

31. US House of Representatives. (1988). Coordination of Genome Projects, in Committee report on H.R. 4502 and S. 1966, the Biotechnology Competitiveness Act (Committee Print No. 138). Subcommittee on Natural Resources, Agriculture Research, and Environment and the Subcommittee on Science, Research, and Technology of the House Committee on Science, Space and Technology. 14 July.

32. FDC Reports. (1988). Biotechnology Act Draws Administration Opposition in Congressional Hearing; Industry Endorses. *The Blue Sheet* (20 July):6.

33. FDC Reports. (1988). Biotech Competitiveness Act Draws Administration Opposition in Subcommittee Hearing on National Policy Commission; Industry Groups Endorse Senate Bill. *The Pink Sheet* (18 July):12.

34. Eckstrand, IA. (1987). Memo to Kirschstein on Meeting of the OHER subcommittee on the Human Genome, 5–6 February. National Institute of General Medical Sciences, National Institutes of Health. 10 February.

35. Ackerman, S. (1988). Taking on the Human Genome. *American Scientist* 76 (January–February):17–18.

36. Crawford, M. (1987). Wyngaarden to Chair Biotech Council. *Science* 238 (11 December):1504–1505.

37. Crawford, M. (1987). Document Links NSF Official to Biotech Firm. *Science* 238:742.

38. Palca, J. (1989). Genome Projects Are Growing Like Weeds. *Science* 245 (14 July):131.

39. Domestic Policy Council. (1987). Subcommittee on Characterization of Complex Genomes, list of questions to be asked of agencies. White House. 5 May.

40. Moscowitz, J. (1987). Note to Dr. Wyngaarden, attached memo. Also used as briefing material by NIH for congressional staff and at genome meetings. National Institutes of Health. 25 August.

41. Office of Technology Assessment. (1987). *Costs of Human Genome Projects* (Mapping Our Genes, Transcript of Workshop "Costs of Human Genome Projects," National Technical Information Service Order No. PB 88-162 813/AS). US Congress. 7 August.

42. Unsigned. (1987). Draft letter to go from the Secretary of Health and Human Services to J. Bennett Johnston. National Institutes of Health. Undated, but after 17 September.

43. Decker, JF. (1987). Letter to James Wyngaarden. Office of Energy Research, Department of Energy. 30 December.

44. Myers, N. (1988). Gene Study Excites Researchers. *Washington Technology* (22 September):29–30.

13. HONEST JIM AND THE GENOME

1. Watson, JD. (1989). *Director's Report* (Annual Report 1988). Cold Spring Harbor Laboratory. 2 August.

2. Davis, B. (1990). Watson Doesn't Use Gentle Persuasion to Enlist Japanese and German Support for Genome Effort. New York: A12. *Wall Street Journal,* 18 June.

3. Roberts, L. (1990). Cold Spring Harbor Turns 100. *Science* 250 (26 October):496–498.

4. Watson, JD. (1988). 60th Birthday Videotape. Taken at "James D. Watson: Celebrating 60 Years" symposium. Cold Spring Harbor Laboratory. 16 April.

5. Bradbury, W. (1970). Genius on the Prowl. 57–66. *Life,* 30 October.

6. Hall, SS. (1990). James Watson and the Search for Biology's "Holy Grail." *Smithsonian* 20 (February):40–49.

7. Barry, JM. (1991). Cracking the Code. *Washingtonian* (February):63–67, 180–184.

8. Watson, JD. (1990). The Human Genome Project: Past, Present, and Future. *Science* 248:44–49.

9. Wyngaarden, J. (1991). The Function of HUGO. American Association for the Advancement of Science annual meeting, Washington, DC. 19 February.

10. Watson, JD. (1981). *James D. Watson—Biologist.*

Excellence, the Pursuit, LTV Seminar, Washington, DC. Found in Cold Spring Harbor Laboratory archives.

11. Cairns, J, Stent, G, and Watson, JD, Ed. (1966). *Phage and the Origins of Molecular Biology.* Cold Spring Harbor Laboratory, Cold Spring Harbor, NY.

12. Schrödinger, E. (1967). *What Is Life?* Cambridge University Press, Cambridge, UK.

13. Judson, HF. (1979). *The Eighth Day of Creation: The Makers of the Revolution in Biology.* Simon & Schuster, New York.

14. Watson, JD, Ed. (1980). *The Double Helix: A Personal Account of the Discovery of the Structure of DNA.* Norton Critical Edition. W.W. Norton, New York.

15. Crick, F. (1988). *What Mad Pursuit: A Personal View of Scientific Discovery.* Basic Books. Same quote also found in "The Double Helix: A Personal View," *Nature* 248 (April 26):766–771, 1974; reprinted in Watson, *The Double Helix.*

16. Watson, JD, 1989, Interview, Engelhardt Institute for Molecular Biology, Moscow, June.

17. Schlessinger, D. (1988). *The Start at Harvard: Those Golden Years* (James D. Watson: Celebrating 60 Years). Cold Spring Harbor Laboratory. 16 April.

18. Crick, F. (1974). The Double Helix: A Personal

View. *Nature* **248** (26 April):766–769.

19. Watson, JD. (1988). *Reflections on My Forty Years in Science.* Harvard University: Department of Biochemistry and Molecular Biology, 13 May.

20. Knox, R. (1973). Nobel Winner Calls Nixon's Medical Research Policy "Lunacy." *Boston Globe* (7 March):67.

21. Gilbert, W, Capecchi, M, Burgess, R, Hendrix, R, Schaffner, W, Wigler, M, and Steitz, J. (1988). 60th Birthday Videotape. Taken at "James D. Watson: Celebrating 60 Years" symposium. Cold Spring Harbor Laboratory. 16 April.

22. *Life* Magazine. (1962). The Take-Over Generation: A Red-Hot Hundred. (14 September):4 and foldout.

23. *Life* Magazine. (1990). The 100 Most Important Americans of the 20th Century: James D. Watson. (Special Issue, Fall):83.

24. Watson, JD. (1965). *The Molecular Biology of the Gene,* 1st Edition. W.A. Benjamin, New York.

25. Watson, JD, Hopkins, NH, and Roberts, JW. (1987). *Molecular Biology of the Gene.* Benjamin, Menlo Park.

26. Levinson, R, 1991, Interview, Office of Science and Technology Policy, Old Executive Office Building, 9 July.

27. Cantor, C, 1991, Phone interview, 2 August.

28. Cantor, C, 1991, Interview, Lawrence Berkeley Laboratory, August.

29. Wyngaarden, J, 1991, Interview, Office of International Affairs, National Academy of Sciences, July.

30. Wyngaarden, J, 1991, Interview and follow-up conversation, Office of International Affairs, National Academy of Sciences, July and October.

31. Roberts, L. (1988). Watson May Head Genome Office. *Science* **240** (13 May):878–879.

32. Fox, JL. (1988). Watson to Head Genome Efforts. *Bio / Technology* **6** (November):1274.

33. Zurer, P. (1988). Watson to Head NIH's Human Genome Effort. *Chemical and Engineering News* (3 October):7.

34. Thompson, L. (1988). Gene Pioneer Will Head Mapping Project. James Watson, Who First Described the Double Helix, Takes on a New Challenge. Washington, DC: *Washington Post.* 27 September.

35. Schmeck, HM. (1988). DNA Pioneer to Tackle Biggest Gene Project Ever. New York: C1, C16. *New York Times.* 4 October.

36. US Senate. (1988). Departments of Labor, Health and Human Services, and Education and Related Agencies Appropriation Bill, 1989, Report (100–399, pp. 83–84). Senate Committee on Appropriations. June 23, 1988.

37. Olson, MV, Hood, L, Cantor, C, and Botstein, D. (1989). A Common Language for Physical Mapping of the Human Genome. *Science* **245**:1434–1435.

38. Roberts, L. (1989). New Game Plan for Genome Mapping. *Science* **245**:1438–1440.

39. Palca, J. (1989). Gene Mappers Meet on Strategy. *Science* **245**:1036.

40. Zinder, N, 1989–1990, Phone conversations, Rockefeller University,

41. Jordan, E, 1990, Interview, National Center for Human Genome Research, National Institutes of Health, 8 March.

42. Fink, L. (1990). Human Genome Office Attains Center Status. *The NIH Record* **42** (9 January):1,4.

43. US House of Representatives. (1990). Departments of Labor, Health and Human Services, Education, and Related Agencies Appropriations for 1991 (Part 4B, pp. 887–960). Committee on Appropriations.

44. Committee on Policies for Allocating Health Sciences Research Funds. (1990). *Funding Health Sciences Research. A Strategy to Restore Balance.* Institute of Medicine, National Academy of Sciences, National Academy Press.

45. Jordan, E. (1990). Reply to Syvanen and Davison postings, BIONET electronic bulletin board. National Center for Human Genome Research, National Institutes of Health. April.

46. Hamilton, DP. (1990). House Prunes Genome Budget. *Science* **249** (10 August):622–623.

47. Roberts, L. (1990). Tough Times Ahead for the Genome Project. *Science* **248** (June 29): 1600–1601.

48. Greenberg, DS. (1991). Programs and Plans Discussed by New NIH Director. *Science and Government Report* (1 May):3–5.

49. Luria, SE. (1989). Human Genome Program (letter). *Science* **246** (17 November):873.

50. Kozak, L. (1990). Letter to Senator William Cohen. Jackson Laboratories, Bar Harbor, Maine. 9 January.

51. Rechsteiner, M. (1990). Letter to William Raub, Alan Bromley, Senator Albert Gore, Jr., and Senator Edward Kennedy. University of Utah. 26 February.

52. Roberts, L. (1990). Genome Backlash Going Full Force. *Science* **248**:804.

53. The Blue Sheet. (1990). Human Genome Project Characterized as 'Terrible Science Policy' in Letter-Writing Campaign to NIH. *F-D-C Reports, The Blue Sheet* (30 May):8–9.

54. Syvanen, M. (1990). Letter signed by Chuck Turnbough, University of Alabama; Michael Syvanen, University of California, Davis; Richard Calendar, University of California, Berkeley; Ryland Young, Texas A&M University; Cathy Squires, Columbia University; and Marlene Befort, New York Public Health. BIONET electronic bulletin board. March through May.

55. Martin, RG. (1990). We GNOMES Find the PROJECT an Atlas but No Treasure. *The New Biologist* **2** (May):385–387.

56. Davison, D. (1990). Bitnet communication. Stanford University. 1 May.

57. Angier, N. (1990). Vast, 15-Year Effort to Decipher Genes Stirs Opposition. *New York Times* (5 June):C1, C12.

58. Kolata, G. (1990). Beginning Scientists Face A Research Fund Drought. *New York Times* (5 June):C1.

59. Sarfaty, M, 1990, Phone interview, Subcommittee on Health, Committee on Labor and Human Resources, US Senate, August.

60. Nelson, M, 1990, Interview, Subcommittee on Science, Space, and Technology, Committee on Commerce, Science and Transportation, US Senate, August.

61. Levinson, R, 1990, Phone interview, Office of Science and Technology Policy, the White House, August.

62. Lokken, P, 1990, Conversation, National Center for Human Genome Research, National Institutes of Health, August.

63. Raines, L. (1987). *Results of IBA Membership Survey on Mapping the Human Genome.* Industrial Biotechnology Association. 1 September.

64. Gage, LP. (1988). *Human Genome Mapping and Sequencing: Why We Should Do It—Now!* Annual meeting, Industrial Biotechnology Association, Charles Hotel, Cambridge, MA. 12 May.

65. Poste, G, and Christoffersen, RE. (1987). Biotechnology and Human Health Care, Session on the Human Genome. Pharmaceutical Manufacturers' Association meeting, Mayflower Hotel, Washington, DC. 13 October.

66. Sundaram, SC, Waters, WS, Abouelnaga, A, and Gabuzda, PG. (1990). *UCLA Field Study Project on the Human Genome Project.* John E. Anderson Graduate School of Management, University of California, Los Angeles. 30 April.

67. Rivers, LW. (1989). *Prioritization of Federal R&D*

Megaprojects. Industrial Research Institute, Inc. 5 December.

68. Davis, BD, and Colleagues. (1990). The Human Genome and Other Initiatives. *Science* 249 (27 July):342–343.

69. Davis, B. (1990). Testimony before the Senate Subcommittee on Energy Research and Development. Committee on Energy and Natural Resources, US Senate. 11 July.

70. Galas, D. (1990). Statement before the Subcommittee on Energy Research and Development. Committee on Energy and Natural Resources, US Senate. 11 July.

71. Roberts, L. (1990). DOE to Map Expressed Genes. *Science* 250 (16 November 1990):913.

72. Mervis, J. (1990). On Capitol Hill: One Day in the Hard Life of the Genome Project. *The Scientist* 4 (#16, 20 August):1, 4, 14.

73. Davis, BD. (1990). Human Genome Project: Is "Big Science" Bad for Biology? Yes, It Bureaucratizes, Politicizes Research. *The Scientist* 4 (#22, 12 November):13, 15.

74. Hood, LE. (1990). Human Genome Project: Is "Big Science" Bad for Biology? No: And Anyway, the HGP Isn't Big Science. *The Scientist* 4 (#22, 12 November):13, 15.

75. US House of Representatives. (1990). Departments of Labor, Health and Human Services, and Education, and Related Agencies Appropriation Bill, 1991 (101–516). Committee on Appropriations. 10 October.

76. US House of Representatives. (1990). Departments of Labor, Health and Human Services, Education, and Related Agencies Appropriations for 1991 (Part 4A, pp. 83–84 and budget tables). Subcommittee on Labor, Health and Human Services, Education and Related Agencies of the Committee on Appropriations. March 20.

77. US Senate. (1990). Departments of Labor, Health and Human Services, and Education, and Related Agencies Appropriation Bill, 1991 (101–591). Committee on Appropriations. 12 July.

78. US Congress. (1990). Making Appropriations for the Departments of Labor, Health and Human Services, and Education, and Related Agencies, for the Fiscal Year Ending September 30, 1991, and for Other Purposes (101–908). Conference Committee. 20 October.

79. Marshall, E. (1990). Genome Center Grants Chosen. *Science* 249 (28 September):1497.

80. Holden, C. (1991). New Genome Centers. *Science* 251 (15 February):742–3.

81. Hedetniemi, JN. (1992). Letter and budget table sent to author in response to letter requesting information on NIGMS Genetics Branch and GenBank funding 1987–1993. National Institute of General Medical Sciences, National Institutes of Health. 8 July.

82. US Department of Health and Human Services, and US Department of Energy. (1990). *Understanding Our Genetic Inheritance: The First Five Years, FY 1991–1995. DOE / ER-0452P.* National Technical Information Service, Springfield, VA.

83. US House of Representatives. (1992). Departments of Labor, Health and Human Services, Education, and Related Agencies Appropriations for 1993 (Part 4, pp. 607–698). Committee on Appropriations. 25 March.

84. Budget tables for NIGMS genetics program, 1992, The year NIH first had a special genome budget, the Genetics Program Branch at NIGMS saw a budget increase from $188.9 million (1987) to $213.8 million (1988). In 1989, when the genome project left NIGMS, the Genetics Program Branch budget went down to $209.6 million, and it dropped again to $207.8 million in 1990, before rebounding to $218.3 million in 1991 and an estimated $229.5 million in

1992. During these same years, the NIH genome budgets were: $17.2 million (1988, within NIGMS); $28.2 million (1989); $59.5 million (1990); $87.4 million (1991); and $104.9 million (1992 estimate). The 1993 increase was even larger than shown, as the GenBank contract amount of roughly $4 million was transferred to the National Library of Medicine.

85. Budget tables for NIGMS genetics program, 1992. By this reasoning, $25 million, or just over 10 percent of the NIGMS genetics program, moved away with the genome center. Some unknown fraction of the science supported by the genome center would likely have been supported by NIGMS, but certainly not all. It seems likely that some would instead have gone for study of gene regulation, recombination, and other basic genetics.,

86. Stegner, W. (1982). *Beyond the Hundredth Meridian.* University of Nebraska Press, Lincoln, NE.

87. Roberts, L. (1990). A Meeting of the Minds on the Genome Project? *Science* 250 (9 November):756–757.

88. Syvanen, M. (1991). Views of the Genome Project. *Science* 251 (22 February):855.

89. Brown, DD. (1991). Views of the Genome Project (letter). *Science* 251 (22 February):854–855.

90. Roberts, L. (1990). Whatever Happened to the Genetic Map? *Science* 247 (19 January):281–282.

91. Anderson, GC. (1990). Creation of Linkage Map Falters, Posing Delay for Genome Project. *The Scientist* 4 (8 January):1, 10, 12, 13.

92. Roberts, L. (1990). The Genome Map Is Back on Track After Delays. *Science* 248 (18 May):805.

93. US Congress. (1988). *Mapping Our Genes—Genome Projects: How Big? How Fast?* Office of Technology Assessment, OTA-BA-373, Washington, DC: Government Printing Office; reprinted by Johns Hopkins University Press. April.

94. National Research Council. (1988). *Mapping and Sequencing the Human Genome.* National Academy Press, Washington, DC.

95. Oliver, SG, van der Aart, QJM, Agostoni-Carbone, ML, Aigle, M, Alberghina, L, Alexandraki, D, Antoine, G, Anwar, R, Ballesta, JPG, Benit, P, Berben, G, Bergantino, E, Biteau, N, Bolle, PA, Bolotin-Fukuhara, M, Brown, A, Brown, AJP, Buhler, JM, Carcano, C, Carignani, G, Cederberg, H, Chanet, R, Contreras, R, Crouzet, M, Daignan-Fornier, B, Defoor, E, Delgado, M, Demolder, J, Doira, C, Dubois, E, Dujon, B, Dusterhoft, A, Erdmann, D, Esteban, M, Fabre, F, Fairhead, C, Faye, G, Feldmann, H, Fiers, W, Francingues-Gaillard, MC, Franco, L, Frontali, L, Fukuhara, H, Fuller, LJ, Galland, P, Gent, ME, Gigot, D, Gilliquet, V, Glansdorff, N, Goffeau, A, Grenson, M, Grisanti, P, Grivell, LA, de Haan, M, Haasemann, M, Hatat, D, Hoenicka, J, Hegemann, J, Herbert, CJ, Hilger, F, Hohmann, S, Hollenberg, CP, Huse, K, Iborra, F, Indge, KJ, Isono, K, Jacq, C, Jacquet, M, James, CM, Jauniaux, JC, Jia, Y, Jimenez, A, Kelly, A, Kleinhans, U, Kreisl, P, Lanfranchi, G, Lewis, C, van der Linden, CG, Lucchini, G, Lutzenkirchen, K, Maat, MJ, Mallet, L, Mannhaupt, G, Martegani, E, Mathieu, A, Maurer, CTC, McConnell, D, McKee, RA, Messenguy, F, Mewes, HW, Molemans, F, Montague, MA, Muzi Falconi, M, Navas, L, Newlon, CS, Noone, D, Pallier, C, Panzeri, L, Pearson, BM, Perea, J, Philippsen, P, Pierard, A, Planta, RJ, Plevani, P, Poetsch, B, Pohl, F, Purnelle, B, Ramezani Rad, M, Rasmussen, SW, Raynal, A, Remacha, M, Richterich, P, Roberts, AB, Rodriguez, F, Sanz, E, Schaaff-Gerstenschlager, I, Scherens, B, Schweitzer, B, Shu, Y, Skala, J, Slonimski, PP, Sor, F, Soustelle, C, Spiegelberg, R, Stateva, LI, Sttensma, HY, Steiner, S, Thierry, A, Thireos, G, Tzermia, M, Urrestarazu, LA, Valle, G, Vetter, I, van Vliet-Reedjk, JC, Voet, M, Vol-

ckaert, G, Vreken, P, Wang, H, Warmington, JR, von Wettstein, D, Wicksteed, BL, Wilson, C, Wurst, H, Xu, G, Yoshikawa, A, Zimmermann, FK, and Sgouros, JG. (1992). The Complete DNA Sequence of Yeast Chromosome III. *Nature* **357** (7 May):38–46.

96. Sulston, J, Du, Z, Thomas, K, Wilson, R, Hillier, L, Staden, R, Halloran, N, Green, P, Thierry-Mieg, J, Qiu, L, Dear, S, Coulson, A, Craxton, M, Durbin, R, Berks, M, Metzstein, M, Hawkins, T, Ainscough, R, and Waterston, R. (1992). The *C. elegans* Genome Sequencing Project: A Beginning. *Nature* **356** (5 March):37–41.

97. Dietrich, W, Katz, H, Lincoln, SE, Shin, H-S, Friedman, J, Dracopoli, N, and Lander, ES. (1992). A Genetic Map of the Mouse Suitable for Typing Intraspecific Crosses. *Genetics* **131** (June):423–447.

98. Barrell, B. (1990). DNA Sequencing: Present Limitations and Prospects for the Future. *FASEB Journal* **5** (January):40–45.

99. Edwards, A, Voss, H, Rice, P, Civitello, A, Stegemann, J, Schwager, C, Zimmermann, J, Erfle, H, Caskey, CT, and Ansorge, W. (1990). Automated DNA Sequencing of the Human HPRT Locus. *Genomics* **6** (April):593–608.

100. Gocayne, J, Robinson, DA, FitzGerald, MG, Chung, F-Z, Kerlavage, AR, Lentes, K-U, Lai, J, Wang, C-D, Fraser, CM, and Venter, JC. (1987). Primary Structure of Rat Cardiac Beta-Adrenergic and Muscarinic Cholinergic Receptors Obtained by Automated DNA Sequence Analysis: Further Evidence for a Multigene Family. *Proceedings of the National Academy of Sciences (USA)* **84** (December):8296–8300.

101. DNA Sequencing Conference, I. (1990). Hilton Head, SC. Laboratory on Molecular and Cellular Neurobiology, NINDS. October 1 to 3.

102. Adams, MD, Kelley, JM, Gocayne, JD, Dubnick, M, Polymeropoulos, MH, Xiao, H, Merril, CR, Wu, A, Olde, B, Moreno, RF, Kerlavage, AR, McCombie, WR, and Venter, JC. (1991). Complementary DNA Sequencing: Expressed Sequence Tags and Human Genome Project. *Science* **252** (21 June):1651–1656.

103. Hamilton, DP. (1992). Venter to Leave NIH for Greener Pastures. *Science* **257** (10 July):151.

104. Bishop, JE, and Stout, H. (1992). Gene Scientist to Leave NIH, Form Institute. *Wall Street Journal* (7 July):B1, B4.

105. Venter, JC, McCombie, WR, and McConnell, JB. (1990). Genome Sequencing Conference I: Summary. *Genomics* **8** (September):186–188.

106. Biosystems Reporter. (1990). Wolf Trap Conference Marks Turning Point in Developing Technologies for Sequencing the Human Genome. *Biosystems Reporter* (#7, February):1, 6, 7.

107. Zinder, NL. (1990). The Genome Initiative: How to Spell "Human." *Scientific American* (July):128.

108. Jaroff, L, with Nash, JM, and Thompson, D. (1989). The Gene Hunt. *Time* (March 20):62–67.

109. Bush, G. (1990). *Text of Remarks by the President at the Annual Meeting of the National Academy of Sciences.* Office of the Press Secretary, the White House. 23 April.

110. Culliton, BJ, and Pool, R. (1991). Presidential Address Sets the Tone. *Nature* **349** (21 February):642.

111. Wright, R. (1990). Achilles' Helix. *New Republic* (July 9 & 16):21–31.

14. FIRST STIRRINGS ABROAD

1. Mannarino, E, 1991, Phone interview and letters, Science Attaché to the Italian Embassy, Washington, DC, March.

2. Dulbecco, R. (1986). A Turning Point in Cancer Research: Sequencing the Human Genome. *Science* **231**:1055–1056.

3. US Congress. (1988). *Mapping Our Genes—Genome Projects: How Big? How Fast?* Office of Technology Assessment, OTA-BA-373, Washington, DC: Government Printing Office; reprinted by Johns Hopkins University Press. April.

4. Dulbecco, R. (1990). The Italian Genome Program. *Genomics* **7** (June):294–297.

5. Dulbecco, R. (1991). Letter. Salk Institute. 16 August.

6. Bodmer, WF. (1987). The Human Genome Sequence and the Analysis of Multifactorial Traits. *In 1987 Molecular Approaches to Human Polygenic Disease*, Ed., pp. 215–228. Chichester: Wiley.

7. Bodmer, WF. (1981). The William Allan Memorial Award Address: Gene Clusters, Genome Organization, and Complex Phenotypes. When the Sequence Is Known, What Will It Mean? *American Journal of Human Genetics* **33**:664–682.

8. Bodmer, WF. (1986). Human Genetics: The Molecular Challenge. *Cold Spring Harbor Symposia on Quantitative Biology, the Molecular Biology of Homo sapiens* **51**:1–13.

9. Bodmer, W. (1986). The 1985 Darwin Lecture: Genetics and Cancer. *Biology and Society* **3**:108–117.

10. Bodmer, W. (1986). Genetic Susceptibility to Cancer. In *Accomplishments in Cancer Research, 1985, General Motors Cancer Research Foundation*, JC Fortner and JE Rhoads, Ed., pp. 198–210. J.B. Lippincott.

11. Bodmer, W. (1988). Presidential Address 1988, British Association for the Advancement of Science, Genes and Atoms for Health and Prosperity. *Science and the Public* (September / October):3–8.

12. Bodmer, W. (1979). Molecular and Genetic Organization: The Future. In *Human Genetics: Possibilities and Realities*, pp. 395–400. Excerpta Medica.

13. Maddox, J. (1987). Brenner Homes in on the Human Genome. *Nature* **326**:119.

14. Newmark, P. (1987). *Mapping and Sequencing the Human Genome in Europe* (Mapping Our Genes Contractor Reports, Vol. 2, Order Number PB 88-162 805 / AS). National Technical Information Service, Springfield, VA. October.

15. GTA Note 244. (1989). *UK Human Gene Mapping Project*. Government of the United Kingdom. April.

16. Dickson, D. (1989). Britain Launches Genome Program. *Science* **245** (31 March):1657.

17. Alwen, J. (1990). United Kingdom Genome Mapping Project: Background, Development, Components, Coordination and Management, and International Links of the Project. *Genomics* **6** (January):386–388.

18. McLaren, DJ. (1991). The Human Genome—UK and International Research Initiatives. *MRC Annual Report, April 1990–March 1991*:44–50.

19. Ferguson-Smith, MA. (1991). European Approach to the Human Gene Project. *FASEB Journal* **5** (January):61–65.

20. McLaren, DJ. (1991). *Human Genome Research: A Review of European and International Contributions*. Medical Research Council, United Kingdom. January.

21. Spurr, NK, and Bishop, MJ. (1990). Newsletter of the UK Human Genome Mapping Project. *G-String* (No. 3, February).

22. Spurr, NK, and Bishop, MJ. (1989). Newsletter of

the UK Human Genome Mapping Project. *G-String* (No. 2).

23. Spurr, NK, and Bishop, MJ. (1989). Newsletter of the UK Human Genome Mapping Project. *G-String* (No. 1, February).

24. Spurr, NK. (1990). Newsletter of the UK Human Genome Mapping Project. *G-Nome News* (No. 4, Autumn).

25. Spurr, NK. (1991). Newsletter of the UK Human Genome Mapping Project. *G-Nome News* (No. 5, January).

26. Spurr, NK. (1991). G-Nome News. *G-Nome News* (No. 6, Spring).

27. Spurr, N, and Bates, C. (1991). G-Nome News. *G-Nome News* (No. 7; Summer).

28. Sulston, J, Du, Z, Thomas, K, Wilson, R, Hillier, L, Staden, R, Halloran, N, Green, P, Thierry-Mieg, J, Qiu, L, Dear, S, Coulson, A, Craxton, M, Durbin, R, Berks, M, Metzstein, M, Hawkins, T, Ainscough, R, and Waterston, R. (1992). The *C. elegans* Genome Sequencing Project: A Beginning. *Nature* **356** (5 March):37–41.

29. Bodmer, W. (1990). Social and Political Implications of the Mapping of the Human Genome. *Science in Parliament* **47** (April):87–93.

30. Vinokurova, S. (1990). Academician Interview on Career in Genetic Research. *Pravda* Translation (31 May):4.

31. Early in the 1930s, Bayev took courses from V.N. Slepkov, whose brother was associated with Bukharin, whom Stalin came to see as a dangerous rival. Graduate students who had studied under Slepkov were rounded up and sent to prison. By such tenuous connections were many lives shattered.

32. Graham, LR. (1974). *Science and Philosophy in the Soviet Union*. Vintage, New York.

33. Medvedev, ZA. (1969). *The Rise and Fall of T.D. Lysenko*. Columbia University Press, New York.

34. Petrov, RV. (1990). Biology's Future: Blossoming or Catastrophe? *Kommunist* translated (January):86–93.

35. Soyfer, VN. (1990). Against Lysenko. *Nature* **344** (1 March):14.

36. Gilbert, W, Maxam, A, and Mirzabekov, A. (1976). Contacts Between the *lac* Repressor and DNA Revealed by Methylation. In *Ninth Alfred Benzon Symposium, Copenhagen: Control of Ribosome Synthesis,* NO Kjeldgaard and O Maaloe, pp. 139–143 and discussion 144ff. New York: Academic Press.

37. Maxam, AM, and Gilbert, W. (1977). A New Method for Sequencing DNA. *Proceedings of the National Academy of Sciences (USA)* **74** (February):560–564.

38. Drmanac, R, Strezoska, Z, Labat, I, Radosavljevic, D, Paunesku, T, and Crkvenjakov, R. (1990). *Towards Genomic DNA Sequencing Chip Based on Oligonucleotide Hybridization*. Cold Spring Harbor meeting on Genome Mapping and Sequencing. 2–4 May.

39. Khrapko, KR, Lysov, YP, Khorlyn, AA, Shick, VV, Florentiev, VL, and Mirzabekov, AD. (1989). An Oligonucleotide Hybridization Approach to DNA Sequencing. **256** (October):118–122.

40. Larionov, V, 1989, Interviews and conversations, Leningrad (Institute of Cytology, Moskva Hotel, Neva riverboat, and strolls during white nights), June.

41. Mirzabekov, A, 1989, Interview, Guest Quarters, Bethesda, MD, in association with a meeting at the National Center for Human Genome Research, December.

42. Mirzabekov, A, 1989, Interview, UNESCO-USSR workshop on human genome research, Moscow, June.

43. Mirzabekov, A, 1990, Interview, Englehardt Institute for Molecular Biology, Moscow, in conjunction with a meeting to plan an international conference

on ethical, legal, and social issues in genome research, October.

44. Bayev, AA, 1988, Interview, First International Workshop on Cooperation for the Human Genome Project, Valencia, Spain, October.

45. Bayev, AA, 1989, Interview, UNESCO-USSR workshop on the Human Genome Project, Moscow, June.

46. Bayev, AA, 1990, Interview, UNESCO Workshop on the Human Genome Project, Paris, February.

47. Bayev, AA, 1990, Interview, Englehardt Institute for Molecular Biology, planning meeting for June 1991 international meeting on ethical, legal and social issues related to genome research, October.

48. Balzer, HD. (1989). *Soviet Science on the Edge of Reform*. Westview Press, Boulder, CO.

49. Bayev, AA. (1990). The Human Genome Project in the USSR. *Biomedical Science* **1**:106–107.

50. Bayev, AA. (1989). *The Human Genome, A General Overview (Genom Cheloveka, Obshchii Uzgliud)*. Scientific Council for the State Scientific-Technical Program "Human Genome," Moscow.

51. State Council for the State Scientific-Technical Program 'Human Genome'. (1989). *Information Bulletin "Human Genome" (Infomatsionnyi Biulleten "Genom Cheloveka")*. Government of the United Soviet Socialist Republics, Moscow.

52. Larionov, V, 1991, Phone conversation, National Institute for Environmental Health Sciences, Research Triangle Park, NC, September.

53. Balzer, H, 1992, Conversation following his return from Moscow, Russian Area Studies Program, Georgetown University, 31 December.

54. Eisner, R, 1989, Interview, Englehardt Institute of Molecular Biology and long treks through Moscow in search of an open restaurant, June.

55. Dickson, D. (1988). Focus on the Genome. *Science* **240** (6 May):711.

56. Kevles, DJ. (1992). Out of Eugenics: The Historical Politics of the Human Genome. *In The Code of Codes: Scientific and Social Issues in the Human Genome Project,* DJ Kevles and L Hood, Ed., pp. 3–36. Cambridge, MA: Harvard University Press.

57. Jordan, B, 1990, Letter to the author, Centre d'immunologie de Marseille-Luminy, 22 August.

58. Jordan, BR. (1991). The French Human Genome Program. *Genomics* **9** (March):562–563.

59. Jordan, B. (1991). Letter. ICRF Laboratories, London. 2 September.

60. Dorozynski, A. (1992). Gene Mapping the Industrial Way. *Science* **256** (24 April):463.

61. Concar, D. (1992). French Find Short Cut to Map of Human Genome. *New Scientist* (23 May):5.

62. Anderson, C. (1992). New French Genome Centre Aims to Prove That Bigger Really Is Better. *Nature* **357** (18 June):526–527.

63. US House of Representatives. (1992). *Departments of Labor, Health and Human Services, Education, and Related Agencies Appropriations for 1993* (Part 4, pp. 607–698). Committee on Appropriations. 25 March.

64. Rabinow, P, 1992. Much of the information in this paragraph derives from personal communications from Paul Rabinow, who cites a paper by Jean-Paul Gaudillere, "French Strategies in Molecular Biology," presented at a Harvard conference on the Human Genome Project, 15 June 1990, Department of Anthropology, University of California, Berkeley.

65. Kevles, DJ, and Hood, L. (1992). Reflections. In *The Code of Codes: Scientific and Social Issues in the Human Genome Project,* DJ Kevles and L Hood, Ed., pp. 300–328. Cambridge, MA: Harvard University Press.

66. Rabinow, P, 1992, Letter to the author, Department

of Anthropology, University of California, Berkeley, 3 June.

67. European Science Foundation. (1991). *Report on Genome Research 1991*. European Science Foundation.

68. Müller-Hill, B. (1988). *Murderous Science*. Oxford University Press, New York.

69. Proctor, RN. (1992). Nazi Doctors, Racial Medicine, and Human Experimentation. In *The Nazi Doctors and the Nuremberg Code*, GJ Annas and MA Grodin, Ed., pp. 17–31. New York: Oxford University Press.

70. Proctor, RN. (1988). *Racial Hygiene: Medicine Under the Nazis*. Harvard University Press, Cambridge, MA.

71. Pross, C. (1992). Nazi Doctors, German Medicine, and Historical Truth. In *The Nazi Doctors and the Nuremberg Code: Human Rights in Human Experimentation*, GJ Annas and MA Grodin, Ed., pp. 32–52. New York: Oxford University Press.

72. Adams, MB, Ed. (1990). *The Wellborn Science: Eugenics in Germany, France, Brazil, and Russia*. Oxford University Press, New York.

73. Gallagher, HG. (1990). *By Trust Betrayed: Patients, Physicians, and the License to Kill in the Third Reich*. Henry Holt, New York.

74. Kevles, DJ. (1985). *In the Name of Eugenics*. University of California Press, Berkeley, CA.

75. Lifton, RJ. (1986). *The Nazi Doctors*. Basic Books, New York.

76. Smith, JD, and Nelson, KR. (1989). *The Sterilization of Carrie Buck*. New Horizon Press, Far Hills, NJ.

77. US Congress. (1983). *The Role of Genetic Testing in the Prevention of Occupational Disease* (OTA-BA-194). Office of Technology Assessment. April.

78. Coutelle, C, 1989, Interview at meeting of UNESCO Director General's Scientific Coordination Committee on Human Genome Research, UNESCO Headquarters, Paris, January.

79. Coutelle, C, 1990, Interview at meeting of UNESCO Director General's Scientific Coordination Committee on Human Genome Research, UNESCO Headquarters, Paris, February.

80. Coutelle, C, 1990, Interview at meeting of UNESCO Director General's Scientific Coordination Committee on Human Genome Research, Englehardt Institute for Molecular Biology, Moscow, June.

81. Dickson, D. (1988). Europe Seeks a Strategy for Biology. *Science* (6 May):710–712.

82. Brenner, S. (1986). Map of Man (First-draft communiqúe to Bronwen Loder). Commission of the European Communities. 10 February 1986.

83. Oliver, SG, van der Aart, QJM, Agostoni-Carbone, ML, Aigle, M, Alberghina, L, Alexandraki, D, Antoine, G, Anwar, R, Ballesta, JPG, Benit, P, Berben, G, Bergantino, E, Biteau, N, Bolle, PA, Bolotin-Fukuhara, M, Brown, A, Brown, AJP, Buhler, JM, Carcano, C, Carignani, G, Cederberg, H, Chanet, R, Contreras, R, Crouzet, M, Daignan-Fornier, B, Defoor, E, Delgado, M, Demolder, J, Doira, C, Dubois, E, Dujon, B, Dusterhoft, A, Erdmann, D, Esteban, M, Fabre, F, Fairhead, C, Faye, G, Feldmann, H, Fiers, W, Francingues-Gaillard, MC, Franco, L, Frontali, L, Fukuhara, H, Fuller, LJ, Galland, P, Gent, ME, Gigot, D, Gilliquet, V, Glansdorff, N, Goffeau, A, Grenson, M, Grisanti, P, Grivell, LA, de Haan, M, Haasemann, M, Hatat, D, Hoenicka, J, Hegemann, J, Herbert, CJ, Hilger, F, Hohmann, S, Hollenberg, CP, Huse, K, Iborra, F, Indge, KJ, Isono, K, Jacq, C, Jacquet, M, James, CM, Jauniaux, JC, Jia, Y, Jimenez, A, Kelly, A, Kleinhans, U, Kreisl, P, Lanfranchi, G, Lewis, C, van der Linden, CG, Lucchini, G, Lutzenkirchen, K, Maat, MJ, Mallet, L, Mannhaupt, G, Martegani, E, Mathieu, A, Maurer, CTC, McConnell, D, McKee, RA, Messenguy, F, Mewes, HW, Molemans, F, Montague, MA, Muzi Falconi, M, Navas, L, Newlon, CS, Noone, D, Pallier, C, Panzeri, L, Pearson, BM, Perea, J, Philippsen, P, Pierard, A, Planta, RJ, Plevani, P, Poetsch, B, Pohl, F, Purnelle, B, Ramezani Rad, M, Rasmussen, SW, Raynal, A, Remacha, M, Richterich, P, Roberts, AB, Rodriguez, F, Sanz, E, Schaaff-Gerstenschlager, I, Scherens, B, Schweitzer, B, Shu, Y, Skala, J, Slonimski, PP, Sor, F, Soustelle, C, Spiegelberg, R, Stateva, LI, Sttensma, HY, Steiner, S, Thierry, A, Thireos, G, Tzermia, M, Urrestarazu, LA, Valle, G, Vetter, I, van Vliet-Reedjk, JC, Voet, M, Volckaert, G, Vreken, P, Wang, H, Warmington, JR, von Wettstein, D, Wicksteed, BL, Wilson, C, Wurst, H, Xu, G, Yoshikawa, A, Zimmermann, FK, and Sgouros, JG. (1992). The Complete DNA Sequence of Yeast Chromosome III. *Nature* 357 (7 May):38–46.

84. Anderson, A. (1992). Yeast Genome Project: 300,000 and Counting. *Science* 256 (24 April):462.

85. Commission of the European Communities. (1988). Proposal for a Council Decision Adopting a Specific Research Programme in the Field of Health: Predictive Medicine: Human Genome Analysis (1989–1991) (COM(88) 424 final—SYN 146). Commission of the European Communities. 20 July.

86. Commission of the European Communities. (1989). Modified Proposal for a Council Decision, Adopting a Specific Research and Technological Development Programme in the Field of Health—Human Genome Analysis: (1990–1991) (COM(89) final—SYN 146). Commission on the European Communities. 13 November.

87. Härlin, B, 1988, Conversation on National Institutes of Health campus, Bethesda, MD, 3 October.

88. Economic and Social Committee. (1988). Opinion of the Economic and Social Committee on the Proposal for a Council Decision Adopting a Specific Research Programme in the Field of Health: Predictive Medicine: Human Genome Analysis (1989–1991) (COM(88) 424 final—SYN 146) (CES 1342 / 33—SYN 146). Commission of the European Communities. 14 December.

89. Council of the European Communities. (1990). Council Decision Adopting a Specific Research and Technological Development Programme in the Field of Health: Human Genome Analysis (1990 to 1991) (90 / 395 / EEC). *Official Journal of the European Communities* (L 196, 26 July):8–14.

90. Academia Europaea. (1991). *Research on the Human Genome in Europe and Its Relation to Activities Elsewhere in the World*. Academia Europaea. 20 March.

91. Philipson, L. (1991). Turmoil in European Biology. *Nature* 351 (9 May):91–92.

92. Doolittle, F, Friesen, J, Smith, M, and Worton, R. (1989). *Mapping and Sequencing the Human Genome: A Program for Canada*. 11 October.

93. Worton, R. (1992). Memo to author. Hospital for Sick Children, University of Toronto. 10 July.

94. Inter-Council Human Genome Advisory Committee. (1991). Report to the Granting Councils: Medical Research Council; Natural Sciences and Engineering Research Council; Social Sciences and Humanities Research Council; and the Secretary to the Advisory Committee, Dr. Lewis Slotin, Medical Research Council. 9 May.

95. Brady, D. (1990). Breaking the Code. *McLean's* (27 August):44–45.

96. Advisory Committee on the Human Genome. (1992). *A Genome Program in Canada*. Summary sheet of the committee's recommendations prepared for the Canadian Cabinet by Charles Scriver. Undated.

97. Spurgeon, D. (1992). Canada Commits Money for Human Genome Research. *Nature* **357** (11 June): 428.
98. Scriver, CR. (1992). Letter. Institut de Recherche de l'Université McGill-Hôpital de Montréal pour Enfants. 17 June.
99. Roberts, L. (1988). A Sequencing Reality Check. *Science* **242** (2 December):1245.
100. Grisolia, S. (1989). Mapping the Human Genome. *Hastings Center Report* (July / August):18–19.
101. Roberts, L. (1988). Carving Up the Human Genome. *Science* **242** (2 December):1244–1246.
102. Rosenblith, WA. (1989). *Unesco Science and U.S. Scientific Interests* (Testimony before the Committee on Foreign Relations, US Senate). Massachusetts Institute of Technology. 19 April.
103. Anderson, GC, and Coles, P. (1990). United States Still Angry. *Nature* **344** (26 April):801.
104. Aldous, P, and Anderson, C. (1992). Reforms Win Praise, but Not Patrons. *Nature* **356** (19 March):187.
105. Mayor, 1989. Fellow Spaniard Grisolia was appointed chair, and the committee included representatives from Chile, France, Germany, Japan, Kenya, Thailand, Tunisia, the United States, and the USSR. Organizational representatives were designated by several international scientific organizations in biology, adding observers from Belgium, France, Israel, and Italy. UNESCO Headquarters, Paris, February.
106. Zharov, V. (1990). Letter and summary document: UNESCO's Programme in the Human Genome (SC / SER / 445 / 90-2550). United Nations Educational, Scientific, and Cultural Organization. 16 July.
107. Grisolia, S. (1991). UNESCO Program for the Human Genome Project. *Genomics* **9** (February):404–405.
108. Dickson, D. (1989). Unesco Seeks a Role in Genome Projects. *Science* **243** (17 March):1431–1432.
109. Matsui, S. (1990). Letter and appendix: Unesco / TWAS Fellowship Programme in the Human Genome (SC / BSC / 445 / 90-3129). United National Educational, Scientific, and Cultural Organization. 4 December.
110. Zharov, V. (1990). Letter and Appendix: Institutions and Key-Persons in the Third World Who May be Involved in the Human Genome Project (SC / BSC / 445 / 90-2921). United Nations Educational, Scientific, and Cultural Organization.
111. Allende, J. (1988). Background on the Human Genome Project. Red Latinoamericana de Ciencias Biologicas. 28 June–1 July.
112. Red Latinoamericana de Ciencias Biologicas. (1988). Resolution on the Human Genome Approved by RELAB in Quito, Ecuador, 1988. 1 July.
113. Allende, JE. (1991). A View from the South. *FASEB Journal* **5** (January):6–7.
114. Red Latinoamericana de Ciencias Biologicas.

(1990). Resolution on the Human Genome Approved by RELAB in Santiago, Chile, 1990. June.
115. Wertz, DCa, and Fletcher, JC, Eds. (1989). *Ethics and Human Genetics: A Cross-Cultural Perspective.* Springer-Verlag, New York.
116. Lander, ES, and Botstein, D. (1987). Homozygosity Mapping: A Way to Map Recessive Traits in Humans by Studying the DNA of Inbred Children. *Science* **236**:1567–1570.
117. Lander, ES, and Botstein, D. (1986). Mapping Complex Genetic Traits in Humans: New Methods Using a Complete RFLP Linkage Map. *Cold Spring Harbor Symposia on Quantitative Biology* **51** (Molecular Biology of *Homo sapiens*):49–62.
118. HUGO Originators, 1988, Walter Bodmer (Imperial Cancer Research Fund, London), Sydney Brenner (Medical Research Council, UK), C. Thomas Caskey (Baylor University, USA), Jean Dausset (Centre d'Etude du Polymorphism Humain, Paris), Renato Dulbecco (Salk Institute and Italy), Leroy Hood (Caltech, USA), Kenichi Matsubara (Osaka, Japan), Frank Ruddle (Yale, USA), John Tooze (EMBO), James D. Watson (Cold Spring Harbor, USA), and Harold Zur-Hausen (Germany), letter about founding the Human Genome Organization, 3 May.
119. McKusick, VA. (1988). Memo to Sir Walter Bodmer, Dr. Sydney Brenner, Dr. J. [sic] Thomas Caskey, Prof. Jean Dausset, Dr. Renato Dulbecco, Dr. Leroy Hood, Dr. K. Matsubara, Dr. Frank Ruddle, Dr. John Tooze, Dr. James D. Watson, and Dr. Harold Zur-Hausen. Johns Hopkins University. 3 May.
120. Roberts, L. (1988). Human Genome Goes International. *Science* **241** (8 July):165.
121. McKusick, VA. (1989). The Human Genome Organization: History, Purposes, and Membership. *Genomics* **5**:385–387.
122. Palca, J. (1988). Human Genome Organization Is Launched with a Flourish. *Nature* **335** (22 September):286.
123. McGourty, C. (1989). A New Direction for HUGO. *Nature* **342** (14 December):724.
124. Aldhous, P. (1990). Imperial's Wellcome Support. *Nature* **344** (1 March):5.
125. Blanchard, F. (1990). Press statement: The Howard Hughes Medical Institute Awards $1 Million Grant to the Human Genome Organization. Howard Hughes Medical Institute. 3 May.
126. Cantor, CR. (1990). HUGO Physical Mapping. *Nature* **345** (10 May):106.
127. Cantor, C. (1990). Memorandum to HUGO-Americas *Ad Hoc* Mapping Committee. Human Genome Center, Lawrence Berkeley Laboratory. 24 August.
128. Maddox, J. (1991). The Case for the Human Genome. *Nature* **352** (4 July):11–14.
129. Will, G. (1989). Science Is Making the News. *Washington Post* (23 April):C7.

15. JAPAN: A SPECIAL CASE

1. Kikkawa, C. (1925). *The Autobiography of Baron Chokichi Kikkawa.* Private publisher, Tokyo. Kindly contributed to the author by Akiyoshi Wada.
2. Brown, SD. (1982). Translator's Introduction. In *The Diary of Kido Takayoshi,* Sidney DeVere Brown and Akiko Hirota, Ed., pp. xvi–xxi. Tokyo: University of Tokyo Press.
3. Kido, T. (1982). The Diary of Kido Takayoshi. Volume I: 1868–1871; Volume II: 1871–1874; Volume III: 1874–1877. Sidney DeVere Brown and Akiko Hirota, Ed. Tokyo: University of Tokyo Press.
4. Reynolds, RL. (1963). *Commodore Perry in Japan.* American Heritage, New York.

5. Dyer, H. (1904). Education and National Efficiency in Japan. *Nature* (15 December):151–152.
6. Wada, A . (1985). Letter to Professor Paul Doty upon his retirement. University of Tokyo. 28 May.
7. Wada, A, and Soeda, E. (1987). Strategy for Building an Automated and High Speed DNA-Sequencing System. In *Integration and Control of Metabolic Processes,* OL Kon, Ed., pp. 517–532. New York: Cambridge University Press.
8. Wada, A, and Soeda, E. (1985). Strategy for Building an Automatic and High Speed DNA-Sequencing System. *Proceedings of the 4th Congress of the Federation of Asian and Oceanic Biochemists,* pp. 1–

16. London: Cambridge University Press.

9. Wada, A. (1988). The Practicability of and Necessity for Developing a Large-Scale DNA-Base Sequencing System: Toward the Establishment of International Super Sequencer Centers. In *Biotechnology and the Human Genome: Innovations and Impact,* AD Woodhead and BJ Barnhart, Ed., pp. 119–130. New York: Plenum.

10. Miyahara, M. (1987). R&D on Human Gene Analysis and Mapping Systems Supported by Science and Technology Agency of Japan (Report Memorandum #135). Tokyo Office of the US National Science Foundation. August 31.

11. Rhodes, R. (1986). *The Making of the Atomic Bomb.* Basic Books, New York.

12. RIKEN. (1986). History of RIKEN. In *Frontiers of Research,* p. 13. Tsukuba Science City, Japan: Institute of Physical and Chemical Research (RIKEN).

13. DeLisi, C. (1990). Letter commemorating retirement of Akiyoshi Wada. Boston University. 2 January.

14. Roberts, L. (1989). Watson Versus Japan. *Science* **246:**576–578.

15. Johnstone, B. (1990). The Human Gene War. *Asia Technology* (February):51–53.

16. Sun, M. (1989). Consensus Elusive on Japan's Genome Plans. *Science* **243** (31 March):1656–1657.

17. Swinbanks, D. (1991). Japan's Human Genome Project Takes Shape. *Nature* **351** (20 June):593.

18. Endo, I, Soeda, E, Murakami, Y, and Nichi, K. (1991). Human Genome Analysis System. *Nature* **352** (4 July):89–90.

19. Chijiya, M, 1990, Interview at 2-2-1 Kasumigaseki, Chiyoda-ku, Tokyo 100, Director, Life Sciences, Science and Technology Agency (kangaku jijutsucho), R&D Bureau, 2 August.

20. Ikawa, Y, 1988, Interview, First International Workshop on Cooperation for the Human Genome Project, Valencia, Spain, October.

21. Ikawa, Y, 1989, Conversations, Englehardt Institute of Molecular Biology and Hotel Ukraina, Moscow, June.

22. Matsubara, K, 1988, Interview at first meeting on Genome Mapping and Sequencing, Cold Spring Harbor Laboratory, NY, 27 April–1 May.

23. Matsubara, K, 1988, Author's notes from a meeting, Office of Technology Assessment, US Congress, Washington, DC, 3 February.

24. Matsubara, K, 1990, Interview, Osaka University, 31 July.

25. Matsubara, K, Oishi, M, Chijiya, M, Mine, A, Wada, A, and Kanehisa, M, 1990, Interviews on trip to Japan, 31 July–4 August.

26. Matsubara, K, and Kakunaga, M. (1992). The Genome Efforts in Japan as of 1991. *Genomics* **12:**618–620.

27. US House of Representatives. (1987). Fiscal Year 1988 DOE Budget Authorization: Environmental Research and Development (No. 58). Subcommittee on Natural Resources, Agriculture Research, and Environment of the Committee on Science, Space and Technology. March 19, 1987.

28. US House of Representatives. (1987). Departments of Labor, Health and Human Services, Education, and Related Agencies Appropriations for 1988 (Part 4A). National Institutes of Health.

29. US Senate. (1987). *Workshop on Human Gene Mapping* (Committee Report No. 100-71). Committee on Energy and Natural Resources. August 31, 1987.

30. Ministry of Education Science and Culture. (1989). *On Promotion of the Human Genome Program* (see also accompanying data document, dated March 1989). Government of Japan. 19 July.

31. Council for Aeronautics Electronics and Other Advanced Technologies. (1988). *Comprehensive Strategy for Promoting R&D on Human Genome Analysis.* Science and Technology Agency, Government of Japan. Different copies dated 27 March and 27 June.

32. US House of Representatives. (1989). International Cooperation in Mapping the Human Genome. October 19, 1989, 2325 Rayburn House Office Building: Hearing before the Subcommittee on International Scientific Cooperation, Committee on Science, Space and Technology.

33. Crawford, R. (1990). Under the Microscope: Japan's Lackluster Effort to Contribute to the Human Genome Project. *Far Eastern Economic Review* **149** (No. 36, 6 September):82.

34. Swinbanks, D. (1989). Japan Still Seeking A Role. *Nature* **342** (14 December):724–725.

35. Greenberg, D. (1990). Q&A with James Watson, Genome Project Chief. *Science & Government Report* **20** (#5, 15 March):1–5.

36. Kendrew, J. (1990). Note to James D. Watson. The Old Guildhall, Cambridge, UK. 4 January.

37. Watson, JD. (1990). Letter to Sir John Kendrew. Cold Spring Harbor Laboratory. 10 January.

38. Watson, JD. (1988). *Reflections on My Forty Years in Science.* Harvard University: Department of Biochemistry and Molecular Biology, 13 May.

39. Davis, B. (1990). Watson Doesn't Use Gentle Persuasion to Enlist Japanese and German Support for Genome Effort. New York: *Wall Street Journal* (18 June):A12.

40. Ikawa, Y. (1991). Human Genome Efforts in Japan. *FASEB Journal* **5** (January):66–69.

41. Yoshikawa, A. (1987). *In Search of the "Ultimate Map" of the Human Genome: The Japanese Efforts* (Mapping Our Genes Contractor Reports, Vol. 2, National Technical Information Service Order No. PB 88-162 805 / AS). Berkeley Roundtable on the International Economy, University of California, Berkeley. August.

42. Brown, S, and Woodbury, R. (1989). Battle for the Future. *Time* (16 January):42–43.

43. Domenici, P. (1989). The Government-Industry-University Link. *Roll Call* (10–16 July):3.

44. Fujimura, BK. (1988). *Biotechnology in Japan.* International Trade Administration, US Department of Commerce.

45. US Congress. (1984). *Commercial Biotechnology, an International Analysis* (OTA-BA-218, Government Printing Office). Office of Technology Assessment. January.

46. US Congress. (1991). *Biotechnology in a Global Economy* (OTA-BA-494, Government Printing Office). Office of Technology Assessment. Taken from appendix B and ch. 4, yielding an estimate of 20 percent government funding. October.

47. Saxonhouse, G. (1986). Industrial Policy and Factor Markets in Japan and the United States. In *Japan's High Technology Industries,* H Patrick, Ed., Seattle: University of Washington Press.

48. Yoshikawa, A. (1987). *The Japanese Challenge in Biotechnology: Industrial Policy* (BRIE working paper #29). Berkeley Roundtable on the International Economy, University of California, Berkeley. September.

49. Scheidegger, A. (1988). Biotechnology in Japan: A Lesson in Logistics? Part I: The Political Substrate. *Trends in Biotechnology* **6** (January):7–15.

50. Scheidegger, A. (1988). Biotechnology in Japan: A Lesson in Logistics? Part II: The Research Policy. *Trends in Biotechnology* **6** (February):47–53.

51. Yoshikawa, A. (1989). Japanese Biotechnology: Government, Corporations, and Technology Transfer. *Technology Transfer* (Winter):32–39.

52. Mowery, DC. (1989). Collaborative Ventures Between US and Foreign Manufacturing Firms. *Research Policy* **18:**19–32.
53. Klausner, A. (1989). *An Analysis of Strategic Alliances in Biotechnology Between US Firms and Those in Europe and Japan* (Contract report prepared for the Office of Technology Assessment, US Congress). August.
54. Stokes, B. (1989). Multiple Allegiances. *National Journal* (11 November):2754–2758.
55. Reich, RB. (1990). Who Is Us? *Harvard Business Review* (January–February):53–64.
56. Reich, RB. (1989). The Quiet Path to Technological Preeminence. *Scientific American* **261** (#4, October):41–47.
57. Committee on Japan. (1992). *US-Japan Technology Linkages in Biotechnology: Challenges for the 1990s* (Washington, DC: National Academy Press). Office of Japan Affairs, Office of International Affairs, National Research Council, National Academy of Sciences. Ch. 1 and table 2, yielding an estimate of 24 percent.
58. Narin, R, and Frame, JD. (1989). The Growth of Japanese Science and Technology. *Science* **245** (11 August):600–605.
59. National Science Foundation. (1988). *The Science and Technology Resources of Japan: A Comparison with the United States* (NSF 88-318). National Science Foundation. June.
60. President's Council on Competitiveness. (1991). *Report on National Biotechnology Policy*. The White House. Ernst & Young Estimated that federal funding was $3.7 billion in 1991, with $3.2 billion industrial investment in R&D, as cited in the National Research Council Report by the Committee on Japan. The Japanese corporate contribution was estimated at 276 billion yen ($2 billion) that same year (see footnote 5 in NRC report), while government funding was ¥89.6 billion ($665 million) (see table 2 in NRC report). Yen to dollar rates are calculated at 135¥ / $ in 1991, 150¥ / $ in 1990, here and throughout the chapter. February.
61. Committee on Japan. (1991). *Intellectual Property Rights and US-Japan Competition in Biotechnology: Report of a Workshop*. Office of Japan Affairs, National Research Council, National Academy of Sciences. 18 January.
62. C. W. C. (1877). Engineering Education in Japan. *Nature* (17 May):44–45.
63. US Department of State. (1990). Biotech-Human Genome: Uncoordinated GOJ Efforts May Result in Private Funding for HUGO (Tokyo 4670, State 60879). US Embassy, Tokyo. May.
64. By these same proportionate estimates, Italy's contribution was 29 percent that of the US, the UK's 66 percent, the USSR's 78 percent (at official exchange rates, but only 5 percent at unofficial Fall 1990 "black market" rates). Those without specific genome programs could not be calculated. US figures are for the NIH and DOE programs. Japanese figures include the Monbusho, STA, and Department of Health programs, but not agricultural or related MITI programs. The UK figure is for the combined MRC-ICRF program, and Italy's includes only the Ministry of Science figures. GNP figures for these calculations are taken from the *1990 World Almanac* (NY: Pharos Books) and *Statesman's Yearbook, 1989–1990* (NY: St Martin's). Unofficial conversion rates for rubles taken from black market rate, Leningrad and Moscow, June 1989. My estimates for Japan's funding are higher than those of the Department of State ($3.5 million), which did not account for differences in budgeting practices for personnel and space. My calculations assumed that 40 percent of US research budgets covered salaries and other expenses not included in the European and Japanese budget items. Genome budget estimates for 1990 are taken from various references that follow.
65. Dulbecco, R. (1990). The Italian Genome Program. *Genomics* **7** (June):294–297.
66. Alwen, J. (1990). United Kingdom Genome Mapping Project: Background, Development, Components, Coordination and Management, and International Links of the Project. *Genomics* **6** (January):386–388.
67. Bayev, AA. (1990). The Human Genome Project in the USSR. *Biomedical Science* **1:**106–107.
68. Sun, M. (1989). Japan Faces Big Task in Improving Basic Science. *Science* **243** (10 March):1285–1287.
69. Owens, CT. (1989). Tapping Japanese Science. *Issues in Science and Technology* (Summer):32–34.
70. Anderson, A. (1990). Japanese Papers Top Charts. *Nature* **343** (18 January):199.
71. US Congress. (1988). *Mapping Our Genes—Genome Projects: How Big? How Fast?* Office of Technology Assessment, OTA-BA-373, Washington, DC: Government Printing Office; reprinted by Johns Hopkins University Press. April.
72. European Science Foundation. (1991). *Report on Genome Research 1991*. European Science Foundation.
73. Yanagida, M. (1990). The Grant-Getting Game in Japan. *Nature* **343** (11 January):111–112.
74. Swinbanks, D. (1991). Survey Pans University Labs. *Nature* **350** (18 April):544.
75. Ministry of Education Science and Culture. (1990). *Five-Year Plan for Promotion of the Human Genome Program at Universities*. Bioscience Sectional meeting, Subcommittee for Regional Promotion of Studies Specified by the Academic Deliberative Council, Ministry of Education, Science, and Culture, Government of Japan. 11 July.
76. Watson, JD, 1990, Conversations about Japan following author's return from a trip there, Cold Spring Harbor Laboratory, NY, and National Center for Human Genome Research, Bethesda, MD, August.
77. Genome Research in an Interdependent World, 1991, Workshop, Building 38A, National Institutes of Health, Bethesda, MD, 10–12 June.
78. Matsubara tried to secure a ¥6.5 billion budget ($43 million) to support two five-year projects under Monbusho. The first year's budget was ¥400 million ($2.7 million) for the biology program. A second, separate but related, program focused on informatics. A new Genome Analysis Center would be established at the University of Tokyo, with a national research group and a five-year commitment (first-year budget ¥220 million, or $1.5 million). This was to be the first installment on a much larger program. The Monbusho plan laid out a strategy for genome research, urging establishment of groups in five areas: (1) physical and genetic mapping, (2) analysis of complementary DNA collections (cDNA, representing those parts of DNA transcribed to RNA and representing the expressed parts of genes), (3) pursuit of new techniques to analyze DNA, (4) informatics, and (5) science and society. The Monbusho proposal set aside funds for international scientific workshops and for training of postdoctoral fellows and other research personnel.
79. Swinbanks, D. (1991). Human Genome: Japan's Project Stalls. *Nature* **349** (31 January): 360.
80. This was roughly ¥200 million ($1.5 million) short of Matsubara's aspirations for Monbusho. The five elements remained, led by Matsubara (Osaka University), Mitsuaki Yoshida (Tokyo University), Minoru Kanehisa (Kyoto University), with a smaller program at Kyushu University.
81. Swinbanks, D. (1991). Good News for Universities. *Nature* **353** (12 September): 102.

82. Akira Ooya was the principal scientist for the Ministry of Health and Welfare program, which was budgeted for ¥460 in 1991. Of this amount, ¥100 million ($740,000) supported a cell line repository and the bulk of the remainder was for scientific grants.

83. Swinbanks, D. (1992). When Silence Isn't Golden. *Nature* **356** (2 April):368.

84. Bankowski, Z, and Capron, AM, Ed. (1991). *Genetics, Ethics, and Human Values: Human Genome Mapping, Genetic Screening, and Gene Therapy. Proceedings of the XXIVth CIOMS Conference, Tokyo and Inuyama City, Japan, 22–27 July 1990.* CIOMS, Geneva.

85. Crawford, R. (1991). Gene Mapping Japan's Number One Crop. *Science* **252** (21 June):1611.

86. Ferrell, J. (1991). "Office Ladies" Aid Research. *Nature* **353** (12 September):99.

87. Swinbanks, D. (1991). DNA Research Institute to Open in Japan. *Nature* **349** (21 February):640.

88. Swinbanks, D. (1990). Japan Gets Its Act Together. *Nature* **347** (20 September):220.

89. Mine, A, 1990, Interview, Japan Biotechnology Association (formerly BIDEC), 4 August.

90. Oishi, M, 1990, Interview, Tokyo University (Todai), 3 August.

91. Fujiyama, A, 1990, Interview, Englehardt Institute for Molecular Biology, Moscow, October.

92. Miyakawa, H, 1990, Dinner at author's home, Derwood, MD, 18 October.

16. A NEW SOCIAL CONTRACT

1. Wexler, NS. (1991). Life in the Lab. *Los Angeles Times Magazine* (10 February):12–13, 31.

2. Saltus, R. (1989). Notes from Interview for *Boston Globe* with Milton Wexler. 10 November.

3. Roberts, L. (1990). Huntington's Gene: So Near, Yet So Far. *Science* **247** (9 February):624–627.

4. Bishop, JE, and Waldholz, M. (1990). *Genome: The Story of the Most Astonishing Scientific Adventure of Our Time—The Attempt to Map All the Genes in the Human Body.* Simon & Schuster, New York.

5. Wexler, NS, and Wilentz, JS. (1978). *Report of the Congressional Commission for the Control of Huntington's Disease and Its Consequences (DHEW Pub. No. 78–150). Congressional Commission for the Control of Huntington's Disease and Its Consequences,* National Institute of Neurological and Communicative Disorders and Stroke.

6. Kolata, G. (1983). Huntington's Disease Gene Located. *Science* **222** (25 November):913–915.

7. Wyman, AR, and White, RL. (1980). A Highly Polymorphic Locus in Human DNA. *Proceedings of the National Academy of Science (USA)* **77**:6754–58.

8. White, RL. (1979). *Human Linkage Mapping with DNA Polymorphism* (R01-GM27313). University of Massachusetts, Worcester. Proposal signed and dated 27 February; logged in at the National Institutes of Health 1 March.

9. Wexler, NS, Conneally, PM, Housman, D, and Gusella, JF. (1985). A DNA Polymorphism for Huntington's Disease Marks the Future. *Archives of Neurology* **42** (January):20–24.

10. Wingerson, L. (1990). *Mapping Our Genes.* Dutton, New York.

11. Mary Jennifer Selznick Mini-Workshop. (1983). *Clinical Impact of Recombinant DNA Research on Neurogenetic Diseases* (Summary prepared by Roger Kurlan, Department of Neurology, University of Rochester Medical Center). Hereditary Disease Foundation. 17–18 August.

12. Gusella, JF, Wexler, NS, Conneally, PM, Naylor, SL, Anderson, MA, Tanzi, RE, Watkind, PC, Ottina, K, Wallace, MR, Sakaguchi, AY, Young, AM, Shoulson, I, Bonilla, E, and Martin, JB. (1983). A Polymorphic DNA Marker Genetically Linked to Huntington's Disease. *Nature* **306**:234–238.

13. The Huntington's Disease Collaborative Research Group* (1993). A Novel Gene Containing a Trinucleotide Repeat That Is Expanded and Unstable on Huntington's Disease Chromosomes. *Cell* **72** (26 March):971–983. *Group 1: MacDonald, Marcy E.; Ambrose, Christine M.; Duyao, Mabel P.; Myers, Richard H.; Lin, Carol; Srinidhi, Lakshmi; Barnes, Glenn; Taylor, Sherryl A.; James, Marianne; Groot, Nicolet; MacFarlane, Heather; Jenkins, Barbara; Anderson, Mary Anne; Wexler, Nancy S.; and Gusella, James F.; **Group 2:** Bates, Gillian P.; Baxendale, Sarah; Hummerich, Holger; Kirby, Susan; North, Mike; Youngman, Sandra; Mott, Richard; Zehetner, Gunther; Sedlacek, Zdenek; Poustka, Annemarie; Frischauf, Anna-Maria; and Lehrach, Hans; **Group 3:** Buckler, Alan J.; Church, Deanna; Doucette-Stamm, Lynn; O'Donovan, Michael C.; Riba-Ramirez, Laura; Shah, Manish; Stanton, Vincent P.; Strobel, Scott A.; Draths, Karen M.; Wales, Jennifer L.; Dervan, Peter; and Housman, David E.; **Group 4:** Altherr, Michael; Shiang, Rita; Thompson, Leslie; Fielder, Thomas; and Wasmuth, John J.; **Group 5:** Tagle, Danilo; Valdex, John; Elmer, Lawrence; Allard, Marc; Castilla, Lucio; Swaroop, Manju; Blanchard, Kris; and Collins, Francis S.; and **Group 6:** Snell, Russell; Holloway, Tracey; Gillespie, Kathleen; Datson, Nicole; Shaw, Duncan; and Harper, Peter S.

14. Negrette, A. (1990). La Catira. *Panorama* (Maracaibo, Venezuela). April. Translated by Hereditary Disease Foundation.

15. Wexler, NS. (1979). Genetic "Russian Roulette": The Experience of Being at Risk for Huntington's Disease. In *Genetic Counseling: Psychological Dimensions,* S Kessler, Ed., New York: Academic Press.

16. Saltus, R. (1989). Notes from Interview with Nancy Wexler. *Boston Globe,* 19 September.

17. Jaroff, L, with Nash, JM, and Thompson, D. (1989). The Gene Hunt. *Time* (March 20):62–67.

18. Ager, S. (1989). Geri's Gamble. Health section, *Washington Post* 13 June:12–18.

19. Newman, A. (1988). The Legacy on Chromosome 4. *Johns Hopkins Magazine* (April):30–39.

20. Grady, D. (1987). The Ticking of a Time Bomb in the Genes. *Discover* (June):26–37.

21. Wiggins, S, Shyte, P, Huggins, M, Adam, S, Theilmann, J, Bloch, M, Sheps, SB, Schechter, MT, and Hayden, MR. (1992). The Psychological Consequences of Predictive Testing for Huntington's Disease. *New England Journal of Medicine* **327** (12 November):1401–1405.

22. Hayes, CV. (1992). Genetic Testing for Huntington's Disease—A Family Issue. *New England Journal of Medicine* **327** (12 November):1449–1451.

23. Roberts, L. (1989). Genome Project Under Way, at Last. *Science* **243** (13 January):167–168.

24. Schmeck, HM. (1988). DNA Pioneer to Tackle Biggest Gene Project Ever. New York: C1, C16. *New York Times,* 4 October.

25. Watson, JD. (1988). *The NIH Genome Initiative.* Molecular Biology Institute, University of California, Los Angeles: CL Joint Committee on Science and Technology UC Systemwide Biotechnology Research and Education Program, Lawrence Berkeley Laboratory, and California Department of Com-

merce. The Human Genome Projects: Issues, Goals, and California's Participation.

26. Fletcher, JC. (1986). Memo to Ruth Kirschstein, Director, National Institute of General Medical Sciences. Magnuson Clinical Center, NIH. 6 August.

27. Office of Human Genome Research. (1989). Ethical and Legal Studies Relating to the Program to Map and Sequence the Human Genome (*NIH Guide for Grants and Contracts* 7 (3 March):9–10.

28. Anderson, GC. (1990). Genome Project Spawns New Research on Ethics. *The Scientist* (22 January):20.

29. National Center for Human Genome Research. (1991). *Ethical, Legal, and Social Implications Program, Grant-Making Status Report.* August.

30. National Center for Human Genome Research. (1989). Ethical, Legal, and Social Implications of the Human Genome Initiative. *NIH Guide to Grants and Contracts* 19 (26 January 1990):1–5.

31. US House of Representatives. (1990). Departments of Labor, Health and Human Services, Education, and Related Agencies Appropriations for 1991 (Part 4B, pp. 887–960). Committee on Appropriations.

32. US House of Representatives. (1990). Departments of Labor, Health and Human Services, and Education, and Related Agencies Appropriation Bill, 1991 (101–516). Committee on Appropriations. 10 October.

33. Murray, TH. (1991). Ethical Issues in Human Genome Research. *FASEB Journal* 5 (January):55–60.

34. Krimsky, S. (1982). *Genetic Alchemy: The Social History of the Recombinant DNA Controversy.* MIT Press, Cambridge, MA.

35. Juengst, ET. (1991). *Recent Developments and Plans.* National Center for Human Genome Research. 21 February.

36. Raub, WF. (1991). *The Ethical, Legal, and Social Implications of Human Genome Research: Preparing for the Responsible Use of New Genetic Knowledge.* National Institutes of Health. January.

37. Social, Legal, and Ethical Issues in Biomedical and Behavioral Research Policy Panel. (1991). *Report on Social, Legal, and Ethical Issues in Biomedical and Behavioral Research.* National Institutes of Health. Undated, but completed before December.

38. Kerem, B-S, Rommens, JM, Buchanan, JA, Markiewicz, D, Cox, TK, Chakravarti, A, Buchwald, M, and Tsui, L-C. (1989). Identification of the Cystic Fibrosis Gene: Genetic Analysis. *Science* 245 (8 September):1073–1080.

39. Riordan, JR, Rommens, JM, Kerem, B-S, Alon, N, Rozmahel, R, Grzelczak, Z, Zielenski, J, Lok, S, Plavsic, N, Chou, J-L, Drumm, ML, Iannuzzi, MC, Collins, FS, and Tsui, L-C. (1989). Identification of the Cystic Fibrosis Gene: Cloning and Characterization of Complementary DNA. *Science* 245 (8 September):1066–1072.

40. Rommens, JM, Iannuzzi, MC, Kerem, B-S, Drumm, ML, Melmer, G, Dean, M, Rozmahel, R, Cole, JL, Kennedy, D, Hidaka, H, Zsiga, M, Buchwald, M, Riordan, JR, Tsui, L-C, and Collins, FS. (1989). Identification of the Cystic Fibrosis Gene: Chromosome Walking and Jumping. *Science* 245 (8 September):1059–1065.

41. Rich, DP, Anderson, MP, Gregory, RJ, Cheng, SH, Paul, S, Jefferson, DM, McCann, JD, Klinger, KW, Smith, AE, and Welsh, MJ. (1990). Expression of Cystic Fibrosis Transmembrane Conductance Regulator Corrects Defective Chloride Channel Regulation in Cystic Fibrosis Airway Epithelial Cells. *Nature* 347 (27 September):358–363.

42. Working Group on Ethical, Legal, and Social Issues in Human Genome Research. (1990). *Workshop on the Introduction of New Genetic Tests.* National Institutes of Health and Department of Energy. 10 September.

43. The Cystic Fibrosis Genetic Analysis Consortium. (1990). Worldwide Survey of the ΔF508 Mutation—Report from the Cystic Fibrosis Genetic Analysis Consortium. *American Journal of Human Genetics* 47:354–359.

44. Williamson, R, Allison, MED, Bentley, TJ, Lim, SMC, Watson, E, Chapple, J, Adam, S, and Boulton, M. (1989). Community Attitudes to Cystic Fibrosis Carrier Testing in England: A Pilot Study. *Prenatal Diagnosis* 9:727–734.

45. Wilfond, BS, and Fost, N. (1990). The Cystic Fibrosis Gene: Medical and Social Implications for Heterozygote Detection. *Journal of the American Medical Association* 263 (May 23 / 30):2777–2783.

46. Wilfond, BS, and Fost, N. (1990). *Cystic Fibrosis Heterozygote Detection: The Introduction of Genetic Testing into Clinical Practice.* Program in Medical Ethics, University of Wisconsin. 9 November.

47. Caskey, CT, Kaback, MM, and Beaudet, AL. (1990). The American Society of Human Genetics Statement on Cystic Fibrosis Screening. *American Journal of Human Genetics* 46:393.

48. Workshop on Population Screening for the Cystic Fibrosis Gene. (1990). Statement from the National Institutes of Health Workshop on Population Screening for the Cystic Fibrosis Gene. *New England Journal of Medicine* 323 (July 5):70–71.

49. US Congress. (1990). Project Proposal—Cystic Fibrosis: Implications of Population Screening. Office of Technology Assessment. August.

50. Beaudet, AL. (1990). Carrier Screening for Cystic Fibrosis. *American Journal of Human Genetics* 47:603–605.

51. Roberts, L. (1990). Cystic Fibrosis Pilot Projects Go Begging. *Science* 250 (23 November):1076–1077.

52. National Institute of Diabetes Digestive and Kidney Diseases. (1991). Omnibus Solicitation of the Public Health Service for Small Business Innovation Research (SBIR) Grant Applications. *PHS 91-2* (item CC):56–57.

53. NIH Grant proposal R01 DK 43425-01. (1990). Evaluation of Carrier Screening for Cystic Fibrosis. Grant Application to the National Institutes of Health. Name of the principal investigator withheld on request. 26 January.

54. Mammalian Genetics Study Section. (1990). Summary Statement evaluating grant proposal R01 DK 43425-01. National Institutes of Health. 14–16 June.

55. Unnamed, 1991, Interview with a member of the initial review group (study section) that scrutinized the grant. Name withheld to protect confidentiality. Meeting at Cold Spring Harbor Laboratory, June.

56. Roberts, L. (1990). Hopeful News for CF Pilot Studies. *Science* 250 (14 December):1514–1515.

57. Program Advisory Committee on the Human Genome. (1990). Resolution from the NCHGR Program Advisory Committee on the Human Genome. National Institutes of Health. 3 December.

58. Workshop on Clinical Studies of Cystic Fibrosis Carrier Testing. (1991). Meeting Summary. National Institutes of Health. 31 January.

59. National Center for Human Genome Research, National Center for Nursing Research, National Institute of Child Health and Human Development, and Diseases, National Institute of Diabetes and Digestive and Kidney Disorders. (1991). RFA-HG-91-01. Studies of Testing and Counseling for Cystic Fibrosis Mutations. *NIH Guide to Grants and Contracts* 20 (No. 14, 5 April):1–9.

60. Wexler, NS. (1991). Letter to the ELSI working group. Columbia College of Physicians and Surgeons. 27 June.

61. Ethical, Legal, and Social Implications Program. (1991). Grant-Making Status Report. National Cen-

ter for Human Genome Research, National Institutes of Health. November.

62. Bush, V. (1945). *Science—The Endless Frontier.* Office of Scientific Research and Development. 5 July.

63. Bush, V. (1970). *Pieces of the Action.* Morrow, New York.

64. Koshland, DE. (1989). Sequences and Consequences of the Human Genome. *Science* **246** (13 October):189.

65. Smith, JS. (1990). *Patenting the Sun: Polio and the Salk Vaccine.* Morrow, New York.

66. King, PA. (1991). Ethical and Legal Constraints on Research. In *Preparing for Science in the 21st Century,* DC Harrison, M Osterweis, and ER Rubin, Ed., pp. 116–125. Washington, DC: Association of Academic Health Centers.

67. Watson, JD. (1971). Statement of Dr. James D. Watson, Harvard University. Reprinted as "Potential Consequences of Experimentation with Human Eggs" in *International Science Policy,* House Committee on Science and Astronautics, pp. 149–161. (I thank Alexander Morgan Capron for finding this reference when I feared I was on a cold trail.) Panel on Science and Technology Twelfth Meeting: International Science Policy. Proceedings before the House Committee on Science and Astronautics, 92nd Congress. 26–28 January 1971.

68. Reilly, P. (1977). *Genetics, Law, and Social Policy.* Harvard University Press, Cambridge, MA.

69. Duster, T. (1990). *Backdoor to Eugenics.* Routledge, New York.

70. Buck v. Bell. (1927). 274 U.S. 200. *Supreme Court* **47:**584.

71. Smith, JD, and Nelson, KR. (1989). *The Sterilization of Carrie Buck.* New Horizon Press, Far Hills, NJ.

72. Kevles, DJ. (1985). *In the Name of Eugenics.* University of California Press, Berkeley, CA.

73. Gallagher, HG. (1990). *By Trust Betrayed: Patients and Physicians in the Third Reich.* Henry Holt, New York.

74. Adams, MB, Ed. (1990). *The Wellborn Science: Eugenics in Germany, France, Brazil, and Russia.* Oxford University Press, New York.

75. Lifton, RJ. (1986). *The Nazi Doctors.* Basic Books, New York.

76. Müller-Hill, B. (1988). *Murderous Science.* Oxford University Press, New York.

77. Proctor, RN. (1988). *Racial Hygiene: Medicine Under the Nazis.* Harvard University Press, Cambridge, MA.

78. Annas, GJ, and Grodin, MA. (1992). *The Nazi Doctors and the Nuremberg Code: Human Rights in Human Experimentation.* Oxford University Press, New York.

79. Keller, EF. (1992). Nature, Nurture, and the Human Genome Project. In *The Code of Codes: Scientific and Social Issues in the Human Genome Project,* DJ Kevles and L Hood, Ed., pp. 281–299. Cambridge, MA: Harvard University Press.

80. Kevles, DJ. (1992). Out of Eugenics: The Historical Politics of the Human Genome. In *The Code of Codes: Scientific and Social Issues in the Human Genome Project,* DJ Kevles and L Hood, Ed., pp. 3–36. Cambridge, MA: Harvard University Press.

81. Rothman, BK. (1987). *The Tentative Pregnancy: Prenatal Diagnosis and the Future of Motherhood.* Penguin Books, New York. See especially ch. 7: Grieving the Genetic Defect.

82. Rothman, BK. (1992). All That Glitters Is Not Gold. *Hastings Center Report* **Special Supplement** (July–August):S11–S15.

83. Wikler, D, and Palmer, E. (1992). Neo-Eugenics and Disability Rights in Philosophical Perspective. In *Human Genome Research and Society: Proceedings of the Second International Bioethics Seminar in Fukui, 20–21 March, 1992,* N Fujiki and DRJ Macer, Ed., pp. 105–113. Christchurch, NZ: Eubios Ethics Institute.

84. Asch, A. (1988). Reproductive Technology and Disability. In *Reproductive Laws for the 1990s,* S Cohen and N Taub, Ed., Clifton, NJ: Humana Press.

85. Asch, A. (1986). Real Moral Dilemmas. *Christianity and Crisis* **46** (#10, 14 July):237–240.

86. Cowan, RS. (1992). Genetic Technology and Reproductive Choice: An Ethics for Autonomy. In *The Code of Codes: Scientific and Social Issues in the Human Genome Project,* DJ Kevles and L Hood, Ed., pp. 244–263. Cambridge, MA: Harvard University Press.

87. Cook, RH, Schneck, SA, and Clark, DB. (1981). Twins with Alzheimer's Disease. *Archives of Neurology* **38** (May):300–301.

88. Kevles, DJ, and Hood, L. (1992). Reflections. In *The Code of Codes: Scientific and Social Issues in the Human Genome Project,* DJ Kevles and L Hood, Ed., pp. 300–328. Cambridge, MA: Harvard University Press.

89. Kevles, D. (1992). Controlling the Genetic Arsenal. *Wilson Quarterly* **XVI** (Spring):68–76.

90. Kaye, HL. (1992). Are We the Sum of Our Genes? *Wilson Quarterly* **XVI** (Spring):77–84.

91. US House of Representatives. (1992). Departments of Labor, Health and Human Services, Education, and Related Agencies Appropriations for 1993 (Part 4, pp. 607–698). Committee on Appropriations. 25 March.

17. BIOETHICS IN GOVERNMENT

1. US Senate. (1989). Human Genome Initiative (S. Hrg. 101-528). Subcommittee on Science, Technology and Space, Committee on Commerce, Science, and Transportation. 9 November.

2. President's Commission for the Study of Ethical Problems in Medicine and Biomedical and Behavioral Research. (1983). *Screening and Counseling for Genetic Conditions.* Government Printing Office, Washington, DC.

3. Hastings Center, Institute of Society, Ethics and the Life Sciences. (1972). Ethical and Social Issues in Screening for Genetic Disease. *New England Journal of Medicine* **286:**1129–1132.

4. Hastings Center, Institute of Society, Ethics and the Life Sciences. (1979). Guidelines for the Ethical, Social and Legal Issues in Prenatal Diagnosis. *New England Journal of Medicine* **300:**168–172.

5. Committee for the Study of Inborn Errors of Metabolism. (1975). *Genetic Screening: Programs, Principles, and Research.* National Research Council, National Academy of Sciences.

6. Capron, AM, Lappe, M, Murray, RF, Powledge, TM, Twiss, SB, and Bergsma, D, Ed. (1979). *Genetic Counseling: Facts, Values, and Norms.* Alan R. Liss, New York.

7. US House of Representatives. (1988). Legislative Branch Appropriations, Fiscal Year 1990. (See section on Biomedical Ethics Board, Justification of Estimates.) Subcommittee on Legislative Branch, Committee on Appropriations.

8. Public Law 93-348, 1973.

9. King, Patricia, 1987–1992. Conversations at meetings of the ELSI working group, Institute of Medicine activities and other events, Georgetown Law

School. King was a commissioner on both the National Commission for the Protection of Human Subjects of Biomedical and Behavioral Research and the President's Commission for the Study of Ethical Problems in Medicine and Biomedical and Behavioral Research; she then served on the NIH Fetal Tissue Transplantation Research Panel and the NIH-DOE Working Group on Ethical, Legal, and Social Issues, as well as on several Institute of Medicine committees and other panels.

10. Capron, Alexander Morgan, 1983–1993, Conversations connected with Office of Technology Assessment activities, Institute of Medicine activities, the Biomedical Ethics Advisory Committee, and other events, University of Southern California. Capron was consultant to the National Commission for the Protection of Human Subjects of Biomedical and Behavioral Research and the Ethics Advisory Board (Department of Health, Education and Welfare); Executive Director of the President's Commission for the Study of Ethical Problems in Medicine and Biomedical and Behavioral Research; and chair of the Biomedical Ethics Advisory Committee.

11. Beauchamp, Tom, 1986–1991, Kennedy Institute of Ethics Tuesday seminars, electronic mail communications, and conversations, Department of Philosophy, Georgetown University. Beauchamp was on the staff of the National Commission for the Protection of Human Subjects of Biomedical and Behavioral Research.

12. Jonsen, Albert, 1985–1992, Conversations at Institute of Medicine activities and other events, University of California, San Francisco, and then University of Washington, Seattle. Jonsen was a commissioner on both the National Commission for the Protection of Human Subjects of Biomedical and Behavioral Research and the President's Commission for the Study of Ethical Problems in Medicine and Biomedical and Behavioral Research.

13. Mishkin, Barbara, 1983–1991, OTA workshop on impacts of neuroscience (1983), a meeting on prospects for a bioethics commission convened for Elizabeth McCloskey, staff to Senator Danforth (1989), and various phone and personal conversations concerning federal bioethics commissions, Hogan & Hartson, Washington, DC. Mishkin was on the staff of the National Commission for the Protection of Human Subjects of Biomedical and Behavioral Research, the Ethics Advisory Board (Department of Health, Education and Welfare), and the President's Commission for the Study of Ethical Problems in Medicine and Biomedical and Behavioral Research.

14. Stryker, Jeffrey, 1984–1993, Conversations, Office of Technology Assessment and the University of California, San Francisco. Stryker was a member of the staff for the President's Commission for the Study of Ethical Problems in Medicine and Biomedical and Behavioral Research and subsequently at the Office of Technology Assessment and the National Research Council.

15. Yesley, Michael S., 1988–1992, Interviews and conversations at meetings, Los Alamos National Laboratory, ELSI working group meetings, and elsewhere. Yesley was executive director of the National Commission for the Protection of Human Subjects of Biomedical and Behavioral Research, which operated from 1974 to 1978; he later became associated with the Department of Energy ELSI program while working in the legal office at Los Alamos National Laboratories.

16. Rothman, DJ. (1991). Chapter 9: Commissioning Ethics. In *Strangers at the Bedside,* pp. 168–189. New York: Basic Books.

17. Watson, JD. (1971). Statement of Dr. James D. Watson, Harvard University. Reprinted as "Poten-

tial Consequences of Experimentation with Human Eggs" in *International Science Policy,* House Committee on Science and Astronautics, pp. 149–161. (I thank Alexander Morgan Capron for finding this reference when I feared I was on a cold trail.). Panel on Science and Technology Twelfth Meeting: International Science Policy. Proceedings before the House Committee on Science and Astronautics, 92nd Congress. 26–28 January 1971.

18. Ramsey, P. (1975). *The Ethics of Fetal Research.* Yale University Press, New Haven, CT.

19. Jonsen, AR. (1984). Public Policy and Human Research. In *Biomedical Ethics Reviews,* JM Humber and RT Almeder, Ed., pp. 3–20. Clifton, NJ: Humana Press.

20. US Senate. (1973). *Quality of Health Care—Human Experimentation* (Washington, DC: Government Printing Office). Subcommittee on Health, Committee on Labor and Public Welfare. 21–22 February (Vol. 1); 23 February–6 March (Vol. II); 7–8 March (Vol. III).

21. National Commission for the Protection of Human Subjects of Biomedical and Behavioral Research, US Department of Health, Education and Welfare. (1975). *Research on the Fetus.* Government Printing Office, Washington, DC.

22. National Commission for the Protection of Human Subjects of Biomedical and Behavioral Research, US Department of Health, Education and Welfare. (1976). *Research Involving Prisoners.* Government Printing Office, Washington, DC.

23. National Commission for the Protection of Human Subjects of Biomedical and Behavioral Research, US Department of Health, Education and Welfare. (1977). *Disclosure of Research Information Under the Freedom of Information Act.* Government Printing Office, Washington, DC.

24. National Commission for the Protection of Human Subjects of Bimedical and Behavioral Research, US Department of Health, Education and Welfare. (1977). *Psychosurgery.* Government Printing Office, Washington, DC.

25. National Commission for the Protection of Human Subjects of Biomedical and Behavioral Research, US Department of Health, Education and Welfare. (1977). *Research Involving Children.* Government Printing Office, Washington, DC.

26. National Commission for the Protection of Human Subjects of Biomedical and Behavioral Research, US Department of Health, Education, and Welfare. (1978). *The Belmont Report: Ethical Principles and Guidelines for the Protection of Human Subjects of Research.* Government Printing Office, Washington, DC.

27. National Commission for the Protection of Human Subjects of Biomedical and Behavioral Research, US Department of Health, Education and Welfare. (1978). *Ethical Guidelines for the Delivery of Health Services by DHEW.* Government Printing Office, Washington, DC.

28. National Commission for the Protection of Human Subjects of Biomedical and Behavioral Research, US Department of Health, Education and Welfare. (1978). *Institutional Review Boards.* Government Printing Office, Washington, DC.

29. National Commission for the Protection of Human Subjects of Biomedical and Behavioral Research, US Department of Health, Education and Welfare. (1978). *Research Involving Those Institutionalized as Mentally Infirm.* Government Printing Office, Washington, DC.

30. National Commission for the Protection of Human Subjects of Biomedical and Behavioral Research, US Department of Health, Education and Welfare. (1978). *Special Study: Implications of Advances in*

Biomedical and Behavioral Research. Government Printing Office, Washington, DC.

31. Yesley, MS. (1978). The Use of an Advisory Commission. *Southern California Law Review* **51**:1451–1469.

32. Faden, RR, and Beauchamp, TL. (1986). *A History and Theory of Informed Consent*. Oxford University Press, New York.

33. Code of Federal Regulations. (1989). *Part 46—Protection of Human Subjects*, revised as of March 8, 1983, reprinted July 31, 1989. Code of Federal Regulations, Title 45; available through the Office of Protection from Research Risks, National Institutes of Health. 31 July.

34. President's Commission for the Study of Ethical Problems in Medicine and Biomedical and Behavioral Research. (1981). *Defining Death*. Government Printing Office, Washington, DC.

35. President's Commission for the Study of Ethical Problems in Medicine and Biomedical and Behavioral Research. (1981). *Protecting Human Subjects*. Government Printing Office, Washington, DC.

36. President's Commission for the Study of Ethical Problems in Medicine and Biomedical and Behavioral Research. (1982). *Whistleblowing in Biomedical Research*. Government Printing Office, Washington, DC.

37. President's Commission for the Study of Ethical Problems in Medicine and Biomedical and Behavioral Research. (1982). *Making Health Care Decisions* (with Vols. 2 and 3 appendices). Government Printing Office, Washington, DC.

38. President's Commission for the Study of Ethical Problems in Medicine and Biomedical and Behavioral Research. (1982). *Compensating Research Injury*. Government Printing Office, Washington, DC.

39. President's Commission for the Study of Ethical Problems in Medicine and Biomedical and Behavioral Research. (1983). *Deciding to Forego Life-Sustaining Treatment*. Government Printing Office, Washington, DC.

40. President's Commission for the Study of Ethical Problems in Medicine and Biomedical and Behavioral Research. (1983). *Implementing Human Research Regulations*. Government Printing Office, Washington, DC.

41. President's Commission for the Study of Ethical Problems in Medicine and Biomedical and Behavioral Research. (1983). *Securing Access to Health Care*. Government Printing Office, Washington, DC.

42. President's Commission for the Study of Ethical Problems in Medicine and Biomedical and Behavioral Research. (1983). *Summing Up*. Government Printing Office, Washington, DC.

43. President's Commission for the Study of Ethical Problems in Medicine and Biomedical and Behavioral Research. (1982). *Splicing Life* (Stock Number 040-000-00464-5, National Technical Information Service, Springfield, VA). November.

44. US House of Representatives. (1983). *Human Genetic Engineering* (Committee Print No. 170). Hearings before the Subcommittee on Investigations and Oversight, Committee on Science and Technology. 16–18 November 1982.

45. Public Law 98-158. (1985).

46. Walters, L. (1989). *Gene Therapy* (transcript of a meeting at the Office of Technology Assessment Conference Center). Biomedical Ethics Advisory Committee. 17 February.

47. Watson, JD. (1989). Meeting of the Joint NIH-DOE Subcommittee on Human Genome Research. Wilson Hall, National Institutes of Health. 5 December.

48. US House of Representatives. (1990). Departments of Labor, Health and Human Services, Education, and Related Agencies Appropriations for 1991 (Part 4A, pp. 83–84 and budget tables). Subcommittee on Labor, Health and Human Services, Education and Related Agencies of the Committee on Appropriations. March 20.

49. US House of Representatives. (1990). Departments of Labor, Health and Human Services, and Education, and Related Agencies Appropriation Bill, 1991 (101-516). Committee on Appropriations. 10 October.

50. US House of Representatives. (1990). Departments of Labor, Health and Human Services, Education, and Related Agencies Appropriations for 1991 (Part 4B, pp. 887–960). Committee on Appropriations.

51. US Congress. (1990). Making Appropriations for the Departments of Labor, Health and Human Services, and Education, and Related Agencies, for the Fiscal Year Ending September 30, 1991, and for Other Purposes (101-908). Conference Committee. 20 October.

52. Elmer-DeWitt, P, with Dorfman, A, and Nash, JM. (1989). The Perils of Treading in Heredity. *Time* (March 20):70–71.

53. Foundation for American Communications. (1989). *The New Genetics and the Right to Privacy*. Foundation for American Communications. 30 November.

54. Green, R. (1990). Tinkering with the Secrets of Life. *Health* **22** (January):46–50, 53, 84, 86.

55. Consumer Reports. (1990). The Telltale Gene. *Consumer Reports* (July):483–488.

56. Wright, R. (1990). Achilles' Helix. *New Republic* (July 9 & 16):21–31.

57. Bulger, RJ. (1991). How the Genome Project Could Destroy Health Insurance. *Washington Post, Outlook* (4 August):C4.

58. Henig, RM. (1989). High-Tech Fortunetelling. *New York Times Magazine* (24 December):20, 22.

59. Stipp, D. (1990). Genetic Testing May Mark Some People as Undesirable to Employers, Insurers. *Wall Street Journal* (9 July):B1, B2.

60. Billings, P. (1991). Beware of Insurers Offering Tests. *San Francisco Examiner* (11 July):A17.

61. Hurd, SN. (1990). Genetic Testing: Your Genes and Your Job. *Employee Responsibilities and Rights Journal* **3** (No. 4):239–252.

62. Frieden, J. (1991). Genetic Testing: What Will It Mean for Health Insurance? *Business and Health* (March):40–46.

63. Fine, MJ. (1991). Coping with Consequences of Genetic Testing. *Philadelphia Inquirer* (3 July):1A–14A.

64. Knudson, M. (1991). Scientists "Mapping" Genes Urge Ethical Safeguards. *Baltimore Sun* (21 July):A3.

65. Nelkin, Da, and Tancredi, L. (1989). *Dangerous Diagnostics: The Social Power of Biological Information*. Basic Books, New York.

66. Harvey, P. (1988). *S-S-S-S-S-SH, It's Genetic*. Paul Harvey News. 27–28 August.

67. Cavalieri, L. (1987). Gene Sequencing: No Easy Answers. *The Scientist* **1** (26 January):13.

68. Counts, CL. (1987). Human Genome Sequencing (letter). *Science* **236** (26 June):1613.

69. Hubbard, R, and Wilker, N. (1989). Evaluating the Human Genome Project. Annual Meeting, American Association for the Advancement of Science, San Francisco. 18 January.

70. Council for Responsible Genetics. (1990). Position Paper on the Human Genome Initiative. Council for Responsible Genetics. Accompanied letter from Nachama Wilker to Robert Cook-Deegan, 10 January.

71. Human Genetics Committee. (1990). Position Paper on Genetic Discrimination. Council for Responsible Genetics. Accompanied letter to Rep. Hoyer and Sen. Harkin. 27 June.

72. US Congress. (1984). *Human Gene Therapy—A*

Background Paper (OTA-BP-BA-32, Government Printing Office). Office of Technology Assessment. December.

73. Holtzman, NA. (1989). *Proceed with Caution.* Johns Hopkins University Press, Baltimore, MD.

74. Rothstein, MA. (1989). *Medical Screening and the Employee Health Cost Crisis.* Bureau of National Affairs, Washington, DC.

75. US Congress, and Office of Technology Assessment. (1988). *Medical Testing and Health Insurance.* Government Printing Office, Washington, DC.

76. Pokorski, R, Alexander, W, Battista, M, and Kay, B. (1989). *The Potential Role of Genetic Testing in Risk Classification.* Genetic Testing Committee, Medical Section, American Council of Life Insurance.

77. Kass, N. (1990). *The Ethical, Legal, and Social Issues Concerning the Use of Genetic Tests by Insurers: Toward the Development of Appropriate Public Policy.* Program in Law, Ethics, and Health, Johns Hopkins University. 9 November.

78. Gostin, L. (1990). *Genetic Discrimination: The Use of Genetically Based Diagnostic and Prognostic Tests by Employers and Insurers.* American Society of Law and Medicine. 10 December.

79. Greely, HT. (1992). Health Insurance, Employment Discrimination, and the Genetics Revolution. In *The Code of Codes,* DJ Kevles and L Hood, Ed., pp. 264–280. Cambridge, MA: Harvard University Press.

80. Subcommittee on Privacy Legislation. (1992). *Genetic Test Information and Insurance: Confidentiality Concerns and Recommendations.* Task Force on Genetic Testing, American Council on Life Insurance. February.

81. Task Force on Genetic Testing. (1992). *Report of the ACLI-HIAA Task Force on Genetic Testing, 1991.* American Council on Life Insurance and Health Insurance Association of America. February.

82. Palca, J. (1988). Another Report Smiles on Human Genome Sequencing Project. *Nature* **332** (28 April):769.

83. Fox, JL. (1988). Critics Converge on Genome Project. *Bio / Technology* **6** (June):643.

84. Recer, P. (1988). Group Proposed Citizens' Committee to Control Use of Genetic Knowledge. AP wire report (19 April):

85. Thompson, L. (1989). $100 Million Sought for Study of Genes. *Washington Post,* Health section (10 January):9.

86. Foundation on Economic Trends. (1989). Disability Rights Leaders from Across the Country, Jeremy Rifkin, and the Foundation on Economic Trend [sic] Challenge First Human Genetic Experiment at NIH Meeting January 30, 1989. Foundation on Economic Trends press statement. 30 January.

87. Roberts, L. (1989). Ethical Questions Haunt New Genetic Technologies. *Science* **243** (3 March):1134–1136.

88. Foundation on Economic Trends. (1988). Human Genome Fact Sheet. Press statement. 19 April.

89. Krimsky, S. (1982). *Genetic Alchemy: The Social History of the Recombinant DNA Controversy.* MIT Press, Cambridge, MA.

90. Howard, T, and Rifkin, J. (1977). *Who Should Play God?* Dell, New York.

91. Rifkin, J. (1983). *Algeny.* Viking, New York.

92. Recombinant DNA Advisory Committee. (1984). Minutes of meeting and background documents. National Institutes of Health. 6 February.

93. Foundation on Economic Trends. (1983). The Theological Letter Concerning the Moral Arguments Against Genetic Engineering of the Human Germline Cells. In *Taking Sides: Clashing Views on Controversial Bio-Ethical Issues,* C Levine, Ed.,

pp. 276–280. Guilford, CT: Dushkin Publishing, 1984.

94. Hatfield, MO. (1983). Genetic Engineering. *Congressional Record* **129, No. 82** (10 June):S8202–8205.

95. Dorfman, P. (1983). The Rifkin Resolution: Less Than Meets the Eye. *Genetic Engineering News* (July–August):4, 12.

96. Levine, C. (1984). Postscript: Is Genetic Engineering a Threat to Future Generations? In *Taking Sides: Clashing Views on Controversial Bio-Ethical Issues,* C Levine, Ed., Guilford, CT: Dushkin.

97. McCormick, R, 1984, Interview, Kennedy Institute of Ethics, Georgetown University, September.

98. Nelson, JR, 1984, Phone interview, Baylor College of Medicine, September.

99. Kevles, D. (1986). Unholy Alliance. *The Sciences* (September–October):24–30.

100. Rifkin, J. (1990). Letter to James D. Watson. Foundation on Economic Trends. 24 July.

101. US House of Representatives. (1991). Conyers Introduces Genetic Privacy Act. Committee on Government Operations. 3 May.

102. Ismail, S, 1991, Phone interview, Judiciary Committee, US House of Representatives, July.

103. Dorozynski, A. (1991). Privacy Rules Blindside French Glaucoma Effort. *Science* **252** (19 April): 369–3700.

104. Knoppers, B, 1991, Comments at ELSI working group meeting, National Institutes of Health and Department of Energy working group meeting, Ramada Inn, Bethesda, MD, 12 September.

105. Li, FP. (1992). Draft Recommendations on Predictive Testing for Germ Line p53 Mutations among Health Individuals (presented to ELSI Working Group, 10 February 1992). Dana Farber Cancer Institute, Boston, MA. 31 January.

106. Kevles, DJ. (1985). *In the Name of Eugenics.* University of California Press, Berkeley, CA.

107. Gallagher, HG. (1990). *By Trust Betrayed: Patients, Physicians, and the License to Kill in the Third Reich.* Henry Holt, New York.

108. Adams, MB, Ed. (1990). *The Wellborn Science: Eugenics in Germany, France, Brazil, and Russia.* Oxford University Press, New York.

109. Lifton, RJ. (1986). *The Nazi Doctors.* Basic Books, New York.

110. Müller-Hill, B. (1988). *Murderous Science.* Oxford University Press, New York.

111. Proctor, RN. (1988). *Racial Hygiene: Medicine Under the Nazis.* Harvard University Press, Cambridge, MA.

112. Annas, GJ, and Grodin, MA. (1992). *The Nazi Doctors and the Nuremberg Code: Human Rights in Human Experimentation.* Oxford University Press, New York.

113. Wertz, DC, and Fletcher, JC. (1989). *Ethics and Human Genetics: A Cross-Cultural Perspective.* Springer-Verlag, New York.

114. Kevles, DJ. (1992). Out of Eugenics: The Historical Politics of the Human Genome. In *The Code of Codes: Scientific and Social Issues in the Human Genome Project,* DJ Kevles and L Hood, Eds., pp. 3–36. Cambridge, MA: Harvard University Press.

115. Government of Italy. (1988). *Sequencing the Human Genome, Ethical and Social Issues* (Brescia, Italy: Clas International, 1989). Government of Italy organized conference of Economic Summit nations and the European Commission. 10–15 April.

116. Knoppers, B. (1989). Human Genetics, Predisposition, and the New Social Contract. In *International Conference on Bioethics. The Human Genome Sequencing: Ethical Issues,* Ed., Brescia, Italy: Clas International.

117. Knoppers, BM. (1991). *Human Dignity and Ge-*

netic Heritage (Study Paper, Protection of Life Series). Law Reform Commission of Canada, Ottawa.

118. The British Warnock Report touched on genetics in several of its recommendations. The German Benda Commission proposed a broad set of laws to cover gene therapy, embryo research, and genetic technologies. This was followed a few years later by a report from the Enquete Commission of the German Parliament, and ultimately led to new German statutes governing scientific research. The French government considered a report on ethics and the life sciences, and produced legislation to protect human subjects of research. The Glover Report to the European Commission included a section on handicaps that discussed the role of genetic testing. A January 1990 genome conference in Paris devoted a session to discussion of eugenics, and scientists' responsibilities in genome research. Santiago Grisolia's second conference in Valencia was entirely devoted to discussion of the social implications of genome research. And finally, the Council of Europe's Ad Hoc Committee of Experts on Bioethics (CAHBI) worked towards reports on the use of genetic testing and screening in medicine and on DNA testing in the criminal justice system.

119. Ad Hoc Committee of Experts on Bioethics. (1991). Draft Recommendation on the Use of Analysis of deoxyribonucleic Acid (DNA) within the Framework of the Criminal Justice System and Draft Explanatory Memorandum (CAHBI (91) 17 Addendum II, Revised [ADBI9117.A]). Council of Europe. 17 December.

120. Ad Hoc Committee of Experts on Bioethics. (1991). Draft Recommendation on Genetic Testing and Screening for Health Care Purposes and Draft Explanatory Memorandum (CAHBI (91) 17, Addendum I, Revised [ABI91AD1]). Council of Europe. 17 December.

121. Braibant, P. (1988). *Sciences de la Vie, de l'Éthique au Droit ("Life Sciences and the Rights of Man")*. *Étude du Conseil d'État*, Government of France, Paris.

122. *Commission of Inquiry into Human Fertilisation and Embryology. (1984). Report* (the Warnock report). Her Majesty's Stationery Office. July.

123. Dickson, D. (1989). France Introduces Bioethics Law. *Science* **243** (10 March):1284.

124. Enquetekommission (Enquete Commission). (1987). *Chancen und Risiken der Gentechnologie* (Munich: Schweitzer). Deutcher Bundestag (German Parliament).

125. Glover, J. (1989). *Fertility and the Family: The Glover Report on New Reproductive Technologies*. The European Commission.

126. Human Genome Research. (1990). Strategies and Priorities. Committee on Genetic Experimentation (COGENE), International Council of Scientific Unions (ICSU), European Community (EC), Federation of European Biochemical Societies (FEBS), International Union of Biochemistry (IUB), and the United Nations Educational, Scientific, and Cultural Organization (UNESCO). 29–31 January.

127. II Workshop on International Cooperation for the Human Genome Project. (1990). International Conference. Fundación BBV, Fundación Valencianan de Estudios Avanzados. 12–14 November.

128. Ministères des Affaires Sociales et de la Solidarité. (1991). Protection of Persons Undergoing Biomedical Research (Laws No. 88-1138 (20 December 1988), 90-86 (23 January 1990), 90-549 (2 July 1990) and 91-73 (18 January 1991), translated by the French government). Government of France. March.

129. Working Group on in Vitro Fertilisation, GA, and Gene Therapy. (1985). Report (The Benda Commission report). Federal Ministry of Justice and Federal Ministry for Research and Technology, Federal Republic of Germany.

130. Dickson, D. (1989). Genome Project Gets Rough Ride in Europe. *Science* **243** (3 February):599.

131. Commission of the European Communities. (1988). Proposal for a Council Decision Adopting a Specific Research Programme in the Field of Health: Predictive Medicine: Human Genome Analysis (1989–1991). Commission of the European Communities. 20 July.

132. Commission of the European Communities. (1989). Modified Proposal for a Council Decision, Adopting a Specific Research and Technological Development Programme in the Field of Health—Human Genome Analysis: (1990–1991). Commission of the European Communities. 13 November.

133. Economic and Social Committee. (1988). Opinion of the Economic and Social Committee on the Proposal for a Council Decision Adopting a Specific Research Programme in the Field of Health: Predictive Medicine: Human Genome Analysis (1989–1991). Commission of the European Communities. 14 December.

134. Council of the European Communities. (1990). Council Decision Adopting a Specific Research and Technological Development Programme in the Field of Health: Human Genome Analysis (1990 to 1991) (90 / 395 / EEC). *Official Journal of the European Communities* (L 196, 26 July):8–14.

135. Spurr, NK. (1992). EC Human Genome Analysis Programme 1990–1992. *G-Nome News* (No. 10, June):21–23.

136. Human Genome Analysis Programme, 1992, Announcement of Grants, European Commission, February.

137. Privacy Commissioner of Canada. (1992). *Genetic Testing and Privacy* (Cat. No. IP34-3 / 1992, Minister of Supply and Services, Ottawa).

138. Scriver, CR. (1992). *Letter*. Institut de Recherche de l'Université McGill-Hôpital de Montréal pour Enfants. 17 June.

139. Capron, A, Pellegrino, E, and Cook-Deegan, R. (1990). Letters to Senator Tom Harkin and Steny Hoyer. Biomedical Ethics Advisory Committee. 25 June 1990.

140. Billings, P. (1990). Genetic Discrimination: An Ongoing Survey. GeneWatch, Council for Responsible Genetics, accompanying 27 June letter to Rep. Hoyer and Sen. Harkin. 27 June.

141. US Senate. (1990). Americans with Disabilities Act—Conference Report. *Congressional Record* **136** (#89, 13 July):S 9687 (Hatch).

142. US House of Representatives. (1990). Conference Report on S. 933, Americans with Disabilities Act of 1990. *Congressional Record* **136** (#88, 12 July):H 4623 (Owens), H 4625 (Edwards), H 4627 (Waxman).

143. Hoyer, S. (1990). Letter to Alexander Morgan Capron, Biomedical Ethics Advisory Committee. US House of Representatives. 1 August.

144. Rothstein, MA. (1990). Recommendations for Regulations to Implement the Americans with Disabilities Act, Submitted to the Equal Employment Opportunity Commission. University of Houston Law Center. 20 November.

145. Equal Employment Opportunity Commission. (1991). Equal Employment Opportunity for Individuals with Disabilities; Notice of Proposed Rulemaking, 29 CFR 1630. *Federal Register* **Part VI** (28 February):8578–8603.

146. Rothstein, MA. (1991). Comments on Proposed Regulations to Implement the Equal Employment Provisions of the Americans with Disabilities Act.

Submitted to the Equal Employment Opportunity Commission. 15 April.
147. The ELSI Working Group urged EEOC to circumscribe health inquiries to only those questions that would elicit information about conditions that would affect job performance. The proposed regulations indicated that employers could obtain the full medical record, but could only *use* information that was job-related to reject a job applicant. This shifted power to the employer and took power away from the prospective employee. To assert discrimination, an individual would have to prove that the reason he or she was rejected was *not* job-related, but the job applicant would have supplied vast amounts of information with little knowledge of how it was used by the employer. If, however, the employer never had medical information beyond that relevant to job performance in the first place, then arbitrary decisions could not be masked. This would increase the effort necessary on the part of employers to define what job-related questions to ask, and on those responding to employer requests, as they would have to filter the medical record rather than ship it out whole. It implied a major change in standard practice, but one in harmony with the ADA statute.
148. Joint Working Group on Ethical, Legal, and Social Issues. (1991). Genetic Discrimination and the Americans with Disabilities Act. Submitted to the Equal Employment Opportunity Commission. 29 April.
149. Aldhous, P. (1991). Closing a Loophole in Discrimination Rules. *Nature* **351** (27 June):684.
150. Berg, P, and Wolff, S. (1991). Letter to Evan J. Kemp, Chairman, Equal Employment Opportunity Commission. NIH/DOE Joint Subcommittee on the Human Genome. 10 July.
151. Equal Employment Opportunity Commission. (1991). Equal Employment Opportunity for Individuals with Disabilities; Final Rule. 29 Code of Federal Regulations, Part 1630. 29 Code of Federal Regulations, Parts 1602 and 1627. Record-keeping and Reporting Under Title VII of the Civil Rights Act of 1964 and the Americans with Disabilities Act (ADA); Final Rule, *Federal Register* **56** (144, 26 July):35726–35756.
152. Childress, JF. (1991). Human Fetal Tissue Transplantation Research. In *Biomedical Politics,* KE Hanna, Ed., pp. 215–248. Washington, DC: National Academy Press.
153. Consultants to the Advisory Committee to the Director. (1988). Report of the Human Fetal Tissue Transplantation Research Panel, Vol. 1. National Institutes of Health. December.
154. Consultants to the Advisory Committee to the Director. (1988). Report of the Human Fetal Tissue Transplantation Research Panel, Vol. 2. National Institutes of Health. December.
155. Davis, M, and Hoben, P, 1988, Phone and personal conversations, Office of the Assistant Secretary of Health, Department of Health and Human Services, April–November.

156. Advisory Committee to the Director. (1988). Human Fetal Tissue Transplantation Research. National Institutes of Health. 14 December.
157. Windom, RE, 1992, Letter to President-elect Bill Clinton, from Sarasota Florida, 7 November.
158. Windom, RE, 1993, Letter to Donna Shalala, Secretary-designate of Health and Human Services, from Sarasota, Florida, 20 January.
159. Harrelson, W. (1991). Commentary. In *Biomedical Politics,* KE Hanna, Ed., pp. 255–257. Washington, DC: National Academy Press.
160. King, PA. (1991) Commentary. In *Biomedical Politics,* KE Hanna, Ed., Washington, DC: National Academy Press.
161. US House of Representatives. (1992). Domestic and International Data Protection Issues (Committee Print). Subcommittee on Government Information, Justice, and Agriculture, Committee on Government Operations. Hearings 10 April and 17 October 1991.
162. US House of Representatives. (1992). Designing Genetic Information Policy: The Need for an Independent Policy Review of the Ethical, Legal, and Social Implications of the Human Genome Project (House Report 102-478). Committee on Government Operations. 2 April.
163. Wise, B. (1992). Letter. Government Information, Justice, and Agriculture Subcommittee, Committee on Government Operations, US House of Representatives. 7 April.
164. Mervis, J. (1992). NIH Forms Policy Centre to Study Research Ethics. *Nature* **356** (2 April): 367.
165. Mervis, J. (1992). From Blue-Chip to White Coat. *Nature* **356** (2 April):367.
166. Social, Legal, and Ethical Issues in Biomedical and Behavioral Research Policy Panel. (1991). Report on Social, Legal, and Ethical Issues in Biomedical and Behavioral Research. National Institutes of Health. Undated, but completed before December.
167. Hatfield, M. (1992). Statement on an amendment to the NIH authorization bill (HR 2507), proposed and withdrawn. *Congressional Record* **138** (No. 49, 2 April):S4719–S4720, S4723–S4724.
168. US Senate. (1992). Letters from Senators Frank Lautenberg and Hank Brown, Congressional Biotechnology Caucus; Louis W. Sullivan, Jr., M.D., Secretary of Health and Human Services; Harry P. Manbeck, Jr., Assistant Secretary and Commissioner of Patents and Trademarks. Letter and fact sheet from Stephen A Duzan, Chairman and CEO, Immunex Corp. *Congressional Record* **138** (No. 49, 2 April):S4724–S4726.
169. U.S. Congress. (1993). Biomedical Ethics in U.S. Public Policy. (OTA-BP-BBS-104, Government Printing Office, Washington, DC). Office of Technology Assessment.
170. US Senate. (1992). Letter from Louis W. Sullivan, Jr., M.D., Secretary of Health and Human Services. *Congressional Record* **138** (No. 49, 2 April):S4725.

18. WIZARDS OF THE INFORMATION AGE

1. Ulam, SM. (1976). *Adventures of a Mathematician.* Charles Scribner's Sons, New York.
2. Rota, G-C. (1987). The Lost Cafe. *Los Alamos Science* **Special Issue** (No. 15):23–32. Republished in *From Cardinals to Chaos: Reflections on the Life and Legacy of Stanislaw Ulam* (NY: Cambridge University Press, 1989).
3. The critical problem was to create and maintain temperatures high enough to ignite the same reactions that take place in the sun and other stars. In work

with C. J. Everett, Ulam pioneered the use of the "Monte Carlo" method to model how Teller's scheme for doing this would play out. The Ulam-Everett calculations showed the reaction would fizzle before generating an explosion. The Monte Carlo technique was a way to simulate complex phenomena mathematically. It was useful in characterizing an enormous range of applications, and had vast practical implications well beyond the hydrogen bomb simulations. The Ulam-Everett calculations predated the ready availability of computer technology needed to make Monte Carlo models generally practical. The fusion reaction calculations were nonetheless among the earliest and most historically significant demonstrations of the technique.

4. Blow, M. (1968). *The History of the Atomic Bomb.* American Heritage, New York.

5. Norman, C. (1990). How the Soviets Got the H-Bomb. *Science* **247** (12 January):151.

6. Ulam, SM. (1972). Some Ideas and Prospects in Biomathematics. *Annual Review of Biophysics and Bioengineering* **1**:277–291.

7. Beyer, WA, Stein, ML, Smith, TF, and Ulam, SM. (1974). A Molecular Sequence Metric and Evolutionary Trees. *Mathematical Biosciences* **19**:9–25.

8. Zuckerkandl, E, and Pauling, L. (1965). Molecules as Documents of Evolutionary History. *Journal of Theoretical Biology* **8**:357–358.

9. Fitch, WM, and Margoliash, E. (1967). Construction of Phylogenetic Trees. *Science* **155**:279–284.

10. Dayhoff, MO, and Eck, RV. (1966). *Atlas of Protein Sequence and Structure.* National Biomedical Research Foundation, Silver Spring, MD.

11. Smith, TF. (1990). The History of the Genetic Sequence Databases. *Genomics* **6** (April):701–707.

12. Stormo, GD. (1987). Identifying Coding Sequences. In *Nucleic Acid and Protein Sequence Analysis: A Practical Approach,* MJ Bishop and CJ Rawlings, Ed. Oxford, UK: IRL Press.

13. Needleman, SB, and Wunch, CD. (1970). A General Method Applicable to the Search for Similarities in the Amino Acid Sequences of Two Proteins. *Journal of Molecular Biology* **48**:443–453.

14. Sellers, PH. (1974). On the Theory and Computation of Evolutionary Distances. *SIAM Journal of Applied Mathematics* **26**:787.

15. Beyer, WA, Burks, C, and Goad, WB. (1983). Quantitative Comparison of DNA Sequences. *Los Alamos Science* (Fall):62–63.

16. Smith, TF, and Materman, MS. (1981). Identification of Common Molecular Subsequences. *Journal of Molecular Biology* **147**:195–197.

17. Karlin, S, and Brendel, V. (1992). Chance and Statistical Significance in Protein and DNA Sequence Analysis. *Science* **257** (3 July):39–49. See especially items referenced in footnote 5.

18. DeLisi, C, 1991, Interview, Boston University School of Engineering; interview in Senator Pete Domenici's office, Dirksen Senate Office Building, 18 July.

19. Goad, WB. (1983). GenBank. *Los Alamos Science* (Fall):52–61.

20. Smith, T, 1988, Interview, Harvard Medical School, the Jimmy Fund Building, 14 November.

21. Waterman, M, 1988, Car ride and interview, Xerox PARC facility, Middleburg, VA, 24 July.

22. Lewin, R. (1986). DNA Databases Are Swamped. *Science* **232** (27 June):1599.

23. Goad, WB. (1982). Acquisition, Integration and Analysis of Molecular Genetic Information: A Workshop. Aspen Center for Physics. 30 August–18 September.

24. Goad, WB. (1983). Terminal Progress Report: Workshop on Computational Analysis of DNA Sequences, Aspen Center for Physics, August 30–September 19, 1982. Biotechnology Resources Program, Division of Research Resources, Conference Grant 1 R RR01628-1. 7 October.

25. Smith, TF, and Burks, C. (1983). Searching for Sequence Similarities. *Nature* **301** (20 January):194.

26. Marx, JL. (1989). Bionet Bites the Dust. *Science* **245** (14 July):126.

27. Goad, WB. (1988). GenBank—A Retrospective. *News from GenBank* **1** (No. 3):1–2.

28. Burks, C, Cinkosky, MJ, Gilna, P, Hayden, JE-D, Abe, Y, Atencio, EJ, Barnhouse, S, Benton, D, Buenafe, CA, Cumella, KE, Davison, DB, Emmert, DB, Etnier, CM, Faulkner, MJ, Fickett, JW, Fischer, WM, Good, M, Horne, D, A., Houghton, FK, Kelkar, PM, Kelley, TA, Kelly, M, King, MA, Langan, BJ, Lauer, JT, Lopez, N, Lynch, C, Lynch, J, Marchi, JB, Marr, TG, Martinez, FA, McLeod, MJ, Medvick, PA, Mishra, SK, Moore, J, Munk, C, Mondragon, SM, Nasseri, KK, Nelson, D, Nelson, W, Nguyen, T, Reiss, G, Rice, J, Ryals, J, Salazar, MD, Stelts, SR, Trujilla, BL, Tomlinson, LJ, Weiner, MG, Welch, FJ, Wiig, SE, Yudin, K, and Zins, LB. (1989). GenBank: Current Status and Future Directions (LA-UR-89-1154). Los Alamos National Laboratory. 2 June.

29. Goad, WB. (1979). Proposal to Establish a National Center for Collection and Computer Storage and Analysis of Nucleic Acid Sequences (LASL P-F80-5). Los Alamos Scientific Laboratory. 17 December.

30. International Advisory Committee for DNA Sequence Databases. (1988). Summary. National Institutes of Health. 16 February.

31. Editorial. (1988). Why Sequence the Human Genome? *Nature* **331** (11 February):465.

32. Philipson, L. (1988). The DNA Data Libraries. *Nature* **332** (21 April):676.

33. Maddox, J. (1989). Making Good Data-Banks Better. *Nature* **341**:277.

34. Roberts, RJ. (1989). Benefits of Databases. *Nature* **342** (9 November):114.

35. Koetzle, TF. (1989). Benefits of Databases. *Nature* **342** (9 November):114.

36. Cameron, G, Kahn, P, and Philipson, L. (1989). Journals and Databanks. *Nature* **342** (21 / 28 December):848.

37. Maddox, J. (1989). Making Authors Toe the Line. *Nature* **342** (21 / 28 December):855.

38. Kuhn, TS. (1962). *The Structure of Scientific Revolutions.* University of Chicago Press, Chicago.

39. McKusick, VA. (1962). On the X Chromosome of Man. *Quarterly Review of Biology* **37**:69–175.

40. McKusick, VA. (1966). *Mendelian Inheritance in Man.* Johns Hopkins University Press, Baltimore, MD.

41. McKusick, VA. (1989). Historical Perspectives: The Understanding and Management of Genetic Disorders. *Maryland Medical Journal* **38** (November):901–908.

42. McKusick, VA. (1990). Foreword to the Ninth Edition. In *Mendelian Inheritance in Man,* VA McKusick, Ed., pp. xi–xxix. Baltimore, MD: Johns Hopkins University Press.

43. Pearson, PL, Lucier, R, and Brunn, C. (1991). Databases to Serve the Genome Program and the Medical Genetics Community. In *Etiology of Human Disease at the DNA Level,* J Lindsten and U Pettersson, Ed. New York: Raven Press.

44. McGourty, C. (1989). Johns Hopkins as International Host. *Nature* **342** (23 November):330.

45. Anonymous. (1989). Strategic Plan: Mission Statement. Genome Data Base. 14 September 1989.

46. Anonymous. (1990). GDB Introduced at HGM 10.5. *GDB Forum* **1** (No. 2, Fall):1,3.

47. Pearson, PL. (1991). The Genome Data Base (GDB)—A Human Gene Mapping Repository. *Nucleic Acids Research* **19** Supplement:2237–2239.

48. Hinton, D, Pearson, P, and Watson, JD, 1990–1991, Conversations fall 1990 through summer 1991, Howard Hughes Medical Institute (DH), Johns Hopkins University (PP), Cold Spring Harbor Laboratory (JDW), and National Center for Human Genome Research (JDW),

49. Human Genome News. (1992). New GDB Remote

Node Opens in Australia. *Human Genome News* **3**, No. 6 (March):9.

50. Matsubara, K, and Kakunaga, M. (1992). The Genome Efforts in Japan as of 1991. *Genomics* **12**:618–620.

51. Pearson, PL, Maidak, B, and Chipperfield, M. (1991). The Human Genome Initiative—Do Databases Reflect Current Progress? *Science* **254** (11 October):214–215, also Genome Maps 1991 center section.

52. Courteau, J. (1991). Genome Databases. *Science* **254** (11 October):201–207.

53. Roberts, L. (1992). The Perils of Involving Congress in a "Catfight." *Science* **257** (10 July):156–157.

54. That process was complex and contentious in itself, involving not only the National Institute of General Medical Sciences and NLM, but also the NIH genome office, the DOE genome office, the GenBank group at Los Alamos, and the Howard Hughes Medical Institute.

55. Ott, J. (1990). Cutting a Gordian Knot in the Linkage Analysis of Complex Human Traits. *American Journal of Human Genetics* **46**:219–221.

56. Risch, N. (1990). Linkage Strategies for Genetically Complex Traits. III. The Effect of Marker Polymorphism on Analysis of Affected Relative Pairs. *American Journal of Human Genetics* **46**:242–253.

57. Risch, N. (1990). Linkage Strategies for Genetically Complex Traits. II. The Power of Affected Relative Pairs. *American Journal of Human Genetics* **46**:229–241.

58. Risch, N. (1990). Linkage Strategies for Genetically Complex Traits. I. Multilocus Models. *American Journal of Human Genetics* **46**:222–228.

59. Lander, ES, and Botstein, D. (1986). Mapping Complex Genetic Traits in Humans: New Methods Using a Complete RFLP Linkage Map. *Cold Spring Harbor Symposia on Quantitative Biology* **51** (Molecular Biology of *Homo sapiens*):49–62.

60. Lander, ES, and Botstein, D. (1987). Homozygosity Mapping: A Way to Map Recessive Traits in Humans by Studying the DNA of Inbred Children. *Science* **236**:1567–1570.

61. Bowie, JU, Lüthy, R, and Eisenberg, D. (1991). A Method to Identify Protein Sequences That Fold into a Known Three-Dimensional Structure. *Science* **253** (12 July):164–170.

62. Blundell, TL, Sibanda, BL, Sternberg, MJE, and Thornton, JM. (1987). Knowledge-Based Prediction of Protein Structures and the Design of Novel Molecules. *Nature* **326** (26 March):347–352.

63. Presta, LG, and Rose, GD. (1988). Helix Signals in Proteins. *Science* **240** (17 June):1632–1641.

64. Fetrow, JS, Zehfus, MH, and Rose, GD. (1988). Protein Folding: New Twists. *Bio / Technology* **6** (No. 2, February):167–171.

65. Rooman, MJ, and Wodak, SJ. (1988). Identification of Predictive Sequence Motifs Limited by Protein Structure Data Base Size. *Nature* **335** (1 September):45–49.

66. Lander, ES, and Waterman, MS. (1988). Genomic Mapping by Fingerprinting Random Clones: A Mathematical Analysis. *Genomics* **2**:231–239.

67. Doolittle, RF, Hunkapiller, MW, Hood, LE, DeVare, SG, Robbins, KC, Aaronson, SA, and Antoniades, HN. (1983). Simian Sarcoma virus *onc* gene, *v-sis*, Is Derived from the Gene (or Genes) Encoding a Platelet-Derived Growth Factor. *Science* **221**:275–277.

68. Downward, J, Yarden, Y, Mayes, E, Scrace, G, Totty, N, Stockwell, P, Ullrich, A, Schlessinger, J, and Waterfield, MD. (1984). Close Similarity of Epidermal Growth Factor Receptor and v-*erb-B*

oncogene protein sequences. *Nature* **307**:521–527.

69. Doolittle, R. (1987). *Of URFs and ORFs: A Primer on How to Analyze Derived Amino Acid Sequences.* University Science Books, Mill Valley, CA.

70. Committee on Models for Biomedical Research. (1985). *Models for Biomedical Research.* National Academy of Sciences.

71. Morowitz, HJ, and Smith, T. (1987). Report of the Matrix of Biological Knowledge Workshop. Santa Fe Institute. 30 October.

72. Committee on Computer-Assisted Modeling. (1987). *Computer-Assisted Modeling. Contributions of Computational Approaches to Elucidating Macromolecular Structure and Function.* Board on Basic Biology, Commission on Life Sciences, National Research Council, National Academy of Sciences.

73. Karp, W. (1968). Controversy and Conquest. In *Charles Darwin and the Origin of Species,* Editors of Horizon Magazine, Ed., pp. 120–135. New York: American Heritage.

74. Ziman, J. (1976). *The Force of Knowledge: The Scientific Dimension of Society.* Cambridge University Press, New York.

75. King, M-C, and Wilson, AC. (1975). Evolution at Two Levels in Humans and Chimpanzees. *Science* **188** (11 April):107–116.

76. Cann, RM, Stoneking, M, and Wilson, AC. (1987). Mitochondrial DNA and Human Evolution. *Nature* **325**:31–36.

77. Gibbons, A. (1992). Mitochondrial Eve: Wounded, but Not Dead Yet. *Science* **257** (14 August):873–875.

78. Cavalli-Sforza, LL, Piazza, A, Menozzi, P, and Mountain, J. (1988). Reconstruction of Human Evolution: Bringing Together Genetic, Archaeological, and Linguistic Data. *Proceedings of the National Academy of Sciences (USA)* **85** (August):6002–6006.

79. Gibbons, A. (1991). Systematics Goes Molecular. *Science* **251** (22 February):872–874.

80. Doolittle, RF, Feng, DF, Johnson, MS, and McClure, MA. (1986). Relationships of Human Protein Sequences to Those of Other Organisms. *Cold Spring Harbor Symposia on Quantitative Biology. The Molecular Biology of Homo Sapiens* **51**:447–455.

81. Bodmer, WF, and Cavalli-Sforza, LL. (1976). *Genetics, Evolution, and Man.* W. H. Freeman, San Francisco.

82. Bowcock, A, and Cavalli-Sforza, LL. (1991). The Study of Variation in the Human Genome. *Genomics* **11** (October):491–498.

83. Bowcock, AM, Kidd, JR, Mountain, JL, Hebert, JM, Carotenuto, L, Kidd, KK, and Cavalli-Sforza, LL. (1991). Drift, Admixture, and Selection in Human Evolution: A Study with DNA Polymorphisms. *Proceedings of the National Academy of Sciences, USA* **88**:839–843.

84. Cavalli-Sforza, LL. (1990). Opinion: How Can One Study Individual Variation for 3 Billion Nucleotides of the Human Genome? *American Journal of Human Genetics* **46**:649–651.

85. Cavalli-Sforza, LL, Wilson, AC, Cantor, CR, Cook-Deegan, RM, and King, M-C. (1991). Call for a Worldwide Survey of Human Genetic Diversity: A Vanishing Opportunity for the Human Genome Project. *Genomics* **11** (October):490–491.

86. Cavalli-Sforza, LL, Piazza, A, and Menozzi, P. (1992). Epilogue. In *History and Geography of Human Genes,* Princeton, NJ: Princeton University Press.

87. Roberts, L. (1991). A Genetic Survey of Vanishing Peoples. *Science* **252** (21 June):1614–1617.

88. Roberts, L. (1992). Genetic Survey Plans Move Ahead. *Science* **256** (19 June):1629.

89. Darwin, C. (1985, reprinted from 1858). *The Origin of Species.* Mentor, New York.
90. Whitehead, AN, and Russell, B. (1910). *Principia Mathematica,* Vol. 1. Cambridge University Press, Cambridge, UK.
91. Whitehead, AN, and Russell, B. (1912). *Principia Mathematica,* Vol. 2. Cambridge University Press, Cambridge, UK.
92. Whitehead, AN, and Russell, B. (1913). *Principia Mathematica,* Vol. 3. Cambridge University Press, Cambridge, UK.
93. van Heijenoort, J, Ed. (1967). *From Frege to Gödel: A Source Book in Mathematical Logic, 1987–1931.* Harvard University Press, Cambridge, MA.
94. Gödel, K. (1931). On Formally Undecidable Propositions of Principia Mathematica and Related Systems I. In *From Frege to Gödel: A Source Book in Mathematical Logic, 1879–1931,* J van Heijenoort,

Ed., pp. 596–616. Cambridge, MA: Harvard University Press, 1967.
95. Gödel, K. (1930). Some Metamathematical Results on Completeness and Consistency. In *From Frege to Gödel: A Source Book in Mathematical Logic, 1879–1931,* J van Heijenoort, Ed., pp. 595–596. Cambridge, MA: Harvard University Press, 1967.
96. Gödel, K. (1931). On Completeness and Consistency. In *From Frege to Gödel: A Source Book in Mathematical Logic, 1879–1931,* J van Heijenoort, Ed., pp. 616–617. Cambridge, MA: Harvard University Press, 1967.
97. Quine, WvO. (1981). Kurt Gödel. In *Theories and Things,* pp. 143–147. Cambridge, MA: Harvard University Press.
98. Chaitin, GJ. (1987). Conclusion. In *Algorithmic Information Theory,* pp. 160–161. New York: Cambridge University Press.

19. DNA GOES TO COURT

1. Wambaugh, J. (1989). *The Blooding.* William Morrow, New York. The Leicester investigation is abstracted from this book.
2. US Congress. (1990). *Genetic Witness: Forensic Uses of DNA Tests* (OTA-BA-438). Office of Technology Assessment. July.
3. Nishimi, RY, 1992, Electronic mail notes, Office of Technology Assessment, US Congress, 24 August.
4. *Frye* v. *United States.* (1923). *293 F. 1013.* District of Columbia Circuit Court.
5. Rothstein, PF, Ed. (1979). *Federal Rules of Evidence.* 2nd ed. Clark Boardman, New York.
6. Committee on DNA Technology in Forensic Science. (1992). *DNA Technology in Forensic Science* (Washington, DC: National Academy Press). Board on Biology, Commission on Life Sciences, National Research Council. 16 April prepublication draft. Final draft and summary released in July.
7. Lander, E. (1992). DNA Fingerprinting: Science, Law, and the Ultimate Identifier. In *The Code of Codes: Scientific and Social Issues in the Human Genome Project,* DJ Kevles and L Hood, Ed., pp. 191–210. Cambridge, MA: Harvard University Press.
8. Wolfe, T. (1987). *The Bonfire of the Vanities.* Farrar, Straus & Giroux; paperback Bantam, December 1988, New York.
9. Lander, ES. (1989). DNA Fingerprinting on Trial. *Nature* 339 (15 June):501–505.
10. Lewontin, RC, and Hartl, DL. (1991). Population Genetics in Forensic DNA Typing. *Science* 254:1745–1750.
11. Chakraborty, R, and Kidd, K. (1991). The Utility of DNA Typing in Forensic Work. *Science* 254:1735–1739.
12. Roberts, L. (1992). Science in Court: A Culture Clash. *Science* 257 (7 August):732–736.
13. Kolata, G. (1992). US Panel Seeking Restriction on Use of DNA in Courts. *New York Times* (14 April):A1, C7.
14. Roberts, L. (1992). DNA Fingerprinting: Academy Reports. *Science* 256 (17 April):300–301.
15. Lewontin, RC. (1992). The Dream of the Human Genome. *New York Review of Books* (28 May):31–40.
16. Anderson, C. (1992). Courts Reject DNA Fingerprinting, Citing Controversy After NAS Report. *Nature* 359 (1 October):349.
17. Aldhous, P. (1993). Geneticists Attack NRC Report as Scientifically Flawed. *Science* 259 (5 February):755–756.
18. Devlin, B, Risch, N, and Roeder, K. (1993). Statistical Evaluation of DNA Fingerprinting: A Critique

of the NRC's Report. *Science* 259 (5 February):748–749; 837.
19. US Congress. (1981). *Impacts of Applied Genetics: Micro-Organisms, Plants and Animals* (OTA-HR-132). Office of Technology Assessment. April.
20. US Congress. (1984). *Commercial Biotechnology, an International Analysis* (OTA-BA-218, Government Printing Office). Office of Technology Assessment. January.
21. National Institutes of Health. (1989). *NIH Data Book 1989* (NIH Publication Number 90-1261). National Institutes of Health. December.
22. US Congress. (1993). *Government Policies and Pharmaceutical Research and Development.* Office of Technology Assessment.
23. Cohen, S, and Boyer, H, 1976, US Patent 4,237,224, Stanford University and University of California, San Francisco.
24. 447 US 303, 1980.
25. Rosenfeld, SC. (1988). Sharing of Research Results in a Federally Sponsored Gene Mapping Project. *Rutgers Computer and Technology Law Journal* 14 (No. 2):311–358.
26. Public Law 96-517, 1980.
27. Public Law 98-620, 1984.
28. Public Law 99-502, 1986.
29. Benson, RH. (1986). Patent Wars. *Bio / Technology* 4 (December):1064–1970.
30. Fowlston, BJ. (1988). Biotechnology Patents in Europe. *Bio / Technology* 6 (August):911–913.
31. Klausner, A. (1988). Patent Office Reorganizes Biotech Coverage. *Bio / Technology* 6 (April):368.
32. US Congress. (1988). *New Developments in Biotechnology, 4. US Investment in Biotechnology.* Office of Technology Assessment, OTA-BA-360, Washington, DC: Government Printing Office. July.
33. Koniarek, JP, and Coleman, KD. (1988). Patents and Patent Office Resources in Biotechnology. *BioTechniques* 6 (No. 2):148–153.
34. Marshall, E. (1991). The Patent Game: Raising the Ante. *Science* 253 (5 July):20–24.
35. US Congress. (1989). *New Developments in Biotechnology. 5. Patenting Life* (OTA-BA-370). Office of Technology Assessment. April.
36. Philipson, L, and Tooze, J. (1987). The Human Genome Project—Mapping and DNA Sequence Determination. *BioFutur* (No. 58):94–101.
37. Gilbert, W. (1991). Towards a Paradigm Shift in Biology. *Nature* 349 (10 January):99.
38. Office of Technology Assessment. (1987). *Issues of Collaboration for Human Genome Projects* (Mapping Our Genes, Transcript of Workshop "Issues of

Collaboration for Human Genome Projects," National Technical Information Service Order No. PB 88-162 787 / AS). US Congress. 26 June.

39. Roberts, L. (1987). Who Owns the Human Genome? *Science* 237 (24 July):358–361.

40. Kayton, I. (1982). Copyright in Living Genetically Engineered Works. *George Washington Law Review* 50 (No. 2):191–218.

41. Jones, RH. (1986). Is There a Property Interest in Scientific Research Data? *High Technology Law Journal* 1:447–482.

42. Payne, RW. (1988). The Emergence of Trade Secret Protection in Biotechnology. *Bio / Technology* 6 (February):130.

43. Korn, DE. (1987). Patent and Trade Secret Protection in University-Industry Research Relationships in Biotechnology. *Harvard Journal on Legislation* 24 (Winter):191–238.

44. Eisenberg, RS. (1990). Patenting the Human Genome. *Emory Law Journal* 39:721–745.

45. Karjala, D. (1992). A Legal Research Agenda for the Human Genome Initiative. *Jurimetrics* 32 (No. 2, Winter):121–310. See especially pp. 192–203.

46. Roberts, L. (1991). Report Card on the Genome Project. *Science* 253 (26 July):376.

47. While Venter was confident of his capacity to attack a genome region of several million base pairs, the longest continuous sequence from his laboratory was in the tens of thousands. It had just been completed as his proposal came up for review. At the time, many scientists were skeptical that a further increase in scale was possible without major technical improvements in the technology, especially in the analytical software. A critical question was how short sequences of 300 to 500 base pairs, whether generated by machines or manual methods, would be pooled into long stretches of sequence millions of bases long. This was a problem that no laboratory had yet solved, and many believed it was premature to try on a massive scale, particularly on the human genome. Human DNA had many regions containing long repeated sequences, and these might prove to be impenetrable barriers to analysis.

48. Cook-Deegan, RM, Venter, JC, Gilbert, W, Mulligan, J, and Mansfield, BK. (1990). *DNA Sequencing Conference, II.* Hyatt Regency Hotel, Hilton Head, SC. Unpublished meeting summary. 1–3 October.

49. Venter, JC. (1991). Letter to James D. Watson, Director, Cold Spring Harbor Laboratory. National Institute of Neurological Disorders and Stroke. 23 April.

50. Venter, JC, 1992, Interview, Institute for Genomic Research, Gaithersburg, MD, 6 September.

51. Berg, P, and Gilbert, W, 1986. Comments taped by C. Thomas Caskey at session on the Human Genome project, symposium on the Molecular Biology of *Homo sapiens,* Cold Spring Harbor Laboratory, 3 June.

52. Adams, MD, Kelley, JM, Gocayne, JD, Dubnick, M, Polymeropoulos, MH, Xiao, H, Merril, CR, Wu, A, Olde, B, Moreno, RF, Kerlavage, AR, McCombie, WR, and Venter, JC. (1991). Complementary DNA Sequencing: Expressed Sequence Tags and Human Genome Project. *Science* 252 (21 June):1651–1656.

53. Adams, MD, Dubnick, M, Kerlavage, AR, Moreno, R, Kelley, JM, Utterback, TR, Nagle, JW, Fields, C, and Venter, JC. (1992). Sequence Identification of 2,375 Human Brain Genes. *Nature* 355 (12 February):632–634.

54. Gillis, AM. (1992). The Patent Question of the Year. *BioScience* 42 (No. 5, May):336–339.

55. Venter, JC, Adams, MD, Martin-Gallardo, A, McCombie, WR, and Fields, C. (1992). Genome Sequence Analysis: Scientific Objectives and Practical Strategies. *Trends in Biotechnology* 10 (January–February):8–11.

56. Roberts, L. (1991). Gambling on a Shortcut to Genome Sequencing. *Science* 252 (21 June):1618–1619.

57. Nowak, R. (1991). Gene Researchers Challenge the Patent Frontier. *Journal of NIH Research* 3 (December):25–26.

58. Adler, R, 1992, Interview, Office of Technology Transfer, National Institutes of Health, 11 September.

59. Anderson, C. (1991). US Patent Application Stirs Up Gene Hunters. *Nature* 353 (10 October):485–486.

60. Human Genome Committee. (1991). Position Paper on Patenting of Expressed Sequence Tags. American Society of Human Genetics. 6 November, with attached letter. Reprinted as a letter to *Science* 254:1710 and 1712.

61. Human Genome Committee and Board of Directors. (1991). The Human Genome Project and Patents. *Science* 254:1710–1712.

62. Human Genome Organization. (1992). HUGO Position Statement on cDNAs: Patents. *G-Nome News* (Number 9, March 1992):2–3. The HUGO statement did not "oppose patenting of useful benefits derived from genetic information. HUGO does, however, oppose the patenting of short sequences from randomly isolated portions of genes encoding proteins of unknown function. . . . The filing and approval of genome patents, whether in this particular case or in other types of genome patent cases, must incorporate global perspectives.

"It is HUGO's position, therefore, that the US EST patent applications should not be approved. We join the American Society of Human Genetics in supporting this view.

"The Human Genome Project is a major scientific area that now depends heavily on international cooperation in order to avoid both costly competition and duplication of effort. . . . it is the entire spectrum of international scientific collaboration that may be jeopardised. HUGO therefore urges a quick resolution to the EST case.

"HUGO urges the development of a process that allows for flexible negotiation and also mediation between the potentially conflicting needs of different scientific communities. Scientists, policy-makers and administrators world-wide must work together and accept responsibility for balancing the many competing priorities."

63. Berg, P, and Lerman, L. (1992). Letters to Dr. Frank Press (President of the National Academy of Sciences) and Dr. Bernadine Healy (Director of NIH). Chairman, NIH Program Advisory Committee for the Human Genome; Acting Chairman, DOE Health and Environmental Research Advisory Committee; Co-chairman and Acting Co-chairman, NIH-DOE Joint Subcommittee. 23 January.

64. Roberts, L. (1991). Genome Patent Fight Erupts. *Science* 254 (11 October):184–186.

65. Roberts, L. (1992). NIH Gene Patents, Round Two. *Science* 255 (21 February):912–913.

66. McBride, G. (1991). Dispute in US Over Gene Patents. *British Medical Journal* 303 (23 November):1286.

67. Adler quoted the 1988 OTA report that stated "genome projects raise no new questions of patent or copyright law" and went on to speculate that "contributing to this lack of foresight may have been an urgency to start the genome program." As author of the offending OTA sentence, I admit to embarrassment about such a bold and misleading oversimplification. The unfortunate result was to distract from a subsequent chapter on technology transfer in the 1988 report that con-

sidered the issues in considerably more depth. The lack of foresight was not quite as bad as alleged. OTA urged early filing of patent applications. Failure to do so could "inhibit full exploitation of an invention" and invited "foreign exploitation of research funded at US taxpayers' expense. . . . Penicillin was discovered in England, but the patent was obtained by US corporations. . . . the United Kingdom claimed the Nobel Prize, but the United States reaped most of the economic benefits." OTA also noted "there is a gray area between invention of new methods and the data that result from using them," but did not predict how DNA sequences themselves, of the sort at issue in the NIH patent application, would become the subject of patent controversy. Like scientists, public policy analysis can be humbled by the march of events.

68. Adler pointed to how sequences might be used to identify a tissue of origin. Rebecca Eisenberg noted the NIH application listed uses for forensic identification or as genetic markers. Just as the use of DNA markers for identification were useful only if population frequencies were known, all these uses would also require a great deal more to be known about the population distribution of the sequences, or how different tissues expressed them. Since Venter's laboratory was identifying the genes for the first time, or they would not be novel, such information was by definition unavailable without further work. An added problem was that coming from protein-coding regions, their use for forensic typing would make these precisely the regions most likely to later prove related to a genetic disorder, making them poor candidates for general use because of the ethical problems this would raise. This does not, however, count against the contention that the sequences might someday be useful for something.

69. Adler, RG. (1992). Genome Research: Fulfilling the Public's Expectations for Knowledge and Commercialization. *Science* **257** (14 August):908–914.

70. Mossinghoff, GJ. (1992). Letter to Bernadine Healy, NIH Director. Pharmaceutical Manufacturers Association. 28 May.

71. Healy, B. (1992). Letter to Gerald J. Mossinghoff, J.D., President, Pharmaceutical Manufacturers Association. Director, National Institutes of Health. 25 June.

72. Mossinghoff, GJ. (1991). Letter to Louis W. Sullivan, Jr., M.D., Secretary of Health and Human Services. Pharmaceutical Manufacturers Association. 22 November.

73. Mossinghoff, GJ. (1992). Letter to Bernadine Healy, NIH Director. Pharmaceutical Manufacturers Association. 14 July.

74. Industrial Biotechnology Association. (1992). IBA Position Paper: Recommended Federal Policy Concerning Human Genetic Sequences Discovered by Federal Researchers, Contractors, and Grantees. Industrial Biotechnology Association, Washington, DC. 10 June.

75. Association of Biotechnology Companies. (1992). ABC Statement on NIH Patent Filing for the Human Genome Project. Association of Biotechnology Companies, Washington, DC. 17 May; see also statement of Henry Wixon at a public meeting sponsored by the Genome Patent Working Group, Committee on Life Sciences, Federal Coordinating Council for Science, Engineering, and Technology, at the National Academy of Sciences, Washington, DC, 21 May.

76. Small, WE. (1992). Letter to Jeffrey Kushan, Legislative and International Affairs, US Patent and Trademark Office. Executive Director, Association of Biotechnology Companies. 6 May.

77. Association of Biotechnology Companies. (1992). Letter to President George Bush. Association of Biotechnology Companies, Washington, DC. 18 May.

78. Eisenberg, RS. (1992). Genes, Patents, and Product Development. *Science* **257** (14 August):903–908.

79. Kiley, TD. (1992). Patents on Random Complementary DNA Fragments? *Science* **257** (14 August):915–918.

80. Healy, B. (1992). On Patenting Genes. *New England Journal of Medicine* **327** (No. 9, 27 August):664–668.

81. Martinell, J. (1992). In re application of J. Craig Venter, et al., Serial No. 07/837, 195 (filed 12 February 1992). Patent and Trademark Office, US Department of Commerce. 20 August.

82. Roberts, L. (1992). Rumors Fly Over Rejection of NIH Claim. *Science* **257** (25 September):1855.

83. Gladwell, M. (1992). Clarifying Patent Law for Genes. *Washington Post* (23 September):

84. Stout, H. (1992). Gene-Fragment Patent Request Is Turned Down. *Wall Street Journal* (23 September):B1.

85. Anderson, C. (1992). NIH cDNA Patent Rejected; Backers Want to Amend Law. *Nature* **359** (24 September):263.

86. Venter, JC. (1992). Statement of J. Craig Venter, President and Director, The Institute for Genomic Research, Before the Subcommittee on Patents, Copyrights, and Trademarks. Senate Judiciary Committee. 22 September.

87. Healy, B. (1992). Testimony by Bernadine Healy, M.D., Director, National Institutes of Health, before the Subcommittee on Patents, Copyrights, and Trademarks. Senate Judiciary Committee. 22 September 1992.

88. Astrue, MJ. (1992). Letter to the Honorable Harry F. Manbeck, Jr., Patent and Trademark Office, US Department of Commerce. General Counsel, Department of Health and Human Services. 18 June 1992.

89. Richman, BS. (1992). Decision on Petition to Suspend Prosecution; Specification of Correspondence Address. Examining Group 180, Patent and Trademark Office, US Department of Commerce. 6 August.

90. Comer, DB. (1992). Letter to Michael J. Astrue, General Counsel, Department of Health and Human Services. Patent and Trademark Office, US Department of Commerce. 26 June.

91. Roberts, L. (1992). Top HHS Lawyer Seeks to Block NIH. *Science* **258** (9 October):209–210.

92. Anderson, C. (1992). Gene Wars Escalate as US Official Battles NIH Over Pursuit of a Patent. *Nature* **359** (8 October):467.

93. Israelsen, NA. (1992). Letter to James Haight, Office of Technology Transfer, NIH. Knobbe, Martens, Olson & Bear. 9 September.

94. Kolata, G. (1992). Biologist's Speedy Gene Method Scares Peers but Gains Backer. *New York Times* (28 July):C1, C10.

95. Bishop, JE, and Stout, H. (1992). Gene Scientist to Leave NIH, Form Institute. *Wall Street Journal* (7 July):B1, B4.

96. Hamilton, DP. (1992). Venter to Leave NIH for Greener Pastures. *Science* **257** (10 July):151.

20. EXODUS: THE END OF THE BEGINNING

1. Metheny, B, and Haley, S. (1992). Dr. Watson's Resignation Raises Questions on Future Commitment to HGP. *Genetic Engineering News* **12** (No. 7, 1 May):3, 24.
2. Culliton, BJ. (1985). Watson Fights Back. *Science* **228** (12 April):160.
3. Gladwell, M. (1992). New Face for Science. *Washington Post Magazine* (21 June):8–13, 23–25.
4. Nowak, R. (1991). Gene Researchers Challenge the Patent Frontier. *Journal of NIH Research* **3** (December):25–26.
5. Healy, B, 1992, Interview, Building 1, Room 126 (Director's Office, National Institutes of Health), 2 September.
6. Charles, D, and Coghlan, A. (1992). Ministers Move to Limit Genome Patents. *New Scientist* (14 March):9.
7. Gillis, AM. (1992). The Patent Question of the Year. *BioScience* **42** (No. 5, May):336–339.
8. Roberts, L. (1991). MRC Denies Blocking Access to Genome Data. *Science* **254** (13 December):1583.
9. Swinbanks, D. (1992). Japanese Researchers Rule Out Gene Patents. *Nature* **356** (19 March):181.
10. Coghlan, A. (1992). US Gene Plan "Makes a Mockery of Patents." *New Scientist* (22 February):16.
11. US Department of State. (1991). Diplomatic cables: 91 Paris 13315 and 33581; 92 Paris 04248, 05678; 92 London 07705, 04119, 02510, 02118, 01637. US Embassies in London and Paris; US Representative to OECD. October 1991–March 1992.
12. McKusick, VA. (1992). First South-North Human Genome Conference. *Genomics* **14**:1121–1123.
13. Anderson, C. (1992). US to Seek Gene Patents in Europe. *Nature* **357** (18 June):525.
14. Curien, H. (1991). The Human Genome Project and Patents. *Science* **254**:1710.
15. Adler, RG. (1992). Genome Research: Fulfilling the Public's Expectations for Knowledge and Commercialization. *Science* **257** (14 August):908–914.
16. Healy, B. (1992). On Patenting Genes. *New England Journal of Medicine* **327** (No. 9, 27 August):664–668.
17. Gore, A. (1992). Statement on proposed Hatfield amendment to HR 2407. *Congressional Record* **138** (No. 49, 2 April):S4726–S4727.
18. Bourke, FA. (1992). Letter to Bernadine P. Healy, Director, NIH. 25 February.
19. Roberts, L. (1992). Sequencing Venture Sparks Alarm. *Science* **255** (7 February):677–678.
20. Anderson, C, and Aldhous, P. (1992). Genome Project Faces Commercialization Test. *Nature* **355** (6 February):483–484.
21. Brown, P. (1992). A Wellcome Injection of Cash. *New Scientist* **135** (No. 1832, 1 August):12–13.
22. Mundell, I, and Anderson, C. (1992). Sweeter Offer by Wellcome Keeps Genome Scientist in Britain. *Nature* **357** (14 May):99.
23. Watson, JD, 1992, Interview, Cold Spring Harbor Laboratory, November.
24. Watson, JD, 1992, phone conversation, September.
25. Eckholm, E. (1992). A Tough Case for Dr. Healy. *New York Times Magazine* (1 December):67, 118, 122, 124, 126.
26. Watson sat on the scientific board of two companies (Diagnostic Products Corporation, Pall Corporation), a research institute (Roche Institute of Molecular Biology), consulted for Diagnostic Products Corp., and held stock in Genetics Institute, Oncogene Science, Amgen, Beecham, Smith-Kline, Diagnostic Products Corp., Glaxo, Eli Lilly, Merck, Pall, and other companies.
27. Watson, JD. (1988). Disqualification from Activities That Would Affect My Financial Interests. Memo

to William F. Raub, Deputy Director, NIH. 31 October.
28. Watson, JD. (1989). Disqualification from Activities That Would Affect My Financial Interests. Memo to William Raub, Acting Director, NIH. 19 October.
29. Watson, JD. (1989). Disqualification from Activities That Would Affect My Financial Interests. Memo to William F. Raub, Deputy Director, NIH. 2 June.
30. Schneider, J, 1993, Interview, Building 1; discussing a February 1991 electronic mail message from Rob Lanman, NIH Counsel, to William Raub, then acting director, NIH, and a letter from William Raub to James Watson, Office of the Director, National Institutes of Health, 5 January. The February 1991 memo did not assert a conflict of interest, but rather referred to a possible "appearance of conflict," based on stock purchases since Watson's previous disclosure statement. The argument of "appearance of conflict," while retaining surface plausibility and potential for public relations mischief, was quite different from the kind of conflict associated with a clinical trial. A clinical trial centered on a drug or device nearing market approval. The results of a trial bore directly on prospects for Food and Drug Administration approval of a new drug, and such approval for a major new drug could make the stock of even a pharmaceutical giant rise dramatically, if successful, or plummet, if unsuccessful. Genome research, in contrast, was in most cases quite distant from commercial application. There were a few exceptions, in the areas of research instrumentation or DNA diagnostics. It was nonetheless difficult to imagine how the results of a grant decision or research initiative on gene mapping could have any substantial impact on the stock of any but the smallest and most targeted genome research company. Conflict might indeed have arisen if a small instrumentation firm or DNA forensics company sought Small Business Innovation Research funding, or a similar scenario. Watson's holdings, however, were not in such companies, but in large pharmaceutical houses whose stock would be almost entirely unaffected by the fate of federal genome research decisions in the short run. They might benefit from the general knowledge emanating from NIH genome research, but this would not fall into the usual definition of conflict of interest. In retrospect, the February 1991 memo appears to be one lawyer's cautious and broad interpretation of the possibility of an "appearance" of conflict, coupled with a narrow definition of permissible behavior.
31. Watson, JD, 1992, Interview, Cold Spring Harbor Laboratory, November. Watson recalled receiving a letter from William Raub that sought information about certain stock purchases. He and Raub had then met with Jack Kress in May of 1991, after Healy had become director of NIH and before Raub left NIH to join the White House Office of Science and Technology Policy. Kress had indicated he would follow up if Watson's activities required a response.
32. Schneider, J, 1992, Memorandum to the author, Office of the Director, NIH, December.
33. Anderson, C. (1992). US Genome Head Faces Charges of Conflict. *Nature* **356** (9 April):463.
34. Hilts, PJ. (1992). Head of Gene Map Threatens to Quit. *New York Times* (9 April):A26.
35. Watson, JD, 1992, phone conversation, March 26 and various conversations April–September.
36. Roberts, L. (1992). Friends Say Jim Watson Will Resign Soon. *Science* **256** (10 April):171.
37. Schneider, J. (1992). Quoted in "Nobel Prize Biol-

ogist Watson Plans to Resign US Position" by David Brown and Malcolm Gladwell. *Washington Post* (9 April):A3.

38. Brown, D., and Gladwell, M. (1992). Nobel Prize Biologist Watson Plans to Resign US Position. *Washington Post* (9 April):A3.

39. Institute of Medicine. (1988). *A Healthy NIH Intramural Program: Structural Change or Administrative Remedies?* (Washington, DC: National Academy Press). National Academy of Sciences.

40. Institute of Medicine. (1984). *Responding to Health Needs and Scientific Opportunity: The Organizational Structure of the National Institutes of Health* (Washington, DC: National Academy Press). Institute of Medicine, National Academy of Sciences. October.

41. Roberts, L. (1992). Why Watson Quit as Project Head. *Science* **256** (17 April):301–302.

42. US House of Representatives. (1992). *Departments of Labor, Health and Human Services, Education, and Related Agencies Appropriations for 1993* (Part 4, pp. 607–698). Committee on Appropriations. 25 March.

43. Watson, JD, 1992, phone conversations, March–December.

44. Anderson, C. (1992). Watson Resigns, Genome Project Open to Change. *Nature* **356** (16 April):549.

45. Gladwell, M. (1992). Biologist Watson Quits Position at NIH. *Washington Post* (11 April):A9.

46. Palca, J. (1992). The Genome Project: Life After Watson. *Science* **256** (15 May):956–958.

47. Roberts, L. (1992). A Standing Ovation from the Troops. *Science* **256** (15 May):957.

48. Maddox, J. (1992). Healy in a Hurry. *Nature* **356** (16 April):547–548. The editorial might have been more persuasive if it had gotten Healy's name right.

49. Watson, JD. (1992). *Annual Report, 1991.* Cold Spring Harbor Laboratory. pp. 1–25.

EPILOGUE

1. Thompson, L. (1993). Healy and Collins Strike a Deal. *Science* **259** (1 January):22–23.

2. Botstein, D, 1988, Interview, Genentech, 19 August.

3. Healy appointed a committee to assist in the search for Watson's successor. The committee roster was included in a press packet distributed at a press conference down the hall from the NIH Director's office. The committee was co-chaired by Ruth Kirschstein, director of the National Institute of General Medical Sciences, and George Vande Woude, a former National Cancer Institute intramural researcher and current investigator at Advanced Bioscience Laboratories. The other members were Raphael Daniel Camerini-Otero, National Institute of Diabetes and Digestive and Kidney Diseases; Daryl A. Chamblee, senior policy adviser and counselor to the director, NIH; Gary Felsenfeld, National Institute of Diabetes and Digestive and Kidney Diseases; Martin Gellert, National Institute of Diabetes and Digestive and Kidney Diseases; Jay Moskowitz, associate director for science policy and legislation, NIH; Maynard Olson, Washington University; David Rodbard, Division of Computer Research and Technology, NIH; Phillip A. Sharp, Massachusetts Institute of Technology; Maxine Singer, Carnegie Institution of Washington; Shirley M. Tilghman, Princeton University; and Nancy S. Wexler, Hereditary Disease Foundation and Columbia University, National Institutes of Health, 4 May.

4. Healy, B, 1993, Statement announcing her intention to resign by 30 June, released at a press conference, National Institutes of Health, 26 February.

5. Riordan, JR, Rommens, JM, Kerem, B-S, Alon, N, Rozmahel, R, Grzelczak, Z, Zielenski, J, Lok, S, Plavsic, N, Chou, J-L, Drumm, ML, Iannuzzi, MC, Collins, FS, and Tsui, L-C. (1989). Identification of the Cystic Fibrosis Gene: Cloning and Characterization of Complementary DNA. *Science* **245** (8 September):1066–1072.

6. Rommens, JM, Iannuzzi, MC, Kerem, B-S, Drumm, ML, Melmer, G, Dean, M, Rozmahel, R, Cole, JL, Kennedy, D, Hidaka, H, Zsiga, M, Buchwald, M, Riordan, JR, Tsui, L-C, and Collins, FS. (1989). Identification of the Cystic Fibrosis Gene: Chromosome Walking and Jumping. *Science* **245** (8 September):1059–1065.

7. Kerem, B-S, Rommens, JM, Buchanan, JA, Markiewicz, D, Cox, TK, Chakravarti, A, Buchwald, M, and Tsui, L-C. (1989). Identification of the Cystic Fibrosis Gene: Genetic Analysis. *Science* **245** (8 September):1073–1080.

8. Wallace, MR, Marchuk, DA, Andersen, LB, Letcher, R, Odeh, HM, Saulino, AM, Fountain, JW, Brereton, A, Nicholson, J, Mitchell, AL, Brownstein, BH, and Collins, FS. (1990). Type I Neurofibromatosis Gene: Identification of a Large Transcript Disrupted in Three NF1 Patients. *Science* **249** (13 July):181–186.

9. Viskochil, D, Buchberg, AM, Xu, G, Cawthon, RM, Stevens, J, Wolff, RK, Culver, M, Carey, JC, Copeland, NG, Jenkins, NA, and White, R. (1990). Deletions and a Translocation Interrupt a Cloned Gene at the Neurofibromatosis Type 1 Locus. *Cell* **62** (13 July):187–192.

10. Roberts, L. (1990). Down to the Wire for the NF Gene. *Science* **249** (20 July):236–238.

11. Cawthon, RM, Weiss, R, Xu, G, Viskochil, D, Culver, M, Stevens, J, Robertson, M, Dunn, D, Gesteland, R, O'Connell, P, and White, R. (1990). A Major Segment of the Neurofibromatosis Type 1 Gene: cDNA Sequence, Genomic Structure, and Point Mutations. *Cell* **62** (13 July):193–201.

12. Roberts, L. (1992). The Huntington's Gene Quest Goes On. *Science* **258** (30 October):740–741.

13. Collins, FS, 1990, Speech at the first of two conferences organized by J. Robert Nelson and others, Genetics, Religion, and Ethics: Implications of the Human Genome Project for Medicine, Theology, Ethics, and Policy, MD Anderson Cancer Research Center, March.

14. Collins, FS, 1993, note to author, National Center for Human Genome Research, NIH, 9 March.

15. Cook-Deegan, RM, Venter, JC, Pearson, ML, and Mansfield, B. (1990). *The Human Genome and the Pharmaceutical Industry.* Unpublished report of a conference on Hilton Head Island, South Carolina, 30 September 1990. October.

16. Anderson, C. (1993). Genome Project Goes Commercial. *Science* **259** (15 January):300–302.

17. Dickson, D. (1993). Britain Plans Broad Strategy on Genome, Approves Therapy. *Nature* **361** (4 February):387.

18. US Congress. (1993). *Pharmaceutical R&D: Costs, Risks, and Rewards* (OTA-H-522). Office of Technology Assessment. February.

19. Knox, R. (1973). Nobel Winner Calls Nixon's Medical Research Policy "Lunacy." (7 March):67.

20. Fuchs, VR. (1990). The Health Sector's Share of the Gross National Product. *Science* **247** (2 February):534–538.

21. Schieber, GJ, Poullier, J-P, and Greenwald, LM. (1992). US Health Expenditure Performance: An International Comparison and Data Update. *Health Care Financing Review* 13 (No. 4, Summer 1992):1–15 plus tables.

22. Burner, ST, Waldo, DR, and McKusick, DR. (1992). National Health Expenditures Projections through 2030. *Health Care Financing Review* 14 (No. 2, Fall):1–29.

23. Lazenby, HC, Levit, KR, Waldo, DR, Adler, GS, Letsch, SW, and Cowan, CA. (1992). National Health Accounts: Lessons from the US Experience. *Health Care Financing Review* 13 (No. 4, Summer):89ff, see esp. p. 95 and Table 1. Drugs accounted for approximately $54.6 billion of $666.2 billion total health expenditures in 1990.

24. Cleeton, DL, Goepfrich, VT, and Weisbrod, BA. (1992). What Does the Consumer Price Index for Prescription Drugs Really Measure? *Health Care Financing Review* 13 (No. 3, Spring):45–51.

25. Philosopher Michael Walzer, of Princeton, wrote a classic essay in 1973 about the dilemma of "dirty hands," when pursuit of political power and moral virtue came into conflict. See next reference.

26. Walzer, M. (1973). Political Action: The Problem of Dirty Hands. *Philosophy and Public Affairs* 2 (Winter):160–180.

27. Fredrickson, DS. (1991). Asilomar and Recombinant DNA: The End of the Beginning. In *Biomedical Politics,* KE Hanna, Ed., pp. 258–298. Washington, DC: National Academy Press.

28. Krimsky, S. (1982). *Genetic Alchemy: The Social History of the Recombinant DNA Controversy.* MIT Press, Cambridge, MA.

29. Wright, S. (1993). The Social Warp of Science: Writing the History of Genetic Engineering Policy. *Science, Technology and Human Values* 18 (No. 1, Winter):79–101.

ACKNOWLEDGMENTS

1. US Congress. (1988). *Mapping Our Genes—Genome Projects: How Big? How Fast?* Office of Technology Assessment, OTA-BA-373, Washington, DC: Government Printing Office; reprinted by Johns Hopkins University Press. April.

2. Watson, JD, Ed. (1980). *The Double Helix: A Personal Account of the Discovery of the Structure of DNA.* Norton Critical Edition. W. W. Norton, New York.

3. Crick, F. (1988). *What Mad Pursuit: A Personal View of Scientific Discovery.* Basic Books. Same quote also found in "The Double Helix: A Personal View," *Nature* 248 (April 26):766–771, 1974; reprinted in Watson, *The Double Helix.*

4. Cook-Deegan, RM. (1991). The Human Genome Project: Formation of Federal Policies in the United States, 1986–1990. In *Biomedical Politics,* K Hanna, Ed., pp. 99–168. Washington, DC: National Academy Press.

5. Cook-Deegan, RM. (1991). The Genesis of the Human Genome Project. In *Molecular Genetic Medicine,* Vol. 1, T Friedmann, Ed., pp. 1–75. San Diego: Academic Press.

Index